D1257516

BEYOND THE BLUE HORIZON

BEYOND THE BLUE HORIZON

Myths and Legends of the Sun, Moon, Stars, and Planets

Dr. E. C. Krupp

Photographs are by the author unless otherwise credited.

FIRST EDITION

Designed by Irv Perkins Associates

Library of Congress Cataloging-in-Publication Data

Krupp, E. C. (Edwin C.), 1944–
Beyond the blue horizon : myths and legends of the sun, moon, stars, and planets / E. C. Krupp.
p. cm.
Includes index.
ISBN 0-06-015653-8
1. Sky—Mythology. 2. Astronomy. I. Title.
BL325.S5K78 1991
291.2′12—dc20 90-55542

91 92 93 94 95 DT/CW 10 9 8 7 6 5 4 3 2 1

for Jane Jordan Browne,
one of those celebrated agents
of cosmic order

Contents

Acknowledgments

EVERY storyteller owes something to those who told the stories before, and I owe plenty. Some of the stories are so rich with artful and logical connections between human lives and events in the sky that it is pleasurable just to repeat them. I am beholden, therefore, to anthropologists Johannes Wilbert and Gerardo Reichel-Dolmatoff, who listen carefully to indigenous people in the swamps and forests of South America. With distinctive combinations of personal style and relentless scholarship, both of these professors inspire admiration.

Arlene Benson and Floyd Buckskin have shared unusual material from northern California. John Rafter's ability to recognize the importance of astronomical themes in the mythology and ritual of southern California's Luiseño and Chemehuevi Indians prompted me to try the same approach with other traditional narratives. Professor Edward Schafer's treatment of the astronomical component of the culture of Tang China is attractive and elegant. I also enjoy Dr. Robert L. Hall's ability to detect cross-cultural meaning in Native American symbol and myth and appreciate the maps and detailed advice he provided to make possible my own pilgrimage to the summit of Chicoma Mountain. Roger Bingham's thoughts and comments about story telling, science, and myth were persuasive enough to help clarify a key theme of this book.

I am grateful to Vernon Hunter, who continues to discover celestial gems buried in the literature of folklore and pass them on to me. Dr. Katalin Barlai of Budapest's Konkoly Observatory took my interest in old Hungarian sky lore seriously and forwarded important studies that were unknown to me. Years ago, Gibson Reaves, Professor of Astronomy at the University of Southern California, told me about the Ph.D. thesis Theony Condos prepared in the Department of Classics—an English translation and commentary on the *Katasterismoi*, the only surviving collection of constellation myths in ancient Greek literature. This is the only version in English known to me, and Dr. Reaves generously and quickly made a copy of it available to me. No serious study of ancient star lore can afford to ignore the *Katasterismoi*, and I am fortunate to have the benefit of Dr. Condos's good work.

I have been lucky, as well, to travel throughout the world to places that reflect our ancestors' involvement with the sky. These expeditions provided some of the pictures that appear in this book along with firsthand knowledge of now more than 950 ancient sites. Dr. Eve Haberfield at U.C.L.A. Extension has continued to encourage me to organize field study tours under the U.C.L.A. banner, and the experience derived from these trips has found its way into this book. I thank, too, those who chose to travel with me on these unusual itineraries, for they have made much of my field research possible.

Dr. B. G. Sidharth, Director of the Birla Planetarium in Hyderabad, generously provided an opportunity to visit India and to gain preliminary insight into its very ancient astronomical tradition. On the same trip, Dr. A. G. Kulkarni, Director of the Birla Planetarium in Jaipur, hosted my visit there and made it possible for me to see the magnificent historical instruments of Jai Singh in operation. Ted Pedas invited me to lecture on several Sun Line cruises, and one of these, in the Mediterranean, proved essential in the preparation of this book. Yvette Cloutier of ETA/Piuma Travel organized two trips to Turkey for me, and in Turkey, Senduran Doğansoy, my good friend and knowledgeable advisor, guided me to many sites off the beaten track, including the 2000-year-old Sabian planet shrine at Sogmatar.

I am mindful of my debt to many professional colleagues, especially Anthony F. Aveni and John B. Carlson, who have told me many funny stories over the years. With humor, sincere interest, honesty, and hard work in the library and in the field, they knowingly and unknowingly have prompted me to pursue a hunter/gatherer strategy in the quest for ancient astronomy.

The late George O. Abell, professor of astronomy at U.C.L.A., continues to inspire all of my professional activity. He was an observational cosmologist who really loved the night sky and turned me into the same kind of missionary.

Although my undergraduate years at Pomona College now seem as remote as the Cretaceous-Tertiary boundary, that liberal education continues to have impact. With one foot planted firmly on a foundation of physics and astronomy, I am still tempted to muddy the waters between science and the humanities.

When I was quite young, my parents gave me a book about the stars. Full-page color illustrations throughout the book introduced me, as they introduced the young boy featured in the story, to the nine worlds of our solar system—and to the stars as well. When the boy was transported to each of these planets, he found a world that matched our best scientific understanding of the conditions that prevailed there and also encountered the god for which it was named, all costumed in his symbols and carrying the instruments of his commission. Astronomical reality and the symbolism of myth, packaged as a complete collection of planets and stories of the stars, proved irresistible. I memorized the names of the nine planets in order and their vital statistics and decided to become an astronomer. The book is long lost. I don't know what it was called. I've never seen another copy. But the story haunts me, and my parents, Edwin and Florence Krupp, get my thanks for doing me a favor that seems to be lasting a lifetime.

Griffith Observatory and the management of the Department of Recreation and Parks of the City of Los Angeles have again generously allowed me to use illustrations from the observatory's extensive collection of diagrams, plans, and artwork. These have been used in observatory programs and publications, especially the observatory's monthly magazine, the *Griffith Observer*. Joseph Bieniasz, a member of the observatory's technical staff, is responsible for many of them. Nancy Mazzie, Photographic Assistant at Griffith, prepared many of the photographic prints.

Once again, I owe more than a contract percentage to my agent Jane Jordan Browne of Multimedia Product Development, Inc. Ever on the lookout for the inevitable slings and arrows that threaten safe passage from the author's Macintosh to the reader's armchair, she assists without intrusion and advises without pressure to protect all interests, especially the reader's. When this book was big enough to anchor the Goodyear blimp, her take-no-prisoners approach to my bumper word crop helped bring the book back to the market. No manuscript should leave home without her.

Everybody needs editing. I was fortunate that Craig Nelson, now Senior Editor at Harper Collins, edited *Echoes of the Ancient Skies*, and I count myself remarkably lucky to be, once more, the beneficiary of his critical judgment, his hard-nose editing instincts, and his encouraging advice on this project. His professionalism sets the industry standard. He saw the value and flaws in this enterprise and invested his own time and effort to enhance the former and eradicate the latter.

Robin Rector Krupp, my wife, has interrupted work on her own books and projects to provide unusual photographs and drawings for this book. She has also endured, with grace and affection, the unsettling priorities I have imposed upon our lives. Books in production have no respect for family and friends. With impunity they command all of an author's disposable time. It's not pretty, and it takes a generous partner to put up with it.

Despite intentions otherwise, I have continued to provide my son Ethan with a distorted picture of what constitutes a balanced family life in late-twentieth-century America. He has no illusions, however, and what sometimes slips through our fingers at home is often recaptured on the trail. He's good company in the canyons, in the desert, and in the jungle, and I thank him for tolerating the siblings these books inevitably become.

BEYOND THE BLUE HORIZON

I

About the Book

FOR tens of thousands of years, we've been asking ourselves the same questions:

Who are we?
From where do we come?
Where are we going?
What is the right thing to do?

We hunger for answers, as did our ancestors. The ice age cave painter in Europe, the stone age rice farmer in the central China, the ancient Egyptian scribe assigned to watch for the Nile's flood, the Maya noblewoman who drew sacrificial blood from her own tongue—all came to grips with these questions. The sky was one place where they found some answers.

VITAL CONCERNS

Sky stories tell us what some of these answers were. Many of these tales were told to give the world meaning and sense. Of course, there are lots of stories about the sky, and they don't all say the same thing. Different peoples in different places at different times may talk about the same celestial object but tell completely different tales for entirely different purposes. There are, however, some enduring themes that emerge nearly everywhere because of their universal importance. They include, for example, world order, the cycle of life, creation, spiritual transcendence, the complementary opposition of male and female, and the conflict between polarized forces. Stories based on these themes are read in the appearance and movement of objects seen in the sky, and the sky's behavior is also used to describe what happens when these themes are revealed in the things that happen on the earth and in people's lives.

Did various groups of ancient people talk to each other and share their symbols over the millennia? Did they travel long distances to do so? Not necessarily. We shouldn't be too surprised that the tales of so many different people often include the same themes. Their skies may vary, but those skies speak nearly the same language.

For similar reasons, British anthropologist Sir James George Frazer came to the conclusion that the recurrent and seemingly universal theme of the Dying and Reviving God was not necessarily the result of cultural diffusion and borrowing. In his pioneering and massive study of magic, religion, and folk custom, *The Golden Bough*, he attributed it to "the effect of similar causes acting alike on the similar constitution of the human mind in different countries and under different skies."

This doesn't mean that everybody thinks alike and sees the world in the same way. Far from it. Local traditions reflect local landscapes, local interests, and local resources. In temperate latitudes the seasons are hot and cold. In the tropics the climate oscillates between wet and dry. Behind such differences, however, reside some elementary concerns and some universal responses. Both in tropical jungles and in temperate woodlands, people have defined the period when food is scarce by the appearance of the Pleiades and have called that conspicuous cluster of stars a small crowd of hungry children. Despite the sideshow variety we encounter of costumes and props in the world's collection of sky tales, there is a worldwide unity of essential themes and an elementary vocabulary of celestial devices to dramatize them.

SKY TALES

Some sky tales are just accounts of what takes place overhead. They encapsulate the knowledge obtained through observation of the sky into a coded narrative that preserves the information and keeps it accessible. A tale of the moon's birth, growth, decline, death, and resurrection documents its monthly cycle of phases.

That's one kind of sky story, but we have also told stories that explain what is happening on the earth by invoking the power brokers in the sky. People see, for example, that the behavior of wild animals and cultivated plants is synchronized with the seasons, and the seasons, in turn, change in time with the annual journey of the sun and with the cyclical appearances of stars. For that reason, both hunters and farmers contrive stories about the availability of food that feature celestial objects and events.

Finally, the logic of analogy prompts people to sense parallels between the behavior of celestial objects and patterns in their lives. They exploit these resonances and tell stories that sound as if they are about the sky but are really about human psychology and community goals. According to Joseph Campbell, a renowned authority on world mythology, the hero's perilous quest is really a story about the crucial passages to maturity our psyches must survive in our lifelong encounter with the world. Because the sun, too, seems to undertake an adventure of danger and triumph, its yearly itinerary is partly grafted to the story of the hero, who is, in fact, our own aspiring soul.

WHY WE TELL THEM

Sky tales operate symbolically to reveal how the cosmos works and what our place in the universe may be. That is their main function. By telling these stories, we became conscious participants in the actions of the universe. First, just by looking at the sky—by taking in its sweep, its majesty, and its mystery—we marveled, and the wonder and awe we experienced made us feel a connection with the cosmos. We liked that feeling. We required it. It gave us an inkling of who we were, from where we came, and where we were going. Second, by watching the sky, keeping careful track of its rhythms and patterns, we understood relationships, scientifically and poetically. That understanding helped us feel at home in the universe. We still can't do without it.

In a practical way, our sky tales let us comprehend and interact appropriately with our environment. We timed the arrival of salmon and the planting of corn by the sun's progress on the horizon and the return of certain stars. In a less tangible but no less indispensable way, the old sky tales gave our brains a stable framework. With that orienting anchor provided by the sky, our imaginations were free to contrive the new tools and stories needed to deal with a changing world.

Penetrating the cores of the tales we have told about the sky allows us to find out what is in them and to understand why it is there. It is not possible, however, to inventory and retell every story about the sky in a book of finite length. The tales tallied or retold in these pages have been selected to illustrate the points that many more like them also convey. The deck is stacked in favor of their similar content and function. What they have in common helps reveal the universal themes people in all ages have plucked from the sky. By refamiliarizing ourselves with the language of the sky—by mastering its symbolic vocabulary and grammar—we shall see how it reveals cosmic order, how it validates the natural and moral order we attribute to the world, and how it integrates us into the cosmos by clarifying the pattern of life with celestial analogies. We shall understand the role the sky has played in our minds through the stories we have told about it.

THE LINK BETWEEN BRAIN AND SKY

We have to work to understand what the sky meant to our ancestors. They watched it. We do not. We are now too preoccupied with daily events to notice the sky and what takes place there. People in ancient and prehistoric times, on the other hand, used it everyday. It was their clock, calendar, and compass. They had to go out and have a look at it. Because we rely on other sources, we no longer need to get practical information from the sky. When we do see the real sky, we treat it as a novelty.

Without the same intimacy that our ancestors maintained with the sky, it is hard for us to appreciate their celestial metaphors. When we encounter the old sky stories, we find it difficult to understand what they are about and why they were told. We can get some additional help, however. Not everyone has quit looking at the sky. There are some who still tell symbolic stories about it and who can explain what their symbols mean.

The Desana Indians are one of those traditional peoples who still look at the sky and still talk about it. They live in the tropical rain forest of the northwest Amazon, and most of what we know about them is due to anthropologist/archaeologist Gerardo Reichel-Dolmatoff, who has done field research among the Desana and other Indians of Colombia. The Desana don't build giant pyramids or monumental stone temples. They have no written language. They now farm a little bitter manioc, but they much prefer to hunt. They don't wear many clothes. For all practical purposes, they are a stone age people, although contact with the outside world is transforming—and obliterating—their traditional culture. There are only six hundred Desana left, but they have adapted well to the jungle environment and have evolved a rich and complex culture.

The Desana speak of the sky as a brain. Like the human brain, the sky is bicameral, divided into two hemispheres, and the fissure between them is the Milky Way. The Milky Way, they say, is, in turn, a pair of entwined serpents—an anaconda and a rainbow boa. These two snakes represent many complementary pairs of aspects of the world. The rainbow boa is associated with things male, with land, and with light and color. The anaconda is female, water, and darkness. The snakes spiral and sway in rhythms that keep time to the twenty-four-hour cycle of day and night, the monthly march of the moon, and the yearly dances of the seasons, the stars, and the sun.

The Desana also say that two such serpents reside in the fissure of their own brains and facilitate the interaction between the left side of their brains and the right. They don't know anything about our scientific portrait of the specialized character of the two halves of the brain and don't locate the nonverbal, intuitive, analog mind in one cerebral hemisphere and the speech- and language-oriented, rational and analytic, digital mind with the other. But they do say that their brains must resonate with the sky, must dance to the same tune, for consciousness to exist.

With a detailed and comprehensive knowledge of their environment, the Desana know what they need to know to live where they do. Their knowledge is classified and organized into a subtle cosmology that tells them what the world is and more important how they fit into it. They know their ability to survive depends on proper integration of their behavior with this knowledge: They are guided by their own experience and by the insights of their shamans, whose direct, mystical experiences with the spirit world provide valuable understanding and power. The Desana shamans even say they go to the sky for some of this power and knowledge.

The knowledge they manipulate is, of course, their picture of the universe. One of its fundamental attributes—and the Desana are explicit about this—is a link between the brain and the sky. What the Desana mean by that link is the relationship they see between the celestial cycles that order time and the world and their use of those cycles to organize their lives in response to the natural environment. They forge that link through the stories they tell, the sacred myths that put meaning into the world.

The Desana, for example, say that the world was first populated where Sun Father set a perfectly vertical rod into the earth. The stick cast no shadow in the same way the vertical rays of the noontime sun penetrate the earth on the equinoxes, when the sun is at the zenith, without casting a shadow. The first people arrived in great anaconda canoes, and when the equinox rains come, great black anacondas swim upriver to mate.

For the Desana, the sky is filled with answers. They and other people all over the world, since the stone age, have seen a correlation between what takes place in the sky and what is happening on earth and in their lives. The tales they have told about the sky are extraordinarily diverse, but they all make use of the vocabulary and grammar of the sky to convey messages about the universe and their place in it.

SPEAKING THE SKY'S LANGUAGE

The sky speaks in celestial objects; the sun, the moon, the planets, and the stars are its vocabulary. The sky's grammar is what these objects do. They rise and set; come and go with the seasons; meet and separate; they circle or tumble or dawdle through heaven and time. The whole sky is a stage, and the things we see there are players. They make their entrances and exits, take their bows, and return for repeat performances.

Our ancestors reached for the sky and pulled meaning from it that persists in their stories and in our language today. With every dawn the sun was reborn, and each night it died another death. They saw creatures on the face of the moon and time's passing in its phases. The seasons were a cyclical quest made by the sun, a perilous journey through the zodiac that started over again as soon as it was done.

In the past, nearly everybody talked about the planets, the Pleiades, Orion, and the Big Dipper. The Babylonians populated the night sky with stellar gods and made a familiar landscape out of their twinkling light. Sky gods of the Amazon breathed life into the wind and spoke in the thunder. Hidden behind a curtain of rain, the Navaho saw Father Sky impregnating Mother Earth, and from her body new life grew. The Siberian sky was seamed along the Milky Way. A nail held the hub of the Lapp universe in place at the sky's north pole.

We once saw antagonism between the Hunter and the Scorpion in the stars of winter and summer on the opposite sides of the night. We once saw stability in the pole star and a highway to heaven in the Milky Way. These sky tales told our ancestors what the world was like, what rules it obeyed, what place was in it for them. Now, to us, they reveal our ancestors' beliefs. These stories deal with themes that are very old and very deep but never out-of-date. They still charm us because we share the same instinct for order and the same desire for heaven. There is still romance in the moon and mystery in the stars.

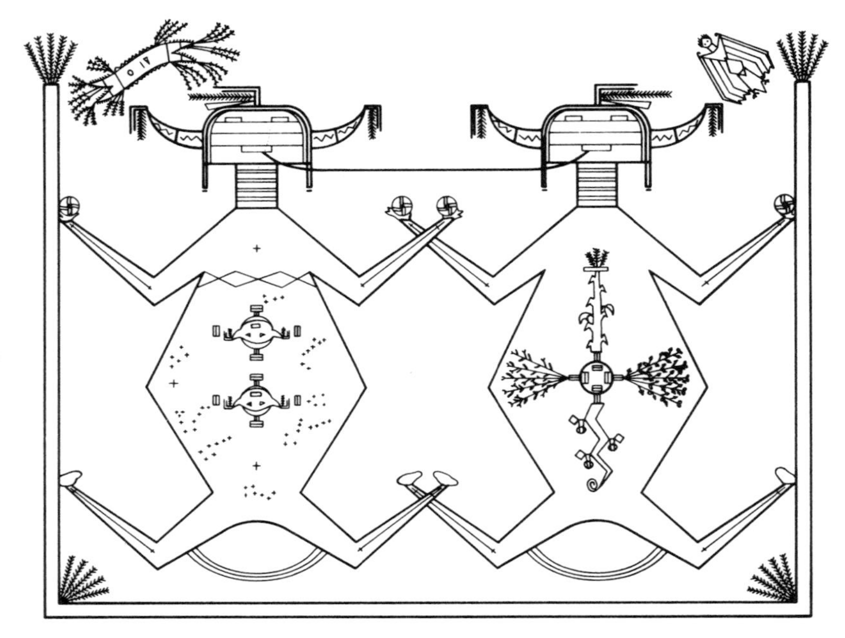

Father Sky (left) and Mother Earth (right) are sometimes paired in sacred Navaho sand paintings. A line of pollen connects their mouths and emphasizes the fertility of the sky and the fecundity of the earth. Several patterns of stars—including the Big Dipper—accompany the faces of the sun and moon on the body of Father Sky. Mother Earth carries the emblem of the world's central lake, which is equivalent to the Navaho and Pueblo concept of the world's center and point of emergence, or creation. Four sacred cultivated plants grow from that lake: corn, beans, squash, and tobacco. (Drawing: Joseph Bieniasz, Griffith Observatory)

The sky tales told in the chapters to come concern the sun, the moon, the seasons, and the stars. Eclipses make an entry, and the planets find their place. Some of their stories are familiar favorites; others follow less well-known scripts. They carry us through the clouds, up the Milky Way, and to the top of the sky. The Great Bear, the Dog Star, the Seven Sisters, the zodiac, and many more stars and constellations are here in their various disguises, their messages disclosed.

It is possible to see how this still works by looking at a modern sky tale told upon a commemorative patch designed by NASA for crew uniforms on mission STS-26, the first space shuttle flight to follow the *Challenger* disaster on January 28, 1986. The emblem shows the exhaust trail of a successful launch against a dark blue sky. A rising sun appears on the horizon to symbolize a new beginning, and the seven stars of the Big Dipper shine in the darkness. They stand for the seven astronauts who died in the explosion of the *Challenger*. Here they are immortalized. Like the unchanging stars of the Big Dipper, they are bright in our memory and inspire us to continue our quest for a future in space. In this symbol of a difficult and uncertain step in our exploration of space, we have taken age-old sky images and fashioned a new story.

The rebirth of America's manned exploration of space is symbolized by a rising sun on the emblem for the first shuttle mission after the Challenger *catastrophe, and the seven astronauts who died in that explosion were remembered as the seven stars of the Big Dipper. Because these stars never set, the ancient Egyptians considered them immortal. (Photograph: Griffith Observatory)*

It is easy enough to understand the sky tale told on the Space Shuttle mission patch. We still remember some of the sky's language, even though we don't see it as frequently as our ancestors did, and we know the meaning of our own story. The meanings of our ancestors' stories are not always so clear. By the time their stories reach us, much may have already been lost. If, however, they are sky tales, we can penetrate them, at least partly, because they are told in the sky's words and by the sky's rules. For example, a study of mesolithic burials at Wadi Halfa, at the Nile's Second Cataract in the Sudan, showed a clear preference for orienting the dead to face the sunrise. We don't have any other fact beyond the apparently deliberate orientation of skulls between the north and south limits of the rising sun, but it is not so farfetched to think the hunters of prehistoric Nubia believed the sun died when it set in the west and was reborn when it rose in the east. If so, their burials are an eight-thousand-year-old parable of solar rebirth. Perhaps they fortified a story about the rebirth of the dead with their celestially oriented burials. Most of the story, of course, is beyond our grasp. We are only guessing the theme. Plot, characters, and setting are all lost. The sky's language, however, lets us ever-so-slightly inside the minds of a people we know only by bones and tools.

STORIES TODAY

Current stories about the sky reflect the concerns and circumstances of our own times. We don't necessarily recognize them for what they are because

Still a stage for heroic celestial adventures, the sky is now big at the box office as the scene of interstellar drama. This model of the Enterprise *is displayed with honor at the National Air and Space Museum in Washington, D.C.* "Star Trek" *'s popular television starship eventually lent its name to the first Space Shuttle test vehicle.*

our concept of the sky has changed; the sky is now outer space, and Luke Skywalker fills our movie screens with adventures in a galaxy far, far away. The timeworn tale of order versus chaos now wears the uniforms of the Rebel Alliance and the Galactic Empire and stages firefights in the vacuum of space. Once upon a time the night sky was the setting for the drama of cosmic order; now the footlights must illuminate the depths of space.

During the Great Depression people looked up to the sky to not a bird, not plane, but—Superman. To escape the explosive death of his home planet Krypton, Superman as an infant was cradled through space in a rocket aimed toward the earth. There he grew to become a champion of justice. Our heroes continue to come from the sky.

Now, decades after the arrival of Superman, "Star Trek" 's Captain Kirk doesn't come from outer space, he "just works there," piloting the crew of the *Enterprise* from adventure to adventure the way Odysseus propelled his warriors around the Mediterranean.

We have also found new mysteries in the sky. UFOs and alien abductions may not be taken seriously by everyone, but they indicate that worlds beyond our own and extraterrestrial life are now lodged in our belief system. Enigmatic and possibly hostile aliens, instead of ghosts and demons, are now what go bump in the night. The sky, in a new disguise, is still speaking to us about the same old themes.

Today's stories about the sky, like those of the ancients, are actually stories about the earth. We still make the sky recite our prayers and our concerns back to us. The stories we now tell about the universe do what the old sky tales did. They help tell us who we think we are, where we think we are going, and what we think is the right thing to do.

BLINDED BY THE LIGHT

The sky still influences us, but not the way it once did. That's because we scarcely see the sky our ancestors saw. We've flooded the sky with artificial light and lost the original meaning of night. The planets, the stars, and even the moon sometimes seem to get lost in the artificial glare. When we somehow manage to escape from our cities and rediscover the stars, that old black magic of night cures our amnesia. We remember instantly the power, mystery, and beauty that have always been part of our interaction with the sky, and are glad to do so. But we now have to make a special effort to experience what used to be an everyday encounter with a marvel.

It has also gotten harder to see the sky in the daytime. In our need to make a living on this planet we've extracted a healthy harvest of energy from fossil fuels. With that power we have cooked up all kinds of entertaining schemes and useful devices, but we've also seasoned our sky with smoke and discolored the day. Sometimes we can't see as far as we used to see, or as clearly, because the sky is less transparent. And the hole our chlorofluorocarbons have punched through the sky's ozone doesn't help us see the celestial landscape any better. It just makes us worry about our future. Around our cities, we are more likely to see an airplane than a star. Landing lights approaching Los Angeles International Airport outshine Venus, the brightest celestial object in the sky after the sun and the moon.

Today we see a different sky from the one our ancestors saw, and we no longer see the same kind of cosmic order there either. Our expanding universe of ever lonelier galaxies is a very different cosmos from the one that was circled by the horizon and domed by heaven. We knew where we stood in that universe. Now it is harder to tell.

In some parts of the world, shamans once flew to heaven, drawn by the lodestone of the sky. They knew where the center was, but we have lost it. And the Desana would marvel that our addled brains can function at all with the Milky Way lost to our sight in our light-drenched nights and with the center gone from the cosmos. But we manage,

Despite losing much of the night sky to the lights of Los Angeles, Griffith Observatory still transports its visitors to other worlds at the eyepiece of its telescope and beneath the dome of its planetarium theater. (Photograph: Michael Copeland, Griffith Observatory)

and it's also clear that the sky still means something to us. It now turns up in our dreams, in our visions of life after death, in reports of UFO sightings and encounters with aliens, in contemporary fairy tales (motion pictures and science fiction), and in our efforts to reach the sky—the rocket-propelled vehicles we send into space. These launches and journeys are not just technological stunts or attempts to collect prosaic knowledge. For many of us, they have an emotional dimension and a spiritual meaning. Our ancestors yearned for heaven. We do, too.

We may still yearn for heaven, but we have forgotten the sky we once knew. Confidences we kept with it for thousands of years got boxed up with other family traditions. They were marked "mythology," "folktales," "fables," and "stories," and left in a storage facility—the library. I intend to retrieve that ancient packet of love letters and reveal the stories they tell about our past affairs with the heavens; this is one of those kiss-and-tell books about the sky.

2

Behind the Stories

IN the fourth century B.C., Aristotle outlined many fundamental principles of storytelling in his *Poetics*. Although his remarks refer in detail to the poetic literature of his own culture, and in particular to dramatic tragedy, some of what he said holds true for stories told by other people at other times. According to Aristotle, the most important element of a story is the structure of the incidents, or the plot, which must have a beginning, a middle, and an end, all of which are connected by cause and effect. This structure is what gives events meaning and gives the story sense.

A story, then, is one thing leading to another, and the link of cause and effect binds the incidents together. We see this happen in everyday life, of course. All kinds of things happen, and some of them are directly related to each other. A telephone call from the office generates an order for merchandise, which puts people to work to provide it, and their work ripples through the economy of a nation, which may, in turn, influence the course of world events. On the other hand, many more events take place every day and have no real bearing on each other. A shopper searching for auto parts pulls into a mini-mall in West Lost Angeles, and a hiker in Glacier National Park pauses at the call of a loon. There is no story there, unless someone puts one together.

When someone does put a story together, we get relationships and connections. We understand not only what happened, but why it happened. The story makes sense. If the story is rational and conforms to Aristotelian standards of logic, it can be very powerful. Science, for example, tells effective, persuasive stories about nature with rational logic. They are powerful because they can predict what will happen next. Scientific stories are also testable. When evidence contradicts their plotlines, they are discarded in favor of better stories. The new stories make better sense.

Making sense, however, is not always the same thing as being rational. It is possible to tell a useful story that obeys a different kind of logic by making use of different relationships devised by our brains. With them, too, our brains can create a sense of congruence and mental propriety. These relationships are not products of cause and effect but of analogy and metaphor. We have told both kinds of stories about the sky.

For example, the sun is said to sustain all life on earth, and in fact it does. When, however, someone says—as the Guajiro Indians of South America say—that the moon has many women and takes every woman, something else is going on. The Guajiro tell a perfectly good story about the moon and women, but it will never find a home in a university astrophysics text. Such beliefs may find no allies in science, but they can still render events in the physical world comprehensible.

To understand the value of storytelling, we should know something about consciousness, but consciousness itself turns out to be a rather slippery concept. Although it is hard to pin down exactly what consciousness is, we are familiar with some of its attributes and consequences. Consciousness contrives the world we perceive, and it does so by letting us organize in space and time the events our senses record. Space and time comprise a framework for what happens in the world. As abstract as the ideas of Einstein's theory of relativity and as immediate as a business appointment kept by calendar, clock, and street address, space and time provide a stage in our minds for the world's events.

Consciousness also edits the world. There is far too much going on, far too much information collected by our senses, for our conscious minds to juggle. Unable to keep all of those balls in the air, we throw most of them away and retain and relate the data we need to get along. Information that does make it through the filter of consciousness must be reconciled with what we already know. Any new data must reside among existing facts and must be integrated with the mind's prevailing worldview; it's done by storytelling.

Through language and symbols, we human beings are able to describe the world around us, but we are not strictly speaking unique in this respect. Most organisms respond to their environments, and some describe them. After discovering an exploitable source of floral nectar, a scouting honeybee returns to the hive and performs a "dance" that informs other members of the bee community about the distance and the direction of the find. This dance is description, however, and not a story. The bees are not concerned about how the blossoms got there or why the honey exists. In a sense, every living thing is an observer of nature, but not necessarily a storyteller.

By contrast, human consciousness adds a distinctive difference to the collection and transmittal of information about the world. Consciousness fabricates the world by letting us tell a story about it. We, like the bee, interact with the world around us, describe what happens, and tell others about it. But our consciousness also lets us observe and describe ourselves observing and describing.

Every time the bee locates honey, it draws the right dance from a repertoire preserved by instinct and shares the appropriate message. In our case, however, we tell a story, and the story we tell helps determine what we do next, either in familiar territory or in a new situation. Our self-awareness enables us to write new scripts to adapt to a changing environment, whether natural or created by ourselves.

Even if our understanding is incomplete or inaccurate, a good story makes us feel as if we understand why things are the way they are. Feeling that we understand the world can be as influential

This medieval pilgrim has traveled to the horizon, crashed through the celestial dome, and seen the mechanisms once thought to make heaven move. He has entered a transcendent realm beyond the blue horizon and the confines of normal reality. Although this illustration is often mistakenly identified as a sixteenth-century German woodcut, it is actually a nineteenth-century work, probably drawn by Camille Flammarion, a famous French astronomer and popularizer of astronomy. It attempts to convey an earlier concept of the universe and harbors some of the meaning the sky once held for our ancestors. To understand the meaning they saw in the sky, we have to go behind their stories to understand the symbolic functions performed by the celestial components of their tales. (Griffith Observatory)

as actually understanding it. It gives us confidence in the future and willingness to act, even in the face of the unknown. Storytelling is not, then, just a frothy pastime. It can be a serious enterprise and a practical tool for survival.

TYPES OF TALES WE TELL

Jack Maguire, a writer and producer for cable television, summarizes the types of stories we tell in his book *Creative Storytelling*. His list includes formula tales, fables and parables, fairy tales, folktales, legends, realistic stories, and myths. Each type of story takes a different approach to narration, has a different purpose, and achieves a different effect.

Throughout *Beyond the Blue Horizon* we encounter nothing but sky tales, but they come in all of Maguire's flavors. Listed below is an example of each type of story and also a sky tale that demonstrates what an effective symbol the sky is for what we want to express about the world and our lives.

Formula tales rely more upon manipulating words and logic than on plot for their impact. *Chicken Licken* (known in England as *Henny Penny*) is a formula tale in which Chicken Licken is struck by a vagrant acorn and jumps to the extravagant conclusion that the sky is falling. Her error drives all of the barnyard fowl into panic. Hardly a serious exploration of real catastrophe, the tale seems to be nothing more than an excuse to recite a litany of silly sounds: Chicken Licken, Henny Penny, Cocky Locky, Ducky Daddles. Goosey Loosey, Turkey Lurkey, and Foxy Woxy. The story does exploit, though, a fundamental meaning of the sky: The world as we know it can't exist without sky, and so its collapse is a genuine threat to the established order. The king, who is responsible for the safety and prosperity of the nation, must be told. Too smart to be taken in by all of this nonsense, the fox sees instead an opportunity to enrich his larder. He tells the panicking poultry he knows a shortcut to the king's castle and guides them to the entrance to his den. In their rush to inform the king, the fowl all fall into the paws of the fox and are doomed.

This is the kind of story that is meant to be enjoyed aloud, and it carries a simple message about common sense along with the funny phonetics. Embedded in it, however, are traces of the sky's fundamental significance. Its cave-in challenges the prevailing law of the world, and the king, the agent of celestial order on earth, needs to be informed.

Fables and *parables* usually impart advice—a moral or some other useful tip on sensible behavior. Talking animals often populate fables, while the setting in a parable is more realistic, and the theme more serious. The Zulus of South Africa tell a kind of Aesop's fable about a hyena who, having found a bone, chanced to see the full moon's reflection in a pond. Mistaking the moon for a morsel of meat, the hyena dropped the bone and splashed into the water. Of course, his imaginary meal disappeared as soon as the water was disturbed, but he saw the moon again once he had climbed back upon the shore. His second dive for the moon's reflection brought him nothing but another muzzle full of water, and this time the perfectly good bone he had found and dropped was stolen by a more realistic hyena. When somebody behaves like the silly hyena and throws away something valuable for nothing, the Zulus say he's like the hyena that jumped into the moon's reflection. The moon stands for everything that is real but out-of-reach and able to delude us with a false value.

The term *fairy tale* is sometimes used for all children's stories, but the fairy tale really has its own special character. It involves or takes place in another realm or world, not the one in which we usually reside. Fairy tales are really stories of the supernatural. Other laws prevail in them, and the creatures that inhabit them do not belong to ordinary reality.

Several theories attempt to account for the idea of the fairy land, but it seems to be related to traditional notions about the spirit world, the realm of dreams, and the kingdom of the dead. Modern European children's literature, beginning with Charles Perrault's *Contes du temps passé* in 1697, domesticated, to a degree, what was once dangerous territory into a less deadly but not necessarily more tranquil zone.

Peter Pan, J. M. Barrie's Edwardian children's story, qualifies as a contemporary fairy tale, and it puts Peter's home ground, Neverland, beyond the sky among the stars. Once aloft for Neverland, the route is clear. We are guided by a star: "Second to the right, and straight on till morning." Although the sky may not be the only highway to the fairy realms, it is easy to see why the sky and Neverland might be linked. Both are remote and inaccessible kingdoms beyond human control.

People also tell *folktales*, and these are stories about how people behave. Firmly embedded in the world of those who tell them, folktales invoke universal themes: love, regret, success, failure, freedom, captivity. The sky can play a role in folktales, too. For example, a famous Chinese folktale, the story of the Weaver Maid and the Cowherd, is not just a bittersweet chronicle of two unhappy lovers confined to opposite sides of the river. It is a story about the bright stars Altair and Vega, which shine down upon us from opposite sides of the "river of heaven," the Milky Way. The two stars are high and visible at the time of year when the story says the two lovers have their only chance to meet.

Legends are stories based in historic but embroidered fact, and they, too, may exploit the sky. Alexander the Great conquered most of the known world in a whirlwind fourteen-year campaign. By the time of his death in 323 B.C., he had created an empire that extended from the Adriatic Sea and North Africa to Samarkand and the mouth of the Indus River. Although he was only thirty-two years old when he died, he had guaranteed his place in the *Guinness Book of World Records*. It's no wonder people told stories about him, and in the Middle Ages, the *Romance of Alexander* exaggerated his deeds and turned his life into legend for an enthusiastic audience. He seemed to be on a first-name basis with the gods, and Alexander's destiny was said to have been written in the sky. According to the legend, when Alexander's mother, Olympias, was about to give birth, she delayed the delivery of her son until the astrologer Nektanebos saw a sign of celestial approval in the favorable disposition of planets. When the proper moment arrived, he cued her to "give birth to a ruler of the world," and Alexander's entry into the world was announced by thunder, lightning, and earthquake. Because the sky is a source of power and a home of the gods, it is only fitting that heaven should salute Alexander the Great at his birth. Heaven had conspired with earth, and Alexander—the beneficiary of their intrigue—would, like a god, transform the world.

Realistic stories that reflect real life are also now part of our literature, and their authors make use of the sky and its meaning when it suits their purpose. Thomas Hardy's *The Return of the Native*, a classic nineteenth-century English novel, opens with an atmospheric conjuration of Egdon Heath, the scene of most of the story's action and the real star of the tale. Before the first sentence is over, we see how the stage is lit: "A Saturday afternoon in November was approaching the time of twilight . . ." By the novel's second sentence, the entire heath is the floor of a tent of cloud that shuts out the sky. Through the sky's dark work the heath wears "the appearance of an instalment of night which had taken up its place before its astronomical hour was come." Sky and heath continue to brood through several more sentences until the heath is identified as a "near relation of night." For a contrast, Hardy tells us the sea is "distilled by the sun, kneaded by the moon, . . . renewed in a year, in a day, or in an hour." Egdon Heath, however, has been "from prehistoric times as unaltered as the stars overhead."

Critics may debate just how realistic Hardy's description of Egdon Heath is and may complain about the transparency of his symbolism. But appreciative or not, the reader is propelled into a landscape that foreshadows what is to come. Part of the symbolic meaning of remorseless, unchanging Egdon Heath is borrowed from the sky.

We can find the sky, then, in realistic literature, legends, folktales, fairy tales, fables, parables, and formula tales, but *myths* are probably what first come to mind when we talk about stories that deal with the sky. Gods and heroes are the main characters in myth, and the sky is often their arena. Myths may describe the origins of things, including the universe itself, and help define the way people see their relationships with the gods. In antiquity, the sky was frequently invoked in religious ritual and symbol to authorize, create, and maintain the prevailing world and moral order. These concerns are all part of what we mean by the sacred, and myths are the way we talk about the sacred. Myths also explain natural phenomena, and in this kind of myth divine power is an armature for the plot. Myths reflect the religious beliefs and philosophical principles of a people, and for that reason the sky can play a part in them.

FISHING AROUND WITH HEAVEN

A story from the Ajumawi band of Pit River Indians of northeastern California is a good illustration of how a myth works. Floyd Buckskin, a member of the Ajumawi Tribal Council who has worked to preserve Ajumawi history and traditional lore, first heard this story from his grandmother. When Floyd Buckskin and California archaeologist Arlene Benson published it in 1987, they recognized its theme and called it "How the Seasons Began."

This myth describes how the seasonal order came to be. It involves the First People, original inhabitants of the earth who lived here before humans arrived. Now they are the animals, the plants, and the celestial objects. For the Ajumawi, the myth is not just an intellectual exercise. It is a didactic story of survival. It is really about the migrating sucker fish, which the Indians diverted into pond traps by the lake shore, where they speared or netted them in great numbers.

At the time of Creation, according to the Ajumawi, Sun and Moon, an old married couple, lived on the eastern horizon, and shared the responsibility for providing the world with light, but Sun's heat was too much for the earth. The present world order hadn't been inaugurated then, and there was a crying need to do something about the sun. That is how things stood when Fisher Man and his younger brother Weasel Man wandered by the round house of Sun and Moon. The fisher, by the way, is a large member of the weasel family. Despite its name, it doesn't fish. It can swim, however, and is at home in the trees.

Now, Sun and Moon had two daughters who lived at home with them, and the two brothers decided to court them. Moon Old Man did not trust the pair, however, and he put Fisher Man through a series of trials of skill and stamina with the hope that Fisher Man would not survive. After several attempts had failed to eliminate Fisher Man, Moon invited him to one more contest at "the place of the teetering pole," a rock on the edge of a high cliff by the Eastern Sea. That rock is still there, a large, ridged boulder, its broad flank carved with many round depressions known as cupules. The Indians say these marks are footprints

California's Ajumawi Indians identified this marked boulder, now in Ahjumawi Lava Springs State Park, as "the place of the teetering pole." According to tradition, the moon was launched into the sky from here by Fisher Man at the time of the world's beginning. (Photograph: Stephen D. Moore)

left on the rock by Sun Old Woman, Moon Old Man, North Star, South Star, Fisher Man, and other First People.

Carrying their fish spears, agile Fisher Man and sly Moon hiked down to the rock. Moon Old Man balanced his fishing spear on the boulder and challenged Fisher Man to walk out to the end of it and bounce. The pole reached beyond the edge of the cliff and into the open air above the Eastern Sea. Far below, Fisher Man could see *Pal-aht-ki-sum,* a monstrous, supernatural serpent lurking near the water's surface. It had the fangs, coils, and rattles of a giant rattlesnake, but elkhorns crowned its venomous, triangular head. Ragged weeds, slimy moss, and the flesh of dead things were caught in the tines of its antlers and dropped around its neck. Its wide mouth made a whirlpool out of the water it swallowed. Rolling and weaving, the great viper enjoyed the prospects of a clumsy pole dance on the precipice above.

Fisher Man had a pretty good idea now why Moon Old Man had suggested this game. But he couldn't back out, so out to the tip of the pole he minced. This was not so risky as it looked. Fishers have little trouble negotiating their way around the high limbs of trees. Fast and steady on their feet, and clever enough to escape tight situations, they usually manage to stay out of trouble.

When Fisher Man arrived at the end of the rod, he secretly hooked the World's Cane (what we know as the Big Dipper) over the end of it to keep it steady. Indian shamans—individuals who were believed to have great spiritual and magical powers—often carried hooked canes, or crooks, like the World's Cane. Moon Old Man never noticed Fisher Man locking the bouncing pole into place with the World's Cane, which contained great power. It could even be invisible, and that's why Moon Old Man couldn't see it. Moon Old Man tried as hard as he could to bounce Fisher Man off that pole and into the hopeful mouth below, but with the help of the World's Cane, Fisher Man rode that rod until Moon Old Man got tired of shaking it.

Prancing in quiet triumph to a more solid roost, Fisher Man politely mentioned it was now Moon's turn to perch on the pole. Moon Old Man was not eager to go and walked only a short way out on the stick. Fisher Man insisted he continue to the end. When he reached it, he looked down and observed that the great serpent was still there. A snake like that would shake anyone's confidence, but Moon started to bounce. He realized he had a little talent for the stunt and continued to bounce, a little higher each time. Fisher Man gave him a snap of the pole, and then Moon Old Man was gone. Fisher Man looked for him down in the serpent's maw. Wherever Moon had bounced, it wasn't there.

Returning to the round house of Sun and Moon with the fish spear, Fisher Man told Sun Old Woman he didn't know where her spouse was. Then, as he sat by the ocean and pondered what had happened, Fisher Man heard Moon Old Man's voice above him. Moon was in the sky, and he seemed to like it there. Now that he was up there, he realized it was where he and his family really ought to be. He asked Fisher Man to send his wife and his two daughters to join him. Fisher Man sensed that was the right home for them, too.

By bouncing the pole on the boulder again. Fisher Man catapulted Sun Old Woman and her two daughters into heaven. Fisher Man put one of them in the north, where she is North Star; and the other in the south, where she is South Star, or *Wah Wah* ("snowshoe hare"). Moon Old Man and Sun Old Woman now pursue their proper courses, while North Star holds her post in the one spot in the sky that does not move, the north pole of the sky.

By remaining in place, North Star keeps the world organized and the sky steady. Even though the World's Cane also circles the north pole of the sky, it keeps its grip on the North Star as it kept its grip on the teetering pole. *Din-hin-na-oose* (the place of the teetering pole), the cup-marked boulder where Fisher Man and Moon Old Man took their bounces, faces north. A fishing spear balanced across its ridge is like a compass needle indicating north and south, and in a sense it points toward North Star and South Star. Because the boulder, on which the fish spear was oriented and balanced, played a part in getting Sun, Moon, North Star, and South Star into the sky, it is said to be a place where the present world was created. The spear and the boulder helped to organize the sky, and organizing the sky organizes the world. The establishment of world, or cosmic, order is

After Fisher Man survived a round of bouncing above the waters of the Eastern Sea, it was the moon's turn to perform the same dangerous stunt over the unsettling maw of the great supernatural serpent that hunted there. The fishing spear is propped over the boulder, and after bouncing high off the pole Moon Old Man is catapulted into the sky, where he is soon followed by his wife, Sun Old Woman, and two daughters, North Star and South Star. This story not only accounts for the presence of these celestial objects but links them to a seasonal configuration of the sky and the seasonal return of fish to Big Lake. (Drawing: Floyd Buckskin)

what we used to mean when we talked about Creation.

South Star, unlike North Star, does not stay in one place, but when you can see it, from the end of January to the first of March, it is low on the southern horizon, a modest star that points the way south. It is in the constellation of Puppis the Stern, near the stars of Canis Major the Greater Dog, in which we find Sirius, the night sky's brightest star. We know South Star as ν (nu) Puppis, a third magnitude star that hugs the horizon when it is seen at all from as far north as Ajumawi territory. The snowshoe hare, for which it is named, changes color with the seasons. The hare is white in winter and dark in summer. South Star also changes its appearance seasonally. It shines in winter, and the rest of the year it is invisible.

When you stand—in South Star's season of visibility and at midnight—in front of the rock where Fisher Man and Moon Old Man teetered on the pole, the Big Dipper, or World's Cane, is invisible, hidden by the hill to the north. That's why Moon Old Man couldn't see it. Its power holds it, however, to the North Star and to the bouncing pole. Behind you, South Star skirts the southern horizon.

South Star turns this sky tale into a fish story. Its appearance in winter marks the season when suckers arrive in Ajumawi waters and are fished out of their traps. The star's visibility is mostly restricted to the month of February, and the Ajumawi word for that month, *al-dammit*, means "fish time." Because the sucker was such an important source of food, this season and its herald—South Star—are key components of the story of Creation. For the Ajumawi, a world ordered by the seasonal appearances of celestial objects is essential for survival. Their story of Creation emphasizes the practical importance of celestial phenomena in the symbolic language of myth.

THE MEANING, PURPOSE, AND ORIGIN OF MYTH

This Ajumawi story fulfills several obligations of myth. It explains the origin of the present world order and provides narrative support for whatever activity had to take place at the time of the suckers run. Each year the original Creation was, in a sense, reenacted when the sun, moon, North Star, and South Star took their places again on the seasonal stage. Mircea Eliade, a specialist in comparative religion, emphasizes the importance of this kind of story in *The Myth of the Eternal Return*. In telling the story about these primordial events, the myth reveals their connection with a supernatural power, which installs the world order and establishes its sacred status. The world order has consequences for people, who must live according to the prevailing rules. Their customs, rituals, and beliefs reflect these rules and are taught or illustrated by the myth.

Eliade's ideas explain some aspects of myth, but not everyone who has tried to account for the nature and source of our myths agrees that the evidence confirms his conclusions. Many of these experts take an entirely different approach. Even when the ancient Greeks looked at their own mythology, some of them—the Stoic philosophers—saw the stories as no more than allegories. They couldn't take the all-too-human theatrics of the gods seriously and assumed the gods were really symbols that masked hidden meanings. As symbols, the gods stood for phenomena and forces in nature and in human behavior.

Since the Greeks, scholars have argued at length over the origin and function of myth, and many of their explanations have treated myth as a distortion of reality. The gods may, on the one hand, be nothing more than misremembered mortals, like Alexander the Great, who eventually become deified. This interpretation originated with Euhemerus of Messina, who wrote the *Sacred History*, a popular critical review of myth and sacred tradition, early in the third century B.C.

Max Müller, a nineteenth-century philologist, saw myth as a different kind of distortion. To him, myth was not inspired by muddled memories of historical facts but by naive, sentimental, and distorted descriptions of natural phenomena. He emphasized the way prehistoric people personified the sky and everything in it, particularly the sun. For this reason, his school of thought is sometimes known as "Solar Mythology." In Müller's interpretation of myth, nature—and especially the sun—was transformed into a divine pantheon through what Müller called a disease of language.

For example, in the Indo-European languages, inanimate objects originally had gender. Most of this is now lost in English, but the sun is still male in French. The moon is still a lady. Müller concluded that these arbitrary grammatical conventions slowly but surely led to conferring gender, personality, and divinity upon celestial objects and natural forces.

A more anthropological explanation of myth sees the personification of nature as a consequence of primitive thought. Our savage ancestors, in this view, misread the world around them as a spirit-ridden realm. Everything was alive. Myths, then, were nothing more than uninformed accounts of nature based upon a false premise. This view was formulated by Edward B. Tylor, an Oxford professor of anthropology, in *Primitive Culture*, first published in 1871. Tylor saw a link between the animist beliefs of surviving primitive peoples and the mythologies of past cultures.

Another doorway to myth opened as Tylor's anthropological approach evolved in the hands of others, such as James George Frazer and the Scottish mythographer and folklorist Andrew Lang. Frazer saw myth as an inaccurate and naive perception of natural phenomena or human conduct and linked the stories to ancient customs and rituals. The main product of his studies, the twelve-volume edition of *The Golden Bough*, published in 1914, especially explored the connection between the notion of the Dying and Reviving God and the widespread custom of ritual sacrifice of the king on behalf of the continuing prosperity of the land and the well-being of its people. Lang, celebrated for his long ethnological campaign against Müller and his solar mythology, advocated the case for myth as a body of stories intended to account for the prevailing manners, customs, and rituals of ancient and living, primitive peoples. Like Tylor, Lang advanced his argument with anthropological examples.

Tylor's anthropological approach to myth evolved in the hands of others to the position that myths were really the narrative poetry or drama that accompanied ancient rituals. They were thought to have no meaning beyond those rituals, although they could acquire other meanings once their parent rituals had fallen out of use and disintegrated to leave only the relic myths behind. For scholars such as anthropologist Lord Raglan, classicist Jane Ellen Harrison, and S. H. Hooke and Theodor H. Gaster—both specialists on the literature, mythology, and religion of the ancient Near East—myth came to represent a form of imitative magic. As such, it was still a story that told people what was the right thing to do in the world they inhabited. Through some magical or religious ritual, whose power was ignited or enhanced by an accompanying recitation of the myth, people believed they could influence the world around them and produce some important, desired result. Such magical thinking was judged, of course, to be a response of the less developed, more childlike mind of the human race in its infancy.

It is natural enough for us to imagine the ancients and the world's uncivilized, tribal peoples as children. We, after all, have arrived much later on the scene and therefore benefit from the accumulated knowledge and experience of many generations. So we think we are much more mature, but that is just a myth we tell ourselves. When Australia was first sighted by a Dutch explorer in 1605, its Aborigines already had forty thousand years of cultural tradition behind them. They may not have been technologically advanced, but they were culturally mature.

We aren't really more mature than our ancestors were, but we do know some things they didn't. That makes us see ourselves as significantly more advanced than our ancestors, and in a sense we are. We know, for example, that thunder is not noise from Thor's hammer, but just the sound produced by the rapid expansion of air heated along the path of a lightning bolt. Our scientific understanding of the weather allows us to predict it and protect ourselves from it more effectively than our ancestors did. So we have more knowledge, and we now tell a better story about thunder. But that does not mean that mythical thought is simple, illogical, or unsophisticated.

The French anthropologist Claude Lévi-Strauss treats myth as a kind of logic. He looks at the way elements in a story are combined in order to see how this kind of logical thought works. It is the abstract structure, then, and not the concrete content of the story that he wants. By isolating this structure, he believes it is possible to reconstruct the cognitive landscape on which all myths are mapped. Once this structure is recognized, it reveals how the components of any story relate to each other. What may seem to be arbitrary elements, items of nonsense, off-the-wall events, and arcane details no longer seem out of place. They are necessitated by the underlying structure and by the relationships between participating components. The Structuralist interpretation of myth explains their presence in terms of the relationships that must hold between the various pieces of the narrative and not by some objective meaning.

MYTH AND THE STRUCTURE OF THOUGHT

The ideas of Lévi-Strauss about the purpose and nature of myth are considered by many scholars to be an important contribution to the modern scientific study of myth. He saw an underlying pattern in the way many different traditional stories are told and concluded that it reveals something about the way in which our minds use categories and narratives to preserve relationships and manipulate abstract concepts.

According to this theory, the details of a myth can be and often are exchanged for some other details while the structure remains unchanged. The structure is really a pattern of thought that is applied to many different problems. For this to be true, all human brains have to think in basically the same way, and that pattern or structure of thought must have been the same in the past as it is now. If not, we wouldn't have a chance of recognizing it ourselves.

These grand and sweeping notions are assumptions. In some respects, they are quite reasonable. All human brains probably do operate in the same way at some neurophysiological level. Whether that way of functioning always makes the mind manipulate data and structure every story in a universal and recognizable pattern is, however, another matter. Certainly, Lévi-Strauss and the Structuralists think it does. By extracting the pattern from some myths and successfully applying it to others, they believe they verify the validity of this approach.

To Lévi-Strauss, a myth has no inner core of meaning. It is nothing more than the structural relation of the components in its storyline. These make use of the tangible properties of things to convey relationships. This is possible because our minds construct schemes to classify what goes on in the world around us. These systems of classification, like the periodic table of elements, catalogue our experience in terms of similarities, differences, and relationships.

What, then, is the basic structure of myth? According to Lévi-Strauss, a myth expresses a conflict or contradiction between two opposed but complementary forces. This means that we exploit our ability to see similarities and differences between things. We use that way of organizing information about the world around us to polarize a story. In each case, a story takes advantage of two contrasting qualities in our experience. These are complementary opposites, like night and day, winter and summer, wet and dry, and female and male. Through them, the story explores whatever abstract concept may be on our minds. It does this in terms of imagery and associations.

These are the actual agents in a myth—characters, objects, and circumstances. We get them from the categories we use for classifying the things we encounter in the world. This says that our minds work in some kind of binary language and, at least sometimes, handle data the way information is processed in some computers. Brain research has not, however, reduced the operation of the mind—or even the way it handles information—to digital processing. The ease with which we liken our brains to computers is a modern myth about human nature. It makes us feel as if we have some insight about ourselves that is consistent with the technology in which we are immersed. But just because we make clever use of polarities does not mean that our minds are constructed that way. We do often see the world in terms of these oppositions, however—good and evil, black and white, nature and culture, earth and sky—and our myths resolve them through some mediating agent in the story.

According to Lévi-Strauss, each story we tell resolves a conflict. We combine complementary opposites to establish logical relations. Mediation of a conflict is, then, the purpose of myth in the Structuralist view. This doesn't mean that problems are solved. It means attitudes are developed. By reconciling a contradiction, the myth allows us to live with the opposing forces that created it. We accept the story it tells. That is how our minds deal with the world.

Perhaps the real value of the Structuralist interpretation of myth to our understanding of sky tales is its insistence that these stories that seem to be about the sky are really layered. Like baklava, they carry one theme on top of another and provide a dense confection of meaning for every circumstance. This is, in a sense, how American folklorist John Bierhorst deals with the problem of myth in general in the introduction to *The Red Swan*, a collection of myths and tales of the North American Indians. There, he defines a myth as "an unverifiable and typically fantastic story that is nonetheless felt to be true and that deals, moreover, with a theme of importance to the believer." With this as his starting point, he soon moves into the notion that most myths involve at least one of six types of fundamental transitions. These include

1. the passage from unconsciousness to consciousness (from sleep to wakefulness, from the womb to birth)
2. the passage from sexual innocence to knowledge (puberty, sexual arousal, marriage)
3. the passage into the animal world or into the celestial world (including shamanic transformation)
4. the passage to or from death
5. the passage from nature to culture
6. the passage from chaos to order (creation and cyclic renewal)

In Structuralist terms, any one of these passages is as good as any other, and a myth that seems to be about one of them is really about all of them.

The sky, then, can also be used in stories to illustrate and illuminate anything that matches the nature of something we see there or fulfills the functions celestial phenomena seem to fulfill. Because the stars look like other things, behave as other things behave, and display characteristics other things display, we can expect to see them treated and described the way other things are treated and described. Their stories look like exercises in analogy.

To people in everyday life, however, the message is more important than the thought process that delivered it. For example, we need to know if the last frost of the spring is likely to have passed. If we're wrong, we may starve. So even if the structure of myths truly tells us something about cognition, the content of the myths organizes empirical experience into useful packets of information. Myths tell us something specific about the world, and people, in general, make use of that kind of information to adapt and survive.

MYTH, SCIENCE, AND METAPHOR

Science and myth are different, but they are both storytelling. Roger Bingham, a science writer and the producer and presenter of documentaries for public television, stresses this point. There is an important difference, he adds, between myth and science: Science changes, and myth does not. When complicated and confusing data are imitated to our satisfaction by a scientific model, we feel we have a grasp of the phenomenon. That model is just a story, and we change that story when we acquire new information.

Unlike science, myth does not change in response to the acquisition of new data. The concerns of myth are timeless. Myth is supposed to reveal transcendental mysteries. Some myths eventually die or change but not because they are contradicted. Instead, the rituals, symbols, and ceremonies that convey the myth's truth to the community lose their effectiveness. Myths, then, are not tested by data; they are matters of belief.

As we try to understand the sky tales in the chapters that follow, we shall find that all of these different approaches to myth will be useful and necessary. The problem is a lot like the elephant investigated by the blind men. They each grasped a different part of the beast and reached a different conclusion about what it was. The elephant was actually everything they found and more. Myths, too, are multidimensional. They are closely linked with ritual and drama, with the idea of the sacred, with the unconscious integration of information in our minds, with the dynamics of the human psyche, and with a people's vision of its own cultural identity.

UNIVERSAL THEMES AND STRUCTURING FORCES

Appreciation for the complexity and the metaphorical logic of myth requires an understanding of the psychological component of myth. Believing that myths are stories told by the unconscious part of the mind, Joseph Campbell applied the principles of Jungian psychology to the study of myth in *The Hero with a Thousand Faces* and emphasized what Carl Jung, a pioneer in the exploration of the human psyche, said in *The Archetypes and the Collective Unconscious*, in *Symbols of Transformation*, and in other works: The logic of myth is like the logic of that other handiwork of the unconscious mind —dreams. According to Jung, dreams and myths reveal, in similar imagery, the archetypes, or primary psychological patterns, shared by all people. Jung believed that these archetypes reside in what he called the "collective unconscious" and that they are expressed in images and themes contained in our minds.

We don't know if there really is a collective unconscious or how it makes contact with the neurophysiology of our brains. We do find, however, themes and patterns and symbols that show up in myths from all over the world. In *Primitive Mythology*, the first book in Campbell's four-volume study of comparative world mythology, he traces some of these themes, patterns, and symbols to the forces that act upon all of us and shape not only our imaginations but the nature of all human life on earth.

Campbell lists five such "structuring forces," and they command our interest here because they all have something to do with the sky tales we tell. They include gravity, which until the space age confined us to earth and which is, in fact, responsible for the very idea of *down*. Because we associate gravity's power to keep us earthbound with all of the limits the earth imposes on us, we think of the sky as a realm of freedom and uninhibited power.

The familiar cycle of light and darkness, generated by the oscillating pattern of day and night, is the second of these structuring forces. The daily change between dusks and dawns imposes a daily pattern on our behavior and also provides a symbolic model for all conflicts and polarities. With sunrises and sunsets, with blue daytime skies and black nighttime skies, the sky itself shapes the character of the world in which we live.

Campbell singled out the moon as the third important influence of nature, but he quickly added the Milky Way, the stars, and, in fact, the entire night sky. Objects in the night sky move and change like clockwork. These rhythms also structure human life, even if there is no direct physical link between them and ourselves, as Campbell wrongly suggested for the moon.

The fourth universal force that colors the way we look at the world and ourselves is the complementary opposition between male and female. Unlike the first three, pairing and sex are not something conferred upon the earth by the sky. But the male and female forces and the interaction between the two are so much a part of our lives that we project it all onto the sky and see there the mirror of our own heart.

In the same way, we also see the fifth structuring force in the sky. This is the cycle of our own lives —birth, growth, death, and, as we imagine, rebirth. The same cycle drives all life on earth. We see it in the life cycle of the butterfly, in the seasonal cycle of flowering plants, and in the agricultural cycle of sown and harvested grain. It is also in the daily birth and death of the sun, the monthly growth and decay of the moon, the annual journey of the sun through weakness and strength, the shadowing of the sun or moon during an eclipse, the ascent and fall of the morning star, and the periodic disappearance of certain stars in certain seasons.

All five of these influences find their way back into the sky. We can almost believe, then, the Ashanti of West Africa when they relate a story about the origin of all the stories in the world, which were once kept in a box owned by Nyame, the Lord of the Sky. Spider knew about the box and wanted it. Nyame offered to sell it, and when Spider finally managed to deliver the payment, he hustled back down to earth with the box. Everyone there was anxious to see what was in the crate, and when Spider opened it, all of the stories came pouring out. They ran everywhere. Spider managed to catch some of them, and his wife got a few. The people in his village also caught some stories to tell, but the rest went to live all over the entire world.

3

Up

WE know where up is. It's overhead, high above us. Up is the wild blue yonder and the black canopy of night. It is the realm beyond earth. With up, the sky is the limit. Up is the sky.

Up, however, is a lot more than a direction. It's an idea. One respectable and massive unabridged dictionary provides eighty-one definitions for *up*. We get up in the morning, and it's not just a change of posture. Getting up is a return to consciousness and action. It's the start of something. We stand up for things we believe in, and ethical behavior makes us upright. Informed or in style, we are up-to-date. Upgrade means to improve. Upbeat is faster; it's also optimistic. Cheered and animated, we feel up. Depressed, we're down. Uppers and downers modulate the mind chemically, and experiences are uppers or downers depending on the mood they inspire. If the computer is up, it is powered for operation and results. When it is down, it doesn't work. "Surf's up" announces the arrival of rolling waves and is also a call to catch the accompanying exhilaration. Success grows as you go up the ladder, and high status is upper class. Prices go up, men are supposed to get it up, and Bugs Bunny asks, "What's up, Doc?" *Up* means much more than "overhead." For a two-letter word, it's a rich mine.

We've incorporated the word into a collection of expressions that reflect a net of related ideas: action, initiative, authority, power, and exhilaration. These are what we expect from the gods, and all of these attributes are also associated with the sky. Gods have power, and so does the sky. That is

Sun symbols have been around since the stone age, but the vintage of this image of the sun is considerably more recent and confirms the continued appeal of the sun's face.

why the supreme gods traditionally reside there. They may, as the sun and the moon, observe the scene below from a celestial zone. They may peer over the pillowed edges of clouds, or they may hurl thunderbolts from mountains which themselves seem to support the sky and scrape its bottom. They create and maintain cosmic order and are linked to the power of the storm and the fertility of the rain.

GOD ON HIGH

Because the sky is the domain of the divine, tales about the sky often bring the gods on stage. When these gods get to work, we learn about the powerful forces that order the world and maintain the cyclic passage of time. The sky puts order on display. Out-of-reach, yet visible to all, the sky gods play by the rules. They can afford to do so, for many of them are celestial objects. Although they can not be manipulated by mortals below, their behavior influences events on mother earth. Confined to earth, we can't tamper with the sky, but we must respond to its rhythms.

The celestial pulse beats in time with the eternal return of day and night, with the monthly prosperity and decline of the moon, and with the seasonal circuits of the sun and the stars. Celestial objects make their regular arrivals, put in their accustomed appearances, depart on schedule, and return at the proper time for another performance; the plot for the most basic story we tell about the sky. It's the story we see in the passage of life and the continuity of generations: birth, growth, death, and rebirth. It is the cycle of cosmic order: creation of order, maintenance of order, intrusion of chaos, and reestablishment of order.

The ancient Greeks personified the real, star-studded sky as Ouranos, the husband of Gaia the earth and the father of the Titans. As such, he represents a primordial creative force. The Greek for *sky*, *Ouranos*, became *Uranus* in Latin. Both names are related to *Varuna*, the Sanskrit name for sovereign god of India's Vedic peoples. *Varuna* contains a root meaning "to cover," which probably refers to the sky's canopy over the earth. Although Varuna was not exclusively a sky god, he functioned as a supreme omniscient judge who upheld cosmic law. In the *Rig Veda*, a collection of early sacred hymns, Varuna establishes the structure of earth and sky to its proper proportion and rhythm with his measuring device, the sun. He is the "emperor of order" to whom mortals appeal for protection and forgiveness.

Varuna was the counterpart in Vedic India to Ouranos, the divine personification of remote night sky in ancient Greece. This representation of Varuna on Rajarani Temple, in Bhubaneshwar, India, is an eleventh-century A.D. *Hindu version of the god as the Lord of Destiny.*

Although the Greeks saw the sky in Ouranos, Zeus was their essential powerhouse in the sky. As the supreme ruler of the Olympian gods, he resided on a mountaintop, upheld the law, and issued decrees packed with celestial power. His weapon was lightning. He gathered the clouds to deliver discipline in the storm and fertility in the rain.

Originally Zeus was an Indo-European sky god of the Aryans who invaded India about 1500 B.C. Although we don't know his earliest name, it was probably something like Dieus or Dyevos. In India's Vedic period, they called him Dyaus; its Sanskrit root links him directly with the sky, for it means "to shine," "day," and "sky." *Zeus* is the Greek version of the same name, and his sovereignty originated with the power of the celestial light of the brilliant daytime sky. His name in Latin, Jovis, shares the same Indo-European pedigree. The Romans also called him Jupiter, which is just a twist on the Sanskrit *Dyaus pitar*, or "Father Sky." As Jupiter, Zeus was also identified with one of the planets, or celestial "wanderers," known to the ancients. The names of others—Mercury, Venus, Mars, and Saturn—are also the Latin names of classical gods.

Jupiter was not, however, just a planet or a simple personification of the daytime sky and the fury of the storm. He was a complex figure with a variety of divine prerogatives. The same may be said of Odin, one of the sky gods of the ancient Norsemen. Odin is called the High One in Norse literature, and he appears to have supplanted Tyr as the Allfather and chief of the gods. Experts on Indo-European languages have reconstructed the original and northwest European name for Tyr as Tiwaz. By the sound of this, we can tell that Tiwaz was just another incarnation of the original Indo-European sky god Dyevos, but as the cult of Odin grew, the status of Tiwaz diminished. In time, Odin became the prime celestial custodian of cosmic order. Day and night and the seasons made their appointed rounds only because of him.

Among the Chinese, the supreme force in the cosmos was Shang di, the Lord of Heaven. He was the source of cosmic order, and this natural harmony was revealed in the cyclical behavior of the sky. Shang di was represented symbolically by the sky's north pole, the spot around which the entire sky seems to circle in an orderly parade. Through the mandate that the Lord of Heaven conferred upon the emperor, the Son of Heaven, celestial order was conveyed to earth.

Zeus began his celestial career as an Indo-European sky god, but in time he became the ruler of all of the Greeks' Olympian gods. This austere and regal bronze head was found at Olympia, the site of an important shrine to Zeus, and it was cast early in the fifth century B.C. (Object in the National Archaeological Museum, Athens)

The Ewe peoples of Togo and Dahomey in West Africa call the highest god Mawu. He is the spirit of the sky and the sky itself. His name means "sky" and originates from a word that means "to stretch over, to overshadow, and to cover." Sir James George Frazer quotes a Ewe's words about Mawu in another of his books, *The Worship of Nature*:

I have always looked up to the visible sky as to God. When I spoke of God, I spoke of the sky, and when I spoke of the sky, I thought of God.

This same kind of connection between god and the sky was sensed, as well, by the Finns. Jumala, one of their old names for the sky god, came to mean "god" in general.

Aborigines of New South Wales, in southeast Australia, had no doubt that Baiame, the Great Spirit Father, knew everything. He couldn't miss a trick because he lived in the sky. Horus, the sky god falcon of ancient Egypt, watched the world through his two celestial eyes, the sun and the moon. In ancient Mesopotamia, the Sumerians called the supreme sky deity An, which meant "high." This name was closely related to the word for sky. An lived in heaven, but he was not the actual sky. He was a cosmic force, a power in heaven. Altaic peoples of southern and eastern Siberia share a common concept in the names of their high god: Tengri, in Kalmuck and Mongol; Tengeri, in Buryat; and Tangara, in Dolgan—all mean sky. The Mongols say he sees everything from his privileged post on high, which is partly why the sky god is a supreme god.

The visible disk of the sun, according to the Aztec of central Mexico, was the god Tonatiuh. Four previous suns had been created in four previous ages and had died with the end of each cosmic era. Tonatiuh was the first moving sun, and the present era is still his. In the Codex Borgia, *a pre-Conquest screenfold manuscript thought to have been prepared in the Mixteca-Puebla-Tlaxcala area, Tonatiuh is depicted with a rayed disk shield upon his back and consuming the blood of an eagle sacrificed on his behalf. The rabbit in the disklike object suspended in the upper right corner represents the markings on the face of the moon, and the moon is attached to another sky symbol: the rectangle with stars (doughnut shapes) dangling from it. (Akademische Druck-u, Verlagsanstalt, Graz, Austria)*

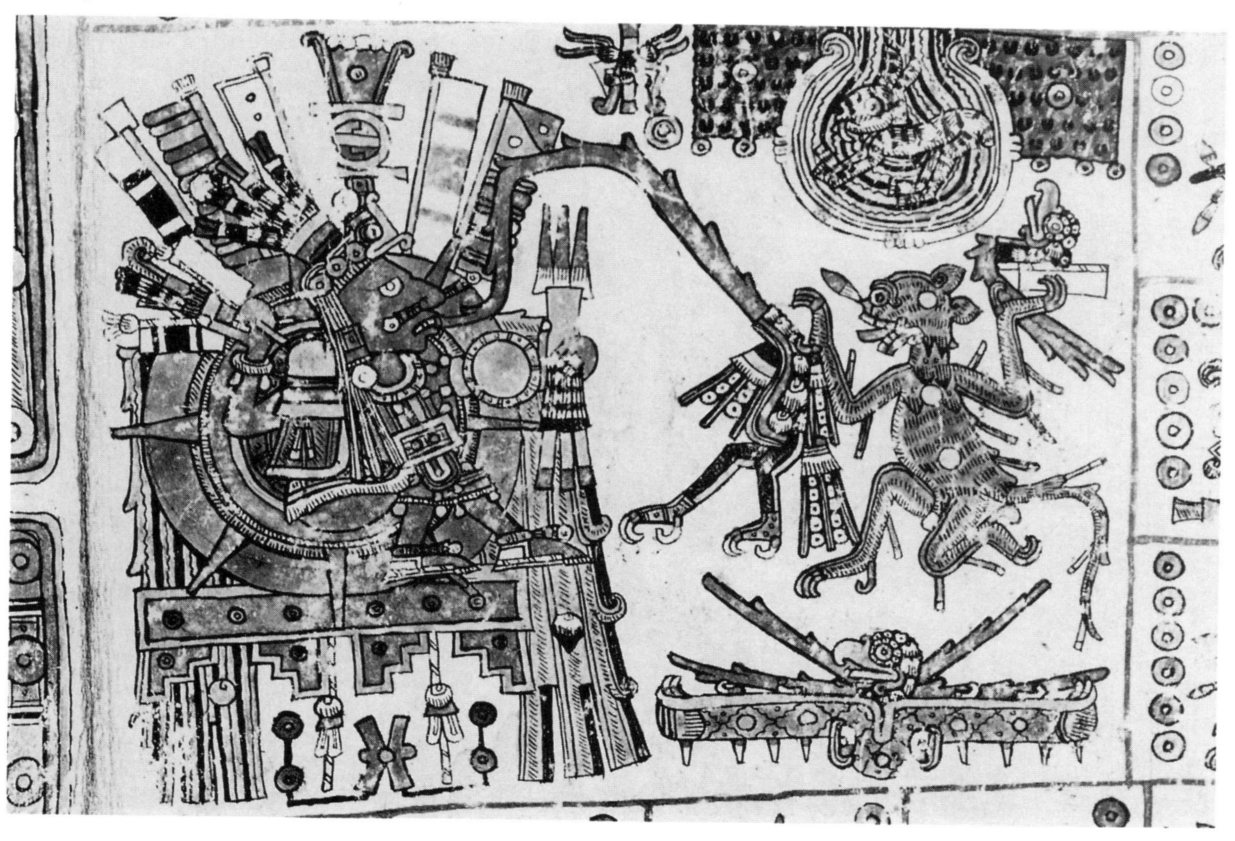

DIVINE LIGHT

To the ancients the light emitted by celestial objects had as much to do with their divinity as their location and movement. They glowed with an inner energy. People could see they had power. Night subjected the world to a darkness only the sun could dispel at dawn. And during the night, despite the unseen dangers that walked abroad, the moon, planets, and stars sailed through the gloom undaunted. They lit their own candles against the darkness.

The brilliant light of the daytime sky is embedded in Zeus's own name, and as a group, the Indo-European sky gods were called the *daevos*, the "Bright Ones." The Yakuts of far eastern Siberia provide many titles for their supreme sky god, including "White Light," and the Ashanti call their high god Nyame, the Shining One. Egypt's sky goddess Nut is "the Great, the Brilliant One." Nanna, the divine moon god of the city of Ur in ancient Mesopotamia, was a "light shining in the clear skies," while Inanna, the Sumerian Venus, burned brilliantly in the evening sky and flared like a "pure torch." To the Aztec, the sun was Tonatiuh, He Who Goes Forth Shining. In the Vedas, the moon is Chandra, "the Bright One." Hieroglyphic texts at Dendera, in Egypt, tell us that the goddess Isis, appearing as the star Sirius,

Two images of the body of Nut, the Egyptian sky goddess, extend across the entire ceiling of the burial chamber of the pharaoh Ramesses VI (ruled 1141–1133 B.C.) in the Valley of the Kings. These two photographs show her head, at one end of the painting, and her hips, legs, and feet, at the other. In the upper image of Nut, she is about to swallow the setting sun (the disk at her mouth), which passes through her body, star-studded like the night sky, and emerges as a reborn disk from her loins (accompanied by the winged scarab, a symbol of sunrise). The stars themselves pass through her body in the lower image. They disappear into her at sunrise and are reborn when night falls.

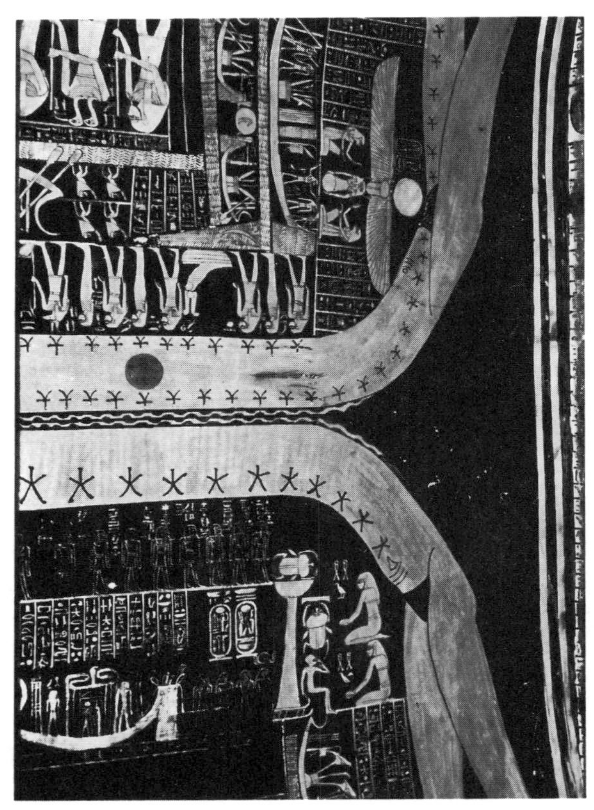

unites with the luminous sun on New Year's Day. These celestial gods dazzled our ancestors with a mystery in open air.

CELESTIAL IMMORTALITY

Another obvious difference between gods and mortals is the gods' immortality, and for that reason they are very old. Itzamná, the high god of the sky in Maya territory, was depicted as an old man. The Tehuelche Indians of Patagonia, now culturally extinct, said that Kóoch, their supreme celestial creator, had always existed. His name means "sky," and the Tehuelche believed he watched over the spirits of the dead Indians, who rose to heaven to become the stars.

According to southern California's Chumash Indians, animals could talk in the world's early days. They were the First People. Then, when death first came to the earth, the First People went to the sky to escape. The Indians said they could still see them up there: Sun, Moon, Morning Star, Evening Star, Sky Coyote, and many more. They resided there, powerful and immortal, while the Chumash had to take their chances down below.

Stars that circled around the sky's north pole enjoyed immortality in ancient Egypt because they never set. Those that fell to earth in the west died there, swallowed by Nut, the sky goddess. Her star-spangled body vaulted the earth, and the stars reborn from her loins rose in the east. Stars like Sirius, the sky's brightest star, also died when they disappeared in the daytime sky, but they were reborn when they reappeared again in the twilight before sunrise.

A dark moon is a dead moon, but in about three days a thin, new crescent shines low in the west. The moon may die when it goes dark, but it, too, is immortal. It returns ever new. Its face may change, but its monthly pulse is constant. Osiris, the dying god of ancient Egypt, was in the same way forever reborn, and in one disguise he was the moon.

Mithras, a solar warrior and an agent of cosmic order, originated in ancient Persia, but his cult spread throughout the Mediterranean during the Roman Empire. He was reborn as *sol invictus*, "the invincible sun," at the time of the winter solstice, when the sun is at its lowest and just prior to its return to the north and summer's strength. In Norse myth, Balder, like Thor, was a son of Odin. Balder died at the summer solstice, when the sun's power seems greatest, and was reborn at the winter solstice, another soldier in the army of the invincible sun. A seasonal cycle of death and rebirth, then, makes an immortal of the sun every year.

COSMIC LAW COMES FROM THE SKY

Sky gods can be sky kings who order the cosmos and keep its laws. Aratus, the third-century B.C. Greek who authored the astronomical poem *Phaenomena*, introduced it with some remarks about the order of Zeus imposed upon the sky. It was he

. . . who set the signs in heaven
Marked out the constellations, and for the year contrived
What stars should best the heralds be
Of seasons to mankind. . . .

In the "Hymn to Zeus," Cleanthes, a Stoic philosopher and a contemporary of Aratus, acknowledged the ruler of the Olympian gods as the "founder of nature, who dost govern all things by law." It is Zeus who knows

. . . how best to make the uneven even,
To order the disorderly. . . .

This can happen only if the architecture of the universe has been put on celestial display and if the cosmic vehicle code is enforced.

Varuna is another one of those defenders of cosmic order. The moon moves by Varuna's decree. He puts the stars where they belong. Each day his will draws the sun forth and prompts it to take its assigned course. He is the "mysterious law of the gods," and that law is the cyclical beat of the sun and the moon and the stars.

To the Egyptians, the natural order of things—the way things should be—was symbolized by the goddess Maat. She, in turn, was the prize won by the sun god Re in his nightly combat with chaos.

Re pursued his proper course across the arched body of Nut, the sky goddess, during the day. After dark, he fought with monsters in the chambers of the night and was reborn each dawn in the company of Maat. Re's journey made sense out of the world. Inevitably, one day followed the next and imposed a rhythm that modulated people's lives.

For the Egyptians, cosmic order was also visible in the return of the goddess Isis as the star Sirius to the predawn sky. At about the same time of year, the Nile, no longer bound to its banks, flooded and refertilized the land. Every year Sirius put in an appearance in the right place at the right time, and the Nile made life possible in Egypt for another year.

These sky gods didn't just inject life into the world and watch it from above. They put celestial power to work. They made the rules of the cosmos and enforced them. In China, it took an entire heavenly bureaucracy to regulate the world. The Chinese saw the universe as a balancing act between two opposed but complementary aspects of nature: yin and yang. Their strength and influence were visible in the oscillation of day and night and in the cycle of the seasons. Yin is passive, and yang is active; yin is dark, and yang is light; yin is feminine, and yang is masculine; yin is winter, and yang is summer. When the Emperor made his annual sacrifice to Shang di on the round, open-air altar in Beijing's Temple of Heaven, the ceremony was intended to help Heaven maintain the balance between yin and yang. A jade tablet inscribed to Imperial Heaven Supreme Emperor was accompanied on the altar by tablets of the sun, the moon, the Big Dipper, the "five planets," the twenty-eight constellations, all of the stars of the sky, the Imperial Ancestors (also associated with the sky),

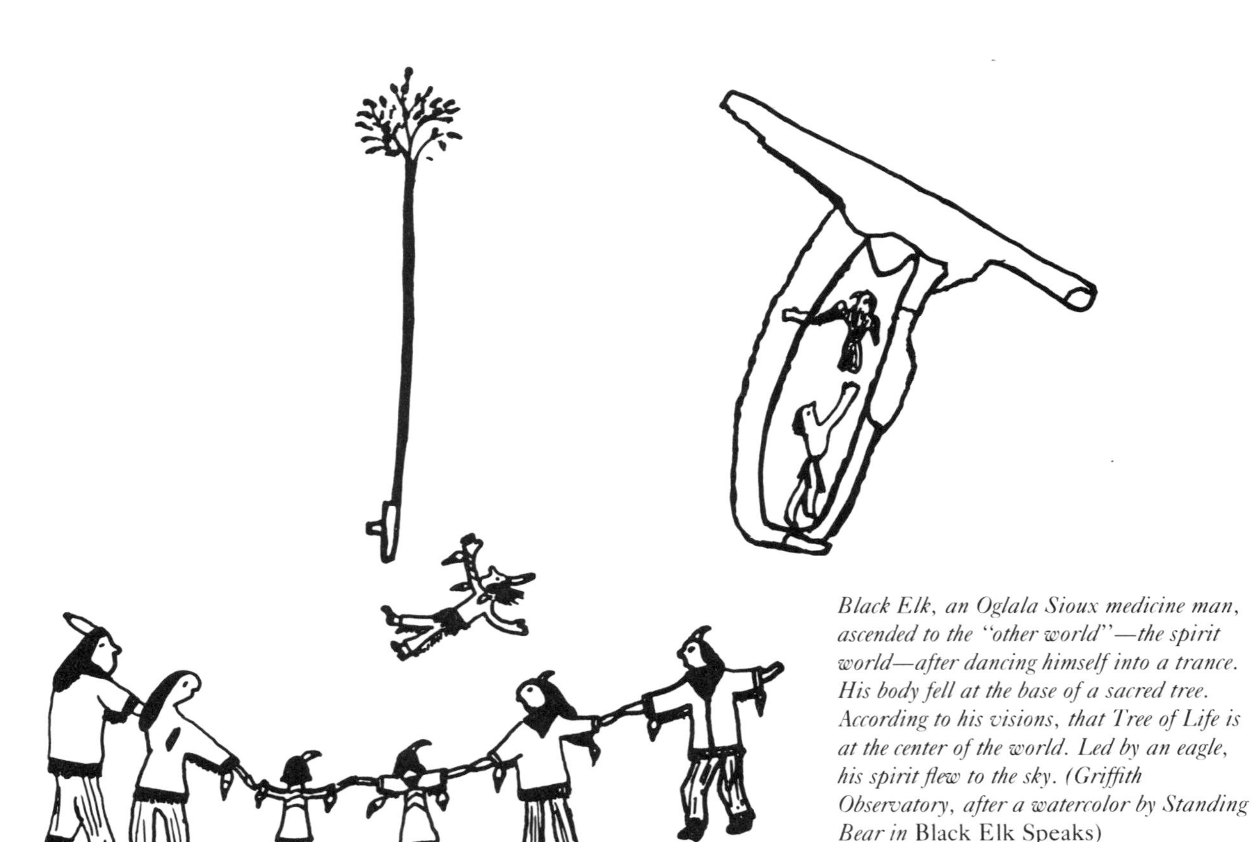

Black Elk, an Oglala Sioux medicine man, ascended to the "other world"—the spirit world—after dancing himself into a trance. His body fell at the base of a sacred tree. According to his visions, that Tree of Life is at the center of the world. Led by an eagle, his spirit flew to the sky. (Griffith Observatory, after a watercolor by Standing Bear in Black Elk Speaks)

and the Masters of the Clouds, Rain, Wind, and Thunder. The entire ministry of Heaven presided on behalf of the normal pattern of nature.

Stories of cosmic order are sky stories because the sky is where that order is seen. It seems to be controlled by all-seeing, all-knowing, powerful gods whose business is law and order. They modify the cosmos. They enrich it with life. They uphold its laws. People felt there was a connection between the cosmic law they saw overhead and human society. In sacred kingships, the ruler was heaven's representative, and the king might trace his lineage back to the sky. In China, the Emperor was the Son of Heaven. Egypt's pharaoh was the son of the sun, and the sun was the divine ancestor of the Inca dynasty in Peru. These sacred kings ruled with a celestial mandate. Because they controlled the calendar and participated in celestial ritual, they conveyed celestial power to earth. They ruled with the force and order of the sky. Heaven had to stabilize the universe, and the kings on earth had to keep order in the kingdom below.

GOING TO HEAVEN

Kings may enforce heaven's commands on earth, but shamans travel to the sky and consort with celestial gods in person. Altaic shamans of Siberia, for example, make a mystical ascent to the sky to chat with Bai Ülgän, the supreme celestial god, and obtain from him inside information on the weather and the harvest. The word *shaman* derives from the Tungus of southern Siberia, who call such an expert on the peregrinations of the soul a *saman*, but the shamanic tradition is found among the Indians of the Americas and in the religions of the Far East, Australia, the Pacific Islands, and even in northern Europe, the Middle East, and the Mediterranean.

Shamans, then, are in touch with the divine. And shamans don't just talk about the sacred. They go there. They may conduct the souls of the dead to their final destinations and may serve as mediators with the gods. A shaman may be a curer or a medicine man, a sorcerer, a performer, or a singer; a rainmaker and weather forecaster; an artist or a poet. Shamans are often responsible for preserving the traditions of their communities and protecting their people from supernatural attack. Among some groups, shamans are the calendar keepers, the skywatchers, and the game wardens who control the timing, frequency, and spirit of the hunt. They may act, in fact, as environmental protection agents who harmonize the behavior of their people with the animals and the plants.

With such a variety of responsibilities, it may seem difficult to define exactly what makes a shaman a shaman. Mircea Eliade, in his comprehensive study *Shamanism,* defines shamanism as a primarily religious phenomenon, calling it an "archaic technique of ecstasy." By "archaic" he means, of course, that shamanism among traditional peoples is, at least in part, tens of thousands of years old.

Eliade's definition identifies shamanism as a "technique." It comprises, therefore, a set of specialized procedures intended to accomplish something. That "something" is what Eliade calls "ecstasy," and ecstasy, in Eliade's terms, is a transcendent state of mind in which the shaman experiences the sensation of contact with the divine. There may be a variety of ways to reach that perception, but it is technique that identifies the shaman as a shaman.

To reach the realm of the sacred, the shaman specializes in putting himself or herself into a trance. According to Eliade, it is the trance that makes the shaman a shaman. The shaman believes that this trance permits his soul to leave his body and journey to distant, spiritual realms where a normal person in a mortal body in a normal state of mind hasn't a prayer of touring. The trance is the shaman's passport to this otherwise foreign country.

Shamans undertake spiritual journeys, and they do it for a reason. Their travels provide them with knowledge and power. In north and central Asia, one of the shaman's most important jobs is magical healing. Because disease is perceived as a malady of the soul, the shaman treats some illnesses with magic and ritual, as well as with a well-chosen herb or two. The soul of a sick person is thought to have wandered from, or been kidnapped from, the body, and it's the shaman's chore to find it. The shaman has to go where the errant soul has gone—perhaps to the underworld, perhaps to heaven.

Shamans are community leaders—very often charismatic community leaders—who negotiate with the gods and spirits, who wrestle with angels, and who enjoy intimacy with the sacred. They are sort of like priests, but whereas priests represent institutionalized religion, the shaman is a freelancer. Priests manipulate symbols of the sacred in ceremonies, sacrifices, and prayers. Shamans do some of these things, too, but in direct, personal, and mystical ways.

Mystical ascent to the sky—through trance—is a fundamental shamanic theme. When the Altaic shamans of Siberia make a sky journey for information from Bai Ülgän (who, like Zeus, is really a storm-rain-fertility god elevated to the position of High God), the ceremony begins with a birch tree stripped of its lower branches and notched with nine steps. It is placed in a new yurt, set up specially for the occasion, with the upper branches extending outside the tent. In the course of the ritual, a horse is sacrificed to allow the shaman to follow the horse's soul to the sky. The shaman plays his drum and dances around the birch tree. On his way, the shaman salutes the moon (sixth heaven) and the sun (seventh heaven) and ascends as close to the realm of the supreme god as he can to converse with him about matters of importance. The sky is the theater for the pageant of cosmic order and is, therefore, a primary source of shamanic power.

Some of the stories people tell about the sky are symbolic accounts of shamanic ascent. The Chumash Indians masquerade the shaman in a tale about a boy named Centipede. He was one of the First People at the time of the world's creation, and so he hadn't turned into a normal centipede yet. He and all of the other boys were highly competitive climbers, but Centipede was the best. The other boys must have grown tired of losing contests with him because one day they conspired with Coyote, who bewitched the climbing pole. Challenged to another contest, Centipede started up the pole with no idea it had been charged with magic. As he climbed toward the top, the pole grew taller. No matter how far he climbed, the top of the pole remained out of reach. Eventually Centipede climbed so high, he passed through the realms of the strong winds and great heat and close enough to the "door to the sky," a light he could see shining above the pole, to make the jump into heaven. There he was attacked by giant mosquitoes. They consumed his blood and left nothing but his bones.

We are tipped off by several details that this story is about shamans. It involves an ascent to the sky at great risk. Also, when a shaman is first initiated and when he undergoes subsequent trances, he is said to die. This death means a complete reduction to bone, which is regarded as the source and essence of life. Then the shaman is reconstituted. His vital organs are often replaced with new ones made of magical substances, such as quartz crystals. This shamanic death and resurrection mirrors the fundamental cycle of cosmic order that is played out in the lives of the plants and animals, in the seasons, in the daily and annual cycles of the sun and stars, and in the phases of the moon. The shaman's spiritual death and rebirth represent the transforming experiences that allow him to pursue his special relationship with the gods. In Siberia, his costume is frequently skeletonized to advertise his transcendent status. The rhythm of the cosmos abides in him. His death and rebirth are additional signs of celestial power.

When we last heard from Centipede, he had gone to the sky and died a shamanic death. Centipede's fate was a little more extreme than Coyote had intended, however, and he was remorseful. He decided to rescue and resurrect Centipede.

Coyote, too, is a shaman. He had the ability to cast a spell on the pole, and his shamanic experience gives him the confidence to contemplate this mission of mercy. Coyote persuaded Xolxol the Condor to lend him some clothes and sticks. With these he would search out Slo'w the Golden Eagle, another one of the Sky People. The eagle, he thought, would fly him to the top of the sky. He would retrieve Centipede's bones and reanimate him. So that's what he did. But as they were all returning to earth, Golden Eagle hit the tip of his wing against the pole, and Coyote fell to earth, where he was dashed to pieces. Centipede jumped to safety on the pole and hustled down to the ground to see what he could do about Coyote. He

After Jacob received the blessing of his father, Isaac, he left Canaan to obtain a wife in Harran. On the way, he slept for a night in the hill country. With a stone for a pillow, he dreamed of a ladder that reached to the sky. Heavenly angels traveled up and down this celestial stairway, and God stood at the top of it. (Gustave Doré, Doré Bible Illustrations, *Dover, 1974)*

reassembled Coyote's bones, and Coyote, like a proper shaman, revived himself.

This story sounds a little like *Jack and the Beanstalk*, and that is not such a farfetched comparison. Jack goes to a dangerous land in the sky to acquire something of great value. The magical beanstalk is a world tree and cosmic axis. Shamans often make their magical ascents by the same route to the north pole of the sky and the North Star that marks it. It is the "sky door" straight above the enchanted pole Centipede scaled to his shamanic death.

The Koran tells us that Muhammad's Night Journey to the throne of Allah began in Mecca. From the mosque there, the Prophet miraculously traveled to Jerusalem and made his ascent from the Dome of the Rock. Part of Muhammad's flight to Heaven was accomplished upon the back of Buraq, the angelic horse led by Gabriel and surrounded by many other angels. The Prophet is veiled and haloed in flame. Clouds swirl through the bottom of the scene, and stars are visible at top. (Shah Tahmasb's Nizami manuscript, Persian, A.D. *1675, displayed in the British Museum, London)*

BOUND FOR GLORY

Many more examples of climbs to the transcendental territory of the sky are recorded in the world's sacred literature. In the Bible (Genesis chapter 28: verses 11–12), Jacob sleeps with stones for a pillow and dreams of a ladder to heaven. Angels file up and down the rungs, and beyond the ladder's summit the voice of God announces his intentions. The dream is a prophecy of the future, and Jacob wakes, terrified by this direct contact with the divine. He sets up his rocky pillow as the foundation stone for the future temple. Oil drips from heaven, and with it Jacob anoints the stone and so consecrates this "gate of heaven" as Bethel, "the House of God."

In Islamic tradition, the Prophet Muhammad embarks upon his mystical Night Journey in the company of the archangel Gabriel and upon the back of Buraq, a miraculous steed with an angel's face. They pass up through the heavens as though these zones were floors on a celestial elevator. From the highest heaven, at the foot of Allah's throne, Muhammad alone rises still higher on a piece of silk brocade. Seated upon the arabesques of this magic carpet, he meets Allah face-to-face.

In many systems of belief the souls of the dead were among those who could travel to the sky. The deceased pharaoh of Egypt rose to the northern undying stars and to Orion, the celestial incarnation of Osiris. On the Great Plains of North America, the Mandan Indians said that newborn children are stars that have descended to earth. When they die, they return to the sky as stars. Following the path of the Milky Way, the spirits of the dead Pawnee made their way to heaven. We still say the souls of the righteous go to heaven and say that our departed friends and relatives are there.

It's not so easy, then, to go to the sky. You have

to be a shaman, a visionary, or dead. Because heaven is so far from earth and so full of power, normal human beings can't just drop in on the gods. Special access is required.

In the old days, heaven was transcendental territory—something beyond ordinary reality—and a sphere of power. That is why we have wished upon stars and yearned to fly over the rainbow. That is why we still tell tales about what we see beyond the blue horizon.

4

From the Darkness

WHEN the Greeks wanted to refer to the entire universe, the word they used was *cosmos*, which meant more than just an inventory of the merchandise in the cosmic warehouse. The universe, they were convinced, had structure. It was ordered—in appearance and operation. In fact, their word *cosmos* actually meant the "ordered whole." The Greeks, like many others, looked for pattern, cycle, and order in the world around them to contrive a picture of cosmic order.

That is what consciousness does. It contrives the world, and it does so by letting us classify phenomena, organize events, and code information. Space and time comprise the framework on which this system of knowledge depends. The Greeks saw an orderly framework of space and time on display in the sky, and they used celestial phenomena to define the character of the universe in which they believed they lived. They saw, for example, the dependable, unending cycle of day and night. It was visible in the oscillation between a bright blue sky accompanied by the sun and a dark black sky populated with stars. They saw the sky rotate. That was evident in the daily movement of the sun and in the nocturnal paths of the stars.

For our ancient Greek ancestors, the sky was a reservoir of cosmic order that put a regular beat into the passage of time and put direction into the landscape. The basic tick in the cosmic clock was that familiar rhythm of darkness and light by which we count the days.

The Greeks, of course, weren't the only ones who paid attention to the regular alternation of day and night. The ancient Maya, for example, in what is now southern Mexico and part of Central America, symbolized the concept of a day with a glyph the Yucatec Maya called *kin*, and they used the same glyph to mean "sun" or "time."

The everyday story of day and night is the first regular pattern of time we are likely to notice as young children. We all soon learn how the pattern works. In general, we are awake when it's light and asleep when it's dark. But the night sheds its own light through its principal agents the stars and the moon. Stars appear to speckle the sheet of shadow rolled out across the sky. Although they glow in the darkness, they really are allies of the night, for that is the only time we can see them. For a few hours, at least, shadow is king, but eventually the sun appears on the horizon and breathes color and life back into the landscape.

All of this is the straightforward result of a cosmic soap opera that plays daily and includes all of us in the cast. The world turns, and as the world turns, our brains forge an ordered cosmos for us out of the complicated and dense sequence of events collected by our senses. We use this steady rhythm of day and night as part of the ground floor of reality and as a symbol for the forces that seem to move and frame our lives.

We grow accustomed to the familiar cycle of day and night long before sophistication and metaphor enable us to give it symbolic meaning. As children, we sometimes fear the dark. We can't see what it hides. It can mean mystery, confinement, and danger. On the other hand, the bright sunshine of a new day brings the promise of renewal, opportunity, and enterprise. In 1789, the English poet William Blake published his *Songs of Innocence* and observed,

The Sun does arise,
And make happy the skies . . .

and almost two centuries later, in 1966, the Beatles kicked off the second side of their *Revolver* album with a similarly cheerful acknowledgment of the day's potential with the song "Good Day, Sunshine."

We make symbols of the light and the dark and extend them to philosophical and ethical concepts—we see the world in binary format, the way a digital computer processes information: The world is either black or white, dark or light. That's why

The rising sun means the promise of a new start and the reawakening of life. Even though electricity has liberated us from the confinement of night, we still take pleasure in the sunrise. Here, the sun is reborn across the water from Egypt's Abu Simbel.

The Queen of the Night sails into a scene from The Magic Flute *on the crescent moon. Piloting through the clouds, she rules the dark and star-sequined sky. Dressed in black, she is the enemy of light. (Karl Friedrich Thiele; set design, hand- and plate-colored aquatint. The Metropolitan Museum of Art, New York)*

the good guys in a canonical Western wear white hats and why Darth Vader, on the dark side of the Force in *Star Wars,* is the appropriately attired agent of evil in an age that transports the scene of traditional conflict to another planet.

CONTENDING WITH THE QUEEN OF THE NIGHT

Even Mozart made use of the accessible symbolism of day and night. His opera *The Magic Flute* invokes the classic conflict between good and evil in the language of darkness and the light and places the action in ancient Egypt, where the hero Tamino falls for Pamina, the daughter of the Queen of the Night. Pamina has apparently been abducted by Sarastro. He is the High Priest of the Temple of Isis, and the Queen's three lady attendants accuse him of being an evil sorceror. After Tamino agrees to rescue Pamina, the Three Ladies present him with a magic golden flute. Its power to control human emotions and to inspire love will aid him in his quest.

Tamino travels to Sarastro's palace in the company of Papageno, a sidekick he picked up in his encounter with the Queen of the Night. Papageno helps unite the lovers, but by that point they are all in the hands of Sarastro and his henchmen. Sarastro denies any evil intentions, however, and it turns out he is telling the truth. All of these events are really part of a larger pageant. Although Tamino and Pamina are destined for each other by the gods, the Queen of the Night, the real villain in this tale, intends to destroy the Temple of Isis.

By facilitating the matchmaking, Sarastro enlists additional allies in his war with the malicious Queen. Trials delay the marriage, however. Both must become worthy and so earn each other's love. Tamino's ordeals make him a member of the temple's elect and prevent the Queen of the Night from regaining her daughter. Because Sarastro also wears the "sevenfold shield of the sun," she can not slay him. After passing a battery of tests and playing a considerable amount of magic flute music, Tamino joins his heart's desire, and the two enter the temple's inner shrine together.

Now the stage is set for the Queen's final raid. Every time she puts in an appearance, thunder roars and lightning tears through the sky. This time, however, the force of heaven drives the Queen and her evil allies into the earth. As they sink into the underworld, they sing, "We are plunged into eternal night." Sarastro, the newlyweds, and a great crowd of members of the Temple of Isis assemble, enveloped by the radiance of intense celestial light. The High Priest confirms that the darkness has been dispersed by the light. Good triumphs over evil. The cosmic order is preserved.

Mozart's plot is a mixed menu of eighteenth-century food for thought. Masonic initiation, political satire, romantic melodrama, and spiritual growth run wild through the libretto. It is not an ancient myth or a traditional tale, but a product peculiar to its own era. That doesn't filter out familiar celestial storylines, however. We never abandon the stories of our ancestors; we just repackage them for our own lives and times.

We often equate life with light, daytime, and the sunrise. Darkness, night, and the setting sun often mean death. When we are awake, we say we are "alive to the world," and being asleep may find us "dead to the world." We bury the dead in the earth, and the setting sun seems to sink into it. We speak of twilight's "dying sunlight," and in slang, "going west" can mean death or decline. We've become a little touchy about old age these days, however, and use the phrase "sunset years" less. Now we favor the euphemism "golden years," but there's still lingering twilight in that image.

There are plenty of examples to show how the setting sun and the west are united with the underworld or with the land of the dead. King Arthur was supposedly wounded at Camlan, in the last battle with his traitorous nephew Modred, and carried off by two beautiful women to the west. There, in the enchanted Isle of Avalon, he waits in immortal sleep until his country needs him again.

The ancient Greeks located the land of the Hesperides in the far west, near the Isle of the Blest and at the edge of the world's ocean. There the Nymphs of the Setting Sun tended the golden apples of immortality.

It's easy enough to see why the dead go west, but the immortals go there too. They confront death as the setting sun faces it, and they emerge unscathed as the sun reappears each day. Anthropologist Peter Roe synthesized the mythical character of the sun according to the Shipibo Indians of Peru and showed that in their tradition, the sun achieves immortality through an underworld transformation and rebirth. The Shipibo say the sun starts the day as a youth—white, hot, and energetic—in the east. He reaches his greatest power high in the midday sky, and then he goes into decline. Descending to the west, he ages and decays and is swallowed by the woman who waits for him at the horizon. Once in the underworld, embraced by the waters of a cold and deep lagoon, he is consumed again, this time by a dragon.

In the dragon's gullet the sun dies. He is reduced to bone and grows younger in the dragon's womb, transformed into the night sun. He moves in darkness, a cannibal moon, paddling down the Milky Way and accompanied by owls, wasps, and vultures, all creatures of death and the night. Fraternizing with evil magicians, ogres, and other evil characters, the night sun spreads disease and death until midnight. Then he begins to change once more. Entering the womb of the World Tree, he is brought to the surface at the eastern horizon, where he is reborn as the young and vital daytime sun. He has, in his way, fought his own battle with the Queen of the Night.

BEGINNING AT THE BEGINNING

In Shipibo myth, every circle of night and day means a new creation of the sun and therefore a reestablishment of cosmic order. According to Mircea Eliade, myth narrates sacred history and so always returns us to the time of creation. That was when the world's order was established, and the natural order of the cosmos is what we mean by the sacred. The story behind all creation stories is simple but absolutely essential: We have what we have now because of what happened then. This is why we value the past and tell stories of the world's beginnings over and over again. We see our existence rooted in the world's origin. If we lose touch with the past, we don't just risk the loss of an engaging, entertaining tale. It's the whole cosmos that's at stake, and we with it. This is why traditionalists and fundamentalists object to scientific scrutiny of the origins of homo sapiens and the world. Literal interpretation of biblical creation may not make any sense in terms of what we know about the age of the earth and the evolution of its life, but for some people the secularization of knowledge has not provided a new and better story that connects them with the cosmos. If they abandon their sense of the sacred, there is no new order to take its place. They don't see a pattern in the universe of expanding galaxies, and they don't feel a bond with the power of natural selection to fill the world with life. Science is not at odds with the sacred, but it can seem to bypass it. Science provides a universe that seems to be accessible only to specialists. In traditional societies, the sacred still binds people together with a shared view of the universe.

Because the cycle of night and day is such a basic part of the world's order, the stories about the world's origin are set at the time of Creation. Many North American Indians describe that era of Creation as the time "when the animals could talk," and the Creeks, one of the Five Civilized Tribes of the southeastern United States tell one of those talking-animal stories to account for the sequence of day and night.

Long ago, then, when the animals could talk, they held a meeting to decide how to divide up day and night. Some naturally preferred the daylight. That's when they got their business done, and they felt they could use all the daytime they could get. They suggested that it would be just as well to let daylight last all the time. Nocturnal animals, on the other hand, sensed a potential problem at this proposal. They said night should just continue uninterrupted. The debate wore on for quite some time, but finally the ground squirrel noticed the raccoon's tail and contrived a solution. "If you look at the raccoon's tail, you'll see that it's divided equally with alternating rings of light and dark. That's how we should split the time between day and night, just like the rings on the raccoon's tail." Ground squirrel's equitable scheme satisfied everybody.

Another story about the origin of the universe exploits fully the polarity of darkness and light. In the beginning, according to the Greek poet Hesiod, only Chaos existed. It was primordial, formless, empty space. Without explanation, Chaos is joined by Gaia, "broad-bosomed Earth." Of course, her bosom is broad. It is the vast surface of the world we inhabit, and it nurtures everything alive. Even without any male assistance, Earth will conceive and give birth to Heaven, or Ouranos, but several other players in the cosmic drama have to appear on stage first. One is black Tartarus, Heaven's counterpart. Located as far below the Underworld as Heaven is above Earth, Tartarus is, in a sense, the bottom of the universe. It is a misty, deep, and remote pit. After Tartarus takes his proper place, Eros is next to arrive. He personifies the force of love, but he means much more than the emotion we know by that name. It was clear to the Greeks that something makes things happen in the world. Some driving desire for creation brings gods, forces, elements, animals, plants, and people together, and through their attraction and longing for each other, a cosmos is fashioned. It looks as if the Greeks figured everything was the product of pairing, but the Greeks were telling themselves a story about the world, describing it through images of procreation. It was a mystery how most things came to be the way they were, but the Greeks could see as well as anyone else that mating brings life into the world. It was a reasonable model, and it described what they knew to be true. For this reason, Eros had to show up early. Without some erotic action, the story of Creation would end much too soon.

Then, Hesoid tells us, from Chaos came black Night (Nyx) and Erebus, or "Darkness." Night we know, but Erebus is a stranger. Not just ordinary darkness, Erebus was deep darkness. As the pervasive gloom of the abyss, he occupied the subterranean chasm between the Underworld and Tartarus. He was the counterpart of his sister Night in the infernal realm below.

The influence of Eros must finally have been felt in the cosmos by the time Night and Erebus arrived, because the two of them coupled and parented Aether and Hemera. Aether was "Space," and by that the Greeks meant the clear upper atmosphere above the air we breathe and below starry Heaven. Hemera was "Daylight." Day, then, is the child of Darkness and Night.

Night indulges in some additional procreating without any male interference, and her dark brood includes "frightful Doom": the black spirits of death; Death itself; Sleep; the "whole tribe of Dreams"; Mockery; sad Distress; the Hesperides, or Nymphs of the Setting Sun; the Destinies; the Fates; Nemesis, the goddess of righteous anger who strikes down those who violate the natural order and are guilty of excess; Deceit; Copulation; Age; and Strife.

Mother Night made her home beyond the earth-encircling Ocean stream, in the far West, at the entrance to the Underworld. There, where the giant Atlas holds up heaven, Night and her daughter Day "approach and greet each other" at the bronze threshold of the horizon and exchange places. They are never home together. When one arrives, the other departs. All-seeing Day brings light to the earth, but "deadly Night". . . hidden in dark clouds, brings Sleep. Sleep and his brother Death live in Mother Night's western chambers, and Helios, the sun, has never seen either of them.

Night and day also figure in the creation myths of the ancient Vikings. Snorri Sturluson, an Icelandic historian, poet, and politician, collected many of the old Norse myths in A.D. 1220 in the *Prose Edda*, and according to him, Night was one of the primordial parents of the cosmos. Night was the daughter of Narfi, one of the giants of Jötunheim. Unlike most Nordic women, who are fair, Night was dark, her hair black. She was first married to Darkling, or Naglfari, which is also the name of the ship of the dead mentioned in the "Völuspà," one of the mythological poems in the twelfth-century Icelandic compilation known as the *Elder Edda*.

In another poem in the *Elder Edda*, the "Allvismál," Thor asks the dwarf Allwise, "What is the night, daughter of Dark, named in each world?" Allwise answers that she is called Night by men and Dark by the gods. The highest gods call her the Hood, a disguise for the world. The giants name her Unlight, the elves call her Joy-of-Sleep, and dwarfs call her the Dream Weaver.

Snorri tell us that Night and Darkling had a child named Aud, or Space, but then the marriage broke up. Night's second husband was Anar, but that's just an alias. It means "Another." Although no one ever says so directly, Anar was probably Odin. Both he and Night are credited with parenting the same daughter—Jörd, the earth. Night, however, seems to have had trouble holding onto her husbands; Anar also left her, but she acquired a third spouse, Delling, or "Dawn," and together they had a son, Day.

So Night married the dark, the high sky god, and the dawn, and had three children: Space, Earth, and Day. Night and Day were equipped with chariots and horses to transport the twelve hours of the night and the twelve hours of the day. Night's horse is Hrimfaxi, or "Frosty Mane." The foam from his bit shows up as morning dew on the grass. Skinfaxi, or "Shining Mane," follows Night's chariot with the vehicle of Day. Light from Skinfaxi's mane adds lustre and light to the daytime sky and to the earth below.

Norse tradition provides more than one explanation for the presence of night and day. In "Völuspà," a sibyl narrates what she knows about the order of the cosmos. In the primordial time of the earth's creation, the sun didn't know which hall of heaven was hers; the stars had no idea where they belonged; and the moon was ignorant of his own power. The gods got together then, and they named night and labeled the phases of the moon; they separated morning, noon, dusk, and evening. Through this process, day and night were clearly established.

IT DAWNS ON US

In both the Greek and Norse accounts of day and night, night is the older. The dark precedes the light, and the light emerges from the dark, usually through the effort of the dawn. At day's end, the night returns, envelops the earth, and extinguishes the daylight, and you can see that finish of the day coming in the twilight. Stories about day and night are, then, often articulated by the dawn or dusk. These transitions between the two realms seem to be what make the cycle of light and darkness possible; you can't get from one to the other without passing through the kingdoms that connect them.

Dawn was one of Night's husbands in the Icelandic *Edda*, and the Greeks called her Eos. Hesiod said she was a child of two of the Titans and therefore the grandchild of Mother Earth and Father Sky. Eos was married to Astraeus (the Starry One), god of the starry night sky, "the ancient father of stars," and he also fathered the four Winds of the cardinal directions and the Morning Star upon Eos. Her brother was Helios, the sun, and her sister was Selene, the moon. Rosy-fingered Dawn drove her own chariot ahead of the sun each morning and parted the gates of the sky for her brother.

The Baltic Slavs made a woman of the sun and said dusk and dawn are her daughters. Their rings are perpetually stolen from their rosy fingers by the evening star and morning star, who run off or disappear when the twilight passes into darkness or day.

In ancient India's *Rig Veda*, hymns are dedicated to the dawn, or Ushas, a name whose root means "to shine." She is the daughter of Heaven, born in the sky. She is also the sister of Night, with whom she exchanges places in an unending cycle of days. The hymns say Dawn is the fairest of all lights. She sheds her brilliance over a vast field, opens the portals of heaven, and makes a path for Surya, the sun.

Immortal and incorruptible, Ushas is always young. She is the "first of endless morns to come" and "follows the path of morns that have departed." Her banner raised in the eastern sky, she dresses herself in sunshine and pours light over her body like tanning lotion. "Like a dancing girl, she puts on bright ornaments; she uncovers her breast . . ." Purple steeds pull her chariot, and in it she carries all the gifts, all the blessings, wealth, and riches that sustain life. And so the devoted pray to her for that brightly colored power that confers children and grandchildren upon us.

LETTING THERE BE LIGHT

Although not always explicit, creation myths often make use of the mechanism of the dawn by describing the first extraction of light from the chaotic darkness. The Bible gets right to the point in the first four verses of the first chapter of its first book, Genesis:

> In the beginning God created the heavens and the earth.
> And the earth was waste and void; and darkness was on the face of the deep: and the Spirit of God moved upon the face of the waters.
> And God said, Let there be light; and there was light.
> And God saw the light, that it was good; and God divided the light from the darkness.

This first light certainly didn't shine from the sun, the moon, or the stars. It was only the first day, and the celestial lamps weren't lit until the fourth day. The true nature of that primordial morning twilight is not clear from the text of the Bible, but later commentaries on the biblical texts tried to come to grips with it. Rabbis in the synagogues of Palestine composed such commentaries as early as the first centuries of the Christian Era, and such an exposition is called a midrash. One midrash identified the first light as the "sort that would have enabled man to see the world at a glance from one end to the other." In medieval times, Jewish mystics equated the first light with the sensation of transcendental light experienced in their visions. However one explains the text, the light summoned in the Bible's third verse is the world's first dawn. In dividing it from the darkness God drew the first pattern upon the world, the eternal exchange of day and night.

THE FIRST DAWN

The Quiché Maya of highland Guatemala tell a story about the time before the first sunrise and before the first Lords of the Quiché set foot upon this earth. These traditions are recorded in a sixteenth-century book written in the language of

Biblical Creation injects light into the world as soon as there is a world in which it can shine. Soon after this moment, God established night and day by separating the darkness from the light. In Gustave Doré's interpretation of this early event in Genesis, the newly created light bursts out of the primordial darkness. (Gustave Doré, Doré Bible Illustrations, *Dover, 1974)*

these descendants of the people who built the pyramids, plazas, and palaces in the great ceremonial centers in the Mesoamerican jungle. This book is known as the *Popol Vuh* ("Council Book"), but the Quiché also call it *The Dawn of Life*.

The Quiché Maya word for world means "earth-sky," and the emergence, or dawning, of sky-earth and everything it contains is the theme of the *Popol Vuh*. This cosmos is orderly and obeys the laws of cyclic time. The four cardinal directions give it four sides and four corners, but in the beginning little of this is evident. Nothing is yet formed. There is only sky above and sea below.

> Whatever there is that might be is simply not there;
> only murmurs, ripples, in the dark, in the night.

A handful of primordial gods dwell in those waters, and they confer with gods from the sky about the creation they intend. Seeds will be sown in the earth, and the sprouting plants will be the dawn of new vegetation. With removal of the water, the dawn of the earth, with its mountains and plains, in the middle of the sky-earth's waters becomes possible. There is still no sun or moon planted in the underworld, which they call Xibalba, to rise above the horizon, as new shoots push above the ground. And there are no people to celebrate, on appropriate days, the gods' efforts and to praise the gods' work. They, too, must be sown in the womb, as plants are sown in the earth, so that they can dawn when they are born.

Three separate attempts to create human life end in failure. Birds and all of the other animals result from the first try, but they can not talk. No living thing can speak the names of the gods. The second time the gods make people out of mud and soil, but these folks are a bigger failure than the first. They can't move, and they can't mate and multiply. They certainly can't glorify the gods. Then they dismantle the mud men and make new people out of wood. At first it looks as if the third version of human life will work, but the wood people turn out to be thoughtless puppets with no feeling for life and no memory of the gods who made them. At the gods' command the wood people, too, are destroyed, by torrential storm, monstrous jaguars, and consuming flood. The world is still dark. There is "just a trace of early dawn on the face of the earth."

In the time before the world's first dawn, Hunahpu and Xbalanque, the Maya hero twins, played ball in the Underworld with the same high stakes that were at risk in the games played at the great ball court at Chichén Itzá in Yucatán, Mexico. The Maya team captain on the right has been decapitated, and streams of blood in the form of snakes spray from his neck and perhaps irrigate the squash vine in the background. The captain on the left holds a stone knife in one hand and his opponent's severed head in the other. The ball itself rests between them and is, as it says in the Popol Vuh, *a skull. (Photograph: Robin Rector Krupp)*

Without sun or moon, no genuine dawn is possible. The rest of the *Popol Vuh* deals with that first dawn. Heroes have to make the world safe before the real human beings can take their place there, and a pair of twin brothers, Hunahpu and Xbalanque, descend to Xibalba to play ball in the court of the Lords of the Night. The father and uncle of the twins were sacrificed by the Lords after they failed a test to spend a night in the Xibalba's Dark House. The hero twins, however, succeed where their father and uncle foundered and beat the Lords at their own ballgame. When they emerge out of the Underworld, the heroes ascend the sky as the sun and the moon. Light comes to the earth and sky, and the world is now prepared for the dawn of human life.

As the two most conspicuous occupants of the sky, the sun and the moon are alternately seen as partners or adversaries in celestial endeavors. Each has its own special character and territory. The moon changes shape and can become a crescent. Except during the unusual circumstances in an eclipse, the sun is always a glowing disk of light. The sun governs the day. The moon, although sometimes visible in the daytime, is said to rule the night. Both were given a face in this detail from the Nuremberg Chronicle, *prepared in 1493. (Griffith Observatory)*

THE SUN IN THE MORNING AND THE MOON AT NIGHT

Day and night do not share the sky. They take turns with it and are on opposite sides of the same coin. One precludes the other, but they are both needed to complete the natural order. This opposed but complementary relationship between day and night is also expressed in the way people have talked about the sun and the moon.

A story told by the Maidu Indians of California's northern Sierra illustrates the way in which people set up the sun and moon in complementary opposition. According to the Maidu, Sun and Moon used to live together in a stone house far to the east, below the horizon. They were brother and sister and never left the house. Despite many attempts to get them to rise, they wouldn't budge. Finally, Gopher and Angle Worm went to see if they could flush out the two celestial stay-at-homes. Angle Worm tunneled under the stone wall and emerged through the living-room floor.

Gopher was right behind, with a bag of fleas. He released them inside the house, and Sun and Moon really started to move. The fleas were so bad, Sun and Moon had to leave the house. Once outside they tried to work out their future plans. Sun said they both couldn't travel at the same time and asked if her brother preferred to make his rounds in the daytime or at night. Moon suggested that Sun go at night, and she did. But it didn't work out. All the stars fell in love with her, and their affections stopped her in her tracks. So Sun returned to her brother the moon and explained the situation to him. He realized he would have to make the night journeys, while Sun would move by day. Ever since then, Moon has moved through the stars of the night, and Sun travels alone during the day.

In some stories, the sun and the moon are personally credited with creating the difference between day and night and express the difference in terms of the presence of stars in the night sky and their absence during the day. The West African Ewe people say the stars are Moon's children. (Tribal people of the Malay peninsula tell almost exactly the same story.) According to the Ewe story, both Sun and Moon had big families, but together they decided to do the kids in. First Sun killed her young and put on a meal for Moon. After cannibalizing her own kids, Sun was supposed to enjoy a second feast of juvenile celestial flesh at Moon's house, but Moon hid her children in a big jar and now releases them only at night. Sun, however, has no children anymore and is alone in the daytime sky.

SUNRISE, SUNSET, AND THE ORDER OF THE WORLD

The pattern that the sun provides doesn't just regulate time. It also organizes space, and it can do this because it doesn't just rise, travel, and set anywhere. There is a zone on the eastern horizon in which it must first appear and a zone in the west where it must take its leave. It's because celestial objects like the sun rise on one part of the horizon and set on another that we make the distinction between east and west in the first place. Although the sun shifts its places of rising and setting a little each day in a seasonal cycle, east and west, in this sense, have broad meaning and refer to the general realms of sunrises and sunsets.

The horizon is just the boundary between the earth and sky, and the word *horizon* means "boundary." It's an interesting place. It is, after all, where we first see the sun, and it's where the sun kisses the earth good night. An anecdote in the *Baba Bathra*, a treatise on the version of the Jewish *Talmud* prepared in Babylonia about A.D. 500, furnishes a nicely compressed description of the way the horizon works. Rabba bar Bar-Hana, a celebrated traveler in antiquity, allows an Arab to guide him to the very edge of the earth. Now, the horizon looks like the place where the earth ends, and that is where the pair travel. They arrive at the horizon at the hour of prayer, and so Rabba sets his basket of bread down on the celestial windowsill. After the prayers, he reaches for his basket and discovers it is gone. "Who has stolen my bread?" he asks his Arab escort. The guide replies that no one has taken it: "The wheel of the Firmament has turned while you prayed. Wait until tomorrow, and you will eat bread again." What the sky ploughs under in the west, it later raises again in the east. Rabba's dinner is transformed into breakfast by the spinning sky.

The spinning sky establishes the world's directions, and establishment of the world's directions is another key event in creation myths. In Snorri Sturluson's account of the Norse creation myth, there are no directions in the very beginning. That's because there is no world. At first, only Fimbulvertr, the mighty winter of nothingness, exists. The universe is a gaping abyss. Within this void, however, a cosmos is created from the corpse of a frost giant whose body has condensed out of the vapor and fog. The giant's name is Ymir, and he is killed by Odin and his two brothers. Together, they fashion a world from his cadaver. The giant killers make the sky out of Ymir's skull and station four dwarfs at the world's corners to hold up the heavens. The names of these dwarfs are the same as the names of the directions where they stand: Austri (east), Vestri (west), Nordri (north), and Sudri (south). The link between the sun's daily movement and these directions is revealed by the original meaning of each name. *Austri* means "glowing bright, burning," a nod to the sunrise and dawn. *Vestri* refers to the evening. *Sudri* seems to have meant "the brilliant," and if so, it probably indicates the sun's high point of the day, when it crosses due south. Finally, *Nordri* means "away, below" and suggests the sun's passage at night through the underworld. The road of the dead that leads to this realm goes north.

The ancient Greeks named the four directions after the prevailing winds rather than the daily stations of the sun, but there is still a bit of light and shadow in two of them. The Greeks called the east (or southeast) wind Eurus, a name linked with brightness. The west wind was Zephyrus, and its name comes from the word *zophos*, which means "the dark" and refers to the darkness that follows when the sun sinks in the west. Roman names for the cardinal directions also transport us to the path of the sun. The east was *oriens*, which means "to rise," and the *occidens*—the west—means "to fall." *Oriental* and *occidental* refer to the two major geographical and cultural divisions of the world.

There's still more astronomy in the Roman di-

Norse cosmography places a dwarf at each corner of the world to support the sky. These four corners are the four cardinal directions, and two of the heaven-holding dwarfs may be what is portrayed on this hogback tombstone. It may be seen in Saint Peter's Church, Heysham, Lancashire, England, and is believed to have been carved in the seventh or eighth century A.D. *Two more world-direction dwarfs are at the other end of the stone.*

rections. Romans didn't look "south" to see the sun at noon. They faced *meridies*, or "midday," because at midday the direction of the sun is due south. The Latin word for north, *septem triones*, means "seven oxen," and the seven oxen are the seven stars of the Big Dipper. They are in the northern sky, and they circle the north celestial pole as the night passes.

In the *Rig Veda*, the god Vishnu personifies the pervasive law or cosmic order that puts cohesion and structure into the world, and in this guise Vishnu preserves the world. As a participant in the world's creation, he separated earth from sky and shored up the space between them by taking three great steps. Vishnu's strides measured out the world and were interpreted by some ancient commentators as the sun's rising in the east, its highest elevation at noon, and its setting in the west. Each day, then, the sun reestablishes world order.

For the Hopi, in the American Southwest, the sun is the "holder of the ways." By following his proper course, he preserves world order. Each day

The sun's daily path is represented by the rainbow in this painting by Hopi artist Fred Kabotie. Emerging from the sun's eastern kiva, the rainbow climbs to the meridian, and at its highest point supports a stool for the sun's brief noontime rest. From there, the rainbow path descends into the sun's western kiva. (Hopi Room of the Watchtower at Desert View, Grand Canyon National Park)

he emerges from his eastern kiva, an underground chamber like the one used by Pueblo Indians for ceremonial activity. After climbing up the ladder and out of the kiva, the sun continues higher in the sky until he reaches his highest station, at noon. He rests there for a little and then descends to his western kiva. Refreshing himself with a little sleep, the sun continues his journey under the ground during the night until he reaches his eastern kiva once more. The Mataco Indians of Argentina's Gran Chaco grass and scrublands tell almost the same story, but with huts instead of kivas.

Shamash, Babylonia's sun god, stepped through the eastern gate and onto the Mountain of Sunrise at the horizon. Continuing his march, he eventually reached the Mountain of Sunset and slipped through the western gate of heaven. A Babylonian hymn addressed to the setting sun reports that Shamash entered the midst, or innermost zone, of heaven at sundown. There, out of sight of the people on earth, he meets his wife, sits down to a feast, and rests until the next day's hike on the celestial sidewalk.

In Egypt, the sun god Re also passed each day through heaven's eastern and western doors. The Vikings said the sun emerged each dawn through Delling's, or Dawn's, door. According to the Bella Coola Indians on Canada's Pacific Coast, the place of sunrise is guarded by a warrior known as the Bear of Heaven, and they say a huge pole holds up the sky where the sun sets.

All of these images of the sun's daily circuit link its entry to and departure from the sky with the world's directions. Such directions provide the fundamental orientation we use to describe the landscape. Most of the time people select four primary directions. In the Quiché Maya's *Popol Vuh*, the world, or sky-earth, has four sides and four corners. Just as the four corners of a building are established by running lines of cord between stakes, sky-earth was measured and aligned with four directions.

Shamash, the sun god of ancient Mesopotamia, sheds undulating rays of light from his shoulders as he emerges from the Eastern Gate, a notch in the mountain horizon. The planet Venus, in the guise of the goddess Ishtar, heralds the sun as a "morning star." The arrows she carries behind her wings may represent the brilliant planet's own beams of light. To the right of the rising sun god, life-sustaining waters, rich with fish, stream from the shoulders of Ea, the god who governed the earth's fresh water. This cylinder seal impression belongs to the Akkadian period (2360–2180 B.C.) and may represent the divine celestial circumstances associated with the akitu, *or New Year ceremony, celebrated at the vernal equinox. (Griffith Observatory, after original in the British Museum, London)*

Hawk-headed Re, the Egyptian sun god, sails through the celestial realm on his solar boat. Crowned with the sun's disk, Re is accompanied by the benu, *his heronlike herald. This bird symbolized the Morning Star. (Tomb of Sennedjem, Deir el-Medina, Dynasty XIX, 1293–1185* B.C.*)*

THE SUN CATCHES A RIDE

Everyone knows that the sun moves from east to west across the sky each day, but the means by which it travels depends on who is talking. Many of the Indo-European peoples—including the Vedic peoples of India, the Vikings, and the Greeks—moved about in chariots, and so it made sense to them that the sun would have wheels, too. The sun's round disk itself suggests a wheel rolling through the sky.

Wheels weren't as important as the river's current in Egypt. There, boats carried most of the goods, and the Nile was the highway. If most things in Egypt moved by boat, so would the sun. Many paintings on papyri show the sun sailing across the body of the sky goddess from one horizon to the other.

In Egypt the sun was also depicted as a winged disk. Since the sun moves through the sky, as a bird does, it certainly could fly on feathered wings. The Assyrians and the Persians had the same idea, and they equipped their winged sun disk with tail feathers, too.

Southern California's Chumash Indians said the sun travels on foot. Bearded and naked, each day he follows a cord that stretches over the world. He carries a torch to light and heat the earth below him. The sun also hoofs it east to west over the west coast of South Australia, where the Karraru aborigines say she's a lady, the Sun Mother, whose first walk over the earth stirs it with life.

CONFINING THE SUN TO ITS PROPER COURSE

It's clear then that the sun gets around, but how does it stay on course? Its movements seem constrained to a preordained pace and path, and they follow a pattern that myth tells us was established at the beginning of the world. Behind Polynesian sun-catcher myths, for example, is the belief that the sun was compelled by the demigod Maui to follow its present and proper course in primordial times. Maui, like the Greek hero Heracles (Hercules), was half-mortal and half-divine, but he was also known as a consummate trickster. As a culture hero, he provided fire, cooked food, and other gifts to people on earth. He fished islands up from the sea to create dry land, and he snared the sun in a net to slow its course. In the version of this tale told on Manihiki, about 650 miles northeast of Samoa, Maui's ropes keep burning up each time he tries to trap the rising sun. When he makes new ropes out of his sister's own hair, however, it is a different story. He places the noose over the hole through which the sun emerges and catches it by the neck when it appears on the horizon. Because the sun agrees to slow down, Maui lets him go, but he refuses to remove the ropes. These are the rays of light that still hang from the sun. They are visible in the morning and at sunset, dangling toward the earth. Drawn up by them from the underworld at dawn, the sun is lowered gently back down at night.

Cahto Indians in northern California's Mendocino County put the sun on course, not by snaring it but by releasing it. In their account of the origin of the basic celestial cycles, Coyote had a hand in making things the way they are today.

In this sun story, Coyote has been sleeping. He actually falls asleep four times, once with his head to the west, once with his head to the north, then with his head to the south, and finally with his head to the east. It sounds as if this story is going to have something to do with sunrise already. By the time Coyote gets to his fourth and last nap, the story starts moving.

While asleep with his head in the east, Coyote's forehead begins to feel hot. Awakened by the feeling, he realizes he has been dreaming about the sun, and decides to go get it and bring it back for people on the earth. They have been doing without it. It must have been pretty dark, for on the way Coyote encounters three mice he is convinced are dogs. Misidentified or not, the mice agree to accompany Coyote to liberate the sun. The rescue party goes straight to the house where the sun was impounded, and Coyote tells the mice that the sun is inside, covered with a blanket, and tied down inside the house. Their job: Gnaw through the straps. They wait, of course, until everyone in the house is asleep, but once the bonds are broken, Coyote will grab the sun and run off with it. It is important, however, to leave the cut straps hanging from the sun. Coyote needs them to carry the orb away.

Coyote then approaches the sun's prison. An old woman comes to the door, and Coyote explains he wants no food, just a place to sleep. She agrees and gives him a blanket. Coyote covers his head with the blanket, and looking a little like the wolf in the nightclothes of Red Riding Hood's grandmother, Coyote sings a lullaby. The old woman nods right off, and the mice get to work. When they finish, Coyote races away, sun in tow.

But there are witnesses. Mole saw Coyote and yells, "He's carrying off the sun." Coyote is still in luck, however. Nobody heard the mole. Then Lizard sees what is happening. He grabs a stick and beats on the old woman's house. "He's carrying off the sun," shouts Lizard. Well, the old lady jumps up and runs after Coyote. She calls out to him, "Why did you take the sun? I was fixing it."

Coyote isn't convinced. He yells back at her, "You were hiding it." Then he tells her to stop where she is and turn to stone. She does. Coyote cuts up the sun, and from the pieces he makes the moon and the stars. What is left becomes the sun we know. Furthermore, he instructs each of them in proper celestial behavior. He tells them when to rise and when to set. Orders are issued to the morning star, which is to come up just before day. The moon, he says, will travel by night and be cold. Finally, to the sun he says, "You shall be hot, come up in the east in the morning, and go down at night." People on earth, very appreciative of Coyote's efforts on their behalf, gave him many presents.

TELLING TALES ON THE SUN

By now it is clear there are many myths about day and night, and in them many natural phenomena, including the sun, are personified. Such stories, especially among Indo-European peoples, inspired the nature myth movement in the nineteenth century. It was spearheaded by Friedrich Max Müller, a leading Sanskrit scholar and professor at Oxford. His theory of Solar Mythology found the sun lurking everywhere.

From nearly every myth, Müller extracted a parable of the sun's trials and adventures. Every time Saint George subdues the dragon, the sun conquers the darkness. Even the nursery tale of Little Red Riding Hood was said to conceal a story about the sun. In fact, it does. The crimson cowl Red wears should make us suspicious. She's a young thing (the morning sun) on a journey (the sun's daily path) to old age—her grandmother's house (the sunset)—where she is devoured by darkness (the wolf). Red Riding Hood emerges (sunrise) from the wolf's gullet, however, and is reborn to carry her basket of goodies another day.

In *Custom and Myth*, folklorist Andrew Lang assaulted Müller and his nature myth enthusiasts with the premise that all peoples share a similar storytelling formula because all peoples experience the same evolution of religious thought. Contemporary anthropologists were reporting examples of totemism and animism among primitive peoples. Totemism traced family bloodlines back to some emblematic creature or natural object. Animism invested every natural phenomenon and feature with spirit and life. We're much too sophisticated to believe in animism these days, but we are convinced sometimes that our automobiles and appliances have independent personalities, especially when they cease to function. Totemism is also in serious decline, but some might argue that baseball and football teams practice a vestigial totemism in the group identities reinforced by their names. There are Bruins and Razorbacks tearing up the field out there.

People who sense spirits in everything the world contains and who align themselves with animal ancestors personify the world's natural phenomena, and that means telling stories—making up myths—about them. Lang said it was this primitive religious approach in the childhood of the race that got people talking in myths. You couldn't, he complained, explain everything in terms of language and the sun.

Some stories, however, are disguised nature myths. The golden apples of immortality in the west really are the immortal setting sun. In another Greek story about a lion and a leopard, the lion stands for the sun and the leopard represents the stars of the night sky. The lion pursues the leopard into a cave. It looks bad for the leopard, but he escapes out of an opening on the other side, circles around, and plunges back into the cave. Now the lion is trapped, and the leopard kills and consumes him.

In many times and places we find the lion equated with the sun. Before declaring itself an Islamic republic, Iran's flag and coat of arms displayed a sword-waving lion with a sun rising over its back. The lion's mane is like a crown of rays; its tawny color is golden as the sun. The majesty of the lion is the majesty of the sun and the majesty of the king. The lion is the King of Beasts, and the sun is the ruler of the day. Both are known for their strength. The leopard, by contrast, is a nocturnal animal. The ancient Egyptians even made its spotted coat stand for the stars of the night sky.

The scientific study of myth began about 150 years ago with Max Müller, but in that relatively brief time, significant controversies have made some explanations fashionable and discredited others. Eventually, Müller's school of Solar Mythology declined like the setting sun; it was "eclipsed," as the American professor Richard M. Dorson wrote. Lang and his accomplices prevailed, but not because there are no nature myths. There are plenty. And lots of them involve the sun. But in attempting to account for myth, Müller narrowed his sights and explained everything in terms of language. Storytelling, we now know, is far more complex than language analysis alone can explain. By now, we also know that the savage mind is not so childlike, naive, illogical, or superstitious as Lang would have had us believe. We have the same brains that our "savage" upper paleolithic ice age ancestors had. Our tools are differ-

Most people recognize the sun in the ray-shedding circles drawn on rocks, pottery, and other handy surfaces since at least the new stone age, or neolithic era. Sometimes there is no doubt that these symbols stand for the sun because accompanying texts or oral commentaries confirm their meaning. Other examples of ancient and prehistoric rayed disks, however, don't come equipped with such confirmation. It is, of course, reasonable to guess that these sunbursts really represent the sun. The sun, after all, looks like a disk. It also shines with light of its own, and its brilliance during the day makes it seem as if light is pouring out of it in every direction. This kind of solar symbolism is found all over the world. Prehistoric Great Basin Indians in California carved this solar petroglyph in Sheep Canyon, near Ridgecrest, California. A bighorn joins the rayed disk and perhaps alludes to some seasonal connotation and some kind of solar myth.

ent, and the vocabulary in the stories we tell about the universe is certainly not the same. But they painted complex and gorgeous animal symbols in their caves and must have told themselves remarkable stories about death and rebirth, night and day, darkness and light.

DEFEATING DARKNESS AND DEATH IN ANCIENT EGYPT

We can only guess that our ice age ancestors had tales to tell about the conflict between the darkness and the light, but ancient Egyptian stories that make a cosmic hero out of the sun have survived. The Egyptians treated the sun as though it were a number of different gods, these various identities emphasized different aspects and attributes of the visible sun that crossed over the Nile each day. At the most fundamental level, the sun was Re, the falcon-headed creator whose cult was originally centered at Heliopolis, or "Sun City." He was engaged in a deadly nocturnal war with darkness and death.

In this myth, Re, crowned with the red disk of the fiery sun on his head, triumphantly navigates the hours of the day in his sun boat until he comes to the western horizon. There, however, the picnic is over. Reaching the mountains at the world's edge and the entrance to *Duat*, the netherworld, the boat Re has been riding since noon carries him into the First Hour of the Night. It is, like twilight, a vestibule preceding the darkness. Now a tourist in the land of the dead, Re himself is dead. He is portrayed with the head of a ram. His allies include a "lady of the boat," who changes each hour as the crew leaves one country, or Hour of the Night, and enters the next. Each lady serves as a local guide to her own Hour of Night.

In the Second Hour of the Night, Re encounters the agent of darkness, one of the great serpents that wars nightly with the sun. In the Third Hour, the netherworld realm where Osiris, lord of life, death, and rebirth, resides, souls of the wicked are hacked and burned. Re's boat is towed through this zone by gods of the netherworld. They must pilot a path through lakes of fire and past an obstacle that may represent a kind of tunnel through the earth.

Re confronts more gigantic snakes in the Fourth Hour of the Night. It's Indiana Jones's worst nightmare: one snake rides a boat; another has three heads, four legs, and wings; a human head protrudes from the tail of a third snake, and other equally strange serpents slither through the night, constantly circling in search of something to eat. This hour is Re's first in the kingdom of Sokar, a hawk-headed ruler of the necropolis affiliated with the slain Osiris and with funereal ritual, and it is much darker and lonelier than the territories through which Re has already passed. Here there are no friendly gods or souls to cheer him on his way, and the darkness is so thick, his light can not cut it. Re changes to a boat whose body is a snake with a head at each end. Flames from this snake help light the way to the night's Fifth Hour, where the infernal city of Sokar is hidden by walls of sand and lakes of fire and is guarded by sphinxes and serpents.

Crossing into the Sixth Hour of the Night, Re travels once more in the boat in which he began this journey. Here he is united with allies, among them nine fire-spitting serpents with knives who are assigned to protect the rising sun, slay its enemies, and cut their shadows to pieces. By the Seventh Hour, Re reaches the hidden home of Osiris, the critical station of the night. Here Re battles his eternal adversary, the serpent of death and darkness—Apophis. Hour Seven is the place of combat, and the forces of the sun, though grimly tested, prevail. Light defeats the darkness on its own turf and is free to head for the horizon. In the tomb of Sety I (ruled 1291–1278 B.C.), a pharaoh of the Nineteenth Dynasty (1293–1185 B.C.), a scene of the Seventh Hour shows Apophis shackled and staked to the ground by six knives. Here, too, the enemies of Osiris, rebels against the Lord of the Netherworld, are decapitated. Their souls are extracted, and only empty, headless husks are left. A crocodile guards the tomb of Osiris, but Re transfixes the creature with his words and gaze. This allows the dead Osiris and his company of souls to look for a brief time upon the shining image of the sun as he continues on his journey and monitors the night.

In the Eighth Hour of the Night reside the dead spirits of those who were mummified properly and buried according to correct ritual. They and the gods who live here are revivified by the sight and light of Re as his boat is towed through the region's ten circles. He commands them to destroy any enemies of his that appear in this hour.

Re's boat is propelled through the Ninth Hour of the Night by divine rowers who also splash water on the souls who stand attentively upon the riverbank. They will help guide Re to the horizon. Re continues toward the east through the Tenth Hour of the Night. He catches here a glimpse of the sacred scarab beetle that pushes the sun's disk into the sky at sunrise, but they are both still in the netherworld. Now many armed gods join Re, and they slay his enemies as he continues forward. In the Eleventh Hour, the darkness begins to diminish. Re feeds the souls of the gods here with his hidden light. Great fire pits are lit to incinerate the enemies of Re. When he reaches the Twelfth, and last, Hour of the Night, his boat carries him—an old and dead sun—into the tail of a snake. The tip of that snake's tail hangs in the darkness that remains at the beginning of this hour. Passing through the body of this snake, the ship exits from the snake's mouth into the night. A dead sun no more, Re is transformed into the scarab beetle he saw earlier. He emerges from the loins of Nut, the

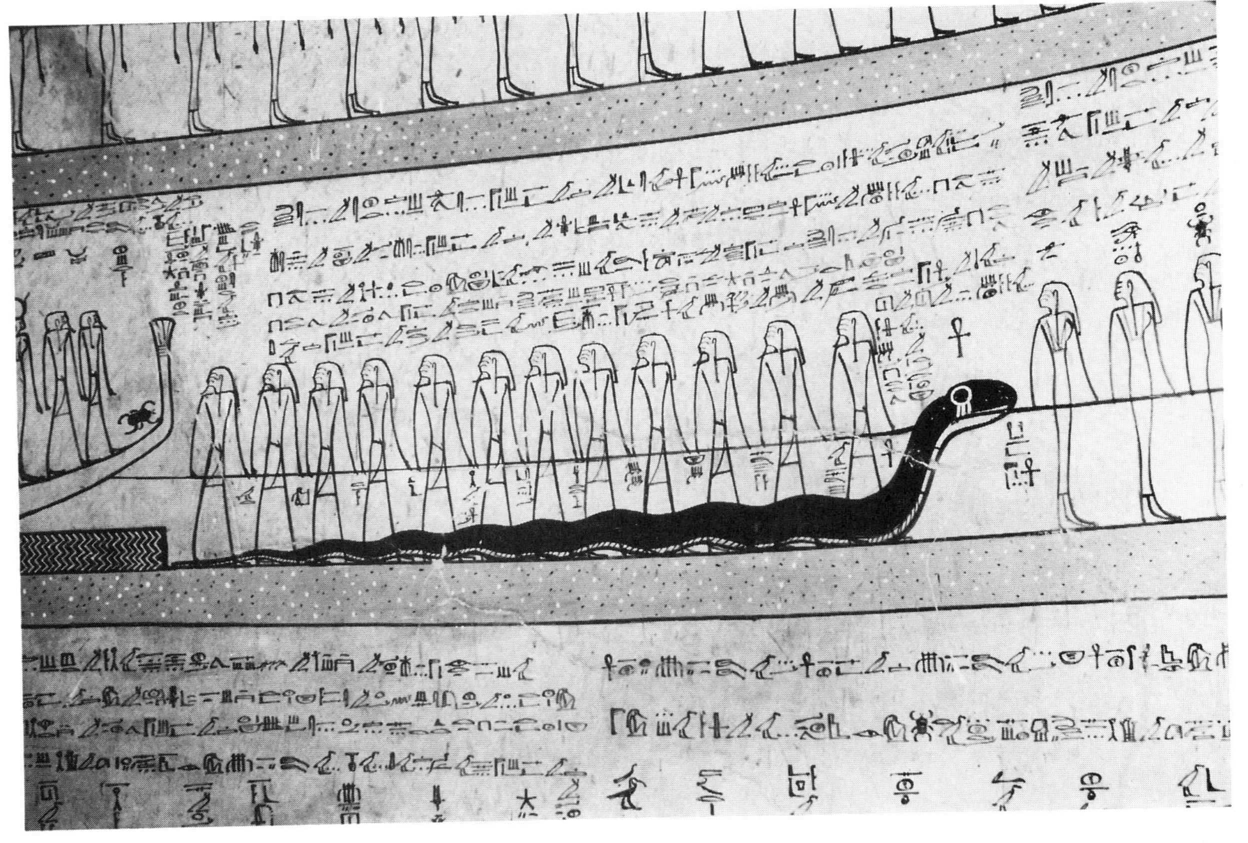

The final trial of the old, dead sun in the Underworld is a passage through the body of a great black serpent. This snake's name is Life of the Gods, and it dominates this detail from the tomb of Thutmose III (ruled 1504–1450 B.C.) in Egypt's Valley of the Kings. A scarab beetle, the symbol of the rising sun, occupies the bow of the sun's boat, on the far left, while twelve gods tug the line that draws the ship into the snake's tail. To the right of the snake stand three of the thirteen goddesses who draw the newborn sun out of the serpent's mouth toward sunrise.

sky goddess, and climbs with the glowing solar disk into the eastern sky.

This version of the nocturnal campaign of Re in the netherworld is given in the Egyptian mortuary text known as the *Book of What Is in the Underworld,* and it appears in a number of New Kingdom (1570–1070 B.C.) pharaonic tombs. There are other descriptions of *Duat* and the Twelve Hours of the Night, and in one of them Apophis is said to be over fifty feet long and armored with flint scales. He roars, hisses, and screams like a dragon and threatens the established order. In the *Book of the Dead,* another Egyptian funereal text, Apophis is defeated in the last hour of the night, just before dawn. The details of the sun's war against the serpent may vary, but the primary theme is clear. The sun, the chief agent of light and cosmic order, heroically engages and subdues his traditional enemy, the agent of darkness and chaos.

THE CYCLE OF COSMIC ORDER

That divine solar scarab served the Egyptians well as a symbol of solar rebirth. Commanded by instinct to gather dung and debris into a ball, the beetle rolls the refuse in front of it. It might as well be rolling the sun, which is also round and which slips across the sky. In a way, the scarab's sphere of waste is, like the sun, a source of heat, nourishment, and new life. Depositing its eggs in a dung ball, the scarab provides its future young with an incubator and with food for their first meal. The beetle then buries the ball in the earth. It drops into its "horizon" and so mimics the way the sun seems to slip into the earth at sunset. The larvae later emerge from the globe, take lunch in the larder their parent has provided, and emerge miraculously from the earth as the sun emerges from the eastern horizon.

The Egyptians didn't know that the scarab beetle laid eggs in the ball of dung it rolled over the ground and into the earth. They thought the ball was like a seed, from which new beetles sprouted. The Egyptian name for the solar scarab—Khepri—means "coming into being" and "self-created." By the time the baby beetles show up, the debris-collecting parent is long gone, and the new scarabs seem to have created themselves the way Re does every morning. This insect was a powerful image for the ancient Egyptians. They tucked stone scarabs into the linen folds of their mummies. It was a symbol of rebirth, and it was said to bear the newborn sun.

We tell ourselves these sun stories about ourselves whenever we rechronicle the triumph of the light over dark. We see ourselves born; we watch ourselves grow; we witness our deaths; and we celebrate the new children that replace us. We also speak of another kind of rebirth, a rebirth of the soul, and the crucial episode in that story is death. That's why the Egyptians put a sun story on the walls of a tomb. They and many other peoples have believed that we complete the cyclical tale and put it in motion again by dying and transcending death through resurrection, reincarnation, or some other transformation of the soul that transports us to heaven, that carries us from the darkness into the light.

Baboons observe the rising sun, rolled forward by a scarab beetle, in this pendant found in the tomb of the pharaoh Tutankhamun (ruled 1334–1325 B.C.). All three are sailing toward the sunrise in the sun's boat. Carried upon wavy waters, they are about to climb the sky, represented by the upper band of five-pointed stars. The rising sun enjoys the company of baboons because like roosters, they have a reputation for being active and noisy at dawn. The crescents on their heads probably stand for the waning crescent moon, which is seen only in the east and in the morning sky. (Object in Egyptian Museum, Cairo)

5

On the Face of the Moon

BECAUSE the moon waxes and wanes between complete darkness and a bright full disk each month it also provides a clear message of the contest between darkness and light. From earth, the changing appearance of the moon is distinctive and eye-catching. Shakespeare called it the "inconstant moon," which "monthly changes in her circled orb." Juliet pressed Romeo not to swear by it, lest his love for her "prove likewise variable." Ironically, when the *Apollo 11* astronauts Neil Armstrong and Edwin "Buzz" Aldrin stepped upon the surface of the "inconstant" moon for the first time in human history, they set foot upon a world whose landscape is really a fossil, scarcely changed in billions of years. Commander Aldrin called the moon's eternal landscape a "magnificent desolation." After its formation about 4.6 billion years ago, the moon experienced a billion years of meteoritic bombardment. Now, except for human footsteps, the surface of the moon is the same as it was 3.6 billion years ago.

Without an atmosphere, there is no weather to sculpt the moon's canyons and wear its mountains. The only real "weather" is the steady "rain" of meteorites that add almost five feet of new soil to the moon's crust every billion years. Even that precipitation will allow the footprints of Armstrong and Aldrin to last for ten million years.

We know the moon far better than our ancient ancestors did, and yet we are far less likely to notice it these days. We don't use it the way our ancestors did to count the days and bundle them into months. Oh, we still stare at it when we see it rising huge, full, and orange and when it hangs as a thin crescent in the western sky, punctuating the twilight. But most of the time we no longer know what the moon is doing day by day or night by night.

If the moon has less influence on our calendar these days, does the moon have less influence on our hearts? Has the intimacy of space travel broken the spell of that old devil moon? Probably not. We still write songs about it. It moves us to music and dances to its own beat.

GOING THROUGH A PHASE

After the sun's cycle of day and night, the moon's cycle was the next most obvious celestial rhythm apparent to our ancestors. The moon goes through a remarkable transformation, a sequence that runs from no moon at all, through a crescent, a half-disk, and an almost-full oval, to an entire disk, and from there, the moon drops backward through a similar set of conversions to invisibility again. The pattern is so familiar and so regular, we even pay our bills in moon intervals. The months of our

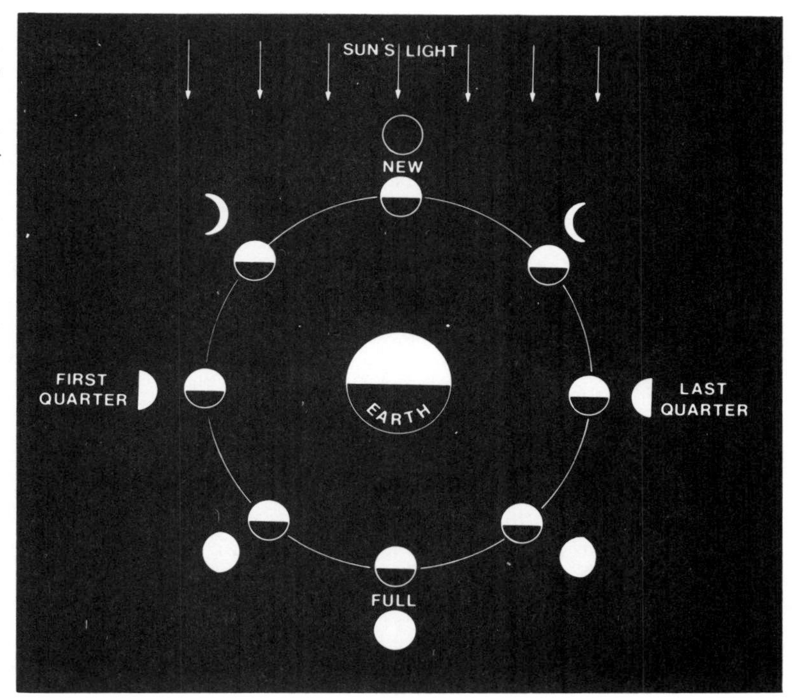

The moon's appearance changes cyclically each month because of its orbit around the earth and because it shines by the reflected light of the sun. Half of the moon is always in sunlight, but here on the earth we do not always see the lighted half. This diagram of the moon's phases shows the location of the moon on the inner ring; the outer ring displays the corresponding shape seen from earth. When the sun and moon are generally in the same direction, we see no moon at all. As the days pass we see the moon grow until it is opposite the sun. Its face then is fully lit. After full moon, the disk we view progressively declines until it disappears. It is gone from sight for two or three days and then reappears as a waxing crescent. (Griffith Observatory)

calendar are "moonths," or the thirty days of one moon.

Originally, the months really were "moons." In the old lunar calendars, the month's start and finish were actually timed by the phases of the moon, but the months we have now are just imitations of the genuine article. The length is about right, but the months in our present Gregorian calendar come and go without regard for what the moon is actually doing anymore.

In the old days, however, each of the moon's packages of periodic time would begin with some easy-to-spot event in the cycle—say the appearance of the first crescent. It shows up low in the west in the early evening right after the brief—roughly three-day—intermission when the moon is invisible. If you look at the moon again at the same time on the next night after first crescent, the crescent will be hanging higher above the western horizon than it did the night before. That means it will set later. If that first crescent reached the western horizon at about seven the previous night, it won't get there until nearly eight the next night. It rose about fifty minutes later, too, which is the average delay in moonrise from night to night.

By the second night, the moon has already grown a bit. At first crescent, the moon can be spaghetti thin, but a day later the moon is not as slim as it was the night before. It looks more like a boomerang, a bow, a boat, a pair of horns—maybe a bison's or a bull's, or perhaps like a banana.

To me, the three- or four-day-old crescent resembles a smile in the sky. It's a smile with no face and reminds me of that disembodied grin of the Cheshire-Cat.

By the time five or six days have passed, the crescent has turned into a half-moon, half of a silver disk that rises at roughly noon, is high overhead at sunset, and sets at about midnight. We are one quarter of the way through the cycle, and the moon in this phase is called the first quarter moon, even though it looks like half a moon in the sky.

These changes in the face of the moon occur because the moon is a spherical world in orbit around the earth and because it reflects light from the sun. It is a "cold" world in space, a celestial mirror that emits no light of its own. Half of the moon is always reflecting sunlight, however, because some half of the moon is always facing the sun. From the earth, though, we don't always see the entire sunlit side of the moon. Just how much of it we see depends on the placement of the moon, the sun, and the earth. When the moon is in the same direction as the sun, we see no moon at all. The half that faces us is the dark half. By the time the first crescent signs on, the moon has shifted east of the sun, and a small fraction of sunlit moon faces us. By first quarter moon, the moon forms a right angle with the sun, with the earth at the corner. That's why we see half of the moon gleaming.

Gravity is what propels the moon on its path through our sky. It is falling around the earth. As the moon circuits the earth from the part of its orbit nearest the sun to the opposite side half a month later, more and more of the moon lights up for us because it is shifting its location compared to the sun.

A few days after first quarter moon, the moon looks bigger and brighter. It's now oval in shape, about three-quarters full, and is called a gibbous moon. *Gibbous* means "humped." The moon has two "humps" in the two curves of its sides. The adjective may be unfamiliar, but we know the shape well. Lemons are gibbous. Gibbous footballs are transferred back and forth, sometimes in gibbous stadiums. Gibbous hips market designer jeans. This gibbous moon is said to be a *waxing gibbous* moon. A waxing moon is a growing moon, and the verb can be traced to Old English for *weaxan* and to German for *wachsen*. Both mean "to grow, to increase in extent."

About two weeks after the moon was located more or less in the same direction as the sun and completely invisible, the moon is opposite the sun. Now the entire sunlit face looks our way, and we see a completely illuminated disk, or full moon. Because it is opposite the sun, we see it on the other side of the sky from the sun. It can only rise when the sun sets, or set when the sun rises.

Even when the moon is full, it is strikingly different from the sun. Its face seems smudged with dark and roughly circular zones. They merge into shapes that some people say make the moon resemble a face. Others detect a rabbit there, or some other creature. (Photograph: Curtis Leseman)

Nearly two weeks before, at first crescent, the early evening moon was low in the west, not far from the sun, which had just set there, too. Now when the sun is setting in the west, the moon is all the way over in the east. Because it rises at sunset, it is up all night, highest at about midnight. We have to get up early in the morning to see this full moon in the west.

It is difficult to tell the difference between the full moon and the nearly full moon. A trained eye can do it, and, of course, careful day-by-day monitoring coupled with counting the days from first crescent or from the invisible new moon can pin it down. But for most purposes, the moon seems full for about three days. Then it begins to shrink, and we see a waning moon. This name derives from the Old English word *wanian*, which means "to lessen." The Old English adjective *wann* means "dark, gloomy," and we still use the word *wan* to mean "pallid," "pale," and "diminished in health, spirit, or power." By the time we get past the period of full moon, our vocabulary is explicit: The moon is in decline.

The waning moon goes through a reverse sequence of phases until it is invisible again. The phases are, in a sense, mirror images of the waxing moon. Whereas before we saw the right side of the moon at first quarter, now it is the left side that's in the spotlight. That first quarter moon rose at noon and set at midnight. You could have seen it in the eastern half of the sky during the daytime, if you had stopped to look. Its appearance marked the end of the first fourth, or quarter, of the month. This last quarter moon fixes the start of the last quarter of the month. It rises at about midnight and sets at about noon. You can still see it in the daytime, but now you have to look in the western half of the sky during the morning. The last visible crescent rises a little before the sun does and is suspended above the eastern horizon for the short time it takes the sun to cross the horizon and outshine its wan light. By the next day, that moon slips out of sight entirely, slain by the sun.

For many ancient and traditional peoples, that disappearance of the moon was really the death of a shiny, silvery celestial god. They could see the moon sicken or succumb to the knife of a solar adversary that sliced away at it in its waning weeks. The Aztec of central Mexico were among those who saw death in the dark of the moon. In fact, they sacrificed the moon on the altar of power politics.

SLICING UP THE MOON IN ANCIENT MEXICO

The Aztec called the actual moon Meztli, but in a story they told about the miraculous birth of their own tribal god Huitzilopochtli ("Hummingbird on the Left"), the moon seems to have been symbolized by Huitzilopochtli's sister. Her name was Coyolxauhqui ("Adorned with Bells"), and the story details her defeat at the hands of her brother. She was not necessarily the moon, but she was perhaps a moon goddess who shared some of the moon's attributes. In a similar way, Huitzilopochtli's high status allowed him to assume some of the sun's character and power. His victory over Coyolxauhqui and the stars of the night sky were, according to the story, what put Huitzilopochtli—and the Aztec—in control of the Valley of Mexico.

As a culture hero and special protector of the Mexica, as the Aztec called themselves, Huitzilopochtli operated under contract. War and sacrifice nourished him, and he enhanced the fortunes of his chosen people. According to legend, the Mexica first emerged from Chicomoztoc, the legendary "Seven Caves" located far to the northwest in the land of Aztlán. Their migration finally brought them to an uninviting little island in the swampland at Chapultepec, near the west shore of Lake Texcoco. That's where they built their capital city Tenochtitlán, the "Place of the Fruit of the Cactus," and Mexico City is now on top of its ruins.

The Aztec started out, then, as nomadic barbarians who had to scratch out a living on land no one else wanted. They were the last to arrive in the Valley of Mexico, indigent and scorned by their neighbors, but in less than two hundred years the Mexica were calling most of the shots. Within another century, however, they were back under the heel, this time the Spaniards'. We know the story of Coyolxauhqui from accounts collected shortly after the Spanish conquest of Mexico in Book Three ("The Origin of the Gods") of the *Florentine Codex*, an important compilation of Aztec life and lore completed in 1580 under the title *Historia de las cosas de la Neuva España* (History of the Things of New Spain).

As far as the Aztec were concerned, they were ordained to rule. The divine favor their tribal god conferred upon them was transformed through war and sacrifice into nourishment for the sun. The sun was sustained by sacrificial blood and the extracted hearts of human victims. By serving the sun and the other gods, the Aztec fulfilled their sacred mission: Their sovereignty preserved cosmic order, and without them, the universe would darken and die.

On the way to the Valley of Mexico and cosmic destiny, the Aztec stopped, they said, at Coatepec ("Serpent Hill"), near the twelfth-century ruins of the Toltec capital Tula, about sixty miles from Mexico City. According to the myth, Huitzilopochtli is born there because that's where his mother, Coatlícue (Serpent Skirt) is confined. Atoning for unspecified sins, Serpent Skirt is sweeping at Serpent Hill.

Coatlícue is Mother Earth, and her previous children include the Centzon Huitzinauhua, or "Four Hundred Southerners." These are stars of the night sky, and they are born each night when they rise out of Mother Earth's body in the east. They are an army on the march.

Now while Serpent Skirt is tidying things up on Serpent Hill, a puff of down softly descends from

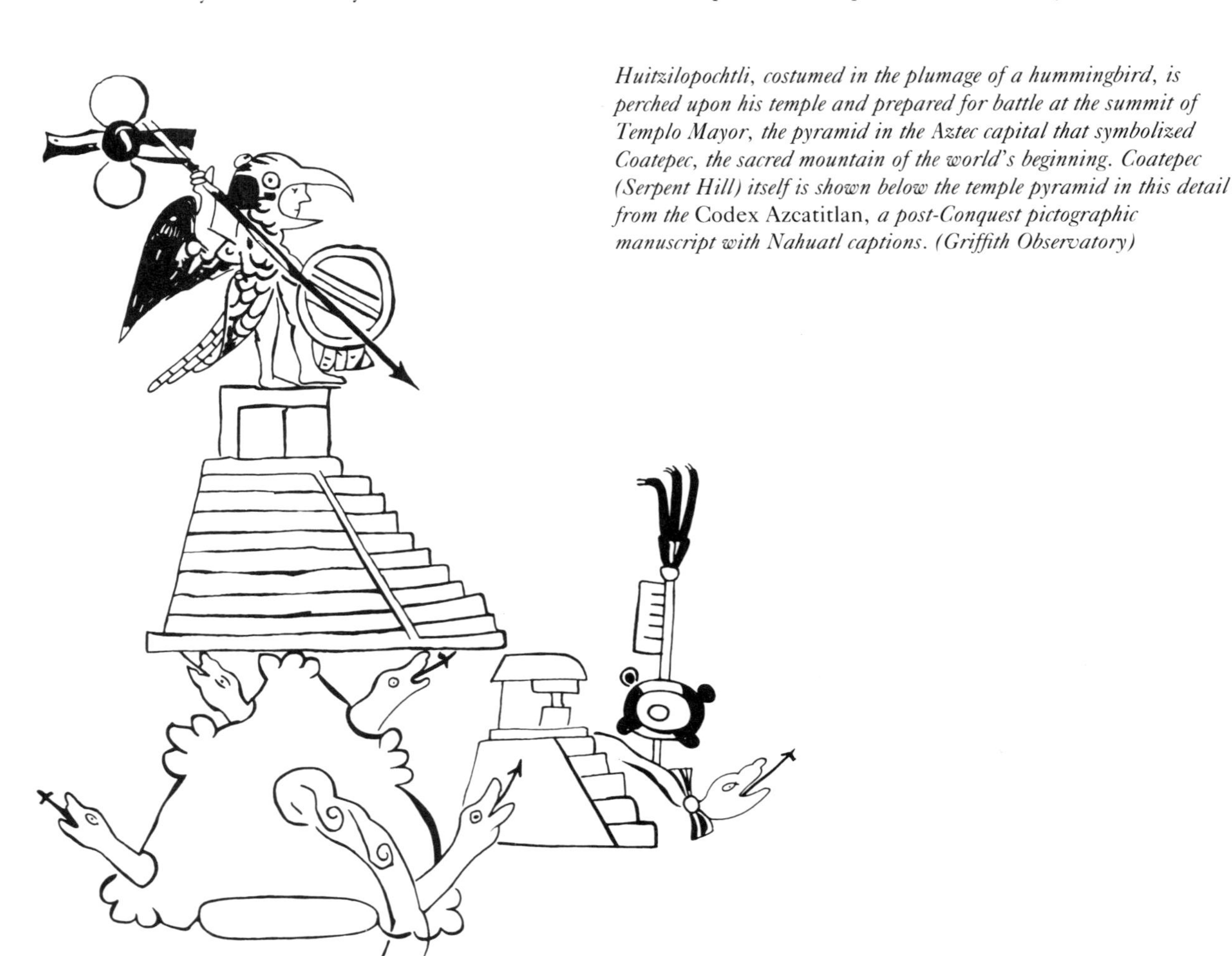

Huitzilopochtli, costumed in the plumage of a hummingbird, is perched upon his temple and prepared for battle at the summit of Templo Mayor, the pyramid in the Aztec capital that symbolized Coatepec, the sacred mountain of the world's beginning. Coatepec (Serpent Hill) itself is shown below the temple pyramid in this detail from the Codex Azcatitlan, *a post-Conquest pictographic manuscript with Nahuatl captions. (Griffith Observatory)*

The lacerated flesh of the dead goddess Coyolxauhqui exposes her bones and splits her neck on this monumental stone relief. Part of the buried remains of Templo Mayor, this disk was discovered in 1978 beneath the streets of Mexico City, at a place that symbolized the spot where Coyolxauhqui's body lay dismembered at the foot of Coatepec. (Instituto Nacional de Antropología e Historia)

heaven. Coatlícue isn't about to allow any debris —even a few feathers—compromise her custodial efforts. She catches the offending feathers in the air and absently tucks them in with the snakes at her waist. She still has some sweeping to do, however, and forgets about the ball of feathers until later, when she reaches under the snaky ribbons of her petticoat. It is then she discovers the feathers gone and herself pregnant. She is surprised, of course, but she accepts it philosophically. She is, after all, Mother Earth, and she is always giving birth. Her children, however, don't take the news so gracefully. The Four Hundred Southerners and Coyolxauhqui are ashamed by the illegitimate child ripening in their mother's womb, and they decide to kill their mother.

Armed for war, the Four Hundred Southerners and Coyolxauhqui advance upon the sanctuary of Mother Earth. Her heart sinks when she hears the terrifying sound of Coyolxauhqui's bells. The doomed Mother Earth and her unborn child have only moments to live before being engulfed by the celestial host. But within her, the voice of her unborn child murmurs comforting words. "Have no fear. Already I know what I shall do."

Reaching the base of the Serpent Hill, the starry troops rush to its summit. Just as they arrive, the mysterious illegitimate son of Coatlícue surges from her womb, fully grown and fully armed. Warrior stripes painted on his face, more stripes on his thighs and arms, the feathers flying from his forehead, ears, and feet, he is dressed for success. He carries an eagle feather shield and flings darts with his throwing stick. He needs no allies. The ranks of stars disperse in confusion. Huitzilopochtli lights his fire serpent and dispatches it upon Coyolxauhqui, his elder sister. Decapitated by the blow, her body tumbles down the hill. She breaks into pieces. Lacerated limbs, bits of bone, fragments of flesh, all strewn across the slopes of Coatepec.

Huitzilopochtli roars down the hillside after his retreating brothers. Though it is one against four hundred, they are clearly outmatched. He chases them around the base of the mountain four times, and they can do nothing but run. The text says: He "destroyed them." He "annihilated them." He "exterminated them."

There is certainly something celestial about this official mythology of the pedigree of Mexica power. Huitzilopochtli banishes and blasts his brothers in the same way the sun slays the stars and disperses the dark. The fate of Coyolxauhqui is like that of the moon. When full, the moon marches with the stars the entire night, and the waning moon dies a little every day at the hands of the sun. The very last crescent is overwhelmed by dawn. Sliced up by sunbeams, Coyolxauhqui's body is torn apart in the same way the waning moon is stripped of flesh, a night at a time, until it dies completely at new moon.

Though Coyolxauhqui is not the actual moon, she does act like it, as Huitzilopochtli acts like the sun. Both gods are players in a drama that tells us how Huitzilopochtli seized control of the cosmos with the naked, unforgiving force of a seasoned warrior. Aztec fortunes were also made with violence and fear. Without lineage or the respect of their neighbors, they had no real claim to territory or power. But what they lacked in legitimacy they made up in strength. In their story of Huitzilopochtli's birth, they mimicked their own history and the cosmic war between the sun and the moon. This moon myth was also a story about the foundation of Aztec power; by exploiting what the sun does to the moon each month, it legitimized Aztec sovereignty.

SLIPPING BACK THROUGH THE STARS

As the moon moves through its phases, it also moves through the stars. Every day it seems to shift farther to the east. The pace is fairly fast. You can see a change in the location of the moon, compared to the background stars, even during the course of a few hours. This is not the same thing as the moon's normal, "daily" progress from east to west, a motion it shares with the sun and the stars because the earth is turning. The west-to-east shift is a product of the moon's own motion. It orbits around the earth every month, and to us it looks as if the moon is moving "backwards" through the stars. It is, in its way, an independent wanderer, for it cuts its own course against the current of the sky's daily turn.

Eskimos all over the Arctic know the moon runs backward through the sky and tell a story about this behavior that pairs the sun and the moon as sister and brother. It also deals with the fact that the face of the sun is pure and bright while the face of the moon is marked and dirty. Moon and sun are often paired in stories about the sky because in some ways they seem very similar. They both look like disks, and they are the same size in the sky; both are large and brighter than anything else. The moon, however, is not as bright as the sun. Mountain ranges, vast "seas" of frozen lava, and impact craters have blemished the moon for billions of years and provided it with a distinctive mask of bright and dark features.

According to the Eskimos, the Moon was a promiscuous fellow who preferred his own pretty sister, the Sun, but she was not about to let her own brother slip into bed with her. To fool her, he disguised his face with ashes and convinced her to dally with him in the darkness. She had no idea who he really was, but she enjoyed his company and invited him for another romp. He complied. And he complied. And he complied again. They kept each other occupied all night long. In the morning, however, when she saw who her partner really was, Sun was very disturbed. Incest was forbidden. Sun warned her brother, "You will never do this awful thing to me again." Not especially remorseful, he replied that he would do as he pleased since he was the male and stronger. Responding to that challenge, she picked up a kitchen knife and cut off her snowy white breasts, which she mixed with blood and urine to make a disagreeable dish. She said, "If you like the taste of me so much, taste this!" Then she ran out of the house with a torch and up into the sky.

Her brother grabbed a torch, too, but he was delayed when he paused to sample the part of her she had left behind. Then he pursued her. This affair and this chase still continue every month. At the new moon Sun and Moon are snuggling together in the dark. But after that Sun races to the west each day with Moon hot on her trail. He's unable to catch her, however, and each day he drops farther back.

By orbiting the earth, the moon does appear to slip eastward from the sun each day. The distance between them increases until full moon. Then the eastward shift carries the moon back to new moon again. If, however, you think of tracing out the line to the moon around the long way from the sun, you just keep having to go farther to catch up with the moon. Because the moon's torch in the Eskimo story burned down to glowing embers, he is less bright, and we can still see the ashes he rubbed on his face as the dark features on the disk of the moon.

THE LUNAR RULER OF TIME

Part of what we choose to see in nature is patterned, cyclic change. That kind of change is what also makes the moon a master of time. It tallies the days with an everchanging face and measures time in cycles of its daily change. Even our name for it includes the idea of measurement. The Indo-European root *me* is embedded in the word *moon*. Transformed into *mami* in Sanskrit, the root means "to measure," and it persists, through Latin *(mensur)*, in the verb *mensurate*, which means "to measure." The moon's connection with measurement involves its use in timekeeping, and the Egyptians certainly recognized the moon's part in timekeeping.

> Moon in the night, ruler of the stars, who distinguishes seasons, months, and years: He comes, ever-living, rising and setting.

The Egyptians also identified the moon with their ibis-headed god, Thoth. As the divine scribe, he was the patron of knowledge and writing and the sun's secretary. Thoth often wore a moon disk cradled in a crescent as his crown, and the beak of the ibis is a crescent. Thoth celebrated literacy, a skill inspired perhaps by the written records that made a calendar out of the moon's phases.

Changing constantly, the moon not only brings us back to a new starting point each month, but also carries us through the year—one cycle of the seasons, which comprises twelve complete cycles of the moon. Sometimes called a lunar year, it lasts 354 days. Here is why. One monthly cycle takes 29 days 12 hours 44 minutes and 2.8 seconds to complete. That's close to 29½ days, but traditional peoples don't count time in half-increments. Some months would therefore be counted as 29 days long, while others would last 30 days. Given

This Western Eskimo mask reveals the face of the moon. He was judged to control the game supply because the availability of game follows the seasonal calendar, which is revealed through the cyclical phases of the moon. His power over the animals is emphasized by the wooden figures of seals and caribou he holds. (Object in American Museum of Natural History, New York)

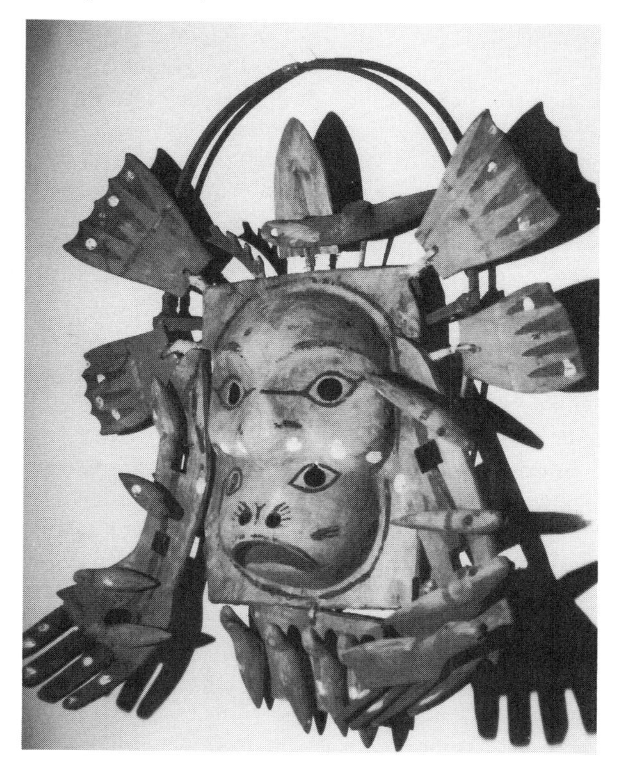

enough lunar cycles, the average would prevail. At 29½ days apiece, they total 12 × 29½, or 354 days. You can keep time reasonably well with a lunar calendar, and lots of people have done so. But there is one disadvantage. The seasons and the lunar calendar get out of synchronization. Instead, the seasons stay in step with the solar year, which is a little over eleven days longer than the lunar year. Without some kind of correction, then, the lunar calendar will slip ahead of the seasons and start sooner. You can add an extra month now and then, however, to keep the moon in tune with the seasons, and that's what people usually did.

In the far northeast tip of Siberia, the Chukchi reindeer herders remember when the moon first agreed to light the sky and measure the year for them with months that signaled seasonal change. It all started one winter when the daughter of one of the reindeer herdsmen was tending the herd far from camp. The tame reindeer who pulled her sled happened to look up, and he noticed that the Moon Man seemed to be getting larger. In a moment, he realized the Moon Man was descending to earth in his own sleigh, pulled by a pair of reindeer. The girl's reindeer friend called to her, "Look out! The Moon Man is coming down to carry you away." The girl was frightened and asked her reindeer what to do. He quickly hoofed a hole in the snow, told her to hop in, and heaped the snow back over her before the Moon Man landed. Only the top of her head remained uncovered, and her hair looked like a small bunch of grass protruding through the mound of snow.

The Moon Man was quite puzzled over the girl's quick disappearance. He decided to leave and come back later. She would have to reappear sometime, and he could catch her and carry her off then.

Once the moon had left, the reindeer kicked away the snow. He and the girl raced back to her father's tent for help, but when they arrived, she discovered her father was still away. She asked the reindeer again what she should do. The reindeer thought a disguise would help. It had worked the first time. He suggested he turn her into a meat-pounding stone, but she said, "No, the Moon Man will know me."

"Well, then, how about a hammer?"

"No, he will know me."

"How about a tent pole?"

"No, he will know me."

"How about a single hair on the hide that hangs over the door?"

"No, he will know me."

The Moon Man, we must guess, is just too bright to be fooled by these transformations. But there the Moon Man's insight must end, for the reindeer then said he would turn the girl into a tallow lamp. The girl agreed this would work, and in a moment she was gone, a glowing lamp in her place.

Disguised not a moment too soon, the girl was safely hidden as the Moon Man swooped down on the camp and tore into the tent. He looked everywhere—among the hairs on the hides, around every tent pole, through every utensil, under the beds, and over every inch of the floor. He didn't see her. She was shining right there in the middle of the tent, but he didn't notice her. His own light was too bright, and he couldn't get too close to the flame, for it would burn him. Genuinely mystified by the girl's disappearance, he decided to return to the sky.

When he went outside to his sled, however, the spunky girl ran out of the tent and baited him. "Here I am! Here I am!"

The Moon Man rushed back into the tent, but by the time he got there, the girl had turned back into the lamp. He looked everywhere for her again, but he failed to spot her. He went back outside and untied his reindeer so he could leave. This time the girl stuck her head through the tent door and shouted again. "Here I am! Here I am!"

Convinced he could catch her, the Moon Man raced back into the tent. But once again the spirited girl was well hidden in plain sight. The Moon Man rummaged through everything. By now he was exhausted. Each time he had run back into the tent, he had grown thinner and weaker. Now he couldn't carry off the girl even if he found her. He couldn't even carry himself back to the sky. He stumbled to the ground, frail and helpless.

The girl realized she could handle the Moon Man now. She tied him up. Because he had threatened to carry her away to the sky, the Moon Man thought the girl would kill him. He admitted he deserved to die at her hands, but he begged her to take him inside the tent and cover him with skins. He was miserably weary and cold.

Not believing what she had heard, the girl asked the Moon Man, "How could you be chilled? How could you want to remain inside a tent? You live in the cold, outdoor sky and wander freely there. You belong outside, and that's where you'll stay."

"If it be my fate to be homeless, set me free, and I'll serve you and your people forever," promised the Moon Man. "I'll be something for people to watch. I'll give them some pleasure. I'll be a beacon in the night to guide them. My light will turn night into day. Set me free, and I'll measure the year for you.

"First I'll be the Moon of the Old Reindeer Stag.
then the Moon of Bitter Cold Udders,
then the Moon of the Full Udders,
then the Moon of New-born Reindeer Calves,
then the Moon of Water,
then the Moon of First Leaves,
then the Moon of Warm Weather,
then the Moon of Shedding Antlers,
then the Moon of Light Frost,
then the Moon of Pairing Reindeer,
then the Moon of the Reindeer's Winter Back,
then the Moon of Shorter days."

The girl still didn't quite trust the Moon Man. "You'll just recover if I let you go. You'll get strong again. And you'll carry me off to the sky."

"No," the Moon Man protested, "you are too clever for me. I'll remain in the sky and give you light."

So she let him go, and the Moon Man kept his promise. He shines and measures the year. And just in case he gets any funny ideas again, he grows weak each month just after he's recovered his full strength.

TRANSFORMATIONS IN MOONLIGHT

By changing so quickly and so obviously, the moon allowed us to craft effective calendars out of its phases. Change is such an essential part of the moon's character, however, the moon came to count for more than just the passage of time. It stood for the process of change itself, and people have said the moon governs all kinds of change: It moves the tides and controls the rain; it presides over the growth of plants and animals and guides the fertility of women; it changes itself, and the rhythm in its celestial walk makes it the model and trigger for all cyclical change. For that reason the moon's own transformation or its alleged ability to transform others became a primary theme in many of the stories we have told—and still tell—about the moon.

There persists, for example, widespread belief in the power of the full moon to induce departures from normal human behavior. Firefighters say there are more fires and false alarms in that phase of the moon. Police swear there are more crimes, more drunkenness, more violence, and more accidents. They are both certain the full moon brings out more general lunacy. The root of that word *lunacy* only confirms what people have felt for quite some time: The moon can make us—at least some of us—temporarily mad. Nurses in hospital psychiatric wards and emergency room personnel are all said to verify that things go a little crazy during the full moon. Surgeons say there are more post-operative hemorrhages. Supposedly more babies are born.

Is all of this possible? Is all of this true? Can the moon really affect us this way? Why do people think so?

Often the argument goes like this: The moon causes the tides. And the tides are water influenced by the moon. Now, the human body is practically all water—80 percent! So if the moon can move the waters of the earth so dramatically to produce two tides a day, it should govern us as well. Is this true? No. Why not?

Ocean tides are the result of the difference in force felt by opposite sides of the earth, but on human beings, the moon's tidal effect is insignificant. The force the earth exerts on a 200-pound person is 200 pounds. The moon's force is one-hundredth of an ounce. So the moon reduces a 200-pound person by only one-hundredth of an ounce. That's not much. The moon's tidal pull is even less. It is equal to one part in 30 trillion of the weight of the fluid on which it is acting. If a

Belief in the moon's power to influence human behavior is evident in the title and in the action in this seventeenth-century French engraving. Beams from an amused and mischievous moon touch crescents above the heads of five women. (Griffith Observatory)

200-pound person is 80 percent water, then the weight of the fluid of concern here is 160 pounds. The tidal force of the moon on this water is about five-trillionths of a pound. This means this book exerts a tidal force on your body thousands of times greater than the moon's.

To test the moon's influence on human births, the late Dr. George Abell, an astronomer at U.C.L.A., reviewed the record of all births during a fifty-one-month period (fifty-one real lunar cycles) in the mid-1970s and demonstrated that they occurred completely randomly with respect to the phase of the moon. In addition, other careful statistical tests have discounted most claims about the behavioral effects of the moon.

People probably believe in a connection between the moon and their bodies because of selective memory. If the moon is full, more people tend to notice it, and when something odd happens, they are more likely to remember both the event and the moon. People who already believe the full moon influences human behavior are more likely to retain the memories that confirm their belief. I have talked on occasion to both medical and law enforcement professionals, who were genuinely puzzled to learn that a recent night of real uproar and lunacy did not coincide with the full moon. Rather than conclude something might be wrong with the belief in full moon, their initial response was skepticism of the astronomical record.

Perhaps the most extraordinary influence of the moon on human behavior is its power to turn an otherwise solid citizen into a snarling werewolf. A particularly colorful victim of moon madness, the werewolf is still a crowd pleaser.

You can find other reflections of popular belief

in the moon's influence on human beings. Marvel Comics, for example, has published the adventures of Moon Knight, a millionaire with a multiple personality whose strength waxes and wanes with the phase of the moon. Appropriately, he made his comic book debut in *Werewolf by Night.* In a more recent series, he died in a remote tomb in the Egyptian desert but was brought back to life by Khonshu, a slight variation in spelling of Khonsu, an Egyptian god of the moon during the New Kingdom (1570–1070 B.C.)

Khonsu was the child of Amun, the supreme god of Thebes who eventually appropriated the divinity of the sun. More than a thousand years earlier, Khonsu's reputation was more savage. He helped catch and kill certain gods the dead king was to consume.

Marvel's Moon Knight is installed by the moon god as the "Fist of Khonshu," and he ruthlessly apprehends evildoers, armed with a crescent-shaped boomerang and scarab darts, their wings outstretched in the shape of a crescent. He pilots a crescent-shaped hang glider and also carries a golden club in the shape of an ankh, the Egyptian symbol of life.

THE LUNAR LORD OF EVERYTHING

Certainly a connection was made between the moon's changes and transformation in the world around us as long ago as ancient Egypt. In Egypt, the god Osiris ruled the transformation from life to death and judged the dead on their way to new life. He was also equated with the moon and known as the Lord of Everything. Texts on the ceiling of the main hall of the Temple of Hathor at Dendera, which probably belongs to the first century A.D. and the time of the Roman Emperor Tiberius, say Osiris is the moon. The ceiling relief shows Osiris in his celestial boat, sailing upon a lintel-like symbol that stands for the sky. The sky symbol is supported by the goddesses of the cardinal directions. Osiris is accompanied by his sister/consort Isis and Nephthys, a funerary goddess who, like Osiris and Isis, is also one of the original children of Father Earth (Geb) and Mother Sky (Nut).

Plutarch, the Greek historian and biographer who lived in the first and second centuries A.D., seems to have preserved at least some of Egypt's traditional lore in his book *On Isis and Osiris.* The Egyptians, he confirmed, saw Osiris in many things, including the moon. Osiris, he said, brought civilization and craft to the people who worshiped him and delivered them from their formerly brutish and poor lives. He gave them laws and instructed them to honor the gods. Marketing world order, he organized agriculture and the foundation of religion. After putting Egypt in order, Osiris franchised himself to the rest of the world. His absence from Egypt, however, provided his chief adversary, the god Set (or Seth) to conspire against him.

Also a child of Nut and Geb, Set stood for the vacant sterility of the desert, lifeless chaos, and the mindless destruction of cosmic order. He is depicted with an animal's head, as are so many Egyptian gods, but in Set's case, the animal remains unidentified. It has some attributes of a donkey as well as characteristics of a canine. He is an enemy of the ordered transformations Osiris orchestrates. In Plutarch, Set is called Typhon, who in Greek myth was one of the monstrous children of Gaia, the earth. His taste for annihilation and darkness is second to none.

With a crowd of co-conspirators, Set tricked Osiris into lying down in a beautifully ornamented box that was really a coffin in disguise. As soon as Osiris lay down to test the fit, Set's gang slammed down the lid, nailed it shut, sealed it with molten lead, and tossed it into the Nile. It floated down to the river's mouth near Tanis, and Plutarch says that even in his day the Egyptians still regarded the Tanitic mouth of the river as "hateful and execrable" because of its part in the death of Osiris.

Now, where does the moon fit into all of this? According to Plutarch, Osiris was murdered on the seventeenth day of the month of Athyr in the twenty-eighth year of Osiris's rule. The number 28 in the death of Osiris symbolizes the time when the month dies, at the last visible crescent. The Egyptians said the moon shines for 28 days each month. They did not include the dark days of new moon in that number. To understand the significance of the number 17, you have to think about the moon's phases. Fourteen days of waxing moon

Fourteen steps on an ascending stairway stood for the fourteen days of the waxing moon in this ceiling relief from the temple of Horus at Edfu in Egypt. The full moon and the fifteenth day of the month are symbolized by the crescent-cradled eye at the top of the stairs. It is attended by Thoth, the ibis-headed scribe god who was responsible for the integrity of the calendar. To the left is the moon's own sky boat. (From the Description de l'Egypte*)*

began with the first crescent, and Ptolemaic-period (332 B.C.–30 B.C.) temples at Edfu and Dendera portrayed these 14 days as 14 tutelary gods upon the 14 steps of a stairway that ascends to a moon disk in the guise of the left eye of the falcon god of the sky. Each month the days climb to the full moon on the fifteenth day. After that the moon wanes, but it is hard to see any decline on day 16. By the seventeenth day, however, the moon has visibly dwindled. Day 17 inaugurates the moon's decay. It was a good day for Osiris to die, but his death didn't take place in any arbitrary month. Osiris died in the month of Athyr, the third month in the season of the Nile's inundation and the month when the Nile reached its greatest height and when its flood started to subside. Osiris died, then, when the river died, when the month died, and when the moon died.

Isis went searching for the body of her husband and collected it at Byblos, a Phoenician city at what is now the Lebanese port of Jubail, where the waves had carried it to shore. She returned with the corpse to the delta of Lower Egypt and hid it among the marsh reeds. Hunting by night in the moon's light, Set by chance discovered the body. Not yet content, he dismembered the corpse into 14 pieces and scattered them up and down the length of the Nile. Isis managed to recover everything except his sexual apparatus and reassembled nearly all of her better half. The penis of Osiris had been dropped in the river, however, and was consumed by the fish. Resurrected but unable to rise, Osiris could not impregnate Isis again. His fertility, now in the Nile, was transferred to the river, and it fertilized new life in Egypt every year. Each day of the waning, or dying moon, was equated with the pieces of Osiris, and all 14 days of the moon's decline also appear in the reliefs at Dendera and Edfu as gods afloat on the moon's boat, piloted by Thoth.

RESURRECTIONS MIDWIFED BY THE REBORN MOON

Osiris, as the Egyptian god associated with the kingdom of the dead and the judgment of souls, embodied the principle of cyclical transformation, and the key mechanism of this change was death and rebirth. For that reason he was equated at times with the moon. Egyptians also played a boardgame in which the action of play followed the same cycle of death and rebirth, and one circuit through the game also imitated the monthly period of the moon. The game was called *senet,* and boards and pieces for playing it were sometimes interred with the dead. Tutankhamun's burial treasure included an elegant set carved from ebony with ivory inlay to mark the thirty squares on which the markers were moved. Evidence of the game can be found as early as 3300 B.C., in the predynastic period.

Usually the spaces on the board were arranged in three rows of ten squares each. The pieces probably followed a serpentine path, up one row and down another. As in most boardgames, some squares were to be avoided if possible. Others were "safe." Although we don't know all of the rules, it appears the player's goal was the successful completion of the passage of his pieces to the end square. Identified as the House of Horus, the final square was sometimes marked with a sun symbol or a falcon, which represented rebirth.

The funerary element of the game is confirmed

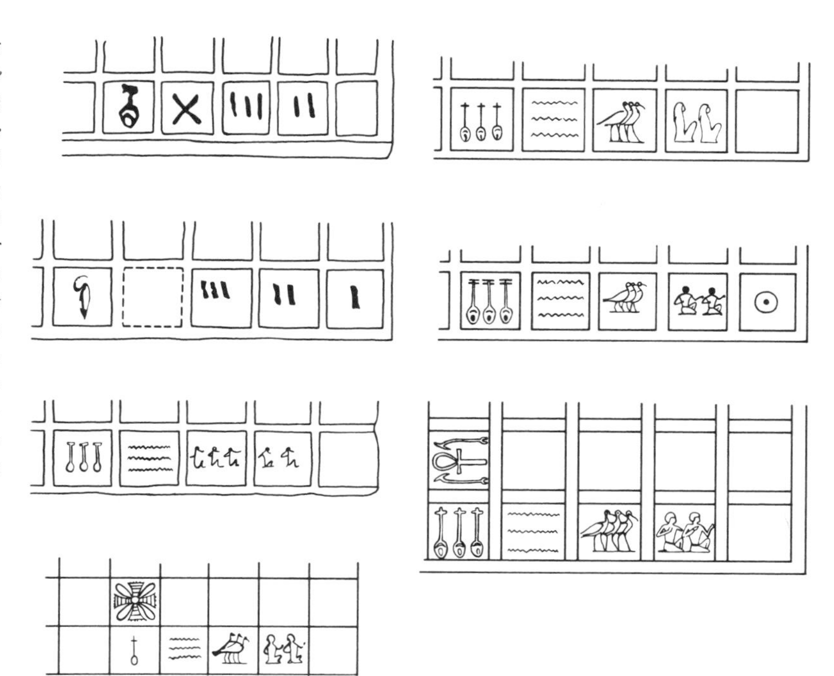

In ancient Egypt, the game of senet *was modeled on the moon's monthly cycle. The last few squares of several Eighteenth Dynasty (1570–1293 B.C.) boards are shown here. In many of them, square 26 (bottom row, fifth from the right) contains one or more* nefer *hieroglyphs. This symbol looks like a spoon with a crossbar at the top of the stem, and it means "good." Zigzag water symbols on the next square represent the risky waters of the Underworld. The last three squares in one way or another count down to the soul's final destination on the board and its successful resurrection. (Griffith Observatory)*

by portrayals of the deceased playing the game in the netherworld. These show up in tomb paintings and papyri accompanied by descriptive texts. Neferteri, the wife of the celebrated New Kingdom, Dynasty XIX, pharaoh Ramesses II (ruled 1279–1212 B.C.) is shown playing *senet* in her elaborately painted tomb in the Valley of the Queens. One version of the Egyptian Book of the Dead, the Papyrus of Ani, shows the scribe Ani and his wife, Tutu, sitting inside a pavilion and at a game of *senet*.

Not just a pastime, *senet* was a dead-serious enterprise. Winning meant the soul's survival. Square 26 with its hieroglyph for "good," was really what the Egyptians called the "beautiful house," the mummy's funeral parlor. Square 27 was the supernatural Nile of the netherworld, the waters that had to be crossed before the tomb in the west and the place of rebirth could be reached. Passing through squares 28 and 29 meant a favorable verdict in the great hall of Orisis. Upon arriving at square 30, the soul joined the sun for rebirth in the east. Magical talismans and religious significance transformed the game into a kind of road map through the netherworld with obstacles, dangers, resources, and goals clearly named and marked.

Play began on the square known as the House of Thoth. Plutarch equated Thoth with the Greek god Hermes, the messenger of the gods and the divine sponsor of commerce, communication, and travel. As the Egyptian Hermes, Thoth ruled the written word, handy for commerce and enterprise, and may have presided over the calendar. He was called the "reckoner of time" at Dendera, and inscriptions at Edfu address him as the "determiner of time," the "scribe of time," and the "divider of time." He "divides seasons, months, and years." Thoth's role in the game and the thirty squares suggest, then, another level of metaphor. In charge of writing and the calendar, Thoth counted out the thirty days of the lunar month. So a game of *senet* also meant going a round with the moon.

In fact, the first lunar month in the Egyptian calendar was called Thoth, and there is more evidence that the game and the calendar were closely related. In the papyrus from Oxyrynchus, there is a fragmentary description in Greek, from the second or third century A.D., that describes an Egyptian game with thirty squares. It seems to identify each square with a day in the lunar month but adds that the sum obtained by adding the numbers of squares 15 through 30 together is 360. This, of course, is close to 365, the number of whole days in a year. Plutarch said Thoth gambled in a boardgame with the moon and won from her one seventieth of each of her periods of illumination. He contrived from that time five extra days to fill out the year from 360 to 365.

Thoth and the moon must have been playing *senet*. The moon plays *senet* every month, and that is the point of the celestial template for the playing board. The moon's death and rebirth were linked to the transformation of the soul in ancient Egypt through a game modeled on the moon.

THE LUNAR LORD OF LIFE

To the Sumerians of ancient Mesopotamia, the moon was Nanna, "a great lord," a "light shining" that establishes the month and brings the year to completion. At new moon, Nanna sat on the bench in the underworld and judged the claims and cases of the gods residing there. His Babylonian name, Sin, is a contraction of Suen, which perhaps originally referred to the crescent moon. When Nanna reappeared as a new crescent, his power to renew himself was evident. He was a

> *. . . lamp appearing in the clear skies,*
> *Sin, ever renewing himself, illuminating darkness,*
> *bringing about light for the myriad people.*
> (Thorkild Jacobsen, *The Treasures of Darkness*, Yale University Press, 1976, p. 126)

He measured time, lighted the night, and revitalized the world. For the sake of the farmers he was "fruit self-grown." In the minds of cattlemen, the new crescent made the moon a "frisky calf of heaven." The boatmen of the marshes at the mouths of the Tigris and Euphrates saw a boat in the crescent's bow. The spring floods, the growth of marsh reeds, the bounty of the fields, and the proliferation of the herds—all proceeded from the moon, and he was in particular credited with guaranteeing an ample supply of the dairy products the herds provided. This food was a foundation of life and was a gift Nanna gave to Ningal, a goddess of the abundant reeds in the marsh. Ningal was the moon's sweetheart, and he said it to her with cheese and milk and cream, not with flowers.

The moon seemed to emerge from the privacy of Ningal's bower each time he rose. In a hymn detailing their courtship, the moon has to satisfy his lady's every whim. He refills the rivers, restocks the fish in the marsh and the game in the forest, and replenishes the reeds. Grain grows full at his command, and garden vegetables ripen. The desert flowers. He puts honey on the table, wine in the cup, and life in the palace. Ningal demands a lot from the moon, but Nanna knows when to call in his chips. After he has completely restocked the world's larder for Ningal, he has her eating out of his hand. She says she'll join him.

> *O Lord Nanna, in your citadel*
> *I will come to live,*
> *Where cows have multiplied, calves have multiplied*
> *I will come to live.*
> (Thorkild Jacobsen, *The Treasures of Darkness,*
> Yale University Press, 1976, p. 126)

And she also agrees to lie down in his bed. It's a match made in heaven.

In their story of Nanna and Ningal, the Sumerians were celebrating the life-enhancing energy they saw in the moon. Elsewhere Nanna was said to be a herdsmen, and the cattle he tended were the star herds he accompanied in the night sky. He milked them and drove them to pasture and won praise from his parents for the dairy goods he delivered to their door.

There is ample evidence in these Mesopotamian traditions for a link between the fertility of the world's plants and animals and the power of the moon. The same kind of notion applied to Osiris. Explicit connections between his death and the sprouting of the next crop of grain prompt us to call him a vegetation god. He is, but that does not make him just a spirit of the plant world. As we have seen, he also represented the fertility of the Nile, death and rebirth, and the moon. His cyclical change represented the principle of birth, growth, death, and rebirth the Egyptians saw in everything. Because the moon can grow, die, and live again, it is said to induce new life into the world along with the inevitability of death.

ENHANCING THE PLANTS

Vegetation follows the same pattern of waxing and waning we see in the moon. From seeds, young plants develop. They grow to maturity, blow to seed themselves, and die. Their deaths, however, set the stage for rebirth in another season of new growth. The seed of the next generation accompanies the death of the last. Because the phases of the moon are like the growth and decay of the vegetation, people saw a relationship between the moon and the cultivation of plants. Some people still do. *The Old Farmer's Almanac* continues to advise us to "plant flowers and vegetables which bear crops above the ground . . . during the LIGHT of the moon; that is, between the day the moon is new and the day it is full. Flowers which bear crops below ground should be planted during the DARK of the moon; that is, from the day after it is full to the day before it is new again." Because underground vegetables are regarded as "growing down," the *Farmer's Almanac* school of plant cultivation links the growth of the subsurface groceries with the downside phases of the moon.

In addition, the *Almanac* provides a table that lists the most favorable planting times based upon the phases of the moon. Other traditional advice counsels pruning vines during the full moon. If you want to retard the growth of a shrub or tree, you are supposed to set or cut it when the moon is dark. On the other hand, if you want it to grow quickly, use the first quarter moon.

Almost two thousand years ago, Plutarch said something about cultivating by the moon. Farmers, he reported, gathered the harvested, or "dead," wheat from the threshing floor when the moon was on the wane. Ohan Mantaguni, an Armenian author of the fifth century A.D., indicated that the moon affects plant growth, and in the seventh century A.D., another Armenian, Anania of Shirag, wrote that the moon had been known centuries earlier as "the nurse of plants." The Chorti Maya, who live in parts of highland Guatemala, El Salvador, and southwest Honduras, say that trees produce fruit under the moon goddess's guidance.

Pointing a finger at the moon will parch the fields in Tzeltal Maya territory on the Chiapas plateau. Nineteenth-century folklore from northern England warned, however, against picking apples during the waning moon; they'll contract and wrinkle. In Devon, on the other hand, most people are convinced that apples harvested at the waning moon would not decay if bruised.

WOMANIZING THE MOON

It's a modest jump from the fertility of the fields to the fertility of women, and so women and their capacity for procreation have been associated with the moon by many peoples. By lasting approximately a month, the menstrual cycle also seduced people into believing there is a connection between women and the moon. The Mataco Indians of northern Argentina actually blame the moon for bringing menstruation and the female fertility cycle into the world.

A long time ago, as the Mataco tell it, when the moon was young, good-looking, and single, he was known as a good hunter. His eating habits, however, were idiosyncratic. After catching a tapir or other game, he would eat continuously and grow larger and larger until he was absolutely full. Then he starved himself, and for fourteen days he grew thinner and thinner, until there was only the slimmest line of light left of him. His flirtation with anorexia finished, he would chase down another tapir and start his food binge all over again. Despite his feast-or-famine life-style, the moon was very popular with the girls.

Finally the moon selected one of them to be his wife, but she lived for only five days after the wedding. The same thing happened with every subsequent wife, and eventually people began to suspect something odd was taking place. Their interrogations of the moon provided no clues, however. All the moon could remember was bedding down with his betrothed, sleeping well, and waking to find her dead.

At last, one girl's family took a cautious approach. After she and the moon were married, they insisted she stay awake all night and keep an eye on her husband. During the day she was not to fall asleep at home but instead return to her family's house and sleep there. At first, she gave the moon a good report card. He had a good disposition and was kind to her. When he finally decided to consummate the marriage, however, his new wife realized her moon man was much too well endowed. She saw that he would split her in two if she allowed him the intimacy he desired, and she understood why his previous wives had died. For her own safety she left him, and word got out about the liability of the moon's love.

From then on, the moon was alone, but years later, he visited his grandchild. (The Mataco don't explain where she came from.) She was fifteen years old and lived in another village. He persuaded her to return home with him, and one night on that journey, while she was sleeping, he moved in on her. Penetrating only partway, the moon

Cruz-Moquen, a Christianized prehistoric megalithic tomb in Brittany, is now associated with lunar folklore that links the full moon with female fertility. (Photograph: Robin Rector Krupp)

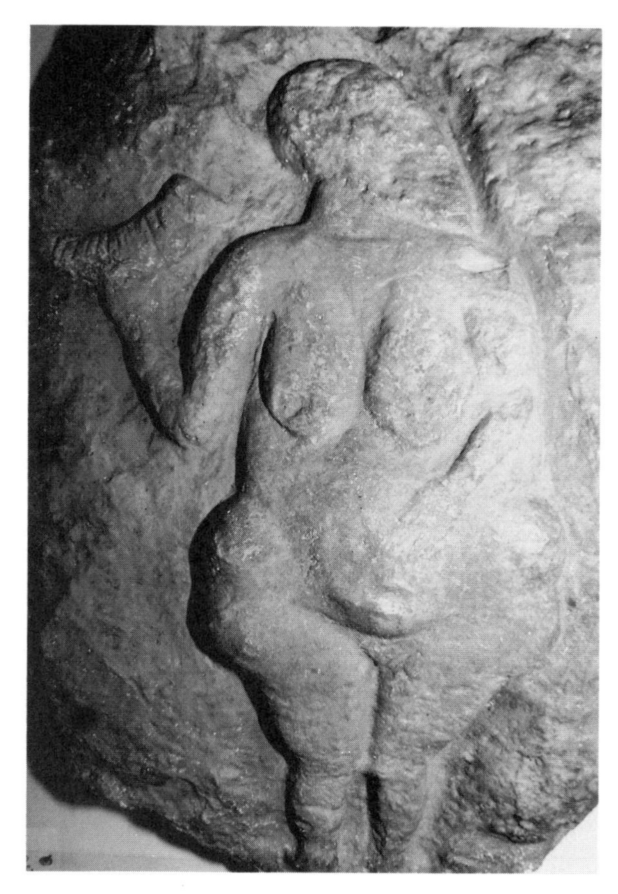

We don't know if this full-figured woman carved perhaps 18,000 to 25,000 years ago during the upper paleolithic period of the ice age hunters and great cave paintings of Europe has anything to do with the moon, but she might. She is known as the Venus of Laussel. Because there was no writing in the old stone age, she carries no inscription to convey what the ice age artist had in mind. She probably represents the goddess of the world's seasonal rebirth. The crescent-shaped bison horn may stand for the moon, whose changes over the course of the year usher in the seasons in sequence. Thirteen marks on the horn may symbolize the nearly 13 cycles the moon experiences in a year or the number of days of visible waxing moon. (Replica exhibited in Musée National de Préhistoire, Les Eyzies-de-Tayac)

woke his granddaughter. She was torn and bleeding, but alive. By forcing the flow of her blood, the moon brought menstruation into the world. Menstruation reminds the Mataco the moon has a way with women.

Peasant women in Carnac Ville, the town next door to the famous prehistoric rows of standing stones in Brittany, used to hitch their fertility to the moon through Cruz-Moquen, a prehistoric monument right in the town itself. Cruz-Moquen is a dolmen, a stone burial chamber that once was covered with an earth mound. It differs from most other dolmens, however, for somewhere along the way Cruz-Moquen was Christianized; it now carries a large stone cross. We have no idea what significance this monument held for the people who built it thousands of years ago, but it didn't involve Christ and may have had nothing to do with the moon. The Breton peasants, however, had their own ideas. Ladies desiring to conceive would go to Cruz-Moquen at the full moon and lift their skirts in front of it.

SCORING IMMORTALITY

Lunar renewal means immortality for moon deities, and the Chinese moon goddess Heng O provides an example of the connection. Heng O's husband was Shen I, the celebrated archer who removed the nine false suns from the sky with his medalist markmanship. After marrying Heng O, Shen I obeyed the orders of his emperor and investigated an unusual path of light that led to the palace of the Queen Mother of the West. She ruled the western paradise. There, in the Kunlun Mountains of Chinese Turkestan, she hosted the Immortals. In Han dynasty times (206 B.C.–A.D. 220), she was believed to dispense immortality, and by building a palace for her, Shen I earned a share of that eternal life. He returned home with the precious pill of immortality the Queen Mother of the West had given him and hid it in the rafters while he began his twelve-month preparation of exercise and diet for everlasting life.

One day, when Shen I was out settling accounts with a notorious criminal, Heng O noticed a beam of white light dropping out of the ceiling. The room smelled delicious. Heng O was intrigued and examined the rafters from a ladder. There she discovered the pill, a luminescent bonbon guaranteed to melt her soul. She consumed it and began to float toward the ceiling, apparently a side effect of immortality. Just then Shen I came home. Untroubled at first by the obvious signs of his wife's weight loss, he appreciated the full gravity of the situation when he could not find the pill. When he asked her about it, she realized she was in trouble and made for the window. Outside, the sky was clear, and the full moon, she judged, would make a handy refuge. As she headed for it, Shen I began his pursuit. He was down before he started, however, blown over by a blast of wind.

Heng O continued on her course to the moon. She arrived to find it a safe, but cold, harbor. Glassy, luminous, and huge, the sphere had nothing growing on it but a huge cassia tree, sometimes called Chinese cinnamon. Not subject to seasonal change, this evergreen is an appropriate emblem of immortality.

As soon as Heng O touched down on the moon, the lady lunar lander coughed up the pill's capsule coating. That husk turned right into a white rabbit, and he is the rabbit that people say can still be seen in the moon. He is also the moon's first pharmacist and continues to grind the elixir of immortality under that shadowless cinnamon evergreen. The bitter taste left in Heng O's mouth prompted her to sip lunar dew and swallow some moon cinnamon. Satisfied with her new surroundings, she took up residence in the silver orb.

Meanwhile, Heng O's estranged spouse was swept by a hurricane to the mountaintop home of the King Father of the East. He was the husband of the Queen Mother of the West and was responsible for compiling the social register of the Immortals. He advised Shen I to be tolerant of his wife and said Shen I would be rewarded with immortality anyway for his acts of heroic service.

Shen I was given a mooncake, flavored with sarsaparilla as the story goes, to protect him from the fierce heat of the sun. That's where he would supervise the cosmic forces assigned to him. The lunar talisman on the mooncake would allow him to visit his wife on the moon. She, however, was not at liberty to travel.

Shen I ate the cake and was about to pack up for the sun when the God of the Immortals reminded him that he was not yet informed about the regulations that govern the sun's arrivals and departures. A bird with golden plumage would therefore accompany him and signal through its songs the proper times for sunrise and sunset. This bird perches in the Fu-sang tree that grows where the sun rises. All ten suns used to take their places in it as ornaments before Shen I shot nine of them out of heaven. Originally a cock, the sun's golden warbler eventually evolved into a three-footed crow, and representations of the sun often show this black bird lodged within the red disk.

After setting up house on the sun, the celestial archer rode a sunbeam to the moon to pass a moment with his long-lost wife. He found her—cold, lonely, and homeless. Although she was apprehensive about seeing him, he reassured her that between fellow Immortals bygones should be bygones. He offered to build her a palace and soon finished the construction of her Palace of Boundless Cold, where he continues to visit her every month at full moon. These circumstances explain why the moon shines as it does: Relying on the sun for its periods of light and dark, the moon glows during marital visits.

This synodic expression of fidelity demonstrates considerable character, for Heng O had, with her first taste of moon dew, turned into a toad. The

Accompanied by a toad, the moon rabbit grinds the elixir of immortality beneath a cassia tree on this lunar disk painted on the hemispherical ceiling of the outer chamber in the Tang dynasty tomb of Prince Zhang Huai (A.D. 654–684), about fifty-five miles northwest of Xi an in north central China.

It is fairly easy to see the rabbit in the moon formed by the dark maria, *or "seas," on the moon's face. His body is on the left side of the disk, and his neck curves over the top to allow his ears to hang down on the right. The toad is more elusive. Although we have no ethnographic data to confirm the toad's image, it is possible to see a toad in the bright highland features that protrude into the* maria *from the southeast limb of the disk. (Drawing: Joseph Bieniasz, Griffith Observatory)*

Chinese aver she is most beautiful on the fifteenth day of each month when her husband comes to call, but even a beautiful toad is still a toad. It is a symbol of water, which the moon is said to influence, and also a symbol of immortality. Toads are supposed to live long. According to the Chinese, both the toad goddess of the moon and the white rabbit can be seen on the face of the moon.

Other people too have seen this moon hare, from the Saxons in northern Europe to the Indians in southeast Asia, the Maya and the Aztec in Mexico, and the Native Americans in the American Southwest. The dark *maria,* solid lunar plains once thought to be seas, define the rabbit's silhouette. He is nestled down on the left side of the moon, as we usually see it. It's a little awkward for him because that side should be down for this reclining hare. His head curves around the top, with his two ears flopping over to the right.

The rabbit is easy enough, shaped by the moon's dark fields, but what about the toad? It's there, but it's a figure in negative space. The shape of a toad shows up as part of the moon's bright zones. You have to look for the toad in the light-colored territory between the rabbit's head and tail. By imagining that as itself a form, and not as the background behind the rabbit, and by seeing the dark rabbit as the background for the toad, you can pick out the toad's head and body emerging from the southeast quarter of the moon. She's standing on her own crescent of bright lunar highlands and looks like some of the representations the Chinese have painted of her on banners and in tombs.

China's lady in the moon is still celebrated every year during the Mid-Autumn Moon Festival, an important traditional holiday. Family gatherings and the consumption of mooncakes are important components of the occasion, and from China to San Francisco, the moon goddess still appears on boxes of mooncakes, floating in heaven with her white rabbit.

The moon's toad and rabbit are included in a detail on a Han dynasty (202 B.C.–A.D. 220) funerary banner recovered from a noblewoman's tomb at Mawangdui, near Changsha, in central China. The toad is standing on the crescent moon and resembles the toad extracted from the bright features in the southwest quarter of the moon. The rabbit is visible just above the toad, but here it doesn't really resemble the rabbit most people can imagine on the moon. (Replica temporarily displayed at the Museum of Chinese Historical Relics, Hong Kong)

When Heng O swallowed the pill of immortality that her husband, Shen I, had hidden, she floated out of the window and up to the moon. Here we see her on the lid of a box of mooncakes, a popular treat during the Mid-Autumn Moon Festival. The rabbit she holds is the rabbit many people say they can see on the face of the moon. (E. C. Krupp; private collection)

MOON MAIDS AND RAINWATER

The moon is so clearly connected with fertility, procreation, and women, it should not surprise us to find that some peoples made a goddess out of the moon. To the Greeks, the moon was female. They called her Selene, a name rooted in the word *selas,* which means "light." She was a young, lovely girl, but notorious for her liaisons.

Although Selene personified the moon, the forever young and virgin huntress and mistress of the wild animals, Artemis (called Diana by the Romans) had clear connections with the moon too. The Romans called the moon itself Luna, and her name persists in the familiar lexicon of moon madness as *lunacy* and *lunatic.*

The Maya represented the moon with a U-shaped symbol, which stood perhaps for the lunar crescent, and in northern Yucatán they seem to have assigned some of the moon's character to Ix Chel, or "Lady Rainbow," a mother goddess who was the wife of the sky god and supreme lord of creation, Itzamná. Healing, fertilization, and childbirth were Ix Chel's special concerns, and in the Postclassic period, her shrine on the island of Cozumel attracted many pilgrims. A picture of her in the Dresden Codex, one of the four screenfold books of Maya hieroglyphic writing known to have survived the Conquest, shows her pouring water out of an upturned base. This portrait suggests she also had something to do with water, rain, and flood.

The moon was believed by many peoples to have some influence on the rain and on the waters of the earth. Perhaps because of its obvious influence on the tides and its association with fertility and vegetal growth, the moon also came to be associated with rain. In detail, the appearance of the moon can change with atmospheric conditions and the transparency of the air. Ancient weather watchers would have noticed this and underscored a connection between the moon and the rain.

Many of the stories about the women said to reside in the moon include details that acknowledge the moon's affinity with water. The Maori tell a story about a woman named Rona the Tide Controller. She was the daughter of Tangaroa, one of the gods who controlled the sea in old New Zealand. She departed one moonlit night with a calabash bucket to carry stream water back home to her three children. On the return trip, when the moon slipped behind a cloud, the path went dark, and she stubbed her toe and kicked her foot against an offending root. Angered for a moment, she unthinkingly vented her spleen on the moon. "You cooked-head moon, where are you anyway?!"

Among the Maori, Rona's intemperate remark constituted a serious curse. The moon heard her. Insulted and angered, it reached down and grabbed her. She seized the branches of a nearby *ngaio* tree. It was ripped out of the ground, and a moment later Rona was in the moon with her water bucket and her tree.

The Haida, the Kwakiutl, and the Tlingit—coastal tribes of British Columbia and Alaska—all believe they see a person in the moon who carries a bucket and a bush. Sometimes they say it is a woman, and when she upsets her bucket, it rains.

The Yakuts, Turkic-speaking folk who live in the basin of the Middle Lena River in northeast Siberia, and the Mongolic-speaking Buryats of southern Siberia also blame the rain on a girl in the moon. The Yakuts say she went out for water one frosty night. Swooping down on her, the moon swept her away. She can be seen there now with a yoke on her shoulders, a bucket suspended from each side. In addition to the girl and her buckets, the Buryats see a willow tree. When the moon descended upon the girl, she, like Rona, clutched the tree, but she wound up in the moon, buckets, willow, and all.

That moon bucket doesn't hold water only in the Siberian tale. It also reaches right into our nursery rhymes. Jack and Jill and their pail of water is really a story about the moon. Originally, it was an ancient Scandinavian tale about two children, a boy named Hjúki and a girl named Bil. According to Snorri Sturluson, who tells us in the *Prose Edda* just about everything we know about them, they were kidnapped by Mane the moon. Returning from the well called Byrgir and carrying a pole named Simul and a bucket known as Soeg, the two kids were conscripted for reasons unknown. There's not much else to go on except the piece of Swedish folklore that identifies the marks on the moon as a boy and girl who carry a bucket of water on a pole between them. S. Baring-Gould, a pioneering British folklorist in the nineteenth century, pointed out the relationship between the names Hjúki and Bil and Jack and Jill. He also said the name Hjúki was related to the verb *jacca*, which means "to pile together, to assemble, to increase." Bil comes from *bila*, "to break up, to dissolve."

Everybody knows how the story goes.

Jack and Jill went up the hill
To fetch a pail of water;
Jack fell down and broke his crown,
And Jill came tumbling after.

But what does it mean? Their familiar climb and catastrophic descent, Baring-Gould argued, were the moon growing to full and slimming down again to the dark new moon. The second verse is also a true lunar tune.

Then up Jack got and home did trot
As fast as he could caper;
And went to bed to mend his head
With vinegar and brown paper.

The moon doesn't bandage itself in vinegar and brown paper, but after its tumble down the hill and consequent disintegration, it's on the mend again, month after month.

Even when the moon is male, it may be said to govern the rain. In ancient India the moon was sometimes called Chandra, or "Luminous," and he scheduled rituals, provided a haven for migrating souls, ruled over the growth of plants, controlled the tides, and influenced and stored the rain. The moon's connection with water extends to a folktale from India, in which the moon is said to be a crystal ball. It contains silver water, and where fish and turtles throw their shadows on the bowl, we see the dark fields upon the milky sphere.

The watery reputation of the moon is well documented worldwide. Ever a ruler of the rain, it served even William Shakespeare with its fluid character. In *A Midsummer Night's Dream,* Titania, the queen of the fairies, explains to Oberon, the fairy king,

Therefore the moon, the governess of floods,
Pale in her anger, washes all the air.

THE MAN IN THE MOON: TALES OF JUSTICE AND SACRIFICE

Shakespeare makes a maiden out of the moon by calling her a "governess" in *A Midsummer Night's Dream,* but elsewhere in the same play he capitalizes on the familiar European tradition of a man in the moon. Shakespeare actually introduces him in "Pyramus and Thisby," the play within the play of *A Midsummer Night's Dream.* The actor playing the moon holds a lantern he identifies as "the horned moon" and says he himself is "the man i' the moon," but Theseus, the Duke of Athens, is an intolerant audience and complains about the error in staging and script. He also generously suggests the problem might be rectified by stuffing the actor into the lantern. Weary of this abuse, the moon impersonator concludes, "All that I have to say is to tell you that the lantern is the moon; I, the man i' the moon; this thornbush, my thornbush; and this dog, my dog." By the time he's finished, he probably figures he's like the King and Queen of the Fairies, "Ill met by moonlight."

Shakespeare invested his play with a little more contemporary European moon lore when he mentioned the man in the moon's thornbush. The way they tell it in Luxembourg, the man was caught stealing turnips out of a neighbor's garden. Suddenly he heard someone yell, "Thief, thief, turnip thief," and felt something pull him up in the air. He grabbed a thornbush to anchor himself to earth, but the force was too strong and uprooted the bush. Now, at full moon, the man and the thornbush are on public display. In many other stories, the man in the moon is seen as a woodcutter carrying a bundle of sticks. He was placed in the moon as a punishment for gathering wood on Sunday.

Those who can see the Man in the Moon often find his familiar face in the moon's bright highlands. His eyes, nose, and mouth are provided by the same dark "seas" of frozen lava that give the moon a rabbit. (Drawing: Neil Passey, Griffith Observatory)

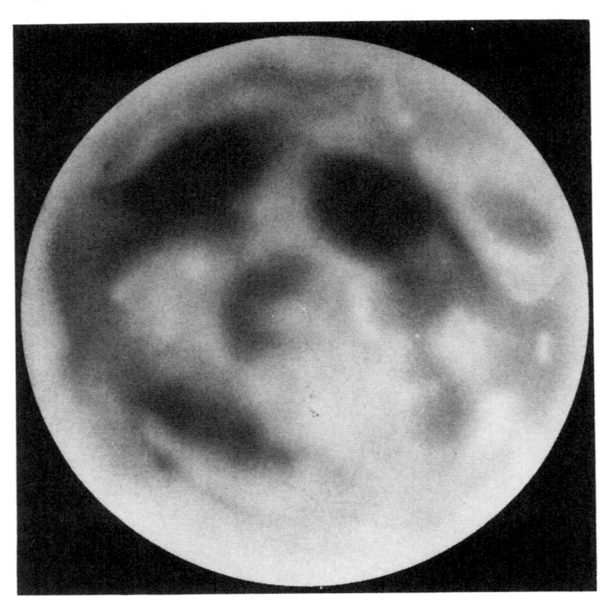

In Poland, they talk about another sinner in the moon—Mr. Twardowski. He was a nobleman who spent all of his money on life's pleasures. Because he had gotten used to having so much money, he had to get rich again, but he didn't want to work for it. Twardowski decided instead to market his soul to the devil, and he asked the Devil what he might get for it. The Devil, of course, laughed and explained that he usually got the souls of noblemen for nothing, but he consented. Twardowski said, "I have been planning to take a trip to Rome for some time now. You must agree not to take my soul until I am in Rome." After they signed the pact, Twardowski said, "Now I have changed my plans. I shall never go to Rome, and I shall live forever." The Devil was not happy, but he kept his bargain until he tired of Twardowski's antics. He managed to trick Twardowski into an inn called Rome. After Twardowski had eaten a splendid dinner, the Devil informed him of his poor choice of restaurants. Twardowski knew he had been tricked, but a gentleman always keeps his word. When the Devil grabbed him, he did not struggle, but as they were flying over the moon on the way to Hell, Twardowski began to feel sorry for his misspent life. He started to sing a hymn his mother had taught him, and suddenly the Devil vanished, which caused Twardowski to fall through space. He was lucky enough to land on the moon, and that is where he may still be seen today, no longer enjoying good times, however, and no longer boasting of his deceit of the Devil.

Chinese folklore also punishes a man—a boy, really—in the moon. He had seen another boy in his neighborhood care for a wounded bird, and when it recovered, the bird gave the boy a magic pumpkin seed; planted, the seed grew into a huge vine with miraculous pumpkins, each filled with gold, silver, and other precious things. The greedy boy deliberately injured a bird, then took care of it; he also received a seed from his bird, who promised him a just reward for his actions. His vine sprouted a pumpkin, too, and the runners reached up to the moon.

Thinking to harvest gold and silver from his pumpkin, the greedy boy opened it. Instead of delivering riches, however, this pumpkin contained an old man who presented the boy with a bill written in red ink. He owed plenty. The old man put him on that pumpkin vine elevator, and the next stop was the moon's floor. The boy was taken to the residence of the immortal moon goddess, the Palace of Boundless Cold, and ordered to chop down the cassia tree that grew there. He could go back home to earth if he succeeded. This tree symbolized immortality, however, and couldn't be destroyed. Every time the boy's ax struck a blow, a white rooster pecked him painfully in the back. By the time he turned around to hit the tree again, the first split in the wood had healed up. So he struck the tree again and felt the rooster's beak again. The cycle never ended. The boy was condemned to chop forever in the courtyard of the white moon's Palace of Boundless Cold.

LUNAR SACRIFICE: ROASTING THE RABBIT AND SPLITTING THE HARE

The lunar themes of sacrifice and immortality may also explain why so many people were able to find a rabbit on the face of the moon. The rabbit, or hare, is itself a symbol of sacrificial death. John Layard's pioneering study, *The Lady of the Hare,* a formidable collection of moon hare lore, clarifies that aspect of the hare's meaning in folklore. Layard related the hare's traditional role as a willing sacrifice to what farmers report seeing a hare do in a burning field. Instead of running, the hare lingers in his hideout in the scrub until the flames come too close. Then it is too late, and the hare is doomed.

Another bond between the hare and the moon can be culled from the pages of *The Old Farmer's Almanac:* Gestation in hares is a thirty-day affair. Hares and women are slaves to the moon. These ideas reemerge in European and North American Easter customs. Easter, of course, is a holiday of rebirth. It celebrates the resurrection of Christ, who was sacrificed on behalf of people and their souls. Its date is regulated by the full moon, and it falls in spring, when the world revives from winter's death and when, through buds and leaves, the earth's vegetation is renewed. We say the Easter Bunny visits early on Easter morning. He may leave presents and sweets, but his true calling card is the Easter egg. Decorated specially for this occasion, eggs at anytime are also symbols of rebirth, and it's a bunny that brings them at Easter. Easter eggs are just the moon rabbit's elixir of immortality in a different costume.

The Aztec tale of the rabbit in the moon is also a story of willing sacrifice on behalf of the world's renewal. The gods had gathered at Teotihuacán (the "Place Where the Gods Were Created") to create the new sun; four previous suns and cosmic ages had each ended in catastrophic destruction. This would be the last. At midnight, in utter darkness, the gods met at the city's sacred fire to determine which of them would sacrifice himself in the hearth of the gods and become the Fifth Sun. At last, two volunteered. Wealthy and fit Taccizté-catl, the arrogant "Conch Shell Lord," and the impoverished, deformed, and diseased Nanahuatzin both agreed to die in an immolation designed to raise the new sun. But when the Conch Shell Lord approached the brink, the flames flared high and scared him back from the edge. Despite shouts of encouragement from his divine colleagues, he could not throw himself into the inferno.

Humble Nanahuatzin stepped up to the pyre. Without hesitation or fanfare he leaped into the blaze. Ignited in the purifying flames, his body sizzled, crackled, and burned. Either inspired or shamed by Nanahuatzin's example, Tecciztécatl followed. The other gods continued to wait in the darkness for signs of the dawn and Nanahuatzin's transformation into the Fifth Sun. At last the sky turned red at the horizon. Every direction was lit, and it was unclear where the sun would rise.

Arguing among themselves, the gods of the creation agreed to place themselves in groups at each cardinal direction to keep a vigil for the new sun. When finally the sun broke into the sky, he appeared brilliant and red in the east, the direction of creation and new life, and swayed from side to side. The Conch Shell Lord, reincarnated as the moon, followed closely on the sun's heels, and it was just as unbearably bright. Neither orb moved properly, however. Both remained still, and in order to get them to follow their ordained paths, the rest of the gods had to submit to sacrifice.

Before they died, however, one of them did something about the moon; he grabbed a rabbit and hurled it into the moon's pristine face. That darkened the moon considerably. The rabbit can still be seen today, a emblem of the Conch Shell Lord's sacrifice on behalf of the cosmos and the badge of his rebirth.

The moon rabbit's connection with sacrificial death is even clearer in a folk tale from India. Three friends were out for a stroll—the ape, the fox, and the hare. It was a fine, sunny day, and they continued walking until they encountered the most destitute and dirty panhandler they had ever seen. He explained he hadn't eaten for several days and was about to starve to death. He wondered if they could help him. The hare explained they didn't have a scrap of food or spare change between them, but the fellow continued to beg. So the hare said they would have to scare something up and advised him to get some rest in the shade while they went hunting for a snack. It took all day, but finally, near sunset, the trio returned. In the ape's long arms was a bunch of ripe mangos, which the beggar accepted gratefully. Then the fox gave him a nest packed with bird's eggs, and again the beggar was thankful. The hare, however, had had no luck and returned empty-handed. He apologized for his failure and then lit a fire, promising the beggar he would provide something for supper. When the flames were good and hot, the hare jumped into them. But the surprised beggar got up and knocked down the flames. In the middle of the ashes stood the hare, unharmed. The ape and the fox were pleasantly surprised to find their friend still alive, and all three were really surprised to see the transformation of the beggar. He appeared in his true form—Indra, the god of the storm and the procreative power of the sky. Indra then honored the hare's willing sacrifice by placing him in the moon, where all can see him and forever remember his gesture.

THE WHITE GODDESS: MAKING PROMISES AND DRESSED TO KILL

Birth, death, sacrifice, rebirth, immortality, women, fertility, water, the growth of vegetation, and fate—all of these themes meet in the moon because the moon, more than anything else, symbolizes the principle of transformation and cyclic change that seems to propel nature and guide its course.

We say the moon dies, but it's not the moon that dies. We die. We tell stories about it because we die, and we describe the moon in terms that are familiar to us, using it to symbolize what matters most to us.

The moon's movement and phases provoke stories that attempt to deal with bottom-line mystery in our existence. Our moon myths are metaphors composed on behalf of the soul. Beguiled by the moon's poetry in motion, the English poet and novelist Robert Graves tried to demonstrate that the myths of the Mediterranean area and northern Europe really represent an ancient metaphorical and symbolic language dedicated primarily to the Moon Goddess and the religious ceremonies performed in her honor. He presented his defense of this notion in *The White Goddess* and made a trinity out of her. As New Moon, the first crescent, she was the white goddess of birth and growth. Dedicated to love and battle, the full moon was the red goddess. And the black goddess of death and divination was the dark Old Moon. According to Graves, the story of the Moon Goddess is a thirteen-act year-long conflict for her favors between the God of the Waxing Year and the God of the Waning Year. Each act is a lunation, and the entire drama embraces the birth, life, death, and rebirth of the God of the Waxing Year and the forces of life.

Most scholars of myth are skeptical about the details in *The White Goddess*, but Graves did understand that the ultimate purpose of sacred myth was facilitating a revelation of reality.

6

Around the Year

JUST as there is a kind of death in the capitulation of the sun each night and in the monthly absence of moonlight from the sky, there is a seasonal death of the earth each year. In temperate latitudes, it occurs in winter. The trees are skeletons of their former selves. The harvest is long past, produce is gone from the garden, and there is nothing to be gathered in the wild. Many animals hibernate in sleep that mimics death. Even the sun is weak. During the day, it almost seems to hug the horizon, and when it is up, it's not up for long. The days are cold and short, and the nights are long and frigid. Life is put on notice by winter. The old, the young, the weak, and the infirm are all at risk of dying.

Then winter relinquishes the year to spring, and in spring the world reawakens. Ice melts, water flows, and thunder is heard again with the return of the rain. Hibernating creatures emerge from their burrows, as if resurrected; birds nest, and flowers bloom. Seeds are planted once the ground thaws enough to cut the first furrow, and through the summer crops grow and mature. Game is plentiful then. The weather can be hot, but if there is no drought, the summer lets nature run loose. Trees wear regal robes of leaves. Melons balloon on the vine, and farmers wage war against the caterpillars that proliferate and prey upon their crops. Expansive in its own abundance, the world in summer lives life to the hilt.

When autumn arrives, however, we see the first signs of the decay that drags us into winter's death. Green sneaks away from the leaves, which burn with the red, yellow, and gold for fall. Then the ice returns. Frozen and ornamented with icicles, the countryside is confined in a white corset of snow, as winter puts on the squeeze and snuffs life out of the landscape once more.

Seasons mean cyclical change. That point is emphasized by the brand name and seasonal imagery on this fruit crate label. The cycle of the seasons provided a pattern of fundamental, ordered change on which prehistoric and ancient peoples modeled some of their stories about the sky.

REASONS FOR THE SEASONS

We wouldn't see seasonal change at all were the earth not tilted in its orbit around the sun. Whatever processes participated in the formation of the sun, and its entourage of planets, also tipped the axis of the turning earth 23½ degrees away from an upright spin. On one side of its orbit, the earth bows its northern hemisphere toward he sun. At that time, the sunrise and sunset occur toward the north. Light falls more directly on the northern hemisphere, and summer there is the result. On the southern half of the earth, however, the rays are less direct and less effective at warming the surface. For people in the southern hemisphere, it is winter.

Seasons in both hemispheres continue to change as the earth continues its travel in its orbit around the sun. Even though the earth moved to the opposite side of its annual circuit in half a year, its axis continues to point in the same direction. The

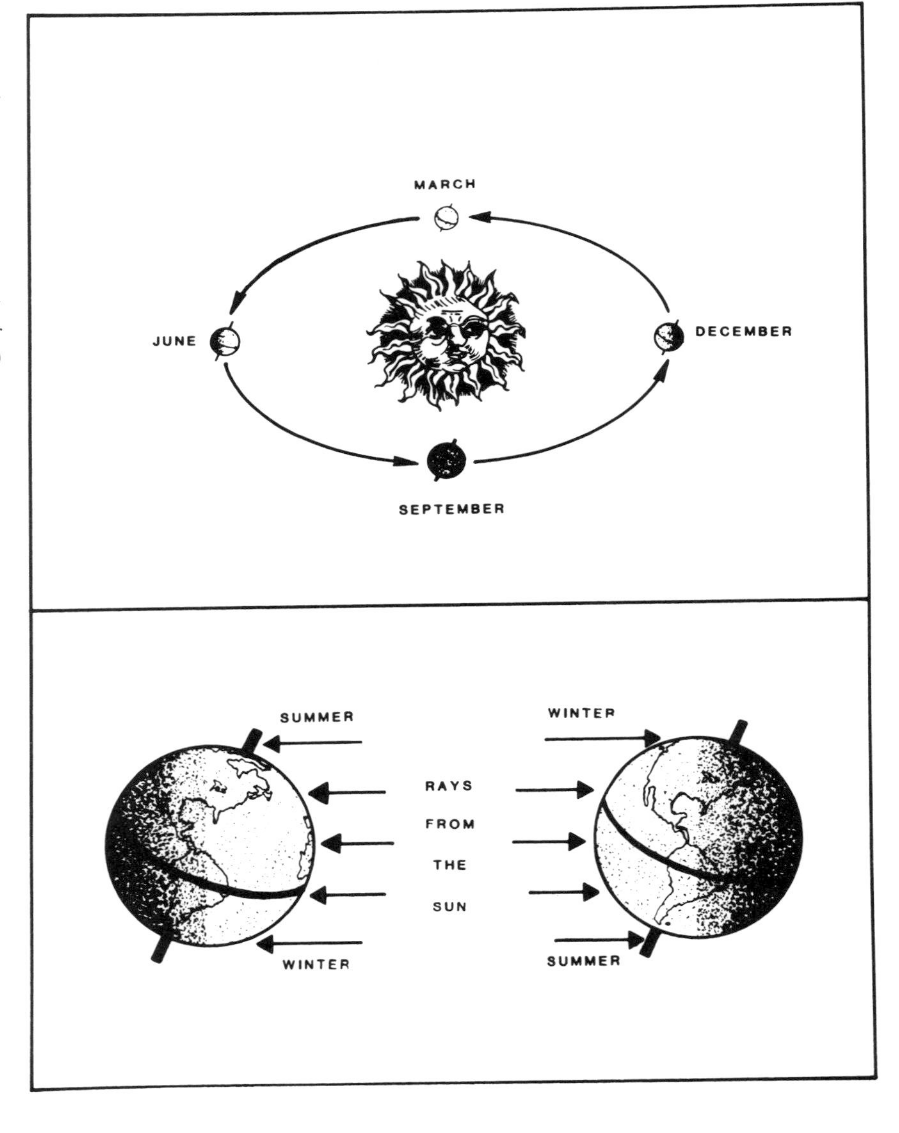

Seasons occur because the spinning earth is tilted in its yearly orbit around the sun. The northern hemisphere is more efficiently heated when it is struck more directly by rays from the sun. At the same time, the southern hemisphere receives less direct light, is less efficiently heated, and endures winter. Six months later, when the sun is in the southern half of the sky, the situation is reversed. (Drawing: Robin Rector Krupp)

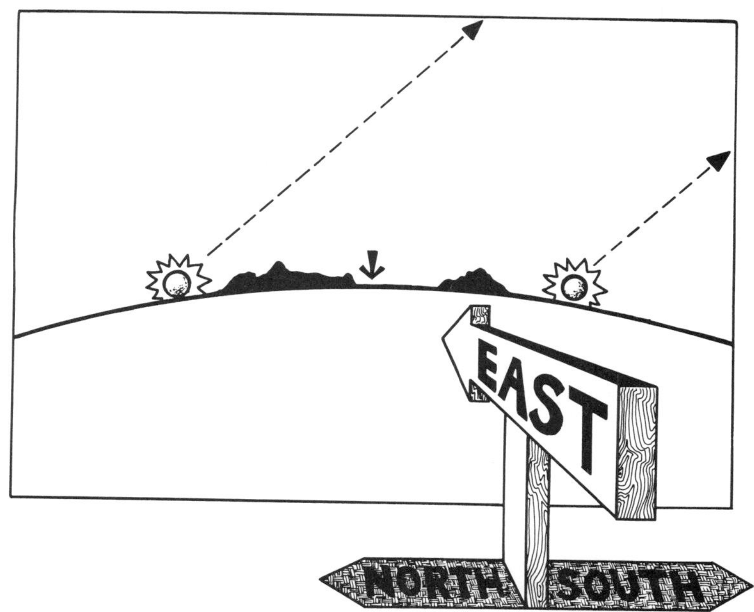

During the year, the rising sun appears in different places along the eastern horizon. The northernmost sunrise (image of sun on the left) occurs on what is called the summer solstice, in June. The winter solstice sunrise (image of sun on the right) is the southernmost limit of the rising sun. The sun reaches this point in December. Throughout the year, the sunrise is seen only between these two points. It moves from one extreme to the other and back again to complete the seasonal cycle. The black arrow indicates due east, the place where the sun rises on the vernal and autumnal equinoxes, in March and September. (Drawing: Robin Rector Krupp)

tilt doesn't change, but when the earth gets to the other side of its orbit, its north pole points away from the sun instead of toward it. Now the southern hemisphere lifts its latitudes more toward the light, and the northern half of the earth has to catch canted sunbeams. Winter now rules the north, but for the folks in the southern hemisphere, the sun is in its "summer place." The seasonal pattern in the southern hemisphere duplicates what goes on in the north, but the timing is half a year out of phase.

The sharp contrast of seasons, from cold to warm and back to cold again, is typical of temperate climates. In the tropics, the seasons oscillate instead between wet and dry. The tropics endure their own brand of seasonal death, however, and it comes with the rain, when the game is gone and there is little else to be foraged. The rain also brings disease. But eventually both disperse with the return of fair weather.

TRAILING THE YEAR WITH THE SUN

Because the earth's axis of daily rotation is tilted with respect to its yearly orbit around the sun, the sun's daily path through the sky seems to shift with the passing of the year. That shift brings us the high-flying sun of summer with its long days and the low-riding sun of winter and its protracted nights. For those of us who live north of the equator, the lowest sun of all is seen at the winter solstice. This is also when the sun appears to rise as late and as far south as we ever see it on the eastern horizon.

Solstice means "sun stand still," but the sun isn't really still at the winter solstice. It rises and sets the way it does every day. But for a few days at the time of the solstice, the sun's walk-on is a repeat performance. If you watch the sunrise for several days in a row, both before and after the winter solstice, you notice that the rising point scarcely changes from day to day. This repetitive rising is what inspired the idea of the solstice. Because the sun runs the same race on several successive days in what the ancient Germanic peoples called the "wet," or winter, season, the event is called the winter solstice; in our present calendar, it takes place on or within a day of December 21.

A fifteen-hundred-year-old Jewish commentary in "Abodah Zarah," a treatise in the *Babylonian Talmud*, tells us that Adam, the first man, observed the consequences of the winter solstice sun. Repentant after being cast out of the Garden of Eden, Adam stood in the Jordan River for forty days. After a while, he noticed that the days were getting

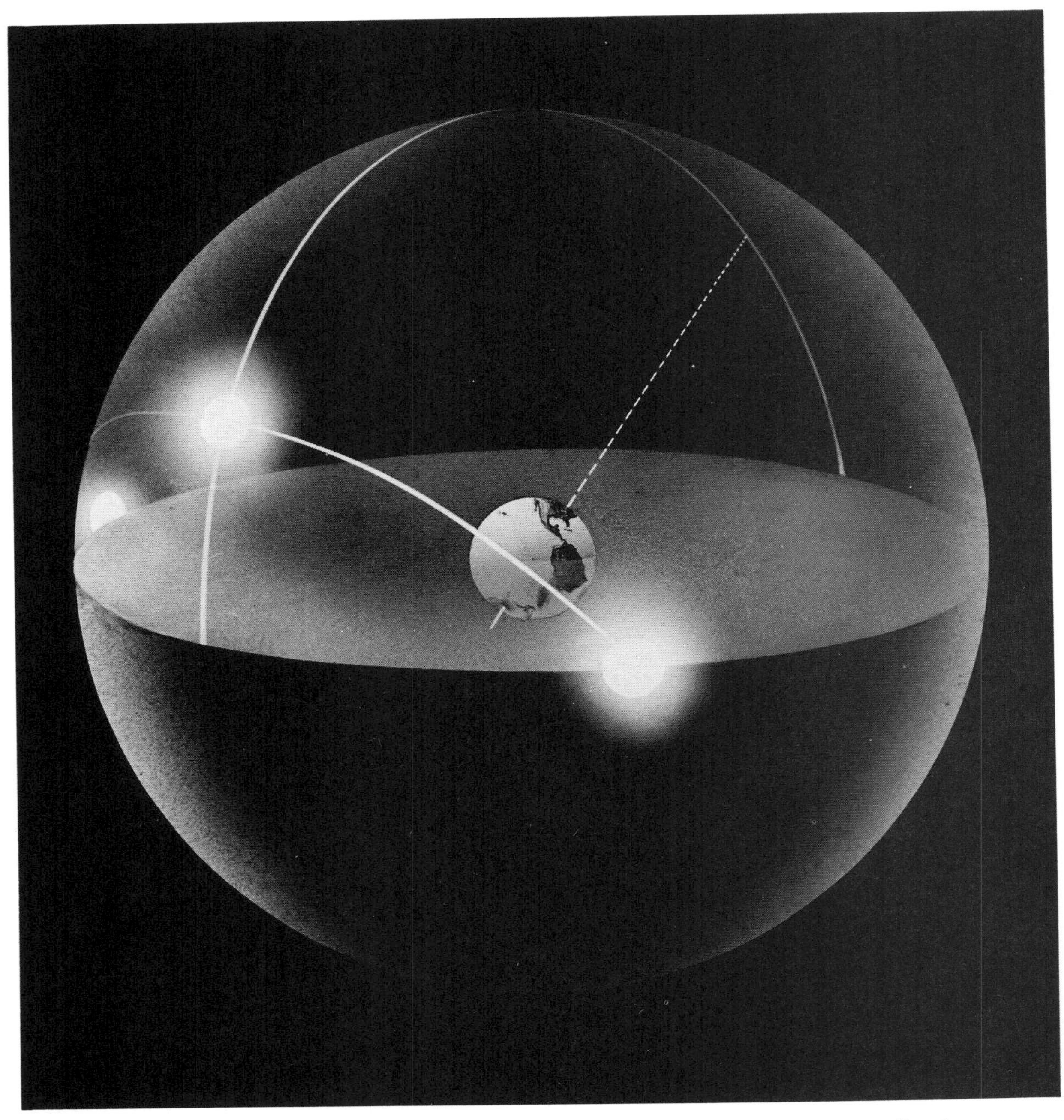

From the ground, the sky looks like a hemispherical vault that domes the earth. The stars, of course, are not all at the same distance from us, but without visual cues of their true distances, they and other celestial objects seem to occupy places on the inside surface of that imaginary celestial dome. For that reason, some of the ancients imagined the sky to be a complete sphere with the earth at its center. Even if the radius of the celestial sphere were the same as the distance to the nearest star, however, the earth in this picture would be too large. Its size is exaggerated here to clarify the relationship between earth and sky. The dotted line is an extension of the earth's axis, which intersects the celestial sphere at its north pole. The glowing disk shows where the sun rises (near side of the celestial sphere), crosses the meridian (highest disk), and sets (far side of the celestial sphere) on the winter solstice. From the latitude represented by the uppermost spot on this view of the earth, this is the sun's shortest and southernmost daily path through the sky. (Drawing: John Lubs, Griffith Observatory)

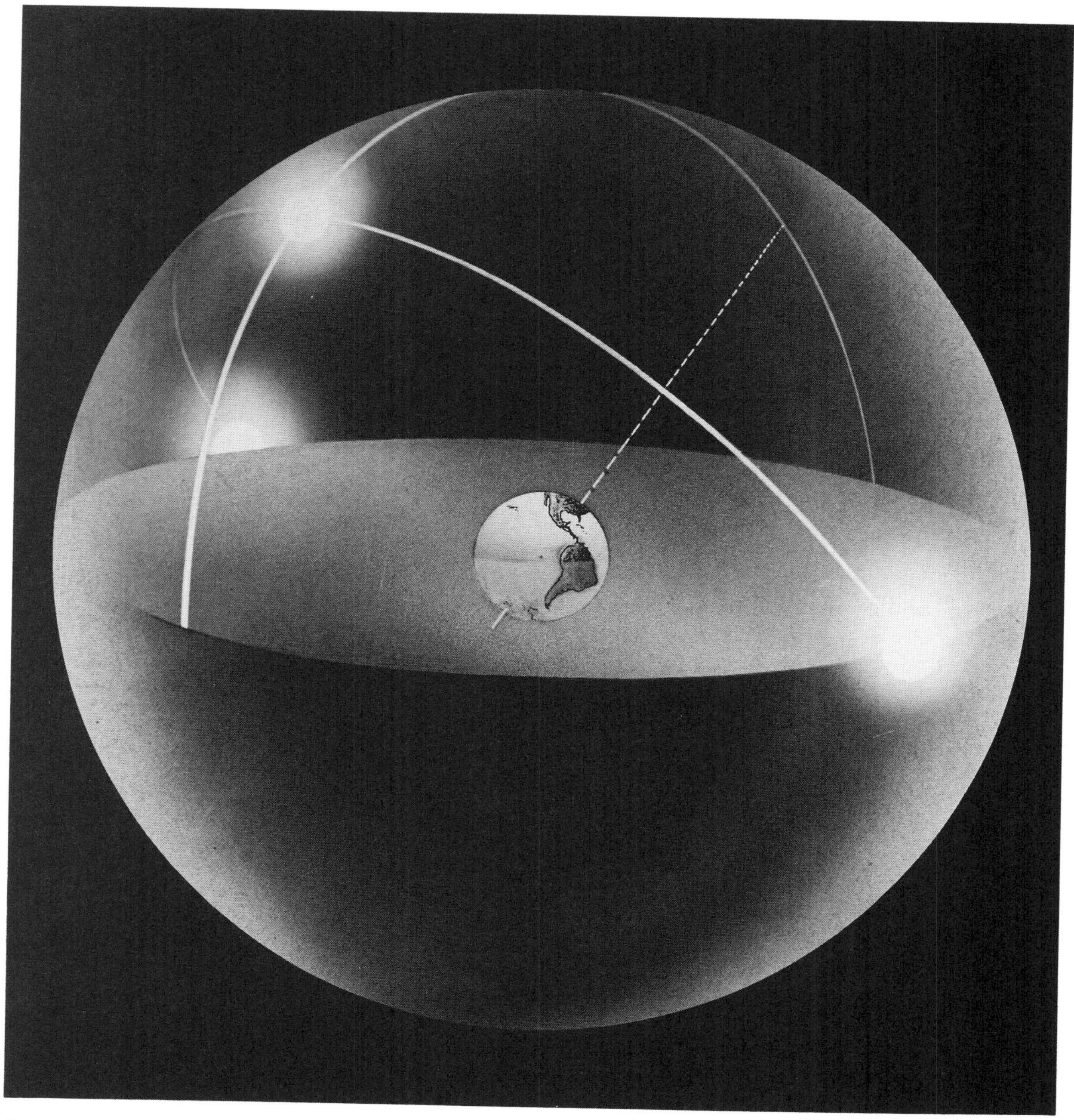

In March, at the vernal equinox, the sun rises due east, crosses higher in the south, sets due west, and follows a path that equalizes the periods of daylight and dark. (Drawing: John Lubs, Griffith Observatory)

shorter. Understandably, this made him uneasy. he was already under a cloud, and shorter days didn't look good. It occurred to him that the world might foreclose entirely, all because of a bit of forbidden fruit, and so he fasted and prayed for eight days. His timing was right, for he prayed right into the winter solstice and then noticed that the days started getting longer again. A year later he remembered the whole business and celebrated the turning of the year just before the solstice and after it too.

After the winter solstice passes, the point of sunrise shifts a little to the north each day, only a small amount at first, but with greater jumps as winter melts into spring. Each day the period of daylight expands, and the nights contract. By March 20 or so, the sun rises due east and sets due west. We've reached the halfway house between the southern and northern limits of the sun's first gleam, and the hours of daylight balance the hours of night. This event and the time when it occurs are called the vernal equinox. *Vernal* means "of spring" in Latin, and our word for this season—*spring*—refers to the emergence of plant shoots as they spring out of the defrosted ground. Translated from Latin, *equinox* means "equal night." At this time of year, the daily migration of sunrise displays its longest leaps. The day-to-day movement is easily observed on the horizon, and the noon sun continues to climb higher. Its rays strike the northern half of the earth more directly. The world warms up to the change.

Continuing its excursion north, in three more months the sunrise pulls into the terminal at the other end of its annual itinerary. Now, usually on June 21, we see the sun come up as far north as it ever shows its first light. Following a steep arc through the sky, it negotiates its highest elevation of the year at noon. Then it continues to a setting point as far to the northwest as it is ever seen on the horizon. The sun now lingers in the north the way it dawdled in the south half a year ago. This is the summer solstice. The sun stands "still" again, and the Sanskrit root of *summer* means "half year." We are halfway home to the new year. Because the daytime path is high and long, the sun is up for more hours than at any other time of the year. Days are long. Nights are brief. Shadows are short. Solar energy nearly hits the northern hemisphere head-on, and the efficient heating cooks the ground and serves a hot season.

After lounging a few days at its northern limit, the sunrise gradually pulls up stakes for the south. In three months, it reaches middle ground again, on September 22 or 23. This time it's the autumnal, or fall, equinox. Night and day share the hours equally again. The sun's path is lower than it was in June. There is a new chill in the air.

People notice the travels of the sun and recognize their link with the seasons. In what is now central California's Sonoma and Mendocino counties and neighboring coastal mountains and inland valleys, the Pomo Indians monitored the solstices, which they called "starting back." An elite specialist of the valley kept track of the sun's progress toward a solstice by watching where it rose over a particular hill on the horizon. When it seemed to rise over the same point four days in a row, the sunwatcher knew the solstice had occurred.

New Zealand's Maori live in the southern hemisphere, and they too are aware of the connection between the solstices and the seasons. During the year, they say, the sun roams from Rangi's head to his toes and back again. Rangi is the sky, and when the sun is near the upper part, or head, of Rangi, it is summer in New Zealand. The Maori also say the sun is then spending time with Hine-raumati, the Summer Maid. He leaves her in December, around the time of their summer solstice, however, and heads toward Rangi's feet, far out to sea, where Hine-takurua, the Winter Maid, resides. The sun enjoys the company of his winter wife until the June solstice, when it's time for him to head back to the land. There, the Summer Maid is cultivating crops and preparing the game of the forest for the summer hunt.

Don C. Talayesva, a Hopi, explains in his autobiography, *Sun Chief*, how important it was for the sunwatcher to keep track of the seasons by observing where the sun appeared and disappeared on the horizon each day. The point of sunrise on the year's shortest day is, he said, "the sun's winter home," and the sun occupies his "summer home" on the longest day of the year. Vital information was provided by the shifting location of the rising and setting sun.

During the summer solstice in June the sun rises as far to the northeast as it is seen throughout the year. It crosses nearly overhead at noon as seen from this latitude and sets far to the northwest; it follows its longest daily path of the year. The duration of daylight is long. Night is short. (Drawing: John Lubs, Griffith Observatory)

Some traditional peoples speak of the solstice horizon points as the sun's summer and winter houses. This petroglyph on a cliff face at San Carlos Mesa in Baja California resembles the type of reed house built by the Tipai Indians of this area. During summer solstice a triangle of midmorning sunlight forms to the left of the house symbol. After eleven minutes, the light glides into the house and shrinks to the triangular spot framed here in the doorway, where it then disappears. (Photograph: Eve Ewing)

> When the sun arose at certain mesa peaks . . . it was time to plant sweet corn, ordinary corn, string beans, melons, squash, lima beans, and other seeds. On a certain date . . . it was too late for any more planting. The old people said there were proper times for planting, harvesting, and hunting, for ceremonies, weddings, and many other activities. In order to know these dates it was necessary to keep close watch on the sun's movements.

Seasonal changes in the amount of food on the table correspond to seasonal changes in the length of the days as far as the Toba-Pilaga Indians are concerned. They live along the middle Pilcomayo River, on the Paraguay-Argentina border, and see the connection between the food supply, the season, and the sun. They say she is a slow-moving fat lady in the warm weather, overloaded with the abundance of summer. The days are long then because she's too fat to run. By winter solstice, however, she's lost all of those excess pounds. Thin and swift, she runs through the sky, and the days are short.

SEASONS IN CONFLICT

The shift in the daily path of the sun shows us a sun on the move, disciplined and determined. Its footprints transform the seasons into destinations, and we tell stories about heroes who walked in the path of the sun on quests that mirrored the order of the year. Glooskap, the hero-trickster of the Micmac, an Algonquian tribe of Nova Scotia and New Brunswick, in eastern Canada, once found himself frozen and exhausted far north in the Land of Ice and accepted the hospitality and warmth of the wigwam of the great giant Winter. Winter welcomed Glooskap with tobacco, stories, and shelter, but his generosity masked his true purpose. Sheltered from the storm and entertained by the giant's tales of the old days, Glooskap was spellbound. As it grew colder and more still, Glooskap gradually slipped into a deep sleep. Trapped by the giant, Glooskap hibernated for six whole months. When, at last, he awoke, he tiptoed out of the wigwam and headed south.

Glooskap's journey to the south brought him back to warmer weather and sunlight. He reached a forest inhabited by tiny people, and their leader was a beautiful, delicate, and diminutive creature named Summer. When Glooskap saw her, he pursued and caught her, securing her with a lasso around her waist. Her people chased after him, but Glooskap carried Summer all the way back to the Land of Ice. Approaching Winter's wigwam, he hid Summer inside his robes and then sat down for another round of the giant's cloying tales. This time, however, Glooskap did not get drowsy; instead, Summer's magic enveloped Winter. He first perspired and then started to melt. His icy wigwam thawed as well. Other signs of life—bird songs and blades of young grass—escaped Winter's grip. The northern Land of Ice was coaxed into summer, and the rest of the little folk caught up with Glooskap, who relinquished Summer to them. She and her people lingered a while in the north, blessing it with warmth, but Glooskap had already set out once more for another journey south.

Glooskap isn't the sun, but he acts like it. He migrates north and south and is the agent of seasonal change. Winter is a giant, and Summer is an elfin lady because summer is brief and winter fierce in the Canadian territory of the Micmac. Despite the brevity of summer, its power is still great enough to disperse winter for at least a few months. By overpowering the Chilly White Giant, Summer is Glooskap's revenge. Through sorcery Winter imprisons Glooskap, but Winter pays for putting him in the cooler. Glooskap will always steal away and spirit summer back north.

ORDERED AROUND BY THE YEAR

The year and its seasons are an essential component of the natural order of things. Our ancestors judged it necessary to act in harmony with the natural order. It was the foundation for all behavior, and they saw the laws, taboos, and rules of conduct of society as extensions of the natural order. For this reason, they sometimes called upon the seasonal component of that natural order to endorse and uphold the social and ethical order of their communities. This process is at work in a familiar Slavic fairy tale, the story of Marouckla, the mistreated stepdaughter of a miserly widow. It was bad enough when Marouckla's own mother had died and her father remarried, but now that Marouckla's father was also gone, she was just excess furniture around her stepmother's house. Far prettier than her halfsister Helen, Marouckla incited only jealousy and hatred at home, and for her good temperament she was assigned all the hardest work. Cleaning the rooms, washing the laundry, cooking the meals, spinning thread, weaving cloth, sewing clothes, carrying the hay, and milking the cow—all kept her busy. Despite her burdens, she grew more attractive each day, while her pouting sister began giving the village's ugliest women a real run for their money. Under those circumstances, Helen's marital prospects seemed grim, and her mother figured something would have to be done to eliminate competition so close to home. So she starved Marouckla, insulted her, and did all she could to make the girl miserable. Despite the accelerating abuse, Marouckla became lovelier and more pleasant.

On the other hand, the more Helen received, the more she demanded, and sometimes her demands were outrageous. In the middle of one year's winter, Helen decided she would like some wild violets and ordered Marouckla to find some in the nearby mountains. This time Marouckla's common sense got the better of her.

She asked, "But, my dear sister, who ever heard of violets blooming in the snow?"

Helen's unambiguous reply left Marouckla with no options. "You wretched creature! Do you dare to disobey me? Not another word; off with you. If you do not bring me some violets from the mountain forest, I will kill you."

Marouckla's stepmother was equally vicious. She and Helen struck the unfortunate girl until they drove her from the house and into the snow, slamming the door behind her.

Hungry, freezing, and wretched, Marouckla wandered into the woods, certain she would die. But far from the cottage and well into the night, she noticed a light on top of a remote mountain. At the top of the peak was a large, inviting fire. Twelve stone blocks arranged in a circle around it provided seats for twelve strange individuals. Marouckla noticed they were not all the same age. The hair on three of them was white with age, while the next three were mature but not as old. Each member of the third trio was young and good-looking, and the last three were the youngest of all.

These men were really the Twelve Months of the Year, and they all stared at the central fire. The oldest of them turned to Marouckla when she asked if she could chase away her chills with a little time by their fire. His name was Setchène (January), and he wanted to know what could have brought her there in such bitter weather. Marouckla explained she was on a quest for violets, and Setchène understandably replied that this was definitely not the season for violets. Marouckla agreed but said that she had been ordered to locate the flowers by her stepmother and stepsister. Setchène then stepped down from his station, which was highest of the blocks, and walked over to Brezène (March). Turning his wand over to Brezène, the old month asked him to move to the high stone. As Brezène shifted chairs, he waved the wand over the fire. It flared up, and the snow started to melt. Buds and young grass sprouted. March had managed an untimely moment of spring and advised Marouckla to gather violets quickly, for winter would have to return in short order. Commandeering a bouquet from the mountainside, Marouckla rushed home and turned it over to her sister. Helen, of course, wanted to know where in the world she got the flowers but expressed no thanks for Marouckla's effort.

Violets in winter made Helen ambitious. On the following day she insisted Marouckla go back out into the snow and scare up some strawberries. If spring violets were a long shot, summer strawberries should be out of the question.

"But who ever heard of snow-ripened strawberries?" Marouckla protested.

Helen's reply was direct: "Hold your tongue, worm; don't answer me. If I don't have my strawberries I will kill you."

Again Marouckla was persuaded to shiver and search on her sister's behalf. Her stepmother pushed her outside and locked the door. With only one choice available, Marouckla headed back up the mountain and explained her situation to white-haired January.

"I'm looking for strawberries," she said.

"Strawberries!" protested Setchène. "Winter is a poor time to shop for strawberries."

Agreeing that the only strawberries likely to turn up in the snow would be frozen strawberries, Marouckla described her stepsister's threat. So Setchène rose from his stony throne and crossed to the other side of the fire. There he gave his wand to Tchervène (June), and asked him to exchange places with him. Tchervène followed Brezène's example, and the wave of his wand stoked the flames. As they climbed into the sky, the icy weather was traded for summer's warmth: Trees attired themselves in full leaf; birds flew and sang up and down the mountain; and strawberries ripened under the luxuriant forest ferns. The hillside flowed red with strawberries, and Marouckla collected an apronful before winter slammed back upon the world. Turning the berries over to her sister, she faced the obvious interrogation.

"Where did you find them?" Helen demanded.

"Just in the mountains. The ones under the beech trees are especially tasty," Marouckla replied.

Well, Helen ate most of the strawberries herself. She certainly didn't give any to Marouckla, and on the next day she had a taste for some apples.

"Apples in winter?" she asked. "The trees are bare!"

"Idle slut," shouted Helen. "Go this minute. If you don't bring back some fresh red apples, we'll kill you."

So Marouckla had to set out in foul and freezing weather again. The snow was deep, her progress slow. By the time she reached the fire and the twelve months on the mountaintop, she was weeping, and her tears turned to icicles. By now Setchène had probably figured the girl required some more unseasonable assistance, but he was still astounded by the object of her trek.

"Look, Marouckla. Winter is in a full blow now. It's a poor time to pine for pippins. Apples are harvested in the fall."

Marouckla spelled out the same old situation. Her life depended on her sister's untimely whims. She wouldn't walk the mountain in winter by choice. If her sister had an eye for apples in a blizzard, it was Marouckla's frostbitten fingers that would have to find them.

So Setchène called over to Zaré (September), one of the older members of the group. Zaré settled into the high throne and waved the wand over the fire. When the flames exploded out of the fireplace, the snow retreated to reveal a mountainside garbed in autumn leaves and the few flowers that embellish the fall. Ferns not yet browned by winter and patches of purple heather paved the way to a single tree reddened not by a change in the leaves but by the ripe apples it offered. A gentle shake prompted the tree to drop one apple, and Zaré let her shake the tree for one more. Marouckla gathered the harvest and headed for home. Her stepmother and stepsister grilled her again about her ability to find fruit and flowers out of season, complained that two apples were not enough, and accused her of collecting more apples and consuming them herself. After more threats, Marouckla ran for refuge in the kitchen. Her stepmother and Helen finally tasted the apples she did bring them and found them tempting enough to turn their attention from Marouckla.

Helen packed off for the mountains herself in pursuit of more fruit. As she reached the mountaintop and the circle of Twelve Months, she stepped right up to the fire without a polite word or permission. When Setchène asked her why she had come, she told him to mind his own business. She then turned away from the fire and headed into the winter-ravaged forest to find Marouckla's apple tree. Behind her, Setchène frowned. Waving his wand over his head, he shut down the fire and summoned heavy clouds, snowfall, and chilling wind to the mountain. Helen cursed her stepsister as she fought through the brutal storm and disappeared in the fog and drifting snow.

When Helen did not return, her mother set out in search for her. The snow continued to drop and erased all signs of life on the mountain. Marouckla's stepmother, lost and frozen like Helen, never returned. Marouckla, however, inherited the house, the field, and the cow. In time she married an honest farmer, and together they lived happily ever after.

In Slavic fairy tales, the year was sometimes personified as it is in the story of Marouckla, with a circle of Twelve Months seated around a fire. It was also said that an old man lived within that eternal fire on top of the mountain. He was bald and had a long white beard. Said to know everything, he favored truth and dispensed justice. He was known as the King of Time. There are already enough clues here to tell us the true identity of the King of Time. He lives in fire and rules time. The twelve months cycle around him. He is, of course, the sun.

The story of Marouckla and the Twelve Months is not, however, about the sun, the calendar, and the seasons. It's a story about the way in which good people should treat each other and about what will happen to the wicked. In it, the calendar and the sun control the seasons, but they do more than that. They sustain virtue. Because Marouckla is innocent, honest, and exploited, the Twelve Months assist her. They also impose a grim verdict on her stepmother and sister, who are greedy and cruel. Their disregard for ethical conduct is underscored by their disregard for the natural order. The story is propelled forward by their refusal to acknowledge the limits of the season.

The Twelve Months are no more than the proper order of the world, but the message is clear. The natural order is aligned with ethical behavior and not with evil. We see the pattern of the year the inescapable limits of our lives. Because the seasons really do require a sensible response from us, the pattern they impose means more than just a change in the weather. It means that some behavior is condoned, that other behavior is proscribed, and that an authority higher than ourselves—here symbolized by the natural order of the year—establishes what is right and what is wrong. Failure to act in congruence with the moral order is like a greedy search for apples in a blizzard. It threatens survival.

PUNCTURING THE YEAR ON HALLOWEEN

In the past, people symbolized the seasonal threat to their survival with myths and rituals that focused the danger to a specific date in the year. The date they chose also usually marked the transition from one season to the next. Those seasonal frontiers were perceived as times when the world, if not always in peril, was at least more exposed to supernatural forces and uncertainty. Even though seasonal change was an inevitable consequence of the natural order, it did carry risks. When the border between seasons was crossed, danger was abroad. One of those hazardous dates is still commemorated in the tradition of Halloween. How could a holiday now celebrated on the last day of October mark the turn of a season? The explanation involves what we mean by the beginning of a season.

Today we usually start the seasons with the solstices and the equinoxes. For example, in North

America we usually call the summer solstice the first day of summer. The ancient Germanic and Celtic peoples of Europe, however, started and ended the seasons on dates that fell about midway between the solstices and the equinoxes. The logic in this custom is easy enough to appreciate; when the sun reaches either of its turning points—either solstice—it is, in a sense, at the apex of a seasonal run, not at the starting block or the finish line. At summer solstice, the sun reaches its greatest height, and as far as the sun is concerned, it's the height of summer even if the seasonal response of the earth lags behind a bit. Winter solstice, by the same logic, catches the sun at the bottom of the year and about to turn again toward the top.

Scholars are not sure whether the ancient Irish considered summer or winter to be the year's first season, but winter makes a fair claim to that distinction. In Celtic tradition, winter began on Samhain, which in our present calendar occurs on the night of October 31 and during the following day, November 1. Samhain is pronounced sah-win, and it means "summer's end."

Halloween, with its customed foragers, tricks and treats, and supernatural trappings, is the harmless descendant of what for the Celts was a genuinely terrifying moment in the annual cycle. As one of the year's two seasonal seams, it was a time when stitches could snap, and through that rip in the protective fabric of ordinary reality, the agents of chaos could emerge. Exploding from the dark otherworld of malicious spirits, the dead and their allies threatened the established natural order. An encounter with the spirits on Samhain night guaranteed traffic with hags, monsters, witches, and fairies. Powerful supernaturals who commanded magical forces, the Celtic fairies could help mortals or lure them to their doom.

THE JUDGMENT OF GROUNDHOGS

Between the winter solstice and vernal equinox the Celts celebrated Imbolg. It survives, like Samhain, as a Cross-quarter Day and is now known as Candlemas. Falling on February 2, it splits winter from spring. Because the old name seems to mean "ewe's milk," the date was probably associated with the lambing season and first signals of spring. In Christian times, the day became associated with Saint Brigit, but in the pagan era, Brigit was an important Celtic goddess. Saint Brigit's time of birth was sunrise, which is congruent with the first spring break of winter's grip.

According to an old riddle, she was born neither inside a house nor outside. That put her on the threshold and is consistent with the idea of seasonal transition and the beginning of spring. Nursed on the milk of a supernatural cow, she matured into a brilliant being who dries her dresses on a clothesline of sunbeams and makes the rooms where she resides glow as if on fire. A perpetual fire dedicated to her at Kildare, in Ireland, was tended by nineteen nuns. In the Christian calendar, Saint Brigit's Day falls just a day before Candlemas, which corresponds to the date when Christ, only forty days old, was first presented in the temple according to custom. Candlemas is also a Quarter Day, one of the days that subdivided the year in Scotland, and Candlemas weather was said to contain clues for the rest of the year.

If Candlemas is fair and clear,
There'll be twa winters in the year.

Even though the Candlemas sun is starting to relieve the world's big chill, too much fine weather too early apparently dooms us to another fierce cold spell before spring breaks up the ice.

A certain amount of prognostication and some fertility associations were also entwined with the theme of spring's advent at Imbolg, and these and the related Candlemas weather watch have survived, in a way, in the American tradition of Groundhog Day on February 2. The hibernating groundhog is obligated to emerge from his burrow in recognition of the year's arrival at the beginning of the slide into spring. If the groundhog sees his shadow, we're in trouble. There must be sunshine, if he has one, and that means we're promised six more weeks of winter. It's the "fair and clear" Candlemas syndrome again. On the other hand, an overcast Groundhog Day means no shadow, declaring winter is finished this time around.

SCREAM OF THE MAY

On May 1, the Celts observed the beginning of the summer half of the year and called the feast they held then Beltine. Calendared about midway between the vernal equinox and the summer solstice as the third Cross-quarter Day, it is now the familiar May Day. For the Celts, Beltine was the second great splice in the year. Huge bonfires were lit, and the name *Beltine* appears to mean "brilliant fire," certainly a reference to the sun.

Beltine's meaning is revealed in the old Welsh story of "Lludd and Llefelys," one of the tales in the *Mabinogion*. According to the story, a great catastrophe occurred every Beltine Eve, when a hideous scream was heard over every hearth in the land. It made the land, water, cattle, and trees all sterile. The stalwart Briton men lost their color and strength on May Eve. Women endured miscarriages; children turned into bubbleheads; May Day was no picnic.

Eventually Lludd, a legendary king of the British Celts, discovered that the terrifying cry was the howl of the Red Dragon of Britain. Locked in battle with a foreign dragon (probably the White Dragon of the intruding Saxons), Britain's talismanic dragon was fighting for the life of its land and was losing. Llefelys, the king of the French Celts and Lludd's brother, advised Lludd to measure the island of Britain with unprecedented accuracy and to excavate a deep pit at the exact center of the land. There he was to deposit a satin-covered vessel of the best mead money could buy. Magically attracted by the mead, the two dragons carried their conflict to the centerpoint. They fought there until too tired to fight anymore and fell into the aperitif and satin. As they consumed the mead, the ferment was stemmed. Lludd rolled the two drunks in the satin sheet and let them sleep it off forever in a secure stone chest at Dinas Emrys, the site of an iron age hill fort in north western Wales. With that modest investment in mead and satin, the threat that accompanied the transition of the seasons was averted, and the fertility of the world was restored in time for summer.

THE FIRST FRUITS AND THE CORN DOLLY

The fourth and final season division in the system of Cross-quarter Days occurs in August, on the first of the month or close to it. It is called Lammas, and it commemorates the beginning of the harvest. Its name is thought to be related to the Anglo-Saxon word *hlafmaesse*, which means "loaf mass" and refers to the first loaf of bread baked from the first harvest.

Lammas continues a calendric tradition the Celts called Lughnasadh, the festival of Lugh, and at this time they held games, celebrated marriages, organized fairs, competed in contests, and prepared feasts. *Lugh* means "the shining one," which suggests a celestial object, and Lugh is sometimes identified with the sun even though explicit evidence for this does not exist. According to one account, he established the festival that bears his name to honor his foster mother, the goddess Tailtiu, and the earthworks at Teltown (Tailtiu), about five miles north west of Navan, Ireland, is the traditional site of her cemetery and the annual Lughnasadh. Perhaps she and Lugh's sisters, Nas and Boi, who were also mourned on Lughnasadh, are alter egos of the Corn Spirit who suffers the first cuts of the harvest at this time. Harvest slays the growing grain, but it is a necessary sacrifice that ensures the continuity of life.

IN SEASON

In *The Golden Bough*, James George Frazer's multivolume monument to magic, nature myth, and the seasonally slain god, Frazer argues that the ritual sacrifice of priests and sacred kings reenacted the annual sacrifice of a fertility god. The god's obligatory death renewed the world and ensured the continuity of life. Because this death—and usually the rituals that celebrated it—were tied to a seasonally significant date in the solar calendar and modeled on the cyclical pattern of vegetation, the story of the god's birth, death, and resurrection parallels the sun's own yearly itinerary. This dying god isn't usually the sun itself, but there is certainly a relationship between what the sun is doing and what is happening to the god.

Sickle in hand, the Sibyl of Cumae holds aloft a sprig of the golden bough. (Engraving: William Turner, British Library, British Museum, London)

The Golden Bough opens with the curious behavior of a Roman priest who tends Diana's woodland shrine near the cratered blue waters of a lake called Nemi. This is where, according to Frazer, we find the mythical golden bough in the title of the book and where we also encounter the ritual sacrifice of an earthly deputy of divine power. The lake is about nineteen miles southeast of Rome, and the priest—the King of the Wood—has to maintain a constant watch for the escaped slaves who come to pick a branch from the bough of the grove's sacred tree. Anyone who collects a branch earns the right to fight the King of the Wood to death. If the intruder manages to kill Diana's priest, he assumes the office himself and holds it until some worthy contender in turn relieves him of his duties.

Continuity of the priesthood of Diana at Nemi mimicked the continuity in the cycle of life, and death was a necessary component of the cycle. Without the death of the old order, there would be no reinvigoration of the world. When the branch was broken from the tree in Diana's grove, the tree's spirit was said to die. When the King of the Wood was killed by the intruder, the consort of Diana—the male to her female—was sacrificed. But the tree was revived, and Diana betrothed once more, when the slayer in the grove himself assumed the mantle of the King of the Wood. Consorts of goddesses and crops must die if there is to be yet another season of life.

Frazer was able to show that the bough in Diana's wood at Nemi was considered by the ancients to be the Golden Bough that Aeneas, the future father of Rome, had to carry to enter the underworld realm of the dead. From the story of Aeneas, we know that the Golden Bough was a talisman of life. When Aeneas entered the cavernous mouth of the underworld on his descent to Hades, he carried a sprig of the Golden Bough to ensure his safe passage. The Golden Bough was said to house the spirit of life of the tree from which it was plucked, and that spirit is what gave it its power. Aeneas needed a passport to travel safely in the realm of the dead, and his branch of the Golden Bough, carried as a gift to Proserpina, the goddess of the underworld, satisfied all requirements.

By the end of *The Golden Bough*, Frazer identifies the magical golden shrubbery in the sacred grove of Diana as mistletoe. Familiar to us as a license to kiss at Yuletide, mistletoe was neither tree nor shrub to the ancients. Not being one thing or the other, it enjoyed an independence that made it a symbol of unrestricted behavior. That idea may have inspired the tradition of authorized kissing.

Although mistletoe is not a tree, it grows on them and is commonly found on oaks. The oak's leaves are long gone in the dead of winter, of course, but the mistletoe remains green. It still signals the presence of life, and so it came to be regarded as the haven where the oak-spirit housed its life. That explains mistletoe's connection with everlasting life, but what, if anything, does mistletoe have to do with gold? Well, the color of mistletoe berries is yellow-white, and a cut bough of mistletoe after a few months turns entirely golden yellow.

Frazer brought his story of the Golden Bough full circle with the Vikings and their myth of the god Balder. His fellow gods in the Norse pantheon held him in such high esteem, they resolved to prevent his death by securing a pledge not to harm him from every single thing in the world.

Presumably invulnerable through the pledge of all creation, Balder entertained the other gods with a parlor trick in the halls of Asgard. He invited them to toss stones, darts, and other missiles at him, and all laughed as the projectiles kept their promises and fell harmless to the floor. Annoyed, however, by the fact that Balder could not be hurt, the trickster god Loki decided to see what could be done to reintroduce pain and suffering into Balder's life.

The mistletoe, Loki discovered, had been excused from making a commitment to Balder because it was too young. Deceptive, spiteful, and sly, Loki collected a branch of mistletoe and encouraged the sightless god Hoder to play pitch-n-toss at Balder with the rest of the crew. Hoder's blind throw could only be aimed with Loki's advice. After targeting Balder's heart, Loki watched the guided mistletoe drop Balder on the spot.

Hel, the goddess of the dead, claimed Balder for

Following Frazer's lead, illustrator H. M. Brock drew the Golden Bough as mistletoe in the sacred grove of Diana. (From Leaves from the Golden Bough, *by Lady Frazer, Macmillan, 1924)*

her dark, subterranean kingdom, but she agreed to release him to live again if, without exception, everything in the world, both dead and alive, wept for him. The whole world did shed tears for Balder. Nearly every one of the gods cried, too, both out of genuine sorrow and in an attempt to fulfill Hel's bargain. But while the entire cosmos cried, there was one who refused. Disguised as a giantess, Loki rejected the appeal of Odin's messengers to weep Balder out of the hands of Hel.

Deprived of Loki's tears, Balder died, and Frazer linked Balder's death with the summer solstice. In Scandinavia, mistletoe was gathered on Midsummer Eve, and bonfires, known in Sweden as "Balder's balefires," were lit the same night. Frazer likened these fires to the funeral pyre of Balder's cremation and argued that the mistletoe was both the agent of Balder's death and the true repository of his life. In this sense, Balder is the tree in Diana's grove, the anxious priest of Nemi slain by the intruder who is next in line, and the god who makes a necessary sacrifice on the altar of the year. With Balder, Frazer closes the circle of his season-inspired interpretation of the place of the slain god in the relationship between human beings and nature.

Solstices and fire certainly make it sound as though mistletoe has something to do with sun. The cut mistletoe's golden color and the living plant's ability to prevail against winter also affiliate it with the sun, whose annual pilgrimage is nothing less than an itinerary of necessary sacrifice and eternal rebirth. If you think of the summer solstice as the beginning of the sun's downhill slide into winter, the time when the sun reaches the high point of its yearly path is like a death. The difference, however, between the sun and seasonally slain gods is important. The sun may "die" each year, but like the moon it is immortal. It is resurrected for another circuit through the seasons. The earth, however, dies a true death as the year passes. It may occur when winter flays the trees to bare bones or when summer withers the fields or when the blade slays the grain at harvest. Life does return to the world at the appropriate season, but it's a new generation that signs in. These are not the same blades of grass, the same apple blossoms, or the same amber waves of grain that were here last year; their deaths were the sacrifices that make the continuity of life possible. Life is handed on like a torch to the young. The fertility gods that energize that flame perish in its passing

DYING GODS: ATTIS

The dying year–god goes by many names. We have already encountered him in Egypt as Osiris. To the Phoenicians of ancient Lebanon and Syria he was Adonis. He can also be traced to Babylon, where his name was Tammuz, and his Sumerian predecessor was Dumuzi. The Phrygians of central Turkey called him Attis.

Attis, according to tradition, died of self-inflicted wounds at the vernal equinox and possibly was resurrected three days later. Attis hadn't intended to kill himself, however. He was actually scheduled to marry Kybele (or Cybele), the daughter of King Midas of Pessinus, the capital of ancient Phrygia, but at the wedding, a savage monster known as Agdistis objected to the match. Long attracted to the boy's good looks, Agdistis wanted Attis for himself. His panpiping drove the wedding party mad, and in a frenzy, Attis castrated himself by a pine tree and died from loss of blood. His blood fertilized the earth, and where it fell, violets blossomed.

The story tells us unmistakably that sacrifice and death bring forth new plant life. In agricultural ritual built around the myth of Attis, this idea was extended from flowers to the crops. At the request of a remorseful Agdistis, Zeus brought Attis back to life as the new seed that fertilizes Mother Earth, Kybele. In this way, Attis fathers the new growth harbored in Mother Earth. His annual sacrifice at the vernal equinox binds the tale to the process of seasonal change and to the yearly journey of the sun.

Priests in the service of the goddess Kybele also castrated themselves and spilled their own blood from other self-inflicted wounds at the equinox ceremonies of Attis. This blood was spattered on the altar and on the sacred tree that represented the dead god in order to revive him. Here, sacrifice again guarantees renewal, but the mysteries of Attis also provide a direct link between this sacrifice and the sexual regeneration of life. Each set of parents dies in its time but renews the world with its own children. The most obvious symbol of parenthood is the apparatus for reproduction, and that's what the priests offered on behalf of the renewal of life. Children become the next set of parents, and they transmit life to the next generation. This story of everyday life is sufficiently profound to prompt people to symbolize it in a myth of a dying god.

AN ADVENTURE IN THE SKIN TRADE

Seasons and planting cycles in the tropics differ from those in temperate climates, but the idea of an annually dying god is still appropriate. The ancient peoples that resided closer to the equator also contrived myths involving divine sacrifice but tailored their scripts of seasonal life and death to suit a climate that cycled through rains and dry spells.

In Aztec Mexico, the god Xipe Totec personified the regeneration of life through sacrificial death. His festival, Tlacaxipehualiztli, coincided with the vernal equinox, a time when the hot and dry season had not yet acquiesced to the rains that usually come in June. At the equinox, the sun rises due east, the place where the sun was born, but vegetation dies, withered in the heat.

Translated, the equinox celebration's name sounds macabre to our ears. It means "flaying of men," and Xipe Totec, whose name means "Flayed Our Lord," was usually depicted as dressed in the skin flayed from a sacrificial victim. This human skin is closely connected with death, of course, but it also represents rebirth. That is because the Aztec likened it to the outer coat of a seed, which contains and confines the next birth of life. When the seed is buried in the ground, its coat cracks open. A young shoot pushes its way through the coat and the soil and into the sunlight, while the discarded husk decays beneath the surface. From the dead skin of the seed, new life emerges. Xipe Totec, then, was the buried seed and the new life that explodes from it.

The Aztec celebrated the festival of Xipe Totec with many human sacrifices. Fray Diego Durán, the Spanish Dominican friar who lived most of his life in and near Mexico City a few decades after the conquest, provided one account of Xipe's equinox sacrifices in his *Book of the Gods and Rites* (Part II of *Historia de las indias de la Nueva España e islas de tierra firma*). After forty days of public glorification, the sacrificial victim who impersonated Xipe Totec in these rituals was taken to the temple, where he was met by six others, costumed as the principal gods. All of them were killed. Their hearts were torn out, and their bodies were flayed. After their hearts were offered to the east—food for the sun about to be reborn—other participants slipped into the flayed skins. In donning them, they became the same gods reborn.

Death alone, however, did not satisfy the liturgy of Xipe Totec. He and his victims had to suffer. For that reason twenty or thirty more prisoners

A sacrificial prisoner, tethered to the top of a small pyramid, defends himself with mock weapons against an armed jaguar knight. Xipe Totec presides over them, and the extra hand dangling from his wrist confirms that he is costumed in the flayed skin of a victim of such sacrifices. (Drawing: Joseph Bieniasz, Griffith Observatory, after Book of the Gods and Rites *and* The Ancient Calendar, *Fray Diego Durán, University of Oklahoma Press, 1975)*

were led, one by one, to the top of a small platform and tied to a stone ring. Each prisoner in turn, armed with a feathered stick—a mock sword—had to defend himself against four warriors, one at a time. The real weapons carried by the warriors guaranteed the outcome: wounds and suffering. The defeated sacrificial victim was then taken to another platform. There, while four men held his arms and legs, his heart was torn out on behalf of the sun.

Hearts of the sacrificed victims were offered to the east as food for the sun, but Xipe, associated with planting, was affiliated with sunset and the west. His connection with the realm of the dying sun echoes the same theme—interment. Consigned in death to a tomb in the earth, the seed corn is reborn as a new crop, only to be sacrificed itself to nourish us. Through the sacrifices of Xipe Totec, the Aztec ritualized what to them was obvious. All life is sustained through sacrifice. Each thing that lives dines on some other living thing and is, in turn, food at some other table. The parade of life is really an interminable meal.

THE SERPENT DESCENDING

A gold disk recovered from the Sacred Cenote, or sacrificial well, at Chichén Itzá, a Maya ceremonial center in northern Yucatán, shows the same kind of human sacrifice performed by the Aztec at the vernal equinox. Four attendants grip the limbs of a victim with a deep cut in his chest. A priest with a stone knife in his hand has removed the heart, while a celestial rattlesnake hovers overhead.

Another kind of celestial serpent is also seen at Chichén Itzá. When the equinox sun moves toward the western horizon, a pattern of light and shadow appears on the west balustrade of the north stairway of the Castillo, the site's tallest and most obvious structure, and this display resembles a descending snake whose head is the monumental stone serpent head at the foot of the stairs. This rippling shadow is cast by the stepped profile of the pyramid's northwest corner. Once the sun gets low enough, about two hours before sunset, the long undulating ribbon of darkness that appears to run from the top of the staircase to the bottom begins to develop a distinctive zigzag pattern that will turn into a sequence of seven well-formed triangles of sunlight. The top triangle appears first. At a stately ceremonial pace, the rest of the luminous triangles show up, one at a time, until the entire north face of the Castillo is dark, save for those seven triangles and the sunlit serpent head at the bottom. By then, the sun is quite low, and as it sinks even more toward the horizon, the triangles disappear very quickly, from the bottom first. This play of light and shadow creates a symbolic serpent of sunlight that slithers down the pyramid—from sky to earth—on the equinox.

If the sun serpent really were part of the original plan of the Castillo, it was probably intended for the March equinox. The rainy season in Yucatán begins in mid-May and continues through late October. At the autumnal equinox, in September, the sun repeats its March matinee, but clouds and rain are more likely to curtain the sky. At vernal equinox, by contrast, the cloud puffs come and go, and the skies are blue. Chichén Itzá's equinox serpent may, then, have something in common with the Aztec equinoctial sacrifice that took place on the festival of Xipe Totec, but we have to know something more about the Castillo and the beliefs of the people who built it to tell if there is a connection.

The part of Chichén Itzá that includes the Castillo was built in the tenth century A.D. by one of the Putún Maya groups that moved from the Gulf Coast into Campeche and Yucatán. They spoke Maya, but they were also strongly influenced by the peoples of central Mexico, especially the Toltec. There are many similarities between the buildings and the style of ornament at Tula, the Toltec capital, and later structures at Chichén Itzá. The ruins of Tula are about fifty-five miles north of what is now Mexico City, and many Toltec traditions were inherited by the Aztec, who carved out the last big empire in central Mexico. We can, therefore, see that some ideas and styles were shared by the Maya of Postclassic Yucatán and the later Aztec.

The Castillo was dedicated to the god Kukulcán. In Central Mexico he was called Quetzalcóatl, and both names meant the same thing, "feathered serpent." But who, exactly, was Kukulcán? The answer is not so clear. Like most of

A serpent of sunlight and shadow slithers down the west balustrade of the north stairway of Chichén Itzá's Castillo near sunset at the time of the equinoxes. A giant sculpted serpent head at the bottom of the stairway makes this seasonal display symbolically convincing.

the gods of ancient Mexico, Kukulcán-Quetzalcóatl is a multilayered, highly textured, complex figure. Said to have brought culture and the arts of civilization to humanity, he was also associated with rainwater, the wind, the planet Venus, and new life.

Feathers make Kukulcán into a kind of sky serpent, but is he the same snake we see in sunlight on his pyramid? The triangular pattern of the equinox serpent matches the side view of a diamondback rattlesnake native to Yucatán, *Crotalus durissus tzab*. All of the feathered snakes in the architecture of Chichén Itzá display rattles on the ends of their tails and tell us Kukulcán was a plumed rattlesnake. So it is the feathered rattler that descends the staircase at Chichén Itzá.

Because the serpent effect is associated with an important date in the solar year, it seems to have something to do with the calendar. Today's Yucatec Maya still say that a rattlesnake gains an additional rattle each year. In fact, rattlesnakes accumulate rattles—and lose them—far less systematically. But people say they pick up a rattle a year, and what people say partially reveals what the rattlesnake means to them. The snake and its rattles are associated with the idea of the year, the passage of time, and, therefore, the calendar. In addition, a new rattle is added when the serpent sheds its old skin. This process renews the rattlesnake; at least it renews its skin. The rattlesnake, then, is a symbol of the renewal of time and of the emergence of new life from old.

*Nearly every feathered serpent depicted in the architecture of Chichén Itzá is a rattlesnake. That confirms that this snake—*Crotalus durissus tzab*—inspired the symbolism of Kukulcán. The diamonds on its back look like triangles from the side and resemble the triangles of light that form upon the Castillo's stairway at the time of the equinox. (Photograph: John H. Tashjian, at Dallas Zoo)*

There is also a link between these ideas and springtime, for that is when rattlesnakes mate. This is not yet the time of new life, however. It is the time when the seed is planted. If anything, the dry, dead vegetation marks the spring as the end of the cycle in Yucatán, not the start. In the dry season of late spring, increasing heat makes rattlesnakes active at dusk, but once the summer rains begin, the snakes become creatures of the day.

There is at work here a network of associations, and it involves the rattlesnake, the vernal equinox, the passage of seasonal time, the transformation from death to new life, and the sky. Xipe Totec, also associated with the vernal equinox and seasonal transformation, carried a ceremonial rattle. Although we cannot be certain this noisemaker was inspired by the rattlesnake, it did have something to do with fertilization, reproductive vigor, and new life. Xipe's costume of human skin—his "golden garment," as the Aztec hymns phrase it—is also like the castoff skin of a serpent. After shedding this skin, the snake looks revitalized. Discarded snakeskin is, then, like the abandoned husk of the seed.

All of the feathered rattlesnake's associations—with the calendar, with the equinox, with death, and with new life—allow the descent of the serpent of sunlight to be a rich and compact symbol. We still can't be sure the effect on the pyramid was intentional and ritually observed. But it does look like something the Aztec would have recognized.

Perhaps the descent of the serpent at Chichén Itzá was a great public event, a visible sign of the sacred. If so, it would have fortified the traditions that spawned it. In traditional societies the mandate of political power, the social structure, the system of economic exchange, and religious life are all interwoven. Each is part of the other. The ceremonial structures of a place like Chichén Itzá are no more exclusively government palaces than they are exclusively churches. They are the skeleton that once supported the ritual life of a people for whom such distinctions were less clear.

In our own world we don't really deal as a community with the sacred. Much of the world is now secularized. Our sense of the sacred is in solitary confinement. And sometimes we feel the loss. Whether or not the Maya of Chichén Itzá made anything of the equinox serpent on the Castillo, it speaks to us today. As many as fourteen thousand visitors now gather there on March 21 each year to witness it. It assembles us on the date when day is balanced with the night, in a time of trial by the sun's fire, before the rains and the new corn, and tells us a story about the renewal of life. For an hour or so, thousands drawn to a thousand-year-old center of Maya power—by curiosity or by the carnival atmosphere—pause to witness a seasonal tale told in sunlight: the seasonal mystery of order and change.

AN EASTER SACRIFICE

When we recite the story of seasonal sacrifice and the emergence of new life, we tell more than the tale of crop cultivation and a narrative of nature's renewal. We outline the foundation of our lives. The growth, death, and rebirth of vegetation provided our ancestors with a glimpse of the real ground rules of life and the world. That's why initiates in the Eleusinian mysteries found epiphany in an ear of grain. It revealed the ordered change that holds the cosmos together. It transcends what's growing in the fields and embraces all that is. It's not surprising that its spiritual value would find its way into Christianity. Christianity's calendar of service and worship annually retells the life story of Christ with the rhythm of the seasons and the guidance of the year.

Easter remembers Christ's death on the cross and rejoices for the Resurrection. It expresses the essential message of Christianity: redemption of the soul. Christmas, on the other hand, celebrates the beginning of Christ's life on earth, the birth of a new age and a new order.

Easter occurs in spring, close to the vernal equinox. It is convenient—and usually correct—to say that Easter falls on the first Sunday after the first full moon after the equinox. Tradition associates the name *Easter* with an Anglo-Saxon goddess, Eostre, whose rites were observed in the spring in Eosturmonath, the Anglo-Saxon month that corresponded closely to our April. If this is true, it suggests that Christianity in Britain borrowed a felicitous name for the most important holiday in the Christian year from pagan tradition. We know that the date of Christmas is a legacy of pagan antiquity. It is, of course, observed in winter, on December 25, about four days after the winter solstice, and many peoples have called the winter solstice the time of the sun's rebirth; Christ's birth nearly coincides with it.

The sun may be born at the winter solstice, but in winter the world is dead. It comes back to life in spring with the sprouting of grass and the budding of trees. In the ancient Mediterranean world and the Near East, springtime also coincided with the beginning of the grain harvest, and that harvest was seen as a sacrificial death. Christ's death and resurrection accompany both the first cutting of the grain and the springtime appearance and flowering of a new generation of plants.

We have already seen how two seasonal phenomena—the annual cycle of vegetation and the annual cycle of the sun—play parts in the traditions of annually dying gods whose seasonal sacrifices guarantee the continuity of life and cosmic order. Christ is a dying god, and Christianity takes advantage of both seasonal metaphors to transmit a message about the destiny of the soul.

Christianity's tradition of springtime sacrifice probably has roots in an ancient Hebrew custom that also became linked with the Jewish Passover. In languages other than English, Easter's connection with Passover is clearer. Easter is *Pascua* in Spanish, *Pâques* in French, *Pasqua* in Italian, and *Pascha* in Latin. All of these names originate from *Pesach*, the Hebrew name for Passover, which begins on the night of the full moon after the vernal equinox.

Passover celebrates the escape of the Israelites from bondage in Egypt. Blood of a lamb slain at sunset (the start of the day in the Jewish calendar) on the fourteenth day—full moon—of Nisan, the year's first month and the month in which the vernal equinox fell, was smeared on the doorframes of their homes. These signs painted in blood protected the Hebrews from the vengeance of God, who appeared in Egypt exactly at midnight. He killed the firstborn in every Egyptian household but passed over the homes of the Israelites.

In the Old Testament, the blood of the first Passover lamb redeemed the Hebrews when death prowled in Egypt. Familiar with the Passover tradition and aware that the Saturday following the Crucifixion coincided with the Passover, Christians portrayed Christ as the sacrificial lamb of Passover.

Despite the direct link of Passover to the account of Moses and the plagues upon Egypt in Exodus, scholars are certain the feast evolved out of an older tradition of a spring sacrifice that accompanied the start of the rainy season and was timed by the vernal equinox and the full moon. Leviticus (chapter 23; verses 5–14) mentions the sacrifice of a young lamb and the "first fruits" of the harvest at the festival of Nisan, which came to be associated with Passover. This was the barley harvest, and the next two verses tell us the Hebrews celebrated the completion of the harvest with Shavuoth, the Festival of Weeks, seven weeks and a day—a total of fifty days—after Passover.

Significant dates in the Christian calendar in the weeks following Easter strengthen the connection between Easter and the ancient Hebrew sacrifice of the first barley grain. Ascension celebrates Christ's ascent to Heaven and marks the end of Christ's appearance on earth. It falls forty days after Easter and signals the return of his human body to its transcendent source. Ten days later, the Holy Spirit descended upon the disciples of Jesus. It consummated Christ's mission on earth by conferring on the Apostles an understanding of the work that awaited them. Pentecost celebrates this event fifty days after Easter, and the name of this church festival derives from the Greek word for "the fiftieth day." Pentecost is called Whitsunday, or White Sunday, in England, and it corresponds to Shavuoth.

Passover and Easter, then, both began the barley harvest. That is why, resurrected, Christ is called the "first fruits" in I Corinthians (chapter 15; verse 20). The "first fruits" of the harvest were sacrificed at Easter to ensure eternal life. The spiritual harvest is gathered in, at the end of the barley season, at Pentecost, fifty days after the Resurrection and the sacrificial offering of first fruits. Through a season ordained by the sun and the moon, the resurrection of Christ and the descent of the Holy Spirit fell in step with the annual sacrifice and harvest of the grain.

The Easter story is not just a symbolic account of the rebirth of life in spring or the cultivation of barley. It is, however, congruent, in narrative and ritual, with both and takes advantage of those metaphors of vegetation to explain the deepest Christian mysteries. Redemption of humanity proceeds from the willing sacrifice of God. Incarnated on earth as Jesus Christ, God suffered the full burden and anguish of life and death. Through his sacrifice on the cross, his human creations are released from physical death to an eternal life of the soul.

For believers, Christ's death makes possible a renewal that redeems the soul from the bondage of time. Decay and corruption may coerce this world toward irreversible change and put direction in the arrow of time, but sorrow over the certainty that things wear out is turned into exuberance for the victory of the soul. From Good Friday's despair over Christ's suffering and death, through the sad and quiet vigil of Holy Saturday, to the liberation of Easter morning, when Mary Magdalene discovered the open tomb, the days of Easter compress the full pageant of death and rebirth.

THE UNCONQUERED SUN

Christmas, of course, celebrates the birth of Christ, and its date is linked to the annual renewal of the sun at the winter solstice. December 25 was the winter solstice in the calendar Julius Caesar established for Rome in 46 B.C. Now the winter solstice usually falls on December 21, and the difference is due to an error in Julius Caesar's calendar. Although the Julian calendar reform provided a year that was 365¼ days long, the year is actually just a little shorter, and that means the solstice arrives a little sooner each year in a 365¼-day calendar. Julius Caesar's approximation missed the mark enough to accumulate a discrepancy of one day in 128 years. Even though the winter solstice arrived 2½ days early in A.D. 274, when the Emperor Aurelian established the sun cult as the state religion of Rome, its traditional date, December 25, was too familiar to abandon. It became *Dies Natalis Invicti*, the birthday of the Unconquered Sun.

Nothing in the New Testament specifies the date of Christmas, and before the fourth century, Christ's birth had been associated with Epiphany, or Three Kings' Day, on January 6. Recognizing, however, that the winter solstice and the Unconquered Sun still commanded an important place in the imaginations of their flocks, the early Church fathers grafted the winter solstice to the story of Christ. Paulinus of Nola, a Christian poet in fourth century Italy, wrote,

> *For it is after the solstice, when Christ born in the flesh with the new sun transformed the season of cold winter, and, vouchsafing to mortal men a healing dawn, commanded the night to decrease at his coming with advancing day.*

Sometime, then, between A.D. 354 and 360 and a few decades after Emperor Constantine's conversion to Christianity, Christmas was shifted to the birthday of the Unconquered Sun.

Deus Sol Invictus, the Invincible, or Unconquered, Sun God, fought the forces of darkness each night to emerge victorious at every sunrise. This eternal champion had been introduced to Rome from Syria in the first century A.D. Legionaries stationed there brought him back as Sol Invictus Elagabal. By the reign of Emperor Commodus (A.D. 180–192), worship of Sol Invictus Elagabal had grown prominent. Through court and royal family intrigue, Elagabalus, the fourteen-year-old high priest of Sol Invictus Elagabal, was installed as Emperor in A.D. 218. Consumed by his religious vocation, he provided little help to the Empire, which was beginning to fray around the edges. He aggravated Rome by inflating the importance of the regional sun god he worshipped and elevating him above the incumbent gods of the Eternal City. His sexual excesses and bloody sacrifices offended Roman sensibilities, especially when he divorced his first wife and married one of the Vestal Virgins in a misguided attempt to unite the cult of the Unconquered Sun with the worship of Vesta, a highly regarded goddess. He, his mother, and his most faithful allies were all killed in the palace gardens, on March 21, 222, the vernal equinox in that year. Fifty years later, under Aurelian, Deus Sol Invictus—stripped of all Syrian predilections—became the chief god of the Roman state. In time, the Fathers of the Church identified Christ as the true sun god, Sol Iustitiae, the Sun of Justice, and informed the faithful that any birthday party for the sun really celebrates the Nativity. As the sun's birthday, every Christmas turns the darkness to light.

That winter darkness is an essential feature of Christmas observances. We can't herald the reinstatement of the light without it. Many Christmas traditions are borrowed from that story: Christmas candles, multicolored lights on Christmas trees, and the burning of the Yule log.

Sweden's Lucia tradition also preserves the link between the winter solstice and light. A crown of lighted candles is donned on December 13 by the youngest girl in the household. She gets up at the first crow of the cock, dressed traditionally in a white robe, with a red sash around her waist and a

Swedish customs are preserved in Los Angeles in an annual Lucia Pageant. Several girls dress in the traditional costume and accompany in song the "Lucia bride," who wears a crown of lighted candles to take some of the edge off of winter's darkness.

crown of evergreens and burning candles on her head. While it is still dark, she rouses her parents from slumber with holiday bread, biscuits, and coffee and sings a special song.

Electric Christmas candles now glow in nearly every window of Bethlehem, Pennsylvania. The city was founded by Moravian Church missionaries from central Europe nearly 250 years ago. Today, visitors and residents can catch the Christmas City Night Light bus. The guide retells the story about why the candles are lit.

> On Christmas Eve, according to the age-old story, the Christ child bearing bundles of evergreens wanders all over the world. Those who long for his coming set a lighted candle in the window to welcome him in their homes and hearts. . . .

Candles even cut through winter's darkness in southern California, where winter has to try harder to command attention in a generally benevolent climate. During the late afternoon service on Christmas Eve in Pasadena's All Saints Episcopal Church, the electric lights are extinguished. What dwindling sunlight remains outside puts pale color into the stained glass windows, and the huge wreaths are transformed into smoky rings on the high stone walls of the nave. In this deep darkness, the story of the first Christmas is told from in front of the choir. The night was long and dark in Bethlehem, too, and it was winter. But the baby born in that darkness would be the light of the world.

A single candle—the Christ Candle—is carried through the darkness, down the central aisle, past the congregation, and up to the altar, where it is installed. It is the first light. Though small and vulnerable, this flame looks as if it could relight the world. Its spark is shared with other tapers, which travel up and down the aisles and light all the candles in the church. Their flickering provides a warm, fragile light that quietly reports that the first footstep out of the darkness has been taken. For a brief time, the eternal moment—the hinge of time—is summoned into the present, and the meaning of Christmas is a revelation in candlelight and carol. We are propelled back on the wings of the winter sun to the first Christmas.

Even the secular aspects of Christmas capture its essential spirit of renewal. All the anticipation on the night before Christmas makes it seem like time is holding its breath for the fresh light of Christmas Day. Children know Santa Claus is coming and can't wait. Going to bed on Christmas Eve is the only way they're going to get to Christmas morning, but it's a profound lesson in patience.

The experience reinjects us with anticipation and allows us each year to feel once again what the Wise Men and the shepherds knew that first Christmas night: Something wonderful is about to happen.

7

Over the Rainbow

FOR an unmistakable display of celestial power, it is hard to compete with the nearby strike of a lightning bolt. It shatters the security of the solid ground with an explosive jolt of savage electricity. In the midst of a rousing storm, even as we expect another thunderclap, each sudden, window-rattling boom arrives without warning and creates an atmosphere of imminent threat. Nerves get frayed. Domestic cats disappear into hidden havens, and dogs look for security in human company. It sounds as if the sky is on a rampage. We feel powerless during the bombardment and can only wait for the fury of the storm to subside. To describe and explain the flames and explosions hurled from heaven, the ancients sometimes said their gods were angry or at war.

Lightning, we know, is not really a grenade thrown by an irate sky god but a natural consequence of the electrical energy in the storm clouds. This electrical energy is generated when electric charge—positive and negative—gets separated in the cloud. Normally the two types of charge are well-mixed, and the material is electrically neutral. Turbulence, however, carries positive charge to the upper part of the cloud and negative charge to the bottom. The polarized storm cloud then provides an awe-inspiring electrical display when the opposite charges reunite.

Any time there is a difference in electric charge in two places, the attractive force between opposite charges will induce electricity to flow. Air is an insulator, however, and is able to inhibit the flow, at least to a point. But when the difference in charge creates a force able to overpower the air's constraints, a rapid and violent release of charge elbows its way through the molecules like a column of storm troopers.

The first lightning in a storm usually travels between one part of the cloud and another, but if a difference in charge develops between the cloud and the ground beneath it, lightning is exchanged with the earth. This starts in the bottom of the cloud, where the preponderance of negative charge generates a temporary difference in charge with the earth below. The earth's surface normally is negatively charged, but a localized zone beneath the storm becomes positively charged with respect to the cloud. If the difference exceeds the insulating capacity of the air, a pencil-thin channel of electricity pushes toward the earth. It takes only a hundredth of a second or so for the negative charge to near the ground, and when it does, positively charged particles fountain back toward it. These opposite charges have a spectacularly fatal attraction, and they meet in this atmospheric wormhole. Their successful union closes the electrical circuit, and electricity flows rapidly back and forth between earth and sky. Although it looks as if the lightning bolt has descended from the sky, the first surge of electricity actually shoots up into the cloud at about 8,000 miles per second.

Inside the lightning bolt the temperature of the charged particles is nearly five times hotter than the surface of the sun. This heat is shared with surrounding air, and it expands so fast, we can hear it coming. Thunder is the sound of this superheated air, and lightning its glow.

Knowing what lightning is doesn't make it any

Lightning is a clear demonstration of celestial power conducted to earth. This bolt was released by a Utah thunderstorm within striking distance of Interstate 70. (Photograph: Robin Rector Krupp)

less impressive. It is a fascinating advertisement for power in the sky even when we understand how it works. The average thunderstorm releases about ten times as much energy as the first atomic bomb did, and every day the earth endures forty-five thousand thunderstorms. Our atmosphere is in constant turmoil, and the turmoil made an impression on our ancestors. Watched from a safe distance, lightning fascinates us. It is a luminous white snake that forks its way to earth. It can splinter trees, set fires, and deafen us with its din. No wonder we once saw it as an unambiguous message from the gods on high.

The eighteenth-century Italian philosopher Giovanni Battista Vico was one of the first to argue that displays of power in nature influenced our ancestors' notions of divinity. According to him, the ancients viewed the sky as divine and imagined that it was trying to say something to them in the thunder. Thunder, lightning, and thick clouds covered Mount Sinai on the day Moses was given the ten commandments by Yahweh.

Some of the old sky gods, then, had a direct hand in the weather. Today we separate atmospheric science from astronomy, but our ancestors saw that the storm performed on the same stage as the stars. Their stories about the sky included the deeds of their weather gods. They could command clouds, thunder, lightning, wind, and rain, and the atmosphere itself was a disclosure of their power.

When the weather is active and violent, there is little doubt that there is power in the sky. Those weather gods purveyed that power into punishment and reward. The storm could be a deadly penalty. Its floods, gales, and lightning bolts were dangerous and destructive. Rain, on the other hand, was a reward that ensured the earth's fertility. Both rain and storm were evidence of celestial power.

THE HIGH AND THE MIGHTY

The Skidi Pawnee, Plains Indians of Nebraska, called their high and powerful god Tirawahat. He was the sky, the universe, everything. He lived at the sky's zenith and was also known as Our Father Here Above. James R. Murie, half-Skidi himself, helped collect and preserve Pawnee traditions at the turn of the century. One of the songs he saved credits Tirawahat with making the heavens and the stars.

Unchanging, eternal, and supreme, he was responsible for the origin of life. Tirawahat caused everything but was himself invisible. He acted through his agents: Lightning, Thunder, Wind, and Clouds. Transcending the action in the atmosphere, Tirawahat's power there was transmitted through Paruksti, a mysterious spirit whose face was the clouds and whose voice was the thunder. Von Del Chamberlain, who compiled and analyzed Skidi Pawnee astronomical lore in *When Stars Came Down to Earth*, provides a powerful image of Paruksti. Paruksti contemplates the creation of life in his lodge at the western horizon and then stands and starts to speak low. His first words are heard in the distant thunder, quiet but with a promise of what's to come. When he goes outside, he stretches his arms, and the blanket of clouds he wears spreads across the sky. Each glance is a display of lightning. Each full breath is a roll of thunder. He shouts thunderclaps.

The Skidi Pawnee said the world was made in the world's first thunderstorm. Now thunder and lightning rarely occur in winter, when the world really seems to be dead, and the first thunder

The Vikings attributed lightning to celestial blows from Thor's hammer (middle figure). In this twelfth-century tapestry from Skog Church, in Hälsingland, Sweden, Odin (left) is recognizable from his single eye, and the god Frey from an ear of grain. (Drawing: Joseph Bieniasz, Griffith Observatory)

heard after winter's death is a herald of spring and rebirth. The year's first thunderstorm is, then, like the world's first thunderstorm. Lightning and thunder, under Tirawahat's command, animated the earth and put life into the trees. They sweetened the waters and vitalized the seeds the gods scattered from the sky. Finally, the storm awakened life in the world's first boy and the world's first girl. They paired, and the world of the Pawnee was born.

A less delicate story told by the Koryak, who live on the eastern coast of Siberia and on the Kamchatka Peninsula, makes the link between thunder, rain, and the world's fertility as explicit as an X-rated movie. The supreme and high god of the Koryak is sometimes known as Thunder-man. His wife is Dampness-woman, and he makes rain by beating on her amputated vulva with his own dismembered member. Because he fastened her genitals to a tom-tom, he performs a thunderous solo with his double-duty drumstick. Every stroke drives water out of the vulva, and it falls to earth as rain.

Thunder-man apparently doesn't believe in too much of a good thing, and it takes Big-Raven and his son to stop the music. When they put the sky's one-man band and his wife to sleep, the Ravens service the instruments. They bake the drum and drumstick over a fire until they are completely dry. When Thunder-man wakes up, he tries to play the same old tune, but every drumbeat brings dryer weather instead of rain.

Brandishing a hammer in one hand and undulating bolts of lightning in the other, the Hittite weather god presided over storms. This portrait of him was carved in the ninth century B.C. and originally came from Zinjirli, in what is now southeastern Turkey. (Object in the Museum of the Ancient Orient, Istanbul)

STORMY WEATHER

The power to populate the world with life has also been embodied in other weather gods who penetrated more personally the earth with rain and dropped bolts out of the blue. Zeus certainly was one of them, but there was a whole gang of Indo-European storm mongers. Viking fortresses shook when Thor's sky chariot roared with thunder. His hammer exploded with lightning. He brought the clouds, dispensed rain, and dispersed the storm. His names in other Germanic languages—*Thunor* in Old English (Anglo-Saxon) and *Thonar* in Old High German—mean "thunder," and they all originated in a common root meaning "to boom, to roar." Celts in Gaul burned prisoners in wood and wicker cages as sacrifices to their version of Thor. His name, Taranis, meant "thunder," and Gallo-Roman statues portray him with a thunderbolt and the symbolic celestial wheel. *Taran* still means "thunder" in Breton and Welsh. In old, pagan Russia, Perun ranked as the highest god. His name, which meant "the striker," is related to common Slavic words for thunder. Like Thor, Perkúnas, the high god of the ancient Lithuanians and related Baltic peoples, carried a hammer, and his name means "thunder." The Hittite weather god's name is lost, but his portrait survives in a relief from Zinjirli in southeastern Turkey. A hammer in one hand and a lightning bolt in the other make him easy to recognize. The Hurrians of northern Mesopotamia called him Teshub.

Ukko, the Finnish god of thunder and storm, is unrelated to the Indo-European tradition, but he, too, hammered his way from spring sowing through the autumn harvest, along with the thunderstorms of summer. The Finns drank to him in spring and poured beer on the ground to assist the fall of rain. Wanton behavior and intoxication accompanied his vernal rituals.

In Vedic India, Parjanya—identified in the *Rig Veda* as the son of Dyaus, who he resembles—personified the rain cloud. He roared like a bull in the thunder, inseminated the plants with raindrops, and was given credit for fertilizing mares, cows, and women. The rain in those clouds over India had to be set free, however, and the *Rig Veda* credits Indra for that deed. His name means "storm," and he was Thor's counterpart among the Vedic gods. He rode the sky in a chariot, in the company of Vayu, the god of the wind. His sceptre was the thunderbolt. With it, he battled Vritra, a monstrous, serpentine cloud dragon, for control of the clouds and rain.

In one passage of the *Rig Veda*, the rain clouds, symbolized as cattle, are kidnapped by the drought demon. Indra trailed the herd to drought's cave, killed the demon at the entrance, and led the rain cows to freedom. Their milk is rain, seed, and semen, and once they are out of drought's dungeon, they shower the earth with fertility.

A better known episode of rain liberation in the *Rig Veda* informs us that when Indra got wind that Vritra had rustled the rain, he went looking for the sidewinder. Armed with his *vajra*, or thunderbolt, Indra hurled it at the evil worm. Sharper than a serpent's tooth, the golden, "hundred-angled" lightning bolt hit Vritra on the back of his neck, in the face, in his soft underbelly that scraped the mountain peaks, and in his vitals. Vritra split a gut and lay open on the mountaintop, releasing the hidden waters he had gorged.

All of the imagery in the second version of the story of rescuing the rain implies that Vritra is a drought dragon castled in clouds and is himself made of clouds. The rain they hold can be freed only by Indra's lightning bolts. Any falling rain is the sign of a rematch, another bout in a never-ending battle, but seasonal drought is the real theme. Winter and spring in India mean fierce heat and no rain; those clouds that manage to appear keep the rain to themselves. But when summer arrives, it brings the monsoon. Indra's penetrating arsenal of guided missiles brings both sex and violence into the atmosphere; drilling Vritra with lightning, then, inseminates the earth with buckets of rain. After the battle, Vritra is likened to a steer, castrated and powerless. Indra, however, is a "bull bursting with seed."

To the Hindus, Indra is the king of the air, and in the Vedic period he was the storm and weather god. He stands here on an east face of Rajarani Temple, a Hindu monument built in the eleventh century in Bhubaneshwar, India. His right hand holds a thunderbolt, and an elephant goad is in his left hand.

Semitic versions of the same kind of storm god

include Adad, who rode a bull and brandished lightning in Assyrian skies, and Baal-Hadad, whose name means Lord Thunderer, did the same thing for the Canaanites. The sculpture of him from Ras Shamra, on the coast of Syria, shows him with a lightning mace in his hand and bull's horns on his helmet. He was also said to bellow like a bull when thunder roared. His father, the Sumerian god Enlil, controlled the atmosphere, and without Enlil, the Canaanites said, "the rain-laden clouds could not open their mouths." Without Enlil, the barley wouldn't sprout, grass would wither, and orchards would fail to bear fruit. Without Enlil, "the wild goats and asses, the four-legged beasts, could not be fertile, could not even copulate." Father and son together represented the power for life and death embodied in the storm.

AXES FROM HEAVEN

We have seen that some kind of ax or hammer was standard issue for licensed thunder masters and weather lords. Thor's hammer, Mjölnir, spit lightning. Its name means "crusher" in the sense of the pulverizing force of a mill. With it Thor smashed frost giants and fought the Midgard Serpent, but the hammer could also give life. Thor's wife, Sif, was connected with the fecundity of the earth and the growth of grain, and as her husband, Thor had something to do with her productivity. The lightning and the rain Thor hammered to earth enhanced the harvest and mirrored the intimacy shared between him and his wife. In northern England, Nordic ideas were preserved in a traditional belief about summer sheet lightning. It was said to ripen and enrich the crops. Thor's hammer, then, symbolized more than lightning and thunder. It had a sexual connotation, and in the story of its theft by the giant Trym, it is obviously the tool of male virility.

When Trym stole Mjölnir and hid it deep within the earth, the goddess Freya in marriage was his price for the hammer's return. Freya refused the dubious honor of cohabiting with a giant, and Thor was forced to disguise himself in bridal linen as Freya. Then, according to the "Trymskvädet," one of the poems in the *Elder Edda*, a thirteenth-century Icelandic compilation of ancient stories, the hammer was laid in the "maiden's" lap to consecrate the bride. The sexual symbolism here is transparent. With his virility back where it belonged—in his own lap—Thor killed the giant and crippled his relatives. If it had been Freya at the bridal shower, and not Thor in drag, having Mjölnir in her lap would have symbolized consummation of the match between her and the giant who commanded the hammer's vigor. When Thor gets it back, however, it is his male strength returned, and he dispenses it like the fury of the storm.

People often thought the god of thunder left other souvenirs of the storm besides the rain. Meteorites, lightning-struck stones, and prehistoric stone axes found in the earth were believed by some to be objects hurled from heaven when the thunder crashed.

Every night, in a dark sky, it is possible to see small interplanetary particles burning up in the earth's atmosphere. These "shooting stars," or meteors, flash for a moment across the sky and disappear. Colliding with molecules of air at high altitude, the particles, known as meteoroids, heat the air through friction until it glows. That trail of incandescent air above the stratosphere is the meteor. In an unusually bright meteor, or fireball, the column of superheated air may be a couple of miles in diameter. The meteoroid responsible for it is usually no larger than a grain of sand or perhaps a small pebble, too small to last long in that kind of abuse. But sometimes stones and metal really do drop from the sky. Known as meteorites, these pieces of interplanetary debris are large enough to survive the vaporization induced by a fiery baptism in the earth's atmosphere.

Some meteorites are made of iron, and so, long before the process of smelting was discovered, iron was known in its meteoritic form. It was rare, and its connection with the sky seems to have been understood by the ancients. Their awareness of its origin is embedded in their names for the metal. The Egyptians called it "stone of heaven" or "produced by heaven." The oldest Sumerian word des-

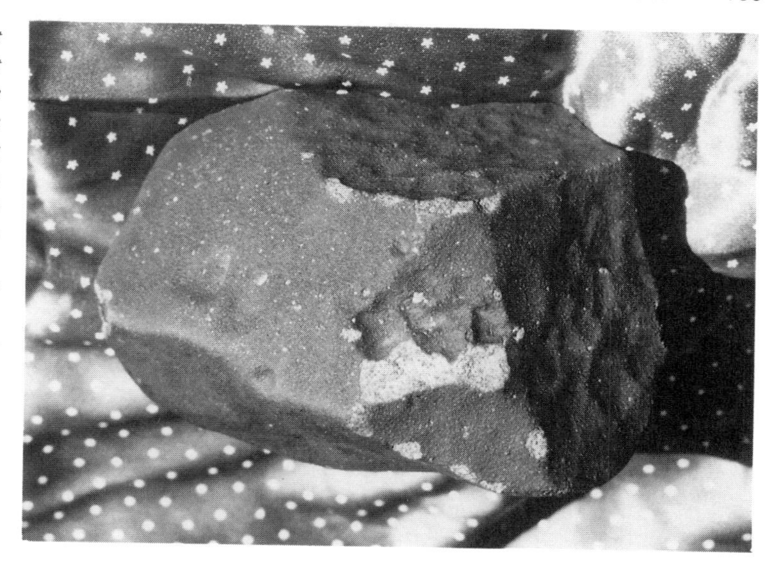

Meteorites are falling to earth every day, but most of them burn up in the atmosphere. Most of those that do manage to reach the surface are lost in the ocean, and of those that hit the ground, few are recovered. The odds of getting hit by a falling meteorite are low, but there is one documented case. This nine-pound meteorite crashed through the roof of a home occupied by Mrs. E. Hulitt Hodges in Sylacauga, Alabama, at 1:00 P.M., Central Standard Time, on November 30, 1954. It bounced off a console radio in her living room and hit her hip, which was bruised enough to require hospitalization. The distinctive dark surface on this rock is its fusion crust, which formed when the stone was heated in its passage through the earth's atmosphere.

ignating iron meant "sky" and "fire." Together the words suggest the "celestial lightning" of a meteor. First in the ancient Near East to use weapons made extensively from smelted terrestrial iron, the Hittites still called the metal "fire from heaven." Iron they said, originated in the sky, unlike copper and gold, whose sources were known to be terrestrial. The Assyrians, too, extracted iron from ore, but they said it was "fragment from heaven," and for them, the very "walls of heaven" —the sky—were made of iron.

Only rarely is a meteorite recovered from an observed fall, for the projectile usually strikes land far from where it is witnessed. Someone actually near the landing zone experiences an entirely different event. A light will suddenly appear. It quickly grows intensely bright, but it hardly seems to move. That is because it is headed right for the witness. Sonic booms and the explosive flames of fragmentation accompany the object's descent. It lands with thunder. Even from a distance, fireballs are sometimes accompanied by sonic booms, and that, too, may have helped associate meteors and meteorites with thunderbolts.

There is evidence that some ancient people saw in meteorites the same procreative powers they attributed to the hammers and axes of the storm gods. Faithful adherents in the Syrian culture of the invincible sun god Gabal probably focused their worship on a meteorite at the shrine at Emesa. This large, black, conical stone was possibly the same one transported by the unpleasant emperor Elagabalus to the temple of the fertility goddess Kybele in Rome. A stone certified to have fallen from the sky was enthroned as the image of the goddess Artemis, or Diana, in the third century B.C. in her famous temple at Ephesus, where she had become equated with the ancient Asiatic fertility goddess.

Pliny the Elder, the Roman writer of the first century A.D., mentioned thunder stones in his encyclopedia and likened them to axes. China's Ming dynasty emperor Kang Xi had an encyclopedia compiled in 1662, and its entry on "lightning stones" indicates that the Mongols picked them up and used them as tools. Some were said to be shaped like knives, others like hatchets and mallets. The book impatiently dismisses the notion that the stones were sent on request by the celestial lord of thunder but confirms they are, in fact, rocks and pebbles vitrified when struck by lightning. It seems likely that some of the lightning stones were just what the Chinese encyclopedia said they were, but the Mongols probably also found prehistoric stone ax blades as well and attributed the buried tools to lightning.

Ax-making was big business in the stone age. Factories are known to have exported their wares over wide areas. For example, the most important ax mill in Brittany, at Sélédin, near Saint-Brieuc, quarried dolerite from a nearby hill, supplied 40 percent of the axes in Brittany, and turned out about five thousand blades per year. It was a blue chip property on the Neolithic Stock Exchange until a new technology—metal—wormed its way into the market about 2700 B.C. Prehistoric axes continue to be found in Brittany today.

The Goldi, or Nanays, of the lower basin of the Amur River in far eastern Siberia also found prehistoric stone implements from time to time. They said they were "thunder axes." The Rengma Naga people of Burma believe that thunder is the voice of Songinyu, the Spirit of the Sky. His words are good for the crops, and the polished stone axes they sometimes find are "sky axes" thrown at the ground or at trees. Aborigines in Australia's Arnhem Land are certain that the Spirit of Lightning, Namarrkon, uses stone axes for the thunderbolts he flings at Australia. The Ashanti people of Ghana in West Africa and the Yoruba of Nigeria called thunderbolts axes; during a storm, these sky hatchets were supposed to plunge to the ground.

BRIDAL CHAMBERS FOR MOTHER EARTH

In many parts of western Europe, prehistoric farmers and herders left monumental shrines of earth and stone. The exact function of these megalithic sanctuaries and tombs is not known, but many are decorated with symbols. The stone axes so frequently carved in them glorify one of the most valuable and respected items in the stone age tool kit, but they may have been intended to invoke the power and presence of the hammer- or ax-brandishing sky god who impregnated the earth in the tremors of the storm with the life-restoring fertility of the spring rain.

Spring means the rebirth of life, and it can symbolize the first birth of life, the time of creation. That theme may echo in Tirawahat's thunder and Thor's hammer, and it may be what is proclaimed in the drawings left by people who could not write their sky tales. A symbol near an ax carving on the Table des Marchands capstone in Brittany remained undeciphered until a second piece of the original menhir was recognized in the extraordinary passage-grave on the island Gavr'inis, near Ile Longe. That match revealed that two bulls once accompanied the ax now seen at the Table des Marchands. These bulls also remind us of the bellowing thunder that may be implied by the axes.

Prehistoric symbols are reticent about their meanings. But there is cohesion and consistency in the ensemble of emblems we encounter in stone age tombs: axes, perhaps bulls, and primordial mother goddesses. This may be a story about Father Sky and Mother Earth. The procreative force of the sky fertilizes the nurturing earth. She pro-

Mané-er Hroek, a prehistoric chambered tumulus near Locmariaquer, in Brittany, contained over a hundred stone axes and this decorated stone. The largest figure on it is bell-shaped and is thought to represent a mother goddess associated with rebirth. Above her may be seen carvings of three hafted stone axes. (Photograph: Robin Rector Krupp)

vides a refuge for the new life she incubates in her underground womb. She is a fierce protector of her young. She shelters and comforts them again when they return, in death, to her bosom. We can't tell if she and the sky god's axes appear on the prehistoric tombs as part of some fertility magic that borrowed power from the spirits of the dead to vivify the world. Instead, she and the storm might stand for some spiritual message connected with the resurrection of the soul. Its career is modeled on the cyclical pattern of nature and would be equally at home in the houses of the dead.

MOTHER EARTH AND FATHER SKY

Hesiod, a Greek poet of the eighth century B.C., documents the old myth of the primordial parents in his *Theogony*. In the beginning of all things Chaos alone existed, but soon after Gaia—"broad-bosomed Earth"—appeared. Her first child was Ouranos, the starry heaven. She mated with her own son, and the two held each other in a close embrace. From this cosmic marriage of earth and sky, children were born: the Titans, the Cyclopes, and more. They were monsters. The Cyclopes were one-eyed giants. Others had fifty heads and one hundred arms. Heaven's procreative power was extraordinary but dangerous.

Ouranos hated his own offspring and quarantined them within Earth. For her, this meant physical pain and heartbreak. Swollen by all she was forced to contain, Gaia conspired with her son Kronos against her husband's brutal hold. She fashioned a stone sickle, gave it to Kronos, and hid him to await the return of Ouranos with the night. Hesiod says he was "longing for love." Spreading out fully, he lay heavy across the body of Earth. Then, however, Kronos reached from his hideout and "harvested his father's genitals" with the saw-tooth blade. Father Ouranos, overthrown by his insolent, rebellious children, withdrew from Earth, and Kronos the Titan assumed the mantle of celestial sovereignty and paternity, only to be deposed in time by his son Zeus. At the shrine of Zeus at Dodona, in Epirus in northwest Greece, the oracles sang,

The Earth is our mother, the Sky is our father.
The Sky fertilizes the Earth with rain
The Earth produces grains and grass.

Many themes meet in this myth, but its foundation is the primeval partnership of earth and sky at the outset of creation. Modeled on human parenthood, they infuse the world with life. Not at all confined to the Greeks, this idea probably has deep Indo-European roots. The *Rig Veda* calls Prthivi, Mother Earth, a dappled milk cow, and Dyaus, Father Sky, is a fertile bull whose seed is the rain. Indra is sometimes said to be their son, as is Surya the sun. Sky is "all-knowing"; Earth "achieves wondrous works." Together they nourish the world.

Nearly half a world away from Hesiod's ancient Greece, New Zealand's Maori quite independently told a similar story about earth and sky. In the darkness of the world's beginning Rangi the Sky descended to Papa the Earth. They held each other so tight and so long, the children of their embrace had to find handholds on the side of Mother Earth or shelter in her armpits. These new gods longed for light and space, and their desire to escape the primeval chaos prompted them to attempt the separation of earth and sky. In succession, four of the sons of Rangi and Papa tried and failed to pry their parents apart. Tane-mahuta, the god of the forest and guardian of birds and insects, then took his turn, but he couldn't crack the divine clinch. Refusing to give up, he propped his head upon the earth, squeezed his feet against the sky, and in a mighty effort stretched to wrench the pair in two. Torn violently from each other, Rangi and Papa roared while wind and light poured into the vault between them. Dozens of concealed progeny ran free across Papa's body and admired the beauty of her figure, curved by hills and visible at last in the world's first dawn. Father Sky, now far removed from Mother Earth, still grieves for his wife's touch and sheds tears of soft rain. She heaves and sighs for him in the mists that rise from her breast.

Navaho sandpainters prepare a drawing of Earth and Sky for the Male Shooting Chant, a ritual performance intended as a cure for certain diseases by the medicine man who sings it. Both Earth and

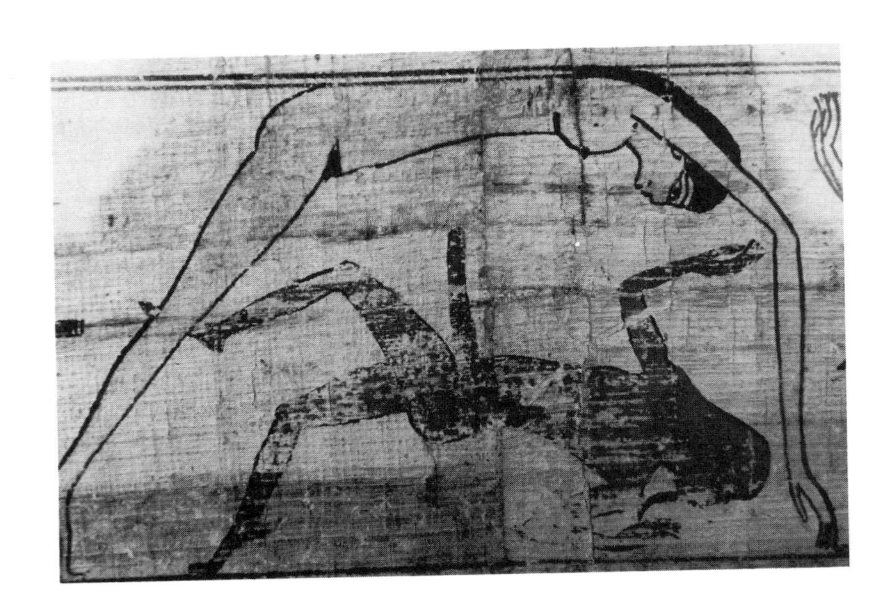

Nut arcs her body over Geb, an obviously male and motivated earth. This sexual role reversal between earth and sky in ancient Egypt is a logical consequence of Egypt's fertilizing water having been delivered by the Nile, not by the sky. (Papyrus of Tameniu, Dynasty XXI, 1069–945 B.C., in the British Museum, London)

Sky are shown as stylized human figures. Sky is black, and his body bears the faces of the sun, moon, the Milky Way, and several conspicuous constellations. Earth is blue with four sacred plants—corn, squash, bean, and tobacco—arranged around a central disk, which stands for the dark and watery Place of Emergence where the first people first entered this world. Earth is a woman, and Sky is a man. Their arms and legs cross, and a line of pollen connects their mouths. They are in intimate contact.

Yuman tribes of the American Southwest also believed the earth to be a woman and the sky a man. When the cosmic pair lay together, a drop of rain fell upon Earth, and from her was born a pair of twins who then lifted Sky to make room on the land for more inhabitants. In West Africa, the Ewe people see a marriage between Mother Earth and Father Sky in the rainy season, when the celestial water induces the growth of plants from the ground.

T'ien lung (Deaf Heaven) and Ti ya (Mute Earth) filled the roles of Mother Earth and Father Sky in ancient China, and it was said that all living things were the spawn of their conjugal bliss. The story is a minor tradition, however, and probably owes a lot to India. In China, the earth is not really a woman, but she is female (yin). The sky is predictably male. The Chinese could see what the rain was doing.

Nearly all cultures made a lady of the earth and a man of the sky. The Egyptians, however, said Nut was the sky, and her husband, Geb, the earth. Both were children of Shu (Air) and Tefnut (Moisture), and once Geb and Nut were born, they, like so many other divine newlyweds, couldn't get enough of each other. Locked in each other's arms, they had to be separated forcibly by Shu, who continued to stand between them. Nut, the "mistress of the gods," is now stretched out over Geb, who is green, like the plants that grow from his ribs. Grammarians judge that these transposed genders of Earth and Sky in ancient Egypt have something to do with the fact that the word for sky in Egyptian was a feminine noun and the word for earth was masculine. It's still fair to ask, however, why the words submitted to a sex change. They reflect the way the world worked in the Nile Valley. There the river, not the sky, brought fertility. It rarely rained in Egypt; the Nile is what quickened the land with life.

Even when the gender of the sky is female, however, the sky is linked with creation and life. Nut gives birth to the sun, the moon, and the stars. She also establishes the courses of their movement, and her hands and feet touch the horizon at the cardinal directions. This makes her arms and legs pillars of heaven and makes her more than a celestial mother. Like most sky gods, she creates and upholds cosmic order. Her arched body shields the ordered cosmos beneath her from chaos outside.

MANY A TEAR HAS TO FALL

Whether people saw their crops watered by the river or by the sky, they worried about whether the next drop would be delivered on schedule. That is not unreasonable. Availability of water is a life-and-death concern. Crops depend on it. Today, however, most of us are so far removed from the sources of our food, we are unconscious of any problems farmers may have in producing it. To us, it comes out of a can or a freezer or a fast food franchise. But the weather still has an impact on the quality and the quantity of the harvest, even though most of us don't notice it until it inconveniences us personally. In 1987, the summer in southern California was a continuation of June, which is usually cool and overcast. Although July, August, and September weren't really unpleasant, they were remarkably cloudy and cool. Southern California didn't really enjoy its customary summer. For beachgoers, it was a disappointment, but the unseasonable weather hardly seemed serious. In midsummer, however, reports began to come in about the tomato crop. The lack of sunshine had diminished the crop. The price of tomatoes was going up. That's how we sense the sky's signals these days—in the wallet.

A long drought plagued much of the United States in 1988, and it hit hard in important agricultural regions: the South, the Midwest, and California. Water in Los Angeles was in short supply from an insufficient snowpack in the mountains and had to be bought from more expensive sources. Water bills went up. Newspapers carried front-page photographs of withered corn crops in Iowa. In the thirties, drought made a dust bowl out of the Midwest and put many farmers out of business for good. Drought-induced famine, aggravated by war, lately has made a calamity out of northern Africa. In Egypt, the level of Lake Nasser dropped far below normal, and the Nile River that made Egypt the breadbasket of the Mediterranean in antiquity turned into a shadow of its former self. Memories of conditions like these once kept the cults of rain gods alive.

One of those ancient rain gods was known as Tlaloc to the Aztec and other Nahua speaking peoples of central Mexico. He toured the mountains, where the rain clouds seem to cling, and his blue face reflected the blue sky in which he operated. His black skin stood for the dark rain clouds he molded. The white clouds seen in fine weather were symbolized by his white heron-feather headdress, and the quetzal-feathered jewel above his forehead signified the life-renewing rain. As the powerful maker of clouds, storm, and rain, Tlaloc sometimes brandished an undulating serpent uncoiling to strike. It stood for the lightning he cast from the clouds. He was also shown flaunting a tomahawk, a jade thunder ax.

Tlaloc was assisted by dwarfs, known as the *tlaloques*. Armed with storm hatchets, they cracked the jars that held the waters of the sky. Rainwater, stored in mountain caves, was released when the rain dwarfs played piñata in the mountains. The sound of their blows also caused the thunder, and the pieces of pottery fell to the earth as thunderbolts.

It was sometimes said there were four Tlalocs, one for each direction. The Maya of southern Mexico and Central America had a similar divine rainmaker named Chac, who had four partners, one for each cardinal direction. Their different colors corresponded to the colors assigned to each quarter, or corner, of the world.

The four Tlalocs controlled the four types of water that could reach the earth. Only one of them —the gentle rain that soon relinquishes the sky to sunshine—was beneficial. It fertilized the earth and fathered the maize. The others put cobwebs and mildew on the grain (lingering dew and moisture), froze the crops (hail), or withered them (the crippling downpour of a tempest).

Mountain shrines were dedicated to Tlaloc, and his powers were enhanced by human sacrifice. Infants were frequent victims. We know something about this unsettling practice from Fray Bernardino de Sahagún, the Franciscan priest who collected information on pre-hispanic traditions from Aztec elders. On the feast dedicated to Tlaloc

> . . . they slew many children; they sacrificed them in many places on the mountain tops, tearing from them their hearts, in honor of the gods of water, so that these might give them water or rain.
>
> The children they slew they decked in rich fi-

nery to take them to be killed; and they carried them in litters upon their shoulders. And the litter went adorned with feathers and flowers. The priests proceeded playing musical instruments, singing, and dancing before them.

When they took the children to be slain, if they wept and shed many tears, those who carried them rejoiced, for they took it as an omen that they would have much rain that year.

Tears are water. They fall like rain, and so, to the Aztec, they induced rain. If tears were a kind of rain, rain could be a kind of weeping. Why does the sky cry? Perhaps it is the child's death that prompts it to tears. The Cashinawa, a Panoan tribe of the selva, the hot jungle lowlands of southeast Peru, say the Creator lives in the sky, rules the Lightning People, and can be heard in the thunder crying for his own lost children. His sobbing booms over the treetops, and his tears fall.

This Olmec carving of a crying child is from Cerro de las Mesas in Mexico's central Gulf Coast region. Like many Olmec images of babies, it has feline features and probably combines the rain and thunder symbolism of sacrificial tears and the jaguar's roar. (National Museum of Anthropology, Mexico City)

Sacrificial tears for rain may also be the meaning of the "crying child" images left by the Olmec, an influential culture from the Middle Preclassic period of ancient Mexico. Between 1500 B.C. and 400 B.C., they constructed monumental ceremonial centers, and their gods, myths, calendar, and ritual all had an impact on neighboring peoples, including the Maya, the Zapotec and Mixtec of Oaxaca, and the anonymous builders of Teotihuacán. Many of the teary child figures have feline attributes—a flattened nose, a snarling mouth with a flared upper lip, and sometimes fangs. More than forty years ago the Mexican artist Miguel Covarrubias built a case to show that the Aztec rain god Tlaloc evolved over centuries from the baby were-jaguar of the Olmec. Tlaloc's fanged jaws may, in fact, symbolize the mouth of a jaguar, whose roar is likened to thunder. Not everyone agrees with Covarrubias, but parts of his arguments are compelling. If he is right, the Olmec weeping child is a source of thunder and rain.

A "weeping eye" also shows up in the symbolism of the Southeastern Ceremonial Complex of the United States. Sometimes called the Southern Cult, its distinctive assemblage of emblems, ritual, and myth spread religious symbols across language boundaries and throughout the area where the prehistoric Temple Mound peoples built their fortified cities and earthen platforms.

This phase of prehistoric America is known as the Mississippian tradition, and evidence for it is found from Tallahassee, Florida, to Oklahoma's share of the Arkansas River, from the southern terminus of the Natchez Trace to the prairies of southern Wisconsin. Although Mississippian centers for trade and ceremony begin to appear as early as A.D. 800, the diffusion of the Southeastern Ceremonial Complex got started later. By the beginning of the thirteenth century, it was well established.

Engraved shells, stone sculpture, carved gorgets, ceremonial stone axes, ceramic jars, and other items carry images of suns, crosses, spiders, birdmen, hands, trophy heads, woodpeckers, arrows with two odd lobes, and more of the typical

A Tiwanakan version of the Andean thunder and sky god may be what looks down from the Gate of the Sun. He hurls thunderbolts with the spear thrower he carries, and tears fall from his eyes. (Photograph: Robin Rector Krupp)

insignia from this fashionable religion. Feathered rattlesnakes suggest influence from Mexico, where the plumed serpent was well known. In addition, many of these same kinds of ritual, or high status, Mississippian luxury goods depict a figure or face with a "forked," or weeping, eye. Sometimes the "tears" look like lightning zigzags, and the weeping-eye figure frequently carries a wand, serpent, or mace in his hand. This is another one of those lost stories. We have some idea about the main character, but without a written record, we can only speculate about the theme and plot. The warrior with the weeping eye does look familiar, however. His weapon may be the lightning, and his tears the rain.

Down by Lake Titicaca, between Bolivia and Peru, the Aymara worshipped a creator/thunder god known alternately as Tiki Viracocha or Thunupa. These forms appear to parallel the Inca versions of the Creator and Thunder. An earlier form of this divinity is probably represented upon the famous Gate of the Sun at Tiwanaku (Tiahuanaco), the southern highland ceremonial center near the southern shore of Lake Titicaca in Bolivia. And it is likely he is the same god depicted on huge ceramic jars from the south coast of Peru. His head is rayed like the sun; there are double-headed serpents in his costume and what look like jaguar or puma heads—thundercats, in any case—in his halo; he has a staff in each arm, and one of them is probably the spearthrower, or "sling," with which he launched his thunderbolts. The other weapon has a double-headed snake on one end, and is probably one of his thunderbolts. He has his weapons. He has his trophy heads. He is accompanied by three rows of winged attendants. Some have heads like his; others look like birds. All are probably the assistants Viracocha was said to have created for himself. Here, then, is one more version of Thunder, and there are tears running from his eyes.

SHAMANS IN THUNDER'S SPOTS

In their quests for supernatural power, shamans affiliate themselves with the spirits and gods in whom that power resides. Because there is evident power in the thunder and rain, the shamans of the Páez Indians of the southern highlands of Colombia say they transform themselves into thunder. To the Páez, this is the same thing as becoming a jaguar, the big spotted cat with thunder in its voice.

The Páez trace their shamanic traditions mythologically to the primordial rape of a Páez woman by a supernatural jaguar—the Thunder Jaguar. They equate this jaguar in their origin myth with the Pijao, warlike rivals who were eventually driven from the lands the Páez now hold. According to them, the Pijao were more animal than human, and the Pijao sorcerors were extremely powerful. Back at the beginning of time, after one of the Páez women was raped by the Thunder Jaguar, she gave birth to Tama, the Son of Thunder, who behaved like Thunder itself. He carried a golden wand that sounded as if it could have something to do with lightning, the sun, or both. Páez shamans now see the Son of Thunder in their visions and receive their supernatural powers from him. They are summoned by the thunder to the shamanic vocation. Shamans can thus turn into jaguars at will and use the spirit of the jaguar and the thunder in their magical endeavors.

Jaguars and thunder also play an important part in the worldview of the Desana Indians who live in eastern Colombia, near the border with Brazil. They say that Sun created the jaguar to act as his agent on the earth. The yellow in his spotted coat is a sign of the jaguar's affiliation with fire and the sun. Sun also put thunder in his roar, for thunder is the voice of Sun.

Lightning is a twinkle in Sun's eye, and when he glances toward the earth, lightning fertilizes the ground. It is a kind of solar semen and is linked to quartz. Its color is the same as that of semen and the sun. Yellow or white quartz pebbles are solidified solar semen. Lightning can scatter pieces of this petrified pollinating fluid where it strikes.

To the Desana, the sun is the essential procreative spirit of the cosmos, and the jaguar represents that same procreative power. Creator Sun is imagined as a Thunder jaguar and is sometimes called by that name. The Desana word for "jaguar," *yé'e*, comes from their word for cohabitation, *ye'éri*, and is related as well to *ye'éru*, or "penis." So the jaguar, already the most powerful predator in the jungle, is a hunter whose genitals are also a weapon, shooting life upon the world.

Responsible for defending Sun's creation on earth from danger and for maintaining the well-being of people and animals, the jaguar guarantees the reproductive capacity of living creatures by acting as an aggressive, sexual, male force in nature. The Desana cosmos is a sexualized universe. The jaguar, the thunder, the lightning, and the rain are all tangible expressions of the erotic momentum that keeps their world alive.

Because the Desana *payé*, or shaman, performs the same job as the jaguar, the shaman and the jaguar are two sides of the same coin. The shaman is said to be able to turn himself into a jaguar. In the Desana dialect, the word for jaguar is also used as the word for the shaman. What the jaguar does for nature, the *payé* does for society. He is a protector and a preserver of social norms who presides over ceremonies and negotiates with the supernatural on behalf of his people. Through magic, he provokes desirable community responses. By enforcing certain prohibitions on hunting and guiding hunters in the proper rituals connected with the taking of game, he is a kind of forest ranger, a gamekeeper responsible for the vitality and population of the wild animals. In this capacity he also communicates with the spirit-masters of the game animals and with shamanic magic strikes the deals that allow the hunters to bring home the bacon. He is a high-stakes power broker—assertive and manipulative—like Sun's jaguar.

All of these bonds between the jaguar and the Desana shaman also make a symbolic sexual aggressor out of the shaman. He wears a cylinder of quartz—said to be Sun's penis—around his neck. The Desana believe he can make lightning by throwing it. In his hands, it is a thunderbolt, with all the fertile and destructive capacities of the storm. He deploys it as needed to fortify the world's life force. Like the jaguar, the shaman is a repository of the same sexual energy that is evident in thunder, lightning, and rain.

THE WRATH OF GOD

Thunder, we have seen, may be a tough customer, but he has a soft spot for the earth. Its needs touch him, and he responds with fertilizing fluids. He sends rain to keep the world alive, but he also demands decorum. People must behave, or they will hear from him.

Several sixteenth-century chroniclers in Peru, including Pedro Sarmiento de Gamboa, Juan de Betanzos, and Joan Santa Cruz Pachacuti Yamqui, confirmed that Thunupa, the Aymara thunder god, was so burned up about the sinful behavior of the people of Cacha, he ignited the town with "fire from the sky" and let it blaze. According to Father Bernabé Cobo, a Spanish Jesuit who came to Peru in 1599, when the Inca storm god Illapa became angry, he barraged the world with thunderbolts.

Thunderbolt justice is certainly not limited to

The uplifted arm of this small bronze figure of Zeus (cast in the fifth century B.C. *and found at the sanctuary of Dodona) is about to send a thunderbolt through the atmosphere. (Photograph: Ethan Hembree Krupp; object in National Archaeological Museum, Athens)*

tribes and empires in South America. The Greeks, for example, were dragged over the coals by Zeus. If he were really disturbed, he let loose with the weather. In the *Iliad,* Homer compares the earth-shaking noise of the Trojan chariots to the drubbing the earth suffers in an autumn storm.

> *. . . when Zeus sends down the most violent waters*
> *in deep rage against mortals after they stir him to anger*
> *because in violent assembly they pass decrees that are*
> *crooked*
> *and drive righteousness from among them and care*
> *nothing for what the gods think.*

In the *Odyssey,* Homer's tale of the long-delayed homecoming of the Greek war hero Odysseus, Zeus destroys his ship and crew after they kill and eat the sacred cattle of the sun:

> *Zeus drew on a blue-black cloud, and settled it over*
> *the hollow ship, and the open sea was darkened beneath*
> *it;*
> *and she ran on, but not for a very long time, as*
> *suddenly*
> *a screaming West Wind came upon us, stormily blowing*
> *and the blast of the stormwind snapped both the*
> *forestays that were holding*

the mast. . . .
Zeus with thunder and lightning together crashed on our vessel,
and, struck by the thunderbolt of Zeus, she spun in a circle,
and was all full of brimstone. . . .

In ancient China, Lei Gong, the Duke of Thunder, and Tien Mu, the Goddess of Lightning, also assaulted the wicked. The Duke had the head of a cock (or a monkey with an eagle's beak), the wings of a bat out of hell, and clawed feet. With a chisel in one hand and a hammer in the other, he beat thunder out of the air, and it reverberated in the halo of fire and drums surrounding him. Tien Mu wore a multicolored robe of rainbows and flashed streaks of lightning with laserlike accuracy from the two mirrors she carried. Together, the Duke and the Lightning Goddess were accurate and dangerous celestial sharpshooters.

There are plenty of ways, then, for the sinful to catch hell from the lords of the air: lightning, thunderbolt, blizzard, storm, deluge, drought, disease, famine, crop destruction, loss of game, and outright death. It takes high crimes and misdemeanors, however, to inspire this kind of choler in a creator. Incest, wasting food, mistreating animals, falsehood, failure to honor oaths—anything that endangers the stability of the community or the lives of its members qualifies as a potential celestial offense.

When the gods decide to dress down the wicked, they do it in high style. In the Old Testament, God threw up the sashes of the windows of heaven and let it rain forty days and forty nights.

The god of the Hebrews was not a personification of the storm or the sky, but he acted effectively there, and at the end of his world-washing judgment, he sent another atmospheric message to earth—the rainbow. God promised Noah and all of the other living things that rode the ark that deluge would not destroy the world again, and the rainbow—the bow he set in the cloud—was the token of his covenant. Every future rainbow would remind those who saw it of this everlasting promise to the world and all its creatures.

FLYING COLORS

The rainbow, a sample of the sky's sleight-of-hand, is a colorful but delicate atmospheric phantom created by sunlight and raindrops. Our word for this appealing luminous arc in heaven tells quite a bit of its story. Shaped like a bow, it only puts in an appearance with the rain. As a result, some peoples have turned it into a weapon in the arsenal of the storm god.

The Moksha are a people who live in the Soviet Union in the Volga River valley. They had a thunder god called At'am and named the rainbow At'am's Bow. The Lapps also equipped their thunder god, Tiermes, with a rainbow, called Tiermes's Bow. A hero in the myths of the Altay Tatars fires lightning arrows with a rainbow bow, and some Ostyaks believed a thunder god bowman fired his thunderbolts with the rainbow. Prehistoric stone ax-heads are said by some to be the points of his arrows. In East Africa the Nyanja, a Bantu people in what is now Malawi, once called the rainbow the Bow of Leza. Leza, the high creator of many east African tribes, authored lightning, thunder, and rain.

Rainbows require more than lightning, thunder, and rain to put in an appearance, however. You don't get a rainbow when the sky is gray and completely overcast and the ground shadowless. Broken clouds for sunlight are also needed. But when sundered rainclouds part for the sun, the rain is also usually nearly over. For that reason, the rainbow is a sign of shower's end and a proper renewal of God's promise about the flood. It says the rain won't keep coming.

The ancient Greeks saw the rainbow as the polychrome messenger between gods and people, between heaven and earth. They called her Iris and dressed her in silk, which has its own respectable measure of iridescence. The word *iridescence* means the play of lustrous, changing colors. Iris was really the personification of the rainbow; her golden wings allowed her to appear and vanish without notice.

Intimacy with the rain and slippery appearances keep us guessing about the rainbow's intent. We can never quite catch it, and so we say "chasing

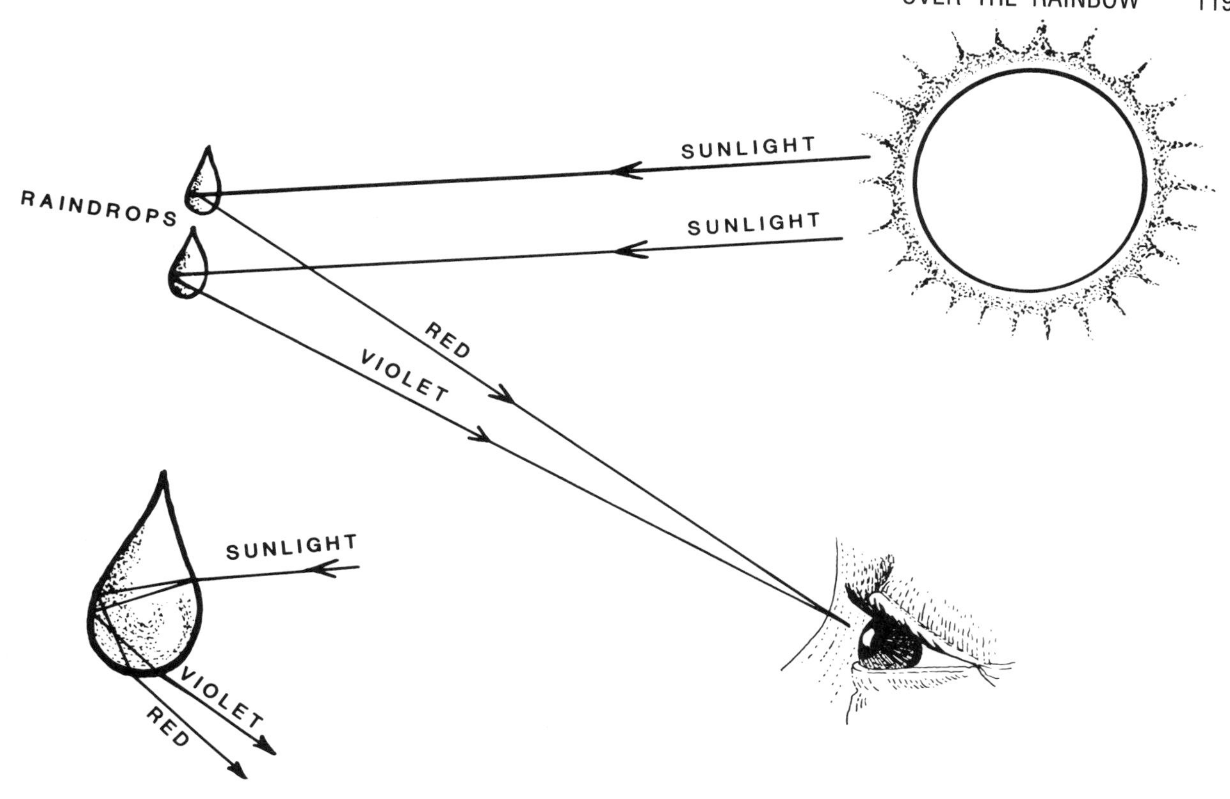

When a ray of sunlight passes into a raindrop, it bends and bounces out of the drop into a minirainbow. Each drop, however, is in position to send rays of only one color into the pupil of an eyewitness. The upper drop in this diagram is responsible for sending redder—and less bent—light into the eye. Violet light is bent more than red light, and so a lower drop directs the violet light into the same eye. It takes, then, numerous drops to produce the full range of color in a rainbow. Because each of us is receiving different light from different drops, we all see a different rainbow. (Drawing: Robin Rector Krupp)

rainbows" when we mean pursuing attractive but unreachable dreams. We generously inform others about the pot of gold at the end of the rainbow, knowing no one is likely to get there before we do. That tradition is at least as old as the mid-nineteenth century, for Jacob Grimm published it in *Teutonic Mythology:* "Where the rainbow touches the earth, there is a golden dish." He also mentions the belief that gold coins drop out of the rainbow and that a treasure is buried where it ends. The rainbow frequently played off the wealth of Uncle Scrooge, the fabulously rich relative of Donald Duck, in comic book gags based on the rainbow's ability to point out the gold. At the end of the rainbow was Uncle Scrooge's money bin and three cubic acres of cold, hard cash.

DRAWING THE BOW

Rainbows form when the sun shines against the rain. Each drop is like a glass fishbowl. It lets the light enter on one side of the drop, and it bounces off the inside of the other side and back out of the drop again. When the sunbeam enters and exits the raindrop, it is also bent a little, refracted by the water. The speed of light is slightly reduced when the light passes from air into water, and it picks up again when it emerges from water into air. That change in speed is what causes the light to bend, and different colors of light are bent different amounts because each color's speed is altered differently. Violet light is bent the most, red the least. White light, which is what we say we see from the sun, is really the mixture of all colors to which our eyes are sensitive.

Distinguishing colors is a subjective proposition: The Chinese, for example, don't really differentiate between blue and green, and Europeans had no word for orange until the fruit with that name —and that color—was imported from the Near East in the seventeenth century. Western tradition now customarily calls out seven colors of the rainbow: red, orange, yellow, green, blue, indigo, and violet. All of these colors can be seen in the rainbow, each as its own bridge of color, nested with the others. Red is on the top, and violet at the bottom.

We all see our own personal rainbow. In fact, each eye sees a different rainbow. To see any rainbow light must travel from the raindrop that is spreading out the light into all of its colors through the pupils of our eyes. The raindrops that send light into the eyes of one person can't aim it into another's. Other drops do that. All of the raindrops are making different rainbows, but each of us only sees the colored rays that manage to bounce and bend into our own.

Because the sky really does manufacture the rainbow with mirrors, the sun is always in the opposite direction. The rainbow is only part of a circle, and just how much circle we see depends on the height of the sun. When the sun is high, the rainbow is low. When the sun is closer to the horizon, the rainbow is high. On the ground, however, we can never see more than a semicircle, except when looking far below from a high altitude, on top of a mountain and looking into a valley, for

The Great Mother Creator Rainbow Serpent lives in water holes and is seen in rainbows. Her flickering tongue is the lightning. Her booming voice is thunder. She is responsible for fertility and rebirth. She guards eggs—symbols of new life—in this 1975 painting by Bardulugubu, a member of the Gunwinggu tribe of Western Arnhem Land. (Painting in the collection of The Kelton Foundation)

example. Where the earth begins, the raindrops end. But it is possible to complete the circle in an airplane; with the sun overhead and raindrops below, a full ring could form under the right circumstances.

The reason we see a circular arch is because in general raindrops have the same shape and make most of the light reflect at a 42-degree angle inside the drop. Any raindrop, then, at this angle from the sun and from the back of my head can send all the colors of the rainbow into my eyes. That means the drops are at a radius of 42 degrees from what looks like the center of the rainbow. If they are all separated by the same distance from a single point, they have to form a circle.

Gorgeous color that appears to shine with its own light and magically transparent sheen against white clouds and blue sky put the rainbow on almost everybody's list of favorite natural wonders. A colored fabric against drops of rain too subtle for us to see, the rainbow forms and fades at the whim of rain, but to our eyes it seems to do as it pleases. That fine sky spectrum is on its own turf and takes its bows with confidence.

THE RAINBOW SERPENT

The rainbow has all the prismatic beauty and sinuous line of a serpent. Peoples in Europe, Africa, and South America have called it a snake, but the best known rainbow serpent is probably the one the Australian Aborigines say helped sculpt the land into its present form during the mythical "Dreamtime" of the world's beginning. As both a creator and a destroyer, the Australian Rainbow Serpent is a transformative force in nature and human society. Known by different names in almost all parts of Australia, the Rainbow Serpent's character, even its sex, depends to some extent on geography and context. In Arnhem Land, the great snake is female and the primordial mother of all things. It is also a powerful male father spirit that keeps a damp house in the streams, billabongs, or water holes in the dry Australian wilderness, and it gets credit for the rain. Initiate shamans in northwest Australia take their first ride to the sky on its back, and in other parts of the continent, shamanic rainmaking involves the idea of being swallowed and regurgitated by the Rainbow Serpent. When a boy is initiated into manhood at puberty, his circumcision is incorporated into a ceremony in which the boy is said to be devoured by, and then vomited from, a giant snake. All of these ideas—creation, birth, initiation, fertility, and rain—are linked with the rainbow in a story the Aborigines told about the two Wawalag sisters.

The sisters had to leave their home because, by having intimate relations with men of the same tribal lineage, they'd broken the incest taboo. Traveling north after a long journey from the Roper River, at the southern boundary of Arnhem Land, the Wawalag sisters stopped for the night, not far from the coast. The older sister had a son, and she was due to have another. The younger sister had just started her first period.

Unknowingly, the Wawalag sisters pitched camp by the water hole of Julunggul, the Rainbow Serpent, and peculiar things started to occur. Each time they tried to cook something, it jumped out of the frying pan and into the serpent's mire. The women realized there was probably some holy snake close by, but so far there was no real problem. They were both uncomfortable, and it was too dark to go any farther; they decided to stay put.

Although it looked as if the night would pass without incident, the expected baby was born right then. That didn't make things any easier for the sisters, and when the sky opened up with thunder, lightning, and a downpour, all they could do was sit tight. Knowing it would be dangerous to have the odor of the bloody placenta linger, they gave the newborn a sponge bath. Then something went wrong.

Maybe some of that blood oozed into the Rainbow Serpent's cold-water flat, or maybe it was some menstrual blood from the younger sister. Whatever it was aroused the snake's ire. It surged out of its pool, erect and angry. Huddled inside their makeshift hut, the two young women tried to stop the rain with ritual singing and dancing. It worked for a while, but they couldn't keep playing the same old song all night long. When they fell asleep, the great snake erected itself again, and like the male sex apparatus probing for a place to plough, it pushed its head into the sisters' hut and coiled his body around them with a constrictor's caress. His voice was the thunder. Water flooded over the ground, and the sisters, the children, and everything else were washed down the snake's throat. It was similar to the beginning of time, when all of the things that would make a real world out of the newly created earth—wild animals, dogs, baskets, stone spearheads and other stone tools, and the ancestors of the Aboriginal people—were contained inside the Great Mother's womb.

Everything now was asleep again inside Julunggul. The situation would have stayed that way, but an ant bit the snake. That bite made it jump, and the jump sent a wave of dyspepsia through its coils. It coughed up the Wawalag sisters and then swallowed them again. Ants continued to nip at Julunggul, and with each bite the serpent, unable to hold its erection, fell back to the ground, spent. Each place where it crashed acquired an imprint that made the site sacred, and these places are where the tribal youths are now initiated.

During the initiation rites of circumcision, the boys of the tribe are "fed" to the Great Mother Kunapipi, who is the Rainbow Serpent, too, to protect the rest of the camp. At the end of the ceremonies, she regurgitates them. They are in this way reborn from the first mother who made everything and had everything inside her in the Dreamtime. Although she swallowed them as boys, they emerge from her as men, the purpose of puberty initiation rituals.

In Arnhem Land, the Rainbow Serpent is the primeval mother of all things, as well as the great father who transforms the boys into members of male society. It is the spirit of water and life's transformations, and the storm that arrives in the rainy season to replenish rivers and pools. Creation and renewal—of the land, of the waters, of the people—are all prerogatives of the rainbow down under.

THE RAINBOW BRIDGE

There is something transcendental about the rainbow. It transports spirits and delivers messages from the gods. It may have its feet on the ground, but it touches the sky. For that reason, the rainbow is a bridge to heaven. In Norse mythology, Bifröst, the Rainbow Bridge, connected this world with the realm of the gods.

Polynesians on the island of Atiu in the Cook Group tell a story about the rainbow and Ina, the goddess in the moon. A mortal she had loved and wed lived with her in the moon for many years, but eventually the man grew old. As death approached, she explained her further sadness: Death has no place on the moon. The aged, infirm man had to return to earth and finish his life there without her help and company. She tried to let him down easy and so built a rainbow bridge to convey him to the ground.

According to the Wyandot Indians of the Great Lakes region, a rainbow conveyed the world's first animals from the earth to the sky. The animals were originally perfectly comfortable on the earth, but when a fight broke out between two brothers, Ice-and-stone and Fire, things took a turn for the worse. Ice-and-stone conjured up Winter, who brought cold death to the world. Fire fought back, and Thunder helped him. Summer, armed with lightning and accompanied by Rainbow, also joined the battle and returned warmth, light, and color to the world.

The animals were grateful but uneasy. Winter, they thought, would demand a rematch. Deer was especially unwilling to deal with the bad weather again and persuaded Rainbow to carry him to the sky. When the rest of the animals saw Deer there, they made up their minds to escape earth and Winter's next visit, too. They marched over the rainbow into heaven. After they crossed, however, they burned their bridge behind them. The rainbow was burnt to ash and spread across the night sky, where we see it as the Milky Way. Many peoples also say the Milky Way, like the rainbow, is a bridge to the sky. After the first animals reached the sky, Rainbow was only a pale image of her original self, but she could still be seen sometimes, weeping in the summer rain. This Wyandot story illuminates some facts about summer, winter, thunder, and the rainbow. Thunder does escort summer into the year, and lightning and rainbows, which have been absent in winter, join them.

OVER THE RAINBOW

To find many of the dreams and folklore of our own times, look at the mass media and popular culture, including movies. One of the best-known rainbow stories of our age was told by Hollywood, the 1939 MGM version of *The Wizard of Oz*, which made America yearn for the treasures of the heart found over the rainbow. Although the film differs in detail from L. Frank Baum's wise and dazzling fairy tale, in spirit it remains true to the book and adds something the book doesn't have—a reminder of the rainbow's power over us. In 1939, the dark clouds of World War II were clotting over Europe. America needed a rainbow.

The film starred Judy Garland as Dorothy Gale, the young girl who rode the whirlwind to Oz. Dorothy Gale needed a rainbow, too. She had been having trouble with Miss Gulch and Auntie Em and with Uncle Henry. It's getting her down, and she tries to talk it out with her little dog Toto. She just wants to find a place where "there isn't any trouble." Before long she'll be going over the rainbow by cyclone and will land in Oz.

Celestial imagery and metaphor are still alive and well in the twentieth century, and our spirits still soar to the sky on the rainbow's flying colors.

8

Through the Zodiac

SEASONAL stations of the year—equinoxes, solstices, and other depots in the calendar—outline and punctuate the sun's annual cycle of interaction with the earth, but the zodiac is the celestial itinerary of the sun mapped in stars. There are twelve constellations, or patterns of stars, in the zodiac, and they form a ring that encircles the sky. We think of them as a complete set. Because there are twelve, we associate them with the year and its twelve months.

The sun appears to travel through all twelve constellations in one year. Its path through these stars is known as the ecliptic. Although the stars are invisible during the day, when the sun is up, by watching what stars appear to trail behind the setting sun or step ahead of the sunrise, it is possible to conclude that the sun is moving systematically through a special set of background stars that straddle the ecliptic. The sun seems to travel on this celestial trail because the earth is moving in orbit on its yearly journey around the sun.

The sun's annual trip through the zodiac shows up in many of the ancient stories told about the zodiac's constellations. But the stories we have for all twelve constellations are not clearly linked to each other, and that suggests that separate stories were told about these groups of stars before they were collected together as a set. By looking carefully at these stories, along with other sources of information on the early development of astronomy, we may be able to understand how some of the stars in what is now the zodiac were first singled out, what the zodiac constellations meant to the ancients who contrived them, and how the zodiac itself was eventually packaged as the even dozen we know so well today.

The ancient Mesopotamians knew as early as 687 B.C. that the sun travels through that belt of stars now known as the zodiac. They could already see the moon moving through the same stars at night and circling back to its starting point in about twenty-seven days, and they called that band "the path of the moon." They also noticed five "stars"—we now call them planets—that could also march through the same zone, but each at a slower pace than the moon. And they realized that the sun completes in a year what the moon can do in a month: one trip through the zodiac.

What we really have, then, in the sun's slow but steady shift along the ecliptic is the earth's orbital flight through the seasons. Because the earth spins at a tilt to its own annual path around the sun, the ecliptic cuts across the line of movement we see in the daily rotation of the sky. Where the sun is on the ecliptic among the stars of the zodiac determines how high the sun flies each day and where it rises and sets. At the winter solstice, when the sun rises and sets in the south and flies low across

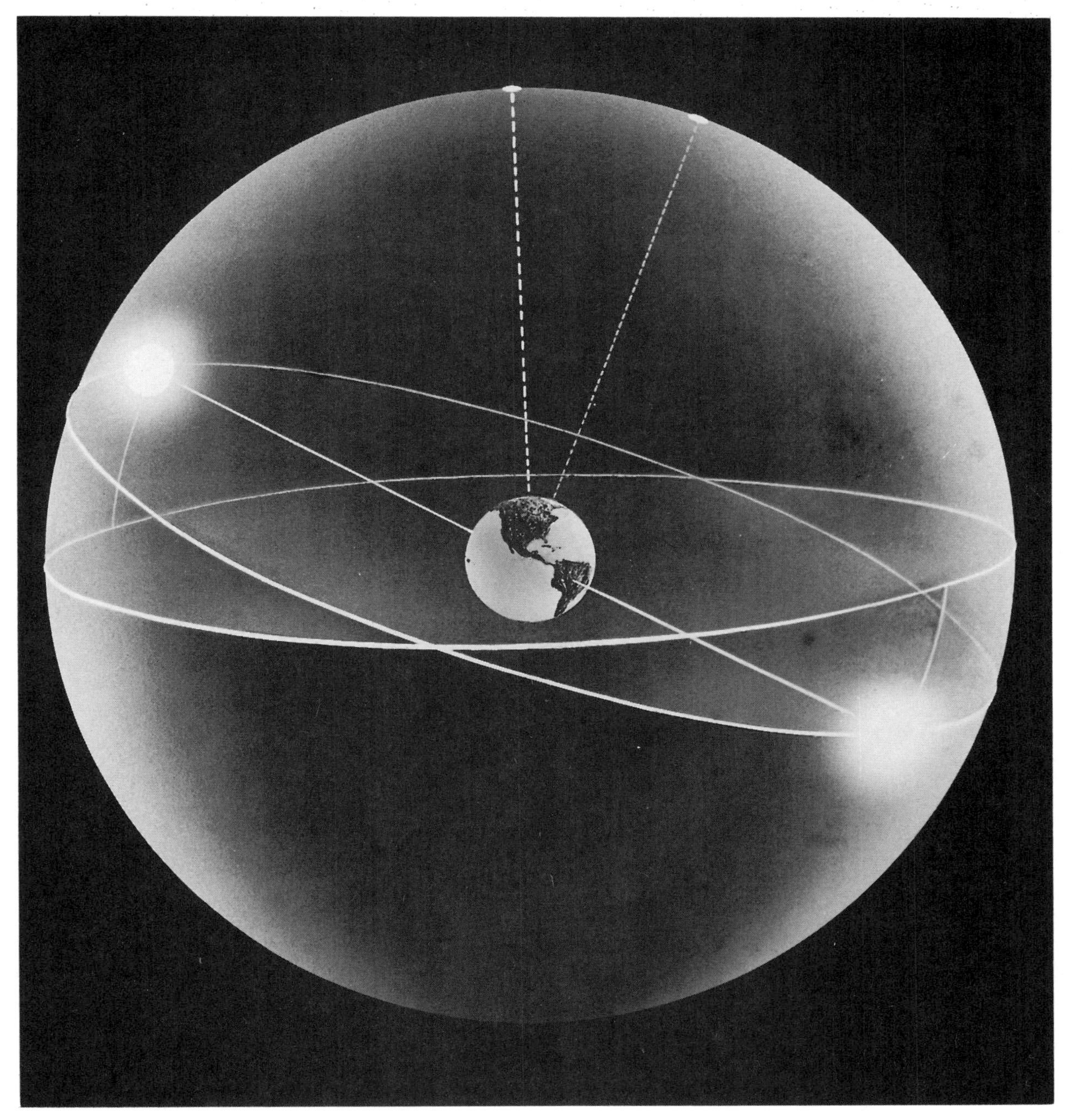

In this view of the celestial sphere, the earth is oriented with its polar axis upright. The terrestrial north pole is on top, and a dotted line continues from there to the north celestial pole, at the top of the celestial sphere. This dotted line is a continuation of the earth's axis of rotation, and it is perpendicular to the horizontal circle, which represents the celestial equator. The celestial equator is the sky's counterpart to the earth's own equator, and these two circles occupy the same plane. A second circle, tilted 23½ degrees to the first, represents the ecliptic, or path of the sun. Two glowing disks on it indicate the placement of the sun at the two solstices. It is farthest above (and north of) the celestial equator at the summer solstice in June and farthest below (and south of) the celestial equator at the winter solstice in December. The ecliptic runs through all of the constellations of the zodiac, and in ancient times those stars helped to establish the location of the ecliptic. The second dotted line is perpendicular to the ecliptic and intersects the celestial sphere at the north ecliptic pole. (Drawing: John Lubs, Griffith Observatory)

the daytime sky, it has just arrived at what is called the *sign* of Capricorn. Six months later, the lofty noon reach of the sun and its appearance and disappearance at its northern limits tell us it is the summer solstice. The sun has also just reached the sign of Cancer. Vernal equinox, in March, coincides with the sun pulling into the sign of Aries, and the sign of Libra balances day against night at the autumnal equinox the following September. As we shall see in a moment, however, these "signs" are not quite the same thing as the stars of the zodiac, but the seasons still come and go in the company of the zodiac signs and constellations occupied by the sun at the same time.

SIGNS ON THE HIGHWAY

Long after the constellations of the zodiac were recognized, ancient astronomers divided the ecliptic into twelve signs of the zodiac. These signs differ from the zodiac constellations, even though they carry nearly the same names. A constellation is like a picture in the sky, a shape defined by the stars that comprise it. Some constellations are large, and others are small. Some have several bright stars, and some have none. Signs, on the other hand, are uniform zones along the ecliptic. Each sign is 30 degrees long, that is 1⁄12 of 360

Proclus, a Byzantine Greek, wrote a treatise on the celestial sphere in the fifth century A.D., and it was still being used as a textbook in the sixteenth century. This title page from a French edition published in 1547 includes an armillary sphere. Its rings mimic the circles of the celestial sphere, with which classical scholars mapped the sky. The north celestial pole is at the top of the sphere, and the south celestial pole is at the bottom. A dark globe at the center stands for the earth, and a horizontal ring around the armillary's middle represents the celestial equator. The conspicuous band adorned with figures and animals and set at an angle to the celestial equator is the ecliptic. Here it is divided into twelve equal compartments. These twelve compartments are not the uneven constellations of the zodiac, but the tidy, uniform signs of the zodiac. (manuscript 95535, reproduced by permission of The Huntington Library, San Marino, California)

degrees, or 1/12 of a complete circle. By convention, the signs are also said to be 18 degrees wide.

When the zodiac signs were first invented, they probably roughly coincided with the constellations that have the same names. Aries the Ram, Taurus the Bull, Gemini the Twins, Cancer the Crab, Leo the Lion, Virgo the Maiden, Libra the Scales, Scorpius the Scorpion, Sagittarius the Archer, Capricornus the Sea Goat, Aquarius the Water Carrier, and Pisces the Fishes. It wasn't a perfect match, however. Two other traditional constellations also held territory in the zodiac—Cetus the Sea Monster and Ophiuchus the Serpent Bearer. They still do, and there are still more differences between the signs and constellations. In two cases, the name of the constellation was truncated when given to a sign. *Scorpius* the constellation became *Scorpio* the sign. *Capricornus* the constellation lost a syllable as the sign of *Capricorn*. In this era, too, discretion has altered the zodiac a little. Because we are uncomfortably familiar with cancer, an often painful and fatal disease, authors of popular horoscopes habitually abandon the sign of Cancer and substitute "Moon Children," but they really mean the Crab.

SIGNS OF THE TIMES

Most people encounter the twelve signs of the zodiac in the daily newspaper horoscope, a skirmish with astrology that usually provides reasonable advice—no matter which sign you read. These twelve signs are allocated according to birthdate, and they refer to the zone in the sky occupied by the sun during the span of birthdates assigned to each sign. For example, the newspaper today advises me and other Scorpios (born October 23–November 21), "Don't get into an argument with a business partner. If you do, serious trouble will result. Your judgment is not at its best right now." Well, I suppose it isn't, but the message for those born under the sign of Taurus (April 20–May 20)—half a year from Scorpio—would be equally helpful if it had been sent to the Scorpios instead. In fact, it's not much different: "Avoid people who like to argue, as they could easily upset you today. Pay special attention to the state of your health and your appearance." Sound advice for any sign in any season.

This celestial self-help column is the tired and pale heir of an ancient belief in the power of the sky to reveal what the cosmos might have in store for us. Based on the principle that what goes on overhead magically frames and guides what occurs on earth, astrology maps the powers in heaven onto that ring of twelve equal signs that circle the sky along the ecliptic.

BULLETINS POSTED ON THE ZODIAC

The whole idea of twelve equal signs is a fairly late invention. They still didn't exist early in the seventh century B.C. The *mul-Apin* texts, compendia of astronomical data from that era, continued, then, to divide the zodiac into four sections that corresponded to the four seasons and were centered on the sun's solstice and equinox positions. Twelve of the eighteen constellations and single stars listed in the *mul-Apin* are on the ecliptic, or "moon's path," and later these twelve provided the names for the twelve signs. They seem to have been contrived in the Neo-Babylonian period, 611–540 B.C., when Chaldaean kings like Nebuchadnezzar II read omens in the zodiac.

Even before the zodiac was separated into signs, its stars and constellations told the Babylonians and Assyrians stories about the destiny of the king and the fate of the land. They called what they read overhead "heavenly writing." A healthy sample of these celestial omens was published in 1900 in R. Campbell Thompson's *The Reports of the Magicians and Astrologers of Nineveh and Babylon*. Thompson, an Assistant in the Department of Egyptian and Assyrian Antiquities at the British Museum, had collected, transcribed, and translated 277 inscriptions. Many of the tablets were written in the court of Essarhaddon, king of Assyria from 681 to 669 B.C., and all but three are astrological. A few examples confirm that the zodiac didn't just map the action in heaven. To the Mesopotamians it meant something.

> When Venus appears in Virgo, rains in heaven, floods on (earth), the crops of Aharrû will prosper; fallen ruins will be inhabited.

> When a star stands at the left rear of the Moon, the king of Akkad will work mightily. When Virgo *(Dilgan)* stands at its left horn, in that year the vegetables of Akkad will prosper. When Virgo *(Dilgan)* stands above it, in that year the crops of the land will prosper. When a star stands at the left horn of the Moon, a hostile land will see evil.

It is clear from the texts that real stars and constellations—not signs—were intended.

SLIDING SIGNS

Differing boundaries is only one reason why the constellations and signs of the zodiac do not coincide. Other changes have also occurred in the zodiac over the last 2600 years. The signs have slided away from the constellations that gave them their names; now when the sun is in the sign of Aries, at the time of the vernal equinox, it is nowhere near the stars of Aries the Ram. It actually occupies the stars of Pisces, and that is why the astrologically-oriented call this the Age of Pisces. When the vernal equinox was back in the stars of Aries, it was the Age of the Ram. Eventually the vernal equinox will slide into the constellation of Aquarius, and we'll be awash in the much-heralded Age of Aquarius.

Those who are discontented with our times have already announced the imminent arrival of the Water Carrier's new age. It can't come soon

Over the centuries, the position of the vernal equinox shifts through the stars of the zodiac. At the present time, the vernal equinox is located in the constellation of Pisces the Fishes and no longer among the stars of Aries the Ram. Astrologers, however, still say that the vernal equinox sun is in the sign of Aries. Signs of the zodiac are uniform in size and do not correspond in extent or position with the constellations whose names they bear. The sign of Aries the Ram is actually among the stars of Pisces the Fishes. (Drawing: Robin Rector Krupp)

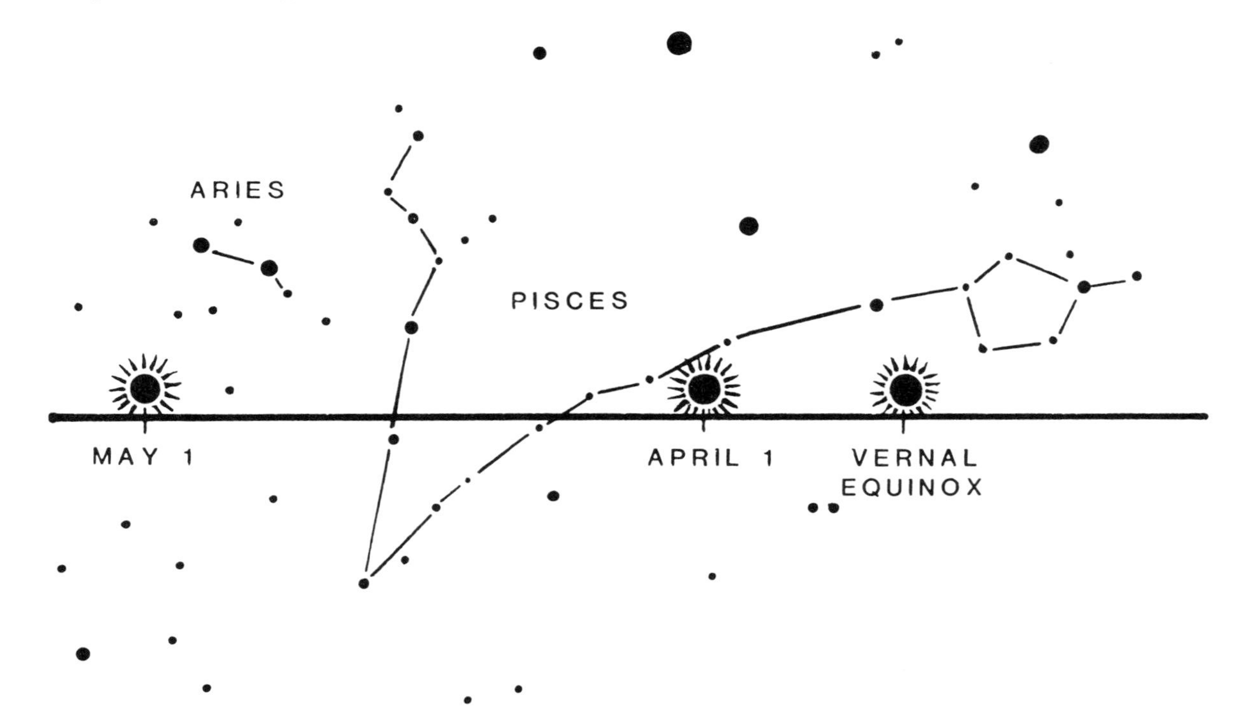

The seasons slip with respect to the stars because the earth is precessing. Precession is the cyclical swiveling of the earth's axis of rotation. This motion is induced by the gravitational influence of the moon and the sun on the earth's equatorial bulge. The axis of the precessing earth swings around the circle shown here once every 26,000 years. The dotted line indicates the present direction of the earth's axis, which at present points to Polaris. In the future, other stars—and sometimes no star—will get the nod. (Drawing: John Lubs, Griffith Observatory)

enough if you're tired of the Fish and figure something better is on the way. Serious investors in the Aquarian Age see a tidal wave of reform in its arrival. A more spiritual, loving, innovative, and progressive era is promised. In *The Six O'Clock Bus, a Guide to Armageddon and the New Age,* English writer Moira Timms contrasts the appealing prospects of the coming Age of Aquarius with what we must endure now in the Age of Pisces, a time of "self-undoing." We have "been trapped in orbit, as it were, around materialism, competitiveness and limitation." The uncertainties and aggravations of our times are the "birth pangs" of a "mighty labor" to birth the new age.

Believers provide various estimates for the onset of Aquarian influence, but guided by the lyrics in a well-known song from the 1960s musical hit *Hair,* we can tell that this "is the dawning of the Age of Aquarius." Despite claims to the contrary, however, it's going to be a long morning twilight for us. The boundary between the constellations Pisces and Aquarius adopted by the International Astronomical Union in 1928 delays the New Age border crossing until A.D. 2614 and gives us plenty of time to get ready. Aquarius has to carry that water six more centuries before he baptizes us into a new era.

What prompts the vernal equinox to plod past the stars? What motivates zodiac signs to migrate through the constellations? Gravity is responsible for allowing the signs to slip along the ecliptic, and gravity in this case acts on the equatorial bulge of the earth to do it.

Although to the uncritical eye the earth looks attractively spherical, it really has midriff bulge. A bit flattened at the poles, the earth is slightly swollen at the equator, and the moon and the sun exert a little extra tug on the extra weight the earth is carrying on its waist. These gravitational forces of the sun and moon try to draw the equatorial bulge toward them. If successful, they would upright the tilted axis of the earth. Because the earth is spinning, however, these forces make the axis swivel, or precess. Something similar happens to a spinning top. Gravity pulls the top toward the floor, but because it is spinning, the top responds by twisting around as it continues to turn.

What the top can do in a matter of seconds takes the earth 26,000 years. As it precesses, the earth's axis slowly shifts direction. Its angle of tilt does not change, but the place where it points gradually moves. We still have seasons, but the stars that mark them play musical chairs in time with a cycle so long it makes recorded history little more than a long afternoon in the 26,000-year "day." If you wait long enough, different stars will appear in different seasons. Stars we now see in summer will light the winter nights in 13,000 years, and our winter constellations will become emblems of summer. In the last 2,600 years or so, since the zodiac signs were established, the stars have drifted forward, a little more than a sign out of season. The zodiac still announces the seasons, but the stars have moved with respect to them. Two and a half millennia ago the sun drove into the stars of Aries in the month of March. Now it doesn't get there until April.

It's a little like wearing a loose belt. When you put the belt on, the buckle is right in front where it should be. Some physical activity may allow it to drift, and then you notice the buckle is off to the side. It is unlikely, though, the buckle will commute through all of the loops and back to the front buttons again. You'll catch it long before that and put it back where it belongs. Nobody, however, stops the zodiac signs from roaming.

A TWELVE-RING CIRCUS, STARTING WITH THE RAM

Despite precession, Aries the Ram has, at least since the time of the ancient Greeks, acted as the leader of the zodiac. Even though its stars do not look much like a ram, they are linked with the ram of the Golden Fleece featured in the story of Jason and the Argonauts. This connection is mentioned by the Pseudo-Eratosthenes, an unknown Greek writer who authored a digested version of a lost work by Eratosthenes of Cyrene, a Greek astronomer, mathematician, and geographer who lived from 276 to 194 B.C. Eratosthenes estimated the size of the earth by observing the shadows cast by the sun at noon on the summer solstice in Alexandria and in Syene, which is now Aswan, in Egypt. His book apparently included names and descriptions of the constellations visible to the Greeks along with accounts of the myths associated with them. A number of these myths were retold two hundred to three hundred years later by the Pseudo-Eratosthenes in the *Katasterismoi*. In the Greek literature passed along to us from antiquity, this summary is the only surviving collection of constellation myths.

It is hard to say just how old these stories are. Eudoxus of Cnidus (408–355 B.C.), the Greek astronomer and mathematician who invented a scheme of concentric spheres to explain the observed motions of the sun, moon, and planets, apparently wrote about the constellations and their myths in fourth century B.C. His book, the *Phaenomena*, is lost, but it was recast in verse by the Greek poet Aratus of Soli, who was born about 315 B.C. Even earlier, in the eighth century B.C., the Greek poet Homer mentioned several familiar constellations in the *Iliad* and the *Odyssey*, and Hesiod recorded them during the same era in his poems. Only fragments of old Greek star lore survive from this time, but it is reasonable to think many of the stories we find in the Alexandrian period, 323–31 B.C., were told at least five or six centuries earlier.

The ram's story begins with a case of child abuse. Phrixus and Helle, the son and daughter of Athamas, the king of Boeotia, were treated cruelly by their stepmother Ino. She also advised her husband to sacrifice the children to Zeus. They escaped, however, on the back of a ram with gold fleece. Sent either by Hermes or by Zeus himself, the ram flew the children across the Aegean toward the Black Sea. Helle lost her grip at the Dardanelles, the strait that separates Europe from Asia and the Aegean from the small Sea of Marmara. Drowned at the crossroads of the continents, Helle lent her name to the Hellespont, or "Helle's bridge," the old Latin name for the Dardanelles. Phrixus and the golden ram carried on to Colchis, on the eastern shore of the Black Sea, where King Aeetes was pleased to meet Phrixus and make him

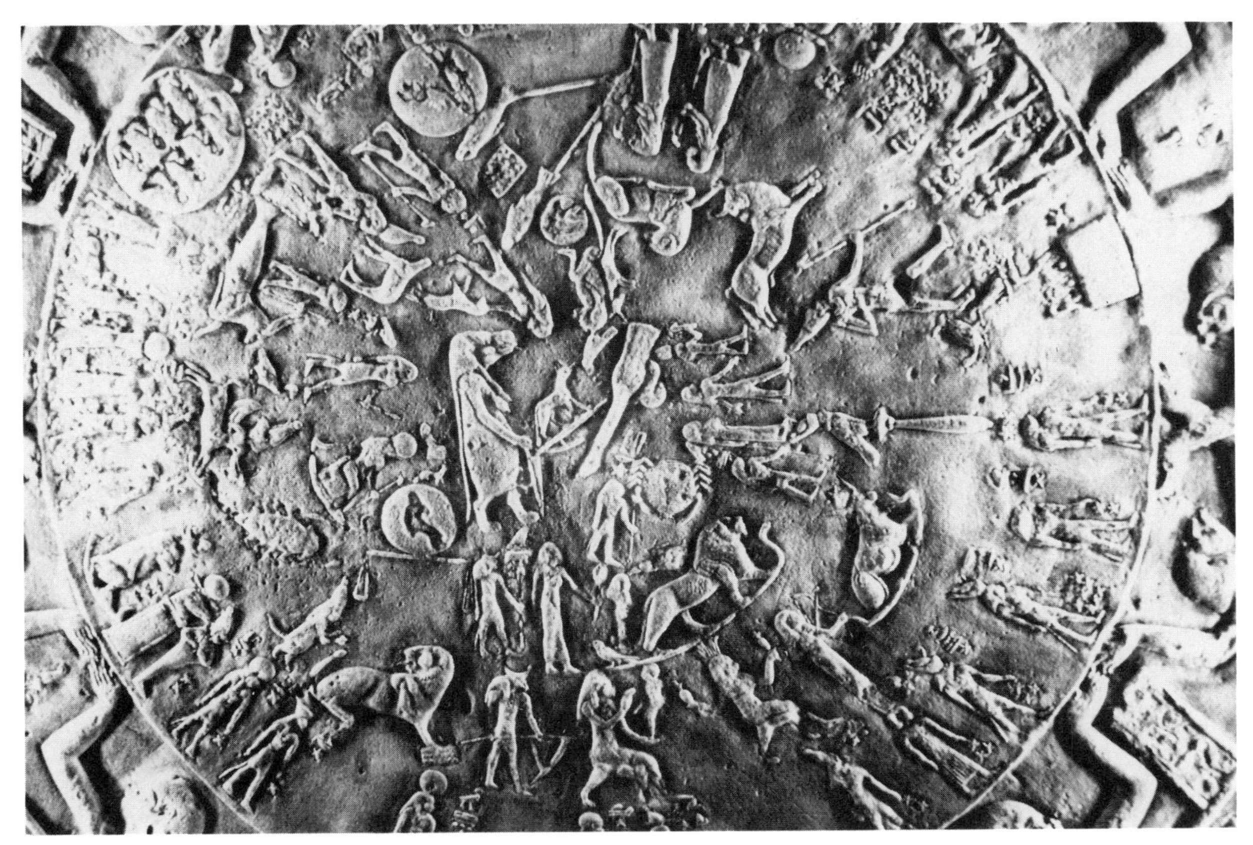

A ceiling in a chapel on the roof of the Temple of Hathor at Dendera, in Egypt, was decorated with this relief in the first century B.C. *Now, however, it is in the Louvre, in Paris. It is a reasonable representation of the constellations as the Egyptians recognized them in the Late Ptolemaic period, and it is called the circular zodiac of Dendera because the constellations of the zodiac are easily spotted on it. Aries the Ram may be seen near the top of this photograph (and upside-down). Taurus the Bull is to the right. Gemini the Twins is shown as a man and woman holding hands. Although the zodiac was imported into Egypt, the Egyptians modified some of its constellations to suit their own taste. A scarab beetle replaces Cancer the Crab. Leo the Lion can be seen below the scarab, and Virgo the Maiden stands behind the Lion with a spike of grain in her hand. To the left of her, we find Libra the Scales. Scorpius the Scorpion is still farther to the left. A centaur archer, sporting an Egyptian crown, represents Sagittarius. Capricornus the Sea Goat, Aquarius the Water Bearer, and Pisces the Fishes complete the circle.*

his son-in-law. In gratitude, Phrixus sacrificed the ram to Zeus and turned the wool, worth its weight in gold, over to the king. Aeetes nailed it to the branch of an oak tree in a grove dedicated to the god Ares and guarded by a dragon. This fleece returns to star in the story of Jason and the Argonauts as the object of the quest, but it already has ample celestial imagery. Retelling an anecdote recorded by the Pseudo-Eratosthenes, Hyginus, the Latin mythographer, mentions the final fate of the ram in his *Poetica Astronomica,* written at about the beginning of the second century A.D. Hyginus says the ram obligingly removed its own fleece, left it with Phrixus, and then made its way to its appointed place in the heavens under its own power.

There's little reason to doubt that Aries the ram is the ram of the Golden Fleece, and in the period these stories were being told, the vernal equinox sun resided in the sign of the Ram. Its ability to fly and its golden coat both imply a connection with the sun. A gold pelt on the branch of an oak in a sacred grove also sounds a lot like the Golden Bough. Although Aries spotlights the equinox and not the solstice, the Golden Bough's immortality may have been appropriated into a symbol of spring renewal here.

SEX ON THE FIRST DATE

Instead of picturing the stars of Aries as a ram, the Assyrians called the first constellation of the zodiac *Lu-Hun-Ga*, the Hired Laborer or Day Laborer. This may seem to be an odd figure to place in the sky, but "hired laborers" were seasonally and economically significant enough to command a commission in the Mesopotamian zodiac. They manually germinated the female palms in a date grove with pollen from the one male tree in the center. The date was an extremely important crop, one of the earliest plants domesticated in southern Mesopotamia and the oldest known crop cultivated from a tree anywhere. At some time the date palms, or their ancestors, must have been pollinated naturally by the wind, but not in human memory. The Sumerians, we know, artifically fertilized the female plants to maximize the yield.

In the *mul-Apin*, the Hired Laborer is said to be the god Dumuzi. Dumuzi was also known as "the one great source of the date clusters," and his bride Inanna, sometimes known as "the lady of the date clusters," embodied the spirit of the storehouse, which she opened for the date harvest in the fall. Her womb was the storehouse, and Dumuzi made sure it was well stocked. That's where the Hired Laborer came into the picture. The constellation of the Hired Laborer originated in the use of its stars to signal the time for pollinating the palms. The female date-palm blossom normally opens in March or April, depending on how cold and wet the winter has been. It must be dusted manually with pollen from the male plant within three or four days of opening. If pollination is delayed longer than that, the sap dries, seeds fail to form, and the fruit does not mature properly. Ancient Mesopotamian date cultivators associated the pollinating season with the heliacal, or first visible predawn, rising of the stars of the Hired Laborer.

Date palms still must be pollinated by hand. A palmero *at the Shields Date Gardens in Indio, California, is dusting the blossoms on a female date palm with pollen from a male tree. The flowers open in the early spring, and in ancient Mesopotamia, this point in the calendar was heralded by the heliacal rising of the stars of Aries the Ram. In Mesopotamia, those stars were known as the Hired Laborer—the date pollinator. This photograph of a date pollinator was taken on 23 March 1988. Precession has shifted the date of the heliacal rising of Aries, but the flowers of the date palm still open in March or April.*

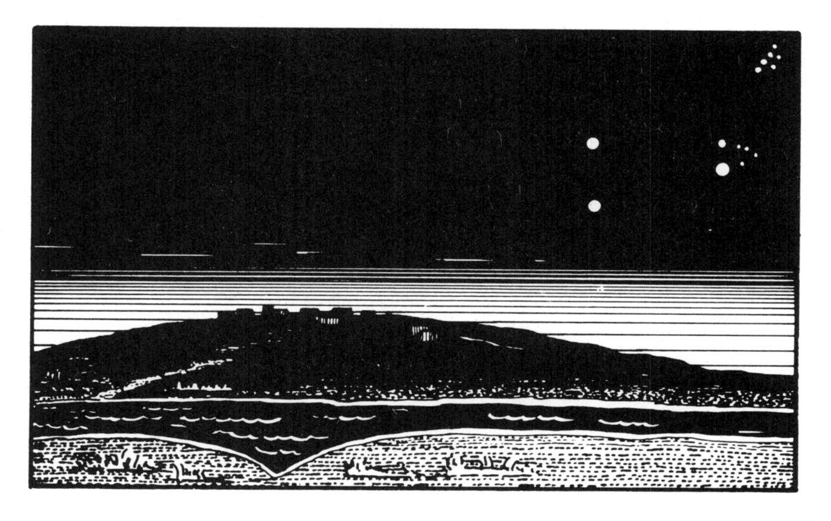

In the weeks before a star rises heliacally, it is lost in the glare of the sun. The sun rises before the star does, and the sky is too bright with sunlight by the time the star is above the horizon to permit the star to be seen. The top panel in this sequence shows that situation for the stars of Taurus the Bull, as seen from somewhere in the Middle East, five or six thousand years ago. After a few weeks have gone by, the orbital movement of the earth has shifted the sun's position with respect to the stars. The sun rises a little later than it did with respect to the stars of Taurus. Taurus still cannot be seen, because the rising sun saturates the sky with enough light to hide the stars (middle panel). After another couple of weeks, the earth moves enough in its orbit to delay the sunrise even more with respect to the stars. A morning will come when the sun has shifted far enough from Taurus to let Taurus be seen for the first time in the predawn sky. This momentary appearance is called an heliacal rising, and that is what we see in the bottom scene here. (Drawing: Robin Rector Krupp)

A few weeks before its heliacal rising, a star can't be seen at all. It is then near the sun, perhaps even east of the sun, and it rises when the sun does or afterward. In either case, it is daytime by the time the star gets into the sky, and the sunlit sky is too bright for the star to make itself known. As the days go by, however, the earth continues to travel in orbit around the sun. This makes the rising and setting times of stars shift. If you could see the stars and the sun at the same time, the sun would appear to move eastward through them just about twice its own diameter each day. Those that had been in its vicinity would be left behind, and, west of the sun, they would rise before the sun does. But even after the sun abandons the company of the calendar star, we won't automatically see it in the dawn. It may rise before the sun, but if it doesn't climb over the horizon early enough, it will still be lost in the glare. Bright stars have to be 12 to 15 degrees west of the sun to be seen rising heliacally.

Eventually the earth journeys far enough in its annual circuit to put some real distance between the star and the sun. Ascending from the horizon, the star is finally seen for a brief moment before the morning twilight overcomes it. That is its heliacal rising, and it pinpoints a day in the calendar. On subsequent mornings, the star rises earlier and earlier and remains visible for a longer time. Its heliacal rising has passed. Eventually the sun circles all the way around the ecliptic and intrudes once more into our star's territory. It's gone again from the night sky, a daylight traveler until its next heliacal rising calls out the date once more.

THE BULL

In Taurus, the next constellation of the zodiac, it is possible to see the face and horns of a bull. They comprise a group of stars known to the Greeks as the Hyades. Aldebaran, the name of its brightest star, may mean "the follower" and may refer to the fact that it and the other members of the Hyades follow the celebrated Pleiades star cluster across the sky.

The Pleiades float on the Bull's shoulder in most representations of Taurus. Pleiades sky lore is found throughout the world, and it is so rich that it receives a chapter of its own, later in this book.

Aldebaran marks the Bull's eye, and the rest of the Hyades has a recognizable V-shape. The point of the *V* represents the bull's nose, and the two "legs" of the *V* extend into horns. On celestial maps, usually only the front half of a bull stands in for Taurus, perhaps a faint echo of the ancients having seen only the head of a bull in these stars.

In Greek tradition, the Hyades were daughters of Atlas, the giant who held up the sky. A springtime disappearance of the Hyades in the west at sunset and their autumn setting at sunrise brought showers in the first season and storms in the second. For that reason they were known as the rainy Hyades. Hyginus linked the rain they brought with their tears of grief for their dead brother Hyas. While hunting, he was killed by a lion. The tears of the Hyades, then, are the spring rains and "sad companions of the turning year," as Manilius styles them in his first-century poem.

The Pseudo-Eratosthenes claimed the celestial bull was really the image of that bull that took a fancy to a ripe Phoenician maiden on the coast of Syria, charmed her into going for a joyride, and carried her off to Crete, where the bull played an important role in myth and ritual. The bull was really Zeus in disguise. The girl was Europa, and she gave her name to the European continent and to European civilization. The story really says that civilization from the Near East was brought to Mediterranean Europe through Crete, which is close to the truth.

The Sumerians also had another name for Taurus. They called it *mul-Is Li-E,* "the bull's jaw," an image suggested by the V-shape of the Hyades. Apparently South American Indians of the Guianas and Brazil thought Taurus looked like a jawbone, too, for they called it the Tapir's Jaw.

THE HEAVENLY TWINS

Gemini the Twins pairs two bright stars as neighbors, and their appearance might inspire calling them twins. The rest of the stars in the constellation drop in two lines from primary stars and provide bodies that let the two bright stars mark the heads of the celestial brothers. The Sumerians said these stars were the Great Twins, and the Greeks, inheriting this notion, identified them with the

The scene on this Ionian-style water jar from sixth century B.C. *Italy shows Heracles in the process of wringing one of the hydra's nine necks. To the right, the crab nips at the heels of Heracles. (The J. Paul Getty Museum, the Eagle Painter, Caeretan Hydria, ca. 525* B.C.*, terracotta, height: 44.6 cm, diameter* [*mouth*]*: 22.9 cm, diameter* [*body*]*: 33.4 cm.)*

Dioscuri (sons of Zeus)—Castor and Pollux—the twin sons of Zeus and Leda. Leda was a married woman when Zeus descended on her in the form of a swan. The boys born of this union hatched from an egg. Both were heroes. Castor liked horses. Pollux liked to box. Pollux was immortal. Castor was not. Both joined Jason on the quest for the Golden Fleece. When Castor was killed in a fight, Pollux begged Zeus to bring him back to life. Zeus agreed to allow Pollux to share his immortality with Castor, but he insisted they alternate shifts; while one walked alive, the other remained in the underworld. Their fraternal loyalty was apparently commemorated by the two stars in Gemini.

In the sky, they are able to enjoy each other's company. There is no more conspicuous and close pair anywhere in the northern heavens. In Greece, their heliacal rising apparently stood for the calm seas of summer and made them a favorable talisman of sailors. Their importance in antiquity stimulated people to ratify oaths by invoking their name. "By Gemini" has evolved into "by jiminy," and that phrase has lent itself as the name of a very famous cricket who helps Pinocchio wish upon a star and is drafted as his conscience. Thus an ancient oath—a pledge of personal honor—survives as a personification of conscience in a Walt Disney cartoon.

AND NEXT THE CRAB

It is difficult to explain the meaning of the Crab. Cancer's stars are faint, and the constellation is relatively small. A rather subtle cluster known as the Bee Hive, or Praesepe (the Manger,) enclosed by four of the Crab's stars, is perhaps its most interesting feature. The Manger informed ancient weathermen about what rain and wind to expect. Aratus, for example, warns us that a darkening of the Manger heralds rain. Moisture in the air could diminish the already dim light from the Bee Hive and serve as a weather sign. The *Katasterismoi*'s account of Cancer claims it is the crab that put the bite on the heel of Heracles (Latinized as Hercules) in the midst of his second Labor when he was slashing and burning the heads of the Hydra. We don't know the meaning of the Crab's name in the *mul-Apin* tablets, but the German Assyriologist Peter Jensen concluded that the Babylonians sometimes identified Cancer as a tortoise. In Ptolemaic Egypt, it took the form of a scarab beetle.

THE LION SHINES

Leo makes a fairly convincing lion in the sky. The "sickle," or backward question mark, that forms his head, mane, neck, and breast, includes the bright star Regulus. Its name means "the Little King," and it locates the lion's heart. An oblique triangle of stars with a bright star at its far corner forms the lion's hindquarters and completes its body. That bright star at the lion's end is Denebola, "the Lion's Tail." Leo, like Taurus, had the same identity in Babylon it enjoyed in Greece. Greek constellation lore represented it as the Nemean lion Heracles strangled in fulfillment of his first Labor.

THE VIRGIN

There is little hope of finding a virgin in the stars of Virgo, but her main attraction, the bright star Spica, offers some hope for explaining the identity of the constellation. The Latin and Greek names mean the same thing: "ear of grain." The Sumerian name of the star was *Ab-Sin*, and the *mul-Apin* says it was the ear of grain of the goddess Shala.

Shala just means "woman." It was a title that might be applied to any of several goddesses, but the consort of the Old Babylonian storm master, Adad, always had this name. In one inscription she is also called "the lady of the field," which sounds as if it might have something to do with grain. The Sumerian name *Ab-Sin* means "furrow," and that appropriate emblem of grain cultivation also tells us where the business about the virgin originated. A furrow is the field's vulva. Before it is seeded, the furrow is a virgin; after seeding, it bears new life. The name suggests the star was the celestial signal for ploughing or planting in the Sumerian era.

The Greeks were a little more specific about what virgin waved the grain. Virgo was said by Hesiod and the Pseudo-Eratosthenes to represent the goddess Dike. She was one of the Horae, three immortals who were in charge of the seasons and the order of nature. Each one stood for a particular season. Dike's was summer. Her name means "justice." Along with her two seasonal sisters—Eunomia (discipline) and Eirene (peace)—she upheld the order of society. The Athenians called her Auxo, which means "growth." She was responsible for the growth, rather than the budding or the ripening, of plants. Those obligations belonged to her sisters, whose Athenian stage names were Thallo (Budding) and Carpo (Ripening). When the Greeks drew pictures of Dike, they put a spike of grain in her hand to make her identity clear.

THE SCALES

Zodiac means "ring of animals" in Greek, and all but one is an animal or a person: Libra the Scales. Scales symbolize justice, and Dike embodied it. Poetic justice, then, puts the Scales next to Dike, our virgin "Justice," on the ecliptic. This doesn't explain why the Scales are here, however, for it was the Babylonians, and not the Greeks, who located a set of scales in this part of the sky. The Greeks adhered to their idea of a "ring of animals" and called these stars the Claws, even after they adopted Libra as a sign of the zodiac. Libra's stars are not particularly bright and don't look much like scales or claws, but scales did make sense in first millennium B.C., when the autumnal equinox balanced the duration of daylight with the length of the night.

To see why the Greeks called these stars the Claws, you have to continue along the ecliptic. If you follow your way east through the Claws, you find they belong to the next constellation.

THE SCORPION

The starry segments of this arthropod form a hook behind a row of three stars that resembles a scorpion's head and the joints where its claws meet its body. The hooked curve ends in a two-star stinger and completes a convincing scorpion image. Greeks and Mesopotamians saw it there. The hook also looks like a hook, of course, and for that reason many Polynesians call this constellation Maui's fish hook. With it, he fished many islands (including Hawaii and New Zealand) to the surface. In the New Zealand version of the myth, he used the jawbone of his ancestress as the fishhook. When he snared the sun to slow it down, he also beat the sun with this jawbone. It all implies that the constellation Scorpius played a role in putting the sun in its proper place. The sun is in Scorpius in the southern hemisphere's summer, and the longer days may be a reminder of the time Maui cited the sun for speeding.

Scorpius, to the Greeks, was the adversary of Orion the Hunter. They said Orion tried to rape Artemis, and she set a scorpion on him. From its sting he died. In another version, repeated by Hyginus, Gaia the Earth was vexed by Orion's boast that he could kill anything she created. A protective mother, she sent the scorpion after the presumptuous giant. After his death, both he and the scorpion were lodged on opposite sides of the sky.

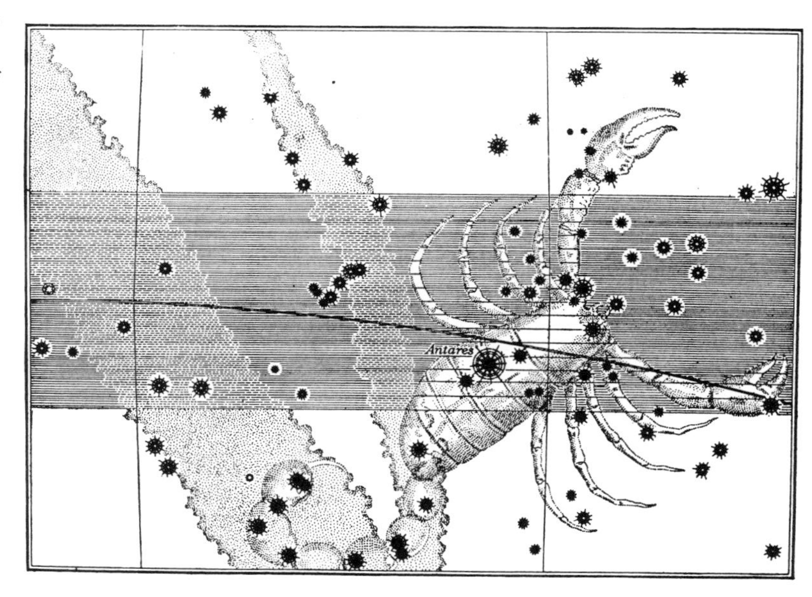

Scorpius probably acquired its name because the hook-shaped pattern of its brighter stars resembles the curving body of a scorpion. Both the Greeks and the Mesopotamians called these stars a scorpion. Originally, Antares, its brightest star, may have been singled out by neolithic Middle Eastern agriculturalists as one of four key seasonal indicators. (From The Stars in Song and Legend *by Jermain G. Porter, Ginn & Company, 1902, after Johannes Bayer's* Uranometria *sky atlas, 1603)*

In our era, Orion is a familiar constellation of winter, and Scorpius claims the summer sky. Tenacious as a pit bull, the Scorpion continues its pursuit of Orion. Whenever Scorpius rises, Orion sets. Fleeing before the claws, the sting, and the poison, he dies behind the western horizon.

THE ARCHER

Sagittarius is usually called the Archer, but he has an unusual aspect not mentioned in his name; he's half horse. Hyginus begins his entry on the Archer by mentioning that some people deny Sagittarius is a centaur at all, and by pointing out that the tail he often sports looks as if it came from a satyr and not a horse. Those who accepted him as a centaur decided he must be Crotus, the son of goat-legged Pan and Eupheme, who nursed the Nine muses in their infancy. Sometimes confused with Chiron, that centaur actually canters in the southern sky as the constellation Centaurus.

We can't translate the Sumerian name used for the stars we call Sagittarius, but we can be reasonably sure the original idea in this constellation came from Mesopotamia. Boundary stones, or *kudurrus*, from the Kassite period of Babylon—from about 1595 to 1157 B.C.—record royal land grants and are known for the astronomical symbols carved on the stones to seal the deal with celestial authority. Obvious emblems of the sun, moon, and the planet Venus, and familiar zodiac constellations—the Scorpion and the Sea Goat—accompany other astronomical symbols. One of the figures on some of these boundary stones resembles our Sagittarius. He is always an archer, and he may have wings. Sometimes he has four legs, sometimes two, but his flanks and legs are never human. There are lower quarters that look as if they've been borrowed from a bird, an insect, a lion, a dog, or a horse. The figure almost always has a scorpion's tail, but he might have a second tail that could belong to a lion or a dog. In at least one example, the Archer's human head shares the neck with the head of a dog. In several figures, his penis is erect and intentionally prominent, and that is consistent with the Greek identification of the figure with satyrs.

When Gilgamesh, the legendary king of the Sumerian city of Uruk and hero of an epic, sets out for the land of the dead, he encounters the Scorpion Man and his mate at the western gate of the mountain that guards both the rising and setting sun. Halting at the portal to the underworld, he is instructed to follow the sun's road through twelve leagues of darkness to its point of rising. If the

Scorpion Man is also the scorpion half-breed archer on the *kudurrus* and if the *kudduru* character stands for Sagittarius, Sagittarius may be the guardian of the door to the dead. The winter solstice station of the sun was a constellation away from Sagittarius in Sumerian times; the Milky Way runs through Sagittarius, and in many traditions, the Milky Way was said to be a path of souls. We don't know what the ancient Mesopotamians made of the Milky Way, but it is possible that it helped the celestial Archer police the passage to the underworld.

THE SEA GOAT

Capricornus has to be the most bizarre beast in the zodiac zoo. Its stars are all fairly dim, provide the rough outline of a triangle, but don't really look like much of anything. Depicted, however, with the head and forelegs of a goat and the tail of a fish, it corresponds to no creature that roamed the earth or swam its seas.

Real or not, the image of the goat-fish, or Sea Goat, is very old. Finding it on Kassite boundary stones, we can be convinced of its Mesopotamian origin. Its Sumerian name means "goat-fish."

In his *Katasterismoi,* the Pseudo-Eratosthenes bypasses the aqueous half of the problem by referring to the constellation as *Aigokeros,* the "Goat Horn," and describes the creature as if, like Pan, it had goat horns and legs but a human torso.

THE MAN THAT BEARS THE WATERING POT

East of Capricornus, Aquarius the Water Bearer pours a trickle of stars into the mouth of Piscis Austrinus, the Southern Fish. Most Greek commentaries agreed that the man with the jug was Ganymede, a mortal youth so attractive that Zeus shanghaied him to Olympus. There he was employed as the cupbearer of the gods while he lit Zeus's fire and quenched his thirst. Alternative identities for Aquarius include Deucalion, survivor of the Great Flood in Greek myth, who presumably was experienced enough with water to qualify him for the job.

Certainly Aquarius also has a Mesopotamian bloodline, for some Kassite *kudurrus* carry a figure who holds a vase from which two streams of liquid flow down to his feet. Mesopotamian tradition doesn't help us much here, however, for it has not been possible to relate the image on the boundary stones to the name usually attached to the stars of Aquarius. The meaning of that name is not known either.

THE FISH WITH GLITTERING TAILS

The last constellation of the zodiac, Pisces the Fishes, was known as "the Tails" in Babylonian, and the name seems to refer to the two fish that are tied together by their tails with a long cord. It is possible to trace out one fish as a circlet of stars on the west side of the constellation. The thread that binds the tails takes the form of a large, deep *V,* but where one longs for a second fish at the other end of the line, it looks like the one that got away. By pirating some stars out of nearby Andromeda, however, it is possible to catch the other fish, and perhaps that is what the Babylonians did. Hyginus explained that these two fish really stand for Venus (Aphrodite) and her son Cupid (Eros), who were just relaxing on the banks of the Euphrates one day when the monster Typhon surfaced in the river. Typhon had the power to panic any goddess—he was huge. Taller than the mountains and able to touch the stars, one hand could knock on the door of sunrise while the other rattled the gate of sunset. His fingers were dragon heads, and he had a hundred of them. A tornado of wings and serpentine coils, he barbecued victims caught by the fire in his eyes. Venus and Cupid didn't waste any time trying to outrace the terror. Instead, they transformed themselves into fish and splashed into the flood. Neither noticing nor recognizing them when they were disguised as fish, Typhon let them escape. This story doesn't tell us much about the meaning of Pisces, but it might explain, at least in part, another Babylonian name for the northern member of the pair, *Anunitu,* or "Lady of the Heavens."

THE TWELVE LABORS OF HERACLES

If the constellations of the zodiac were originally linked as symbols of a single celestial narrative, we might expect the plot to mimic the yearly adventure of the sun, perhaps with each zodiac figure representing a different episode. After reviewing what the Greeks had to say about the constellations of the zodiac, however, we have to abandon that notion. What at first seemed to be a complete and unified set of symbols—twelve figures that encircle the sky, mirror the months, contain the year, and direct the movement of the sun, moon, and planets—is actually an unmistakably unmatched set of stories.

But even if there is no real zodiac epic, there are heroes of myth who take on some attributes of the sun. Because they borrow some of the sun's character, fragments of the zodiac may accompany them as they follow their destinies.

Heracles qualifies for consideration as such a solar hero. Both Macrobius, a pagan Latin grammarian and philosopher of the fourth and fifth centuries A.D., and Nonnus, the fifth century A.D. Greek poet, affirmed he was the sun. Apollodorus, the Greek mythographer guessed to have written in the first century A.D., recounts how Heracles traded places with the sun while seeking the cattle of Geryon, his tenth Labor. Irritated by the heat of the sun, Heracles aimed an arrow at Helios. Because he admired Heracles's grit, Helios allowed him to sail to Erythia in this golden cup. In the sun's boat, Heracles crossed the ocean. Heracles also wore the pelt of a lion, the beast of the summer sun.

His twelve Labors remind us of the twelve signs of the zodiac, the twelve months, and other felicitous groups of twelve that seem richer for the mutual association. Robert Brown, a first-string player on Max Müller's discredited Solar Mythology Team, finds in the original Heracles a "Semitic toiling, warring, voyaging, travelling, man-slaying, at times maniacal Sun-god." In his book, *Semitic Influence in Hellenic Mythology*, Brown adds that Heracles is "the Sun-god, who has a special labour in each month and Sign of the Zodiac." Attempts to link each Labor with a zodiac constellation are unconvincing, however, if we insist on keeping signs and Labors in order. Let's have a closer look at his twelve Labors.

In a fit of madness visited upon him by jealous adversary Hera, Heracles killed his own children. Returning to his senses with remorse, he made a pilgrimage to the oracle at Delphi to learn what penance he should perform for his crime. He was instructed to indenture himself to his cousin Eurystheus, the king of Tiryns and Mycenae, for twelve years. At the successful completion of his trials, he would receive the gift of immortality. So he went to the court of Eurystheus, who was not an especially likable fellow, for his first assignment, slaying the Nemean lion. The other tasks included

destroying the Lernaean Hydra,

capturing the gold-antlered hind, or doe, of Ceryneia,

bringing back the ferocious Erymanthian boar alive,

dispersing the destructive birds of Lake Stymphalus,

cleaning the Augean stables,

returning with the Cretan bull,

bronco-busting the man-eating mares of Diomedes,

acquiring the girdle of Hippolyta, the queen of the Amazons,

herding the cattle of Geryon from the far west back to Greece,

acquiring the golden apples of the Hesperides, and

dragging Cerberus, the three-headed serpent hound of Hades, from the gate of the underworld to the surface of the earth.

The Nemean lion, we already know, was identified with Leo. Although the Lernaean hydra doesn't make a home on the ecliptic, the crab that opened a second front on Heracles is there. We could certainly put the Cretan bull with Taurus, and maybe we could assign Hippolyta's girdle to Virgo. After that, though, nothing else fits.

Jane Harrison, an English classicist and the author of *Themis, a Study of the Social Origins of Greek Legend,* takes a different approach to the solar aspect of Heracles. She analyzes what she calls the "Year-Daimon," the transcendent power and embodiment of the cycle of birth, growth, death, and rebirth. She calls Heracles a seasonal fertility-daimon and a "daimon of the Sun-Year" and fits his twelve labors into a luni-solar calendar cycle.

Heracles was a hero on a quest, and somewhere in the telling and retelling of his story he picked up a zodiacal justification for a Labor of two to punctuate the solar character of his quest. If Jane Harrison is right, Heracles provided regeneration —doing what the sun does.

THE QUEST FOR THE GOLDEN FLEECE

On the other hand, Jason and the Argonauts are clearly questing for the sun and the vernal equinox. That ram with the golden fleece is a pretty good Aries. But what about the rest of the voyage of the *Argo?* Does it navigate through the zodiac? There is certainly at least one correspondence: Gemini is in the crew as Castro and Pollux.

We can probably provide a bull, too. When Jason arrives in Colchis, King Aeetes puts him to a test. He must harness a yoke of bulls, plough a field, and sow the furrows with dragon's teeth. The bulls are a nightmare: They are fierce, and have hooves of bronze and fiery breath. The teeth sprout as men, fully grown and fully armed. Jason tosses a stone into the middle of the army he's cultivated, and the entire crop of soldiers goes to war with itself.

The rest of the voyage is quite interesting, but there is no more zodiac to be found on the itinerary of the *Argo.* The boat makes it into the sky, though; the old southern constellation of the Ship was called Argo. And even without much zodiacal imagery, the quest for the Golden Fleece is a solar quest. Jason travels east, to the land of the sunrise, to secure the fleece. It represents the renewal of the year and the renewal of life collected from the realm of the dawn and the vernal equinox. New life is hanging on that oak branch. Its guardian, the dragon, maintains an unsleeping vigil for the new steward of the talisman of life. The dragon is like the King of the Wood who guarded Diana's grove at Nemi from his own successor and killer.

Each year the sun embarks upon a quest that carries it through the zodiac, into one season after another, around the year. That solar quest, so successfully completed each year, acted as an example for all quests. So when our ancestors put the seasonal meaning of the sun to work in their stories of heroic quest, pieces of the zodiac would tag along. The heroes embarked on great quests, not to circle the year, but to obtain what was of the highest value—a golden fleece, immortality, or the salvation of the soul—and the zodiac might peek through the plotline

BUILDING THE ZODIAC

Mesopotamian names for many of the zodiac constellations tell us that the roots of the zodiac system probably reach back to a prehistoric period when conspicuous stars or constellations on the sun's yearly path proclaimed the end of one season and the start of another. This idea is present in the *mul-Apin* manuals, which distribute the seasons among twelve months according to the sun's annual itinerary in the sky. The right to establish this system of cosmic order was conceded to the Babylonian supreme god Marduk, after his victory over Tiamat, the monster of primordial chaos. Recorded on tablets inscribed at Nineveh, the edition of the *Enûma elish,* or epic of creation, shelved in the library of King Ashurbanipal, is an Assyrian version of a much older story. Ashurbanipal ruled Assyria from 668 to about 626 B.C., but the creation myth was recited annually perhaps as early as 1800 B.C., in the Old Babylonian period, as an essential element of the *akitu,* or New Year's ceremony.

At Babylon, the *Enûma elish* was staged as a sacred drama in a New Year's festival that lasted twelve days. Ceremonial reading of the creation myth was the same thing as reenacting it, and so each year the world was symbolically created all over again. Creation in the *Enûma elish* did not, however, mean manufacturing the components of the cosmos. It meant the establishment of order. The *akitu* ceremony annually mirrored the original creation, or establishment of cosmic order, by re-establishing the terrestrial order. That was accomplished through ritual renewal of the authority of the king at the vernal equinox, which coincided with New Year's in Babylon.

At the start of the festival, Marduk, the divine agent of cosmic order, and his earthly representative, the king, were missing. Somber rites performed in the Esagila, the temple of Marduk in Babylon, lamented Marduk's absence. Confined within the mountain, in the underworld, Marduk's retreat mirrored the dormant, desolate condition of the world before its revival in spring. But on the evening of the fourth day, the words of the creation myth, spoken aloud, were powerful enough to release Marduk from his prison. On the next day, the king made his first appearance in the ceremonies.

Seated in assembly at the time of creation, the gods had granted Marduk kingship over the whole universe. By giving him the emblems of kingship—the scepter, the throne, and the royal robe—and matchless weapons, they legitimized his sovereignty. In the same way, the *akitu* ceremony authorized the mandate of the king and linked his office to the maintenance of order and renewal of life in the land he ruled. Through a ritual battle with wooden representations of Tiamat and her monstrous brood, the king won the Tablets of Destiny, which recorded new proclamations, laws, appointments, promotions, and goals for the coming year. The list of assignments and royal patronage for the new year was read on the ninth day. It corresponded to Marduk's distribution of assignments of the gods. The continued vitality of the land was also a royal responsibility, and so a "sacred marriage" consummated by the king and the priestess of the goddess Ishtar followed a banquet on the tenth day. Male and female were united in order to quicken life in the world womb once more. A second reading of the Tablets of Destiny on the next day informed the populace of the new laws, tasks, and commandments. By the twelfth day, law, society, the life in the land, the gods, and the king were all reinstated. It was time to get down to business. The year, fired by a celestial trigger, was underway.

MAKING THE WORLD SAFE FOR ASTRONOMY

Marduk's reward for defeating Tiamat was the right to organize the universe. He began by setting up the sky. Splitting Tiamat's corpse open like a clamshell, he left half of her in place as the earth, and he raised the rest of her overhead to form the vault of heaven. The great gods—Anu, the god of heaven; Enlil, the divine ruler of the earth and the atmosphere; and Ea, the lord of indispensable fresh water from rivers, rain, and underground wells—were instructed by Marduk to inhabit their proper residences. He also created "stations" for them, the three "ways" or "paths" of Anu, Enlil, and Ea. The *Enûma elish* continues with the constellations of the zodiac, said to be the likenesses of the gods. Marduk set them up, and he

> *. . . made the year, divided its boundaries,*
> *[for the] twelve months three stars each he set.*

Thirty-six stars altogether, then measured out the months on the sky. In *Science Awakening II*, B. L. van der Waerden, a Swiss historian of science, has analyzed Babylonian inscriptions for more information about the use of these thirty-six stars. Each became visible for the first time in the predawn sky on a specified day in the calendar. By watching for that first morning appearance, the Babylonian astronomers kept the calendar. These heliacal risings occurred when the stars in question were far enough from the sun to be seen in morning twilight before sunrise.

Marduk's "paths" for Anu, Enlil, and Ea were like zones of latitude on the earth. They circled the celestial sphere and paraded completely by us here on the ground once every twenty-four hours. The "path of Anu" was a belt of sky centered on the sky's "equator," extending about 16½ degrees above and below. The rest of the sky above it was centered on the sky's north pole and was the "path of Enlil." Its counterpart, below the "path of Anu" in the southern half of the sky, was the "path of Ea."

Statements in the *mul-Apin* tablets plainly tell us the connection between the seasons and the sun's progress on its yearly rounds. From the beginning of the twelfth month to the end of the second month, "the sun is in the path of Anu," and this meant "wind and storm." During the next three months, "the sun is in the path of Enlil"; it was a time of "harvest and heat." From the first day of the sixth month to the last day of the eighth, "the sun is in the path of Anu" once more, and this season also brought "wind and storm." Months nine, ten, and eleven put the sun "in the path of Ea." The weather was cold. Picturing the trail of the sun as a circle inclined to the path of Anu, the Babylonians at this stage divided their zodiac into the four seasonal segments. We know the vernal equinox occurred in the middle of the year's first month. That meant the solstices and equinoxes fell in the middle of each season. When the sun first occupied the path of Anu, it was spring. The harvest and heat of Enlil's path confirmed that it housed the summer solstice. Autumn's wind, storm, and equinox followed in proper course when the sun crossed again through Anu's realm, and winter solstice accompanied the cold as the sun rode south in the path of Ea.

Division of the year and the sun's path into four segments confirms that the twelve-sign zodiac had not yet emerged from the system of thirty-six stars by the time they were incorporated into the *mul-Apin* tablets. In first century B.C., however, the Greek historian Diodorus Siculus provided a report on Chaldaean astronomy, in which he makes the zodiac an essential ingredient of the thirty-six-star system. Twelve of the stars, he writes, are recognized to have higher authority, and to each of them is assigned a month and a sign of the zodiac. In this way, a solar year with twelve artificial "months" was reconciled to the path through which the sun moves each year.

THE FIRST CELESTIAL SEASONINGS

It is likely, then, that the concept of the zodiac emerged out of a few millennia of calendar observations of heliacally rising stars, some of which were also known to occupy the annual path of the sun. Sumerian names for three zodiac constellations—Taurus, Leo, and Scorpius—certainly endorse the notion that these groups of stars were recognized as a bull, a lion, and a scorpion long before the Babylonians fleshed out the zodiac. Each of them looks reasonably like the animal after which it is named, and it is reasonable to guess that such resemblances inspired the early sky-watchers in Mesopotamia to assign these identities. In a paper published in the *Journal for Near Eastern Studies* in 1965, Willy Hartner, a German historian of science, has attempted to trace the symbols of these constellations back even further, to the neolithic farmers of Mesopotamia, Elam, and Persia.

Early in the fourth millennium B.C., neolithic people in what is now southern Iraq began to settle the first cities in what would become the civilization of Sumer. Five centuries before the invention of writing, their drawings of lions, bulls, and scorpions studded with star symbols imply a tradition of organized sky lore.

Realizing that the constellations of Taurus, Leo, and Scorpius were spaced more or less one-fourth of the way around the ecliptic from one another, Hartner suspected that representations of celestial bulls, lions, and scorpions stood for seasonal appearances of these zodiac constellations long before the zodiac itself was imagined. In 4000 B.C., they didn't coincide with equinoxes or solstices, but they did put in special appearances in the predawn sky that would have made them handy markers of seasonal change.

Six thousand years ago, the stars of Taurus, Leo, and Scorpius rose heliacally at significant times of the year, as seen from latitude 30 degrees north. This is close to the latitude of Ur, the Sumerian city founded on the Euphrates River around five thousand years ago. Agricultural settlements with organized irrigation near Ur had been established at least seven thousand years ago. The motivation and circumstances needed to undertake practical, seasonal astronomy were certainly in place by 4000 B.C.

The stars of Taurus, and in particular the conspicuous cluster in it known as the Pleiades, reappeared before dawn at the time of the vernal equinox in 4000 B.C. Regulus, the brightest star in Leo did the same thing at summer solstice, and the heliacal rising of Antares, the "heart" of Scorpius and its brightest star, accompanied the autumnal equinox. The stars of these three constellations were thus conveniently linked to the seasonal progress of the sun. Hartner guessed that there had been a fourth constellation for the winter solstice, and the numerous depictions of a celestial ibex and of lion-ibex conflicts led him to believe that some of the stars we now include in the constellations of Capricornus and Aquarius comprised a star group once known in the ancient Near East as the Ibex. It would have been the only one of the set of four seasonal markers to lack a bright star, but a brighter star, Scheat, in the constellation of Pegasus, although not on the ecliptic, rose heliacally at the same time as stars in his Ibex. Hartner speculated that Scheat might have acted as the agent of the Ibex and signaled the winter solstice for it.

THE LION BRAWLS WITH THE BULL

This combat scene between a lion and a bull was engraved upon a Sumerian goblet around 3000 B.C. The lion seems to represent the stars we still know as Leo the Lion, and the bull, then, is what we still call Taurus the Bull. This bull may be wearing a starry rosette in its horns to confirm its celestial identity. (From Mesopotamian Archaeology *by Percy S. P. Handcock, G. P. Putnam's Sons, 1912)*

Hartner's attention was especially drawn to representations of the lion and bull in combat. Such scenes can be found on an Elamite seal from the fourth millennium B.C., on a Sumerian shell goblet from 3000 B.C., on a monumental staircase constructed in the sixth century B.C. at Persepolis—a capital of the Persian empire, and on a Persian miniature painted perhaps no earlier than the thirteenth century A.D. Five thousand years may separate the oldest and most recent of these lion-bull battle scenes, but the theme remains strikingly consistent.

Seen in the skies of 4000 B.C., the lion-bull combat begins to make sense. The lion kills the bull in early February, at the first ploughing and sowing. At the start of the night, Leo is directly overhead and rules the heavens. Taurus is at the western horizon, about to follow the sun into the land of the dead (evening heliacal setting). The bull dies because in the days that follow, the sun moves into conjunction with Taurus. Slain in the light of the sun, Taurus vanishes from the night sky, a victim of the victorious lion, still high and visible in the early evening sky. Forty days later, the bull comes back to life at dawn, resurrected in the heliacal rising of Taurus at the vernal equinox, traditionally the time of the rebirth of life in the ancient Near East. Depictions of bulls secure in the company of the starry Sacred Tree, an emblem of life, probably symbolize the bull's dawn return at the equinox.

The lion-bull combat on the six-thousand-year-old Elamite seal is convincingly explained by Hartner's hypothesis. But if precession shifts all of these constellations with respect to the seasons, why would the Achaemenid kings of Persia still find the lion-bull combat meaningful three and a half millennia later? Hartner's calculations revealed an interesting circumstance in the skies and seasons of Persepolis. The latitude of Persepolis is 30 degrees north, and from that latitude, in 500 B.C., the Pleiades, in Taurus the Bull, were seen for the last time at sunset in the west again (evening heliacal setting), with Leo in the driver's seat high overhead, but in this era the lion's conquest

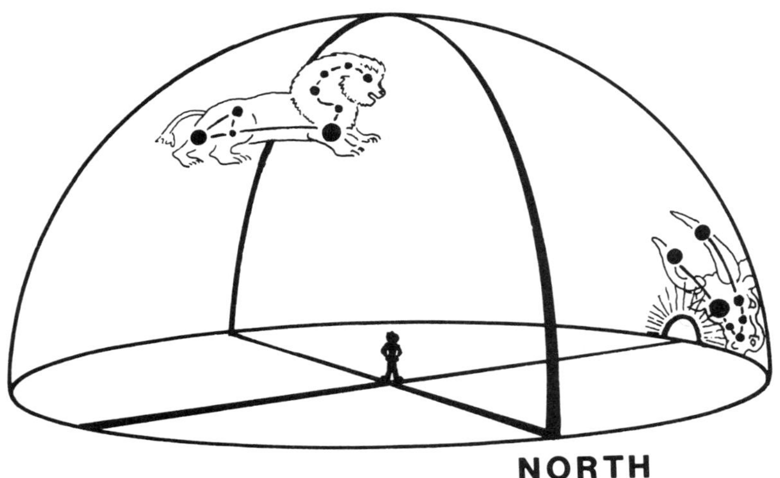

The celestial altercation between the Lion and the Bull has seasonal consequences, for it symbolizes the "death" of the stars of Taurus in the evening twilight during early February, approximately six thousand years ago. As Taurus went down in the west, Leo roared high in the south and ruled the night. (Drawing: Robin Rector Krupp)

occurred on March 28 and not in early February. March 28 falls just a week after the vernal equinox, and that was close enough to prompt the Persians to associate the event with the New Year. In the ancient Persian calendar, the New Year festival of *Nauruz* started the year with the sighting of the first crescent moon after the vernal equinox.

By 500 B.C., the Persian Empire included Babylonia and Assyria, and Assyrian influence on the Achaemenid Persians is evident at Persepolis. *Nauruz*, the Persian New Year equinox festival, incorporated the same themes that activated the Babylonian *akitu:* world renewal and restoration of cosmic order. *Nauruz* New Year ceremonies were performed at the *tachara*, or temple-palace, built by Darius the Great at Persepolis, and the fight between the lion and the bull is displayed prominently at Persepolis on the great staircases of its *Apadana*, or assembly hall.

GOOD AND EVIL IN ANCIENT PERSIA

Conflict between good and evil is the fulcrum of the Persian Zoroastrian religion, and the battle for all of creation is fought in black and white. This cosmic war incites all of the elemental polarities to active opposition: light and darkness, day and night, summer and winter, order and chaos, life and death, truth and falsehood. Ahura Mazda,

Although by 500 B.C. precession had shifted the celestial death of Taurus the Bull and the victory of Leo the Lion to a time close to the vernal equinox, that didn't stop the Achaemenid Persians from extracting seasonal significance from the lion-bull combat. It came close to coinciding with their equinoctial New Year rites, which were staged at Persepolis. (From Media: The Story of Nations *by Zénaïde A. Ragozin, Unwin, 1889)*

whose name means "wise lord," is the creator god and lord of light. His adversary is Angra Mainyu, the author of all evil.

The first reference to Ahura Mazda by name occurs in a list of gods retrieved from the ruins of Ashurbanipal's library at Nineveh, but his roots seem to reach back to Varuna, the celestial architect of cosmic order in the Vedic tradition of the early Indo-Europeans. Ahura Mazda's star-spangled robe demonstrated his bond with the sky. In the Persian prophet Zoroaster's hands, he came down hard on the side of ethical behavior and responsible government. His alliance with light was still remembered in the earlier decades of this century, when a brand of light bulbs was marketed under the trade name Mazda.

In Zoroastrian tradition, a twelve-thousand-year struggle between good and evil began when Angra Mainyu rose mindlessly from the endless dark abyss into Ahura Mazda's kingdom of infinite light. There he encountered the subtle spirits Ahura Mazda had so finely wrought, and infuriated by Ahura Mazda's creations, Angra Mainyu vowed to destroy them all. Ahura Mazda overpowered Angra Mainyu with a prayer, and the evil spirit tumbled back down into the gloom, bewildered but still committed to his adversary's destruction.

While Angra Mainyu slowly recovered from his initial defeat, Ahura Mazda fabricated the universe, a theater for the war to come. Starting with the sky, he manufactured and installed 6,480,000 celestial spirits to fight against evil. Ahura Mazda had a military contingent in each quarter of the world. His general from the west point was the star *Sadwes*, or Antares. *Wanand*, or Vega, the star of the south, also took a commission. *Haftoreng*, the Big Dipper, presided over the north, and in the east *Tishtrya*, the star Sirius, reviewed the troops. A general of generals occupied the north celestial pole and was known as "the peg in the center of the sky."

AHURA MAZDA'S FAITHFUL INDIAN COMPANION

Among those who marched in Ahura Mazda's forces, Mithra also held high rank. Originally he was the god Mitra, in the *Vedas* of India, and his character then evolved into an object of worship in a Roman cult that called him Mithras. His résumé identifies him above all as an emissary of celestial light. As an enemy of evil, his luminous power dispelled darkness and cleared the way for the sun. He was not the actual light, but rather the transcendent spirit of light and its capacity to kindle life.

In Sanskrit, *mitra* means "friend." This is consistent with his Zoroastrian role as mediator, in the middle realm of the upper atmosphere, between Ahura Mazda's eternal celestial light and the bottomless black gloom of Angra Mainyu. From his lofty station, Mithra witnessed all that transpired on earth. The *Avesta* says he has a thousand ears and ten thousand eyes. He never slept. Nothing escaped his notice, and he held everyone accountable. Acting as a judge of both the living and the dead, he was viewed as the divine cosigner of contracts. Integrity, ethical behavior, and truthfulness maintained his friendship. Breaking oaths, promises, and bonds exhausted his tolerance and risked his wrath. And if you crossed Mithra, your soul might well be anxious about crossing the razor sharp bridge to the next world. He stood by, ready to provide a helping hand or a timely shove.

In the first century B.C., Antiochus I, the ruler of Commagene, a small kingdom in southern Turkey, commemorated his reign with colossal sculpture and an artificial mound 164 feet high constructed on the 7,000-foot summit of Mount Nemrut. On this panel from the west terrace of Nemrut Dag, Antiochus (left) shakes hands with a hybrid version of Mithra. His celestial character is evident in the rayed disk that frames his head and the finely carved stars that garnish his Phrygian-style cap. This is the oldest-known Mithraic monument.

The Mithra that came to Rome differed somewhat from the Persian original. Although still closely associated with celestial light and even at times equated with the sun, he appears to have completely displaced Ahura Mazda in the primary role of personalized religion, dedicated not to kings but to the purification of the individual immortal soul. Light and darkness still battled it out, but in Rome they went to the mat in a contest for souls. Mithras continued in his role as "friend" and assisted the soul in its upward progress toward the light by interceding on the soul's behalf. In Rome, secret Mithraic ritual and doctrine incorporated celestial metaphor to convey its message of personal salvation.

THE ZODIAC, THE SEASONS, AND THE COSMIC BULL

Zodiac symbolism also figured in the Mithraic pilgrimage of the soul. When the soul descends from the realm of infinite light to take up residence in a newborn child, it passes through the gate of Cancer and the spheres of seven "planets," where it accumulates impurities. At the end of life, if the soul passes judgment, it ascends back to heaven through the gate of Capricorn. Now, the sun is at the summer solstice when it is in the sign of Cancer, and it starts its descent to the lower half of its annual circuit. Winter solstice finds the sun in the sign of Capricorn, and once it bottoms out on the ecliptic there, it begins to climb back toward the upper portion of its path. Mithraic metaphor borrowed imagery from the sun's journey through the zodiac to illustrate an esoteric doctrine of the destiny of the human soul.

Mithras was born on December 25, the same day as Christmas and the birthday of the Unconquered Sun. At times Mithras was called Sol Invictus and was merged with the sun, but his worship was quite distinct from the state religion of Sol Invictus Elagabal and should not be confused with it. At his birth on the winter solstice, Mithras emerged as a youth from a rocky outcrop on a river bank, dawning on the world the way light breaks out of the eastern horizon before sunrise. He arrived beneath the branches of a sacred fig tree, a symbol of knowledge and of the power to generate life. Wearing his trademark cap, he was already equipped with the two instruments of his office: a torch to provide light and a knife to slay the primordial bull.

Only two young shepherds beheld his miraculous delivery from the rock, and images of them occupied every Mithraic temple. Known as Cautes and Cautopates, the meaning of their names is unknown. Each carried a torch, but while Cautes held the flame up, his companion always tipped the torch toward the ground. Variously interpreted as light and dark, day and night, sunrise and sunset, heat and cold, spring/summer and autumn/winter, and the extinction of the soul's light at birth and its reignition at death, the real meaning of the pair is as uncertain as their names.

Mithras's first encounter, with the Sun, demonstrated his own superior strength. The Sun conceded the match and pledged his loyalty and service to Mithras. From then on they remained close friends and faithful allies.

The next episode in the career of Mithras involved Ahura Mazda's primeval bull of creation. According to Zoroastrian scriptures, written in the Pahlavi language of southwest Iran between the third and seventh centuries A.D., Ahura Mazda brought all of the world's animals into being as a single primordial ox or bull. Obliging as a combination sperm bank and Noah's Ark, the bull warehoused the seed of every species within its own body. He was an obvious target for Angra Mainyu. The Wise Lord tried to protect him from the demon's deadly influence, but Angra Mainyu infected the bull with illness. Its soul left its body for the sky, where it encountered both the sun and the moon, and it is reasonable to conclude from that account that the soul of the cosmic bull became the constellation Taurus.

MITHRAS BUTCHERS THE SPRING BULL

Mithras starred in his own rodeo by bulldogging the cosmic bull. Grabbing it by the horns, he swung himself over the beast and rode the bull bareback. Not about to be broken in by any Persian cowboy, the bull stampeded across the countryside and threw Mithras off of his back. Mithras gripped those horns, however, and was dragged alongside this four-legged locomotive until the bull ran out of steam. Throwing the exhausted bull over his back, Mithras carried it by its hind legs to a cave, where he planned to kill it. Somehow the bull broke out, and Mithras had to pursue the creature again. Taking refuge in the same cave it had just escaped, the bull was trapped by Mithras. He pulled its head back by the nostrils to expose its throat and drove the sacrificial knife in to the hilt.

Because this primordial bull of creation was no ordinary bull, all of the world's useful plants and all of the world's animals grew from its body when Mithras stabbed it. Wheat sprouted from the spinal cord, and its blood became the sacred wine consumed at Mithraic services.

Depictions of Mithras in the act of killing the bull are remarkably consistent in Mithraic temples throughout Europe. Mithraic sanctuaries were often built underground to resemble the cave in which Mithras killed the bull. A curved vault overhead symbolized the sky, and the banquet with the Sun was also usually illustrated. At the far end of the chapel, a scene of Mithras slaying the bull directed attention to the primary episode in the myth. Mithras has the bull by the nose. The knife is buried in the heart. An ear of grain grows from the bull's tail, while a dog and snake appear to mount their own attacks, and a scorpion claws the bull's testicles, the most transparent symbol of the bull's fertility. Sometimes a raven and a lion also attend the bull-killing, and a few scenes also include a goblet.

Most of the players in the bull's death scene also watched from the sky when Taurus the Bull said his lines for the last time on the western horizon before becoming submerged with the sun. The darkness below the western horizon was a cave where the Bull was slain. As Taurus died, Leo the Lion prowled overhead. Hydra the Water Serpent slithered below the Lion, with Corvus the Crow (raven) and Crater the Cup (goblet) lodged in its coils. If Canis Major, the Greater Dog, or its bright star Sirius were symbolized by the dog, then the dog, too, was in the sky just after sunset. Libra the Scales, which was traditionally known as the Scorpion's Claws, was rising in the east and threatening the sinking Bull.

We don't know at what point in the year Mithras slaughtered the bull, and we don't know when this

In Mithraic tradition, the bull still dies, but it is Mithras who kills him. Bull-slaying scenes like this one from the second century A.D. were essential components in Mithraic sanctuaries, and they incorporate numerous celestial references, including here the sun and the moon in the upper corners. (Object in the Louvre, Paris)

mythical act was reenacted in the Mithraic grottoes to baptize initiates in the blood of the bull. But the depictions of the sacrifice displayed in the chapels are entirely consistent with the astronomical circumstances of a specific night, just after sunset, in spring. Although precession shifted the date of this configuration even further from the equinox by the time Mithraism had spread through the Roman Empire, it still occurred in spring and close to the start of the new year. The Latin poet Vergil, writing in the first century B.C., confirmed that "the white bull opened the year with his golden horns." His contemporary Ovid explained in one of his poems when the year began: the month of April. This Roman New Year was timed by the bull's last stand in the west, and that still satisfied the theme of sacrifice and world renewal in spring.

Most of those who worshipped Mithras, however, were not preoccupied with agriculture. Exclusively men, and many of them soldiers, sailors, merchants, and slaves, they were interested in the fate of their souls. A precise seasonal and agricultural meaning was no longer necessary, for the celestial bull's death had transcendental meaning. The purpose of the cult of Mithras was personal salvation. The dying bull poured new life into the soul and triggered a spiritual rebirth.

It seems reasonable to imagine that the zodiac and the astral components of Hellenistic religions like Mithraism, have roots in a tradition of celestial myth that goes back six thousand years to the use of heliacally rising stars as indicators of seasonal change. Neolithic settlers of the fourth millennium B.C. perhaps inherited an even older tradition of sky figures, or devised their own, and exploited seasonal appearances of brighter stars—in the predawn and just after dark—in key locations of the sky. They likely watched for them at either horizon and at their highest elevation above the horizon, or culmination. The meaning and identity of figures and individual stars also likely evolved from what was happening in nature, in agriculture, in domestic life, and in ritual through the seasonal round. Those events were the important things in people's lives. If the stars appeared to herald them, then the stars commanded attention.

Out of that early effort to keep track of the sky, more refined astronomical conceptions may have slowly evolved. As centuries passed, precession could make original meanings of individual stars and constellations obsolete, but adjusting the stories and adding new elements could keep the sky in use. Tradition would still preserve some of the images whose meaning had gone through time's ringer. We at the end of the line would be handed a very mixed bag. That's what we have in the zodiac.

9

Until the Next Date

BOTH the sun and the moon complete appointed rounds through the zodiac, but each takes its own time to do it. They were said to meet in special harmony—some even said in marriage—when the independent cycles of the lunar month and the solar year brought them together in astronomical and calendrical congruence. Such nuptials would be celebrated, for example, when both objects started a new cycle on the same date. A new month—signaled, say, by the first visible crescent in the west—might coincide with the start of a new year.

The circumstances required are straightforward. Both objects must start and end a cycle on the same beat. This means the moon must complete a whole number of months—no fractions—in exactly the length of time it takes the sun to complete a year. This is a very tidy notion, but it doesn't happen. There isn't a whole number of lunar months in a solar year.

It takes the sun 365.242199 days to complete one yearly circuit through the solstices and equinoxes. This period is called a tropical year, from the Greek word *tropikos*, which means "turning." For the moon, 29.530588 days are required to run through a complete set of phases. This is called a *synodic* month, and the name is based on two Greek words, *syn* (with) and *hodos* (way), which when combined mean "a meeting." In that sense, the moon is engaged in the same "meeting," or configuration, with the earth and the sun each time it is in the same phase. Despite the high precision of these quoted intervals of time, both are rounded off at the sixth decimal place. Neither is a whole number, or even a simple fraction, of days, although for convenience we often say that the year lasts 365¼ days and that the moon's phases consume 29½ days. But either way, we run into the same problem: A year that begins together with the start of a synodic month can't end on the beat with the finish of a synodic month. Instead we get 12.36827 cycles of the moon each year. Now 12 lunar cycles appropriate about 354 of the year's days for their "lunar year," and the 11 more days needed to close out the solar year are just left out of the moon's program for those 12 cycles.

GETTING TO THE CHURCH ON TIME

In a single year, there is no way the sun and the moon are going to meet again in wedded bliss, but if they wait patiently through enough of their appointed rounds, they'll finally meet again in some calendrical love nest. Such reconciliations were desirable. They kept the lunar calendar and all of the human activities dependent on it in step with the seasons.

Reconciliation of the sun and the moon is more than just an affair of the heart. It's a ritual requirement for seasonal ceremonies and festivals. Most Greek festivals were scheduled for the full moon, and the names of the months, which varied widely from one city-state to another, tell us that certain ceremonies had to be held in certain months. If the sun and the moon were permitted to wander through time without calendrical recalibration, those Greeks would certainly have found themselves celebrating the harvest at the time of first ploughing and making appreciative offerings to the gods for a great crop before they had ever committed a seed to the soil. So they devised a way to bring the moon back in step with the sun. If the lunar year (12 lunations) is a bit short of the solar year, you can recalibrate the lunar calendar by adding an extra month every so often.

Geminos of Rhodes, writing in first century B.C., confirms that such a system of intercalation was in place some time before the second half of the fifth century B.C. and says that in its earliest form it involved supplementing the normal tally of 96 lunar months over an eight-year period (8 years × 12 months/year = 96 months) with three additional months. The Greeks, then, tried to broker a marriage between the sun and the moon through an eight-year engagement. They called this eight-year cycle the *octaeteris*, and they made "leap years" out of the third, fifth, and eighth years of the count by inserting one of the three extra months into each of them.

CARRYING A TORCH FOR THE SUN AND MOON

Our custom of holding Olympic games every four years, in the same year in which we elect a President and add an extra day to February to keep the calendar in line, goes back to ancient Greece, as, of course, do the Olympics themselves. The earliest recorded games were held in the territory of the Greek city-state of Elis at the sanctuary of Olympia in 776 B.C., and it's quite possible that some type of athletic competition took place there in the northwest region of the Peloponnesian peninsula long before that. The Olympics were said by Pindar, who authored Greek odes during the fifth century B.C., to commemorate the chariot race victory of Pelops, the mythological hero after whom the Peloponnesus, or southern peninsula of Greece, is named.

The tale of Pelops is essentially the story of the inevitable replacement of the older and weaker sovereign by his younger, stronger adversary, whose vitality solidifies his claim to the crown. Although Pelops eventually won the crown, he got off to a bad start in life. His father Tantalus killed him, sliced and diced him, stewed him, and served him up as an offering to the gods. Only Demeter sampled the dish. She immediately realized that the meal violated the dietary code of Olympus, and the gods terminated the banquet and brought Pelops back to life.

Resurrected from a sacrificial death and partially consumed by the hungry goddess of the earth's growth and grain, Pelops has some of what it takes to be a dying year god. He is closely associated with the Olympic games, which at least originally had something to do with the New Year and the inauguration of a new calendric cycle. In this case, however, the "New Year" was really a junction in the longer, eight-year cycle that packaged up the past and got the future off to a proper start in the same way the year rounds up and restarts the seasons. Such an interval of time is like the term of office of a sacred king who must defend his right to rule by demonstrating his continued strength and vigor. If, like the King of the Wood, he is overpowered by an assailant, that challenger takes over, elevated to the throne as the New Year is itself seated.

In time Pelops conceived a great desire to take Hippodamia, the daughter and wife of King Oenomaus. Oenomaus was a sporting king and made a habit of offering his consort to potential suitors willing to compete with him in a chariot race. Agreeing to race, a contestant for Hippodamia's hand would take her with him in a chariot and drive as fast as possible to the altar of Poseidon at the Isthmus of Corinth. Oenomaus would try to catch the fugitive in his chariot before they got there. If they outran him, Hippodamia acquired a new king; should Oenomaus catch the pair, he was

To the Greeks, the sun was Helios, and the moon was Selene. Each traveled independently through the celestial landscape in a separate chariot, but every eight years, their cycles meshed in a calendrical marriage that brought the same phase of the moon back to the same date in the year. This match made in heaven is portrayed on a goblet that halos Helios in a sun disk and crowns Selene with the horns of a crescent moon. (From Themis *by Jane Ellen Harrison, Cambridge University Press, 1927)*

permitted to kill the wife-stealer. In this manner, Oenomaus dispatched twelve potential bride seekers and prominently displayed their severed heads around the palace to impress his guests.

Hippodamia was tiring of this game and looking for a new conquest of her own. Pelops suited her just fine, and so she persuaded Myrtilus, the driver of Oenomaus's chariot, to throw the race. Myrtilus replaced the wheels' bronze linch pins with wax replicas. In the heat of the race, the wax failed, and the wheels started to go their own way without Oenomaus. Myrtilus jumped clear. Oenomaus, fouled in the flying reins, died as he was dragged behind his horses. Pelops won the race, the girl, and the country. In a contest between the old king and the new, the prize was the woman and the land, both of whom were made fruitful by a young and virile king, whose power was secure until the next renewal of time.

With twelve thwarted lovers of Hippodamia already notched on his chariot, Oenomaus became the thirteenth victim of the race. These numbers suspiciously resemble the number of months in a year, with and without the intercalated month, and suggest that the death of Oenomaus and the victory of Pelops has something to do with the calendar. The connection between the *octaeteris* and the Olympic games helps explain what it is.

The games' date was set by the progress of the sun and the moon through the eight-year calendar. From the first full moon after the winter solstice, the Greeks counted eight more full moons. The eighth full moon would be the central day of the first Olympiad of the cycle; it usually fell in August. The next Olympics were then set to take place exactly 50 months after the first, very close to four years later. Finally, the next Olympiad was scheduled for 49 months after that, and because the eight-year/99-month cycle does a fair job of keeping the months where they belong, the Olympics took place when they should, at the eighth full moon after the first full moon after the winter solstice—probably when the New Year was originally observed.

Pelops, driving the chariot, represents the sun, and Hippodamia stands for the moon on this ancient amphora. (From Gods and Heroes *by Gustav Schwab, Pantheon Books, 1974)*

If the Olympics are, in a sense, a reenactment of the triumph of Pelops, they also commemorate the marriage of Pelops and Hippodamia. At the end of one *octaeteris* and the start of the next, the term of the old king is over. The new king marries the queen, and through their union and through the renewal of the calendar, the life and fertility of the land are renewed for another run with the sun and the moon.

In our story, Hippodamia plays the moon, and although she started the run through the *octaeteris* together with her husband, she is much faster than he is. By the last leg of the cycle, the eighth year, she is out in front. The moon is usually ahead of the sun, which plays catch-up each time an extra month is added. Now we can see why the chariot race had thirteen losers. Successful completion of the eight-year cycle occurs in a year that enjoys an extra, or thirteenth, month. Oenomaus and the other twelve contenders for the moon die just in time to be replaced by a new round of time and a new king. At the end, the old king must die because the *octaeteris,* the year, and the month all die, only to be reincarnated in Pelops, who was himself restored to life. His sacred marriage to Hippodamia is really the congruence of the solar year and the lunar month brought together once more. The two start a new life together, celebrated in summer, when the sun is strong enough to handle anyone and when the moon is full, and so also at her best.

RACING AGAINST THE MOON

This idea of a race between the sun and the moon with calendrical congruence at the finish line is also mixed into the story of Atalante. She, like Hippodamia, was pursued by a steady parade of admirers, but she always managed to stay one step ahead of them and maintain her virtue because she could outrun anybody. Stripped naked for every footrace, she promised herself to any man who could outsprint her; she also demanded his life if he failed. Despite the risk, the charms of her running suit inspired numerous challengers. We have no idea how many men lost their lives chasing her tail, but Atalante collected them like Olympic medals. She would still be running men into the ground had Hippomenes (or in some versions, Melanion) not entered the race with three golden apples. Aphrodite, the goddess of love, was just a little tired of Atalante's virginity and lack of sexual appetite, and she gave the apples to Hippomenes to ensure that Atalante would lose the race. Instructed to drop an apple each time Atalante started to overtake him, Hippomenes caught glimpses of her stopping for the golden fruit while he pressed on to the finish line. As the race drew to a close, Atalante, cradling all three precious apples in her arms, exploded toward the ribbon but reached it just after Hippomenes broke through as the winner of the race. Golden apples delayed the lady and delivered her into the embrace of Hippomenes.

Professor Charles F. Herberger, author of *The Thread of Ariadne* and *The Riddle of the Sphinx*, thinks the story of Atalante is a calendar tale. Atalante is the moon, who can outrun everybody. Abandoned as an infant in the mountains and nursed by a she-bear, she is an athletic daughter of the wild, a virgin and a huntress. These characteristics make her a kind of human version of the goddess Artemis and strengthen her link with the moon. The three golden apples are the three extra months added to the *octaeteris* to keep the moon in step with the seasons and the sun. They are dropped for Atalante every time it looks like she's going to pass Hippomenes, who stands for the sun. Just enough apples means the moon will marry the sun at the end of eight years, which is what Atalante does. In the earliest version of the *octaeteris*, the 99 lunar months are just a little longer than the eight 365-day years. That small discrepancy, about a half day, is the amount by which Hippomenes edged out Atalante.

THE METONIC TONIC FOR CELESTIAL CONGRUENCE

By at least the fifth century B.C., Greek calendar calibrators had realized that the eight-year cycle they used was not as accurate as they would have liked. The small difference between ninety-nine lunar months and eight tropical years was cumulative, eventually threw the moon off rhythm, and subverted the marriage of the sun and the moon. They also realized, however, that a longer, nineteen-year interlude known as the Metonic cycle returns the sun to the arms of the moon. Its discovery was attributed to the Greek astronomer Meton, who lived in Athens during the second half of the fifth century B.C., although the Babylonian astronomers had been using it perhaps a hundred years earlier. Because the Greek author Geminos of Rhodes gave Meton credit for this cycle in the first half of the first century B.C., it is still known as the Metonic cycle. It works because, in days, 19 years are almost exactly equal to 235 lunations, or synodic months.

$$19 \times 365.242199 \text{ days} = 6939.6018 \text{ days}$$
$$235 \times 29.530588 \text{ days} = 6939.6882 \text{ days}$$

Of course, the lengths of the year and of the synodic month weren't known with this precision back in the time of Meton, but even the values available to him provided a good match.

$$19 \times 365.25 \text{ days} = 6939.75 \text{ days}$$
$$235 \times 29.53 \text{ days} = 6939.55 \text{ days}$$

Every nineteen years, to the day, the same phrase of the moon occurs on the same date, at least for a century or so. The small discrepancy in the relationship accumulates, however, and over more than a hundred years, it becomes enough to make the moon miss the right date. It is possible that the nineteen-year cycle of the sun and moon was widely known in antiquity. In China it was called the *chang*, but we don't know how early the relationship was recognized. Certainly it was known by the second century B.C., and perhaps much earlier. Two dates for the full moon inscribed on a Shang dynasty (1766–1122 B.C.) oracle bone are separated by exactly eight times the Metonic interval, some think this implies a knowledge of that cycle. There are other hints of its use outside of Greece. Diodorus Siculus, born of Greek descent in Sicily in the first century B.C., wrote a kind of general encyclopedia he called the *Historical Library* and in quoting a lost work by Hecataeus of Abdera, who lived in Thrace at about

300 B.C., mentions a nineteen-year return of the moon known to the Hyperboreans:

> Opposite to the coast of Celtic Gaul there is an island in the ocean, not smaller than Sicily, lying to the North—which is inhabited by the Hyperboreans, who are so named because they dwell beyond the North Wind. . . .
>
> In this island, there is a magnificent grove of Apollo, and a remarkable temple, of a round form, adorned with many consecrated gifts. . . .
>
> It is also said that in this island the moon appears very near to the earth, that certain eminences of a terrestrial form are plainly seen in it, that Apollo visits the island once in the course of nineteen years, in which period the stars complete their revolutions, and that for this reason the Greeks distinguish the cycle of nineteen years by the name of "the great year." During the season of his appearance the God plays upon the harp and dances every night, from the vernal equinox until the rising of the Pleiades, pleased with his own successes.

Perhaps Diodorus and Hecataeus both really knew what they were talking about, but it's not easy to tell. Hyperborea probably was what we know as Britain, and it has even been speculated that the "remarkable temple, of round form" is Stonehenge, the famous prehistoric monument in southern England. Astronomical alignments on the sun and the moon there have prompted some interpreters to regard it as an observatory, although it is more likely that it was a place of ritual and assembly consecrated by celestial and seasonal events incorporated into its architecture.

In any case, Diodorus's description of the nineteen-year interval between Apollo's Hyperborean holidays certainly sounds something like the Metonic cycle, but that's primarily due to his claim that it took nineteen years for the stars to "complete their revolutions." After that remark, Diodorus seems to be playing with less than a full deck. The moon actually would appear to be "very near to the earth" about every nineteen years, but not because the Metonic cycle conducted it there. Another cycle takes about nineteen years—really 18.61 years—and in antiquity it might have been figured, at least descriptively, as a nineteen-year cycle. It occurs because the moon's monthly path does not always follow the same course through the sky. Instead, the moon's orbit slowly swivels, and it takes 18.61 years for it to twist around one complete turn.

THE MOON'S MOVING SIDEWALK

To understand how the moon's orbit shifts, you have to picture it as a circle that goes around the sky. The moon's monthly path looks a lot like the sun's yearly path, the ecliptic, and it almost coincides with the ecliptic. "Almost" means a difference of five degrees. The moon's orbit is tilted about five degrees from the ecliptic—that's about the angle covered by the width of three fingers held at arm's length. The two stars in the front of the bowl of the Big Dipper, the Pointers, are also separated by about five degrees.

Because of this small difference, the ecliptic and the moon's orbit cross each other, twice. These two intersections are called the nodes of the moon's orbit. The line that connects them is known as the line of nodes.

Although the whole sky appears to turn from east to west, from one night to the next the moon moves eastward along its orbit. The sun does the same thing on the ecliptic, but the moon's one-day movement is much greater—on the average, thirteen times greater—and much more obvious than the sun's. For all practical purposes, the moon falls completely around the earth each month in response to the gravitational attraction of the earth, but the sun also influences the motion of the moon. So even as the moon moves eastward through the stars, its orbit actually slips backward a little each month. By perturbing the moon gravitationally, the sun induces the slow but steady pivot in the movement of the moon. After 18.61 years of twisting slowly in the solar wind, the line of nodes at last points again exactly the way it did when the cycle started.

Professor Alexander Thom, a Scottish professor of engineering who carefully studied prehistoric monuments in Britain and France, thought that some prehistoric monuments in Britain were deliberately oriented to the moonrise and moonset at certain times in the 18.61-year cycle. The 18.61-

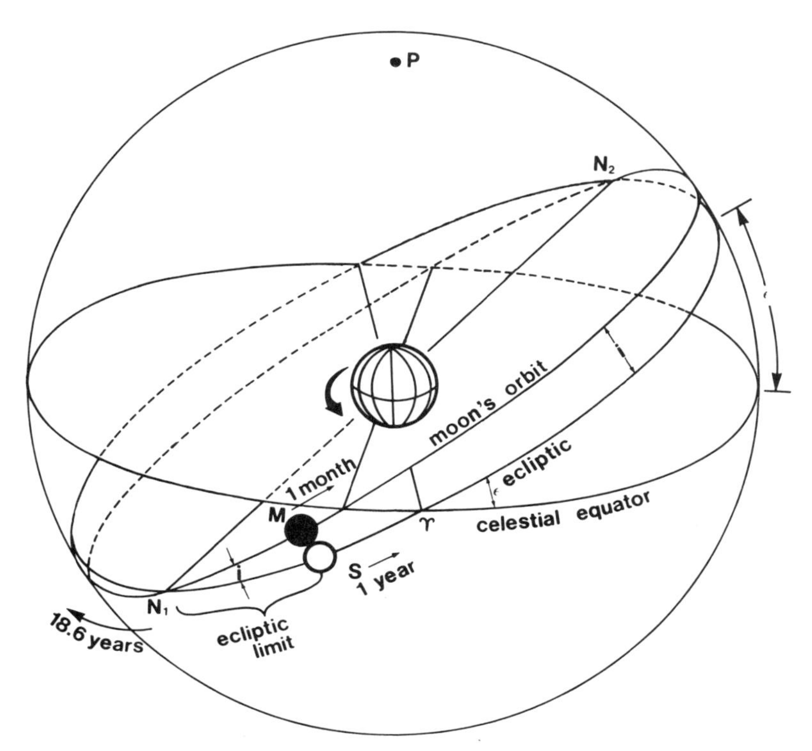

Because the moon's orbit is tilted five degrees with respect to the ecliptic, or path of the sun, the moon appears to follow its own path around the sphere of the sky as it orbits the earth. This circular trail intersects the ecliptic in two places (N_1 and N_2), which are known as the nodes. In this diagram, N_1 is the ascending node, and N_2 is the descending node. The angle i *is the five-degree inclination of the moon's orbit, and the Greek letter* ε *stands for the 23½-degree obliquity of the ecliptic.* P *marks the north celestial pole,* S *is the sun, and* M *is the moon. The symbol that looks like a stylized pair of ram's horns and is located at the intersection of the ecliptic and the celestial equator stands for the vernal equinox. Each month the moon travels completely around its orbit. It takes the sun a year to complete one trip around the ecliptic. Gravity causes the moon's orbit to swivel to the left (clockwise), and for that reason the line that connects the two nodes rotates backward with respect to the motion of the moon in its orbit. This backward movement is called the regression of the line of nodes. One complete regression cycle takes 18.61 years. (Drawing: Joseph Bieniasz, after A. Thom, Griffith Observatory)*

year cycle alters the positions of northernmost and southernmost moonrise and moonset each month. When the greatest monthly excursion in 18.61 years is reached, it is said to be the time of the moon's major standstill. The moon appears to rise and set at the same northern and southern limits for several months in a row, and that's why the phenomenon is called the moon's major standstill. Although there is one month when the absolute limits are reached, the difference is small for at least a year before and after the time of actual major standstill. The summer full moon is always fairly low, but at the major standstill, the summer full moon is more south than ever. From many northern temperate latitudes, the major standstill summer full moon will spend only a few hours above the hills and trees on the southern horizon and disappear below them again. If you go far enough north, above 61½ degrees north latitude, you won't see that summer full moon at all.

The low, major standstill summer full moon may be what Diodorus was talking about when he said that Apollo visits the Hyperboreans every nineteen years, but Apollo was associated with the sun, not the moon. Perhaps that is not really a problem, however. Apollo played a role in the lunar calendar

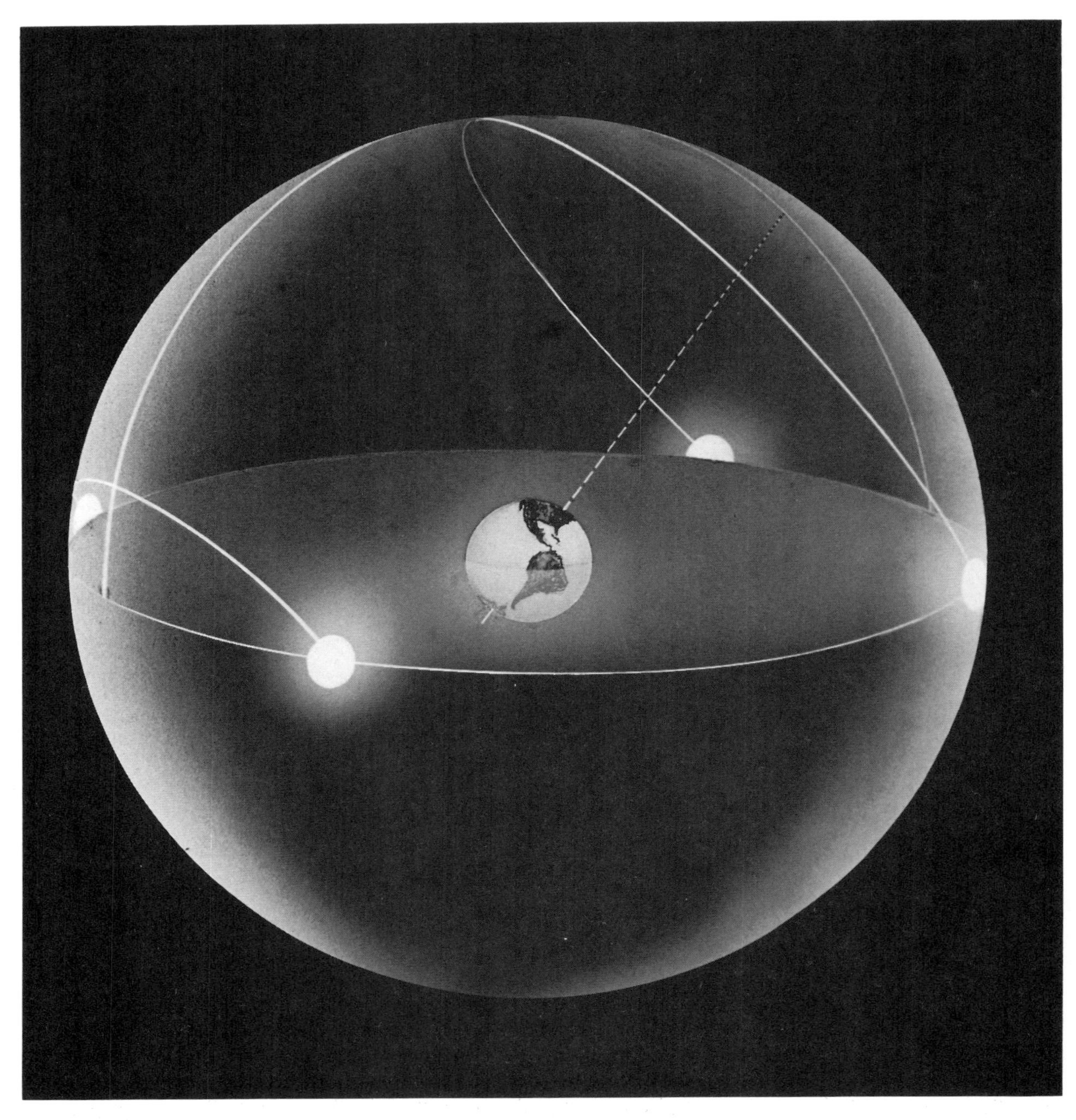

Each month the moon mimics the sun's annual cycle by rising and setting at northern and southern limits on the horizon. Regression of the line of nodes, however, changes the angle between the celestial equator and the path followed by the moon. This cyclical shift in the orientation of the moon's orbit alters the monthly reach of the moon's limits of rising and setting. Every 18.61 years, these monthly limits stretch to their greatest extent. Then the moon is said to have reached its major standstill. For several months in a row, the monthly excursions of the moon will seem to repeat their greatest limits in a way analogous to the sun's repetitive daily path during the days that surround the solstices. We see here the northernmost and southernmost daily paths of the moon at the time of the major standstill in this diagram. When the major standstill occurs, the moon will rise and set as far to the north as we ever see it travel, and two weeks later, the moon will rise and set as far to the south as it can go. (Drawing: John Lubs, Griffith Observatory)

of the Greeks. The seventh day of every lunar month was sacred to him, and his official interests also included the computation of the months.

What, though, did Diodorus mean when he said, "During the season of his appearance the God plays upon the harp and dances every night, from the vernal equinox until the rising of the Pleiades"? It sounds like a description of the moon during the summer months, but it's hard to tell what its nightly performances have to do with either the major standstill or the Metonic cycle. But perhaps we shouldn't be too critical of Diodorus and Hecataeus. The movement of the moon is complicated, and it is easy to get confused. Some ancient skywatchers were motivated to master it anyway though, and we can understand why. If you know exactly what the moon is doing, you can predict eclipses.

10

Within the Shadow

WE have already seen how it was possible for the sun and the moon to follow their own itineraries through the zodiac and get together now and then in calendrical accord. Eclipses, on the other hand, were another story. No matter whether the moon crossed exactly in front of the sun or saluted it in faultless alignment from the opposite side of the sky, a perfect meeting between the sun and the moon meant an eclipse. And eclipses meant trouble.

In general, most ancient peoples were alarmed by the occurrence of an eclipse, or at least anxious about what it might mean. In the *Book of Songs*, or *Shih Ching*, one of the five Confucian Classics assembled during China's Zhou dynasty in the sixth century B.C., solar eclipses are said to be "ugly" and "abnormal." A report in the *Shu Ching*, another Classic, tells us that the "sun and moon did not meet harmoniously" when an eclipse occurred. By the Han dynasty (206 B.C.–A.D. 221), Chinese astronomers understood why eclipses occurred. Despite this knowledge, solar eclipses still were judged to be omens, and their messages depended on when the eclipse was seen.

For the ancient Babylonians, not every eclipse

Total solar eclipses are unusual and commanding. When the moon crosses directly in front of the sun, it looks as if a hole has been punched in the sky. The sun's outer atmosphere, or corona, becomes visible during totality and frames the moon's dark disk with a luminous mane. This total solar eclipse occurred over central Java on June 11, 1983. (Photograph: Allen Seltzer)

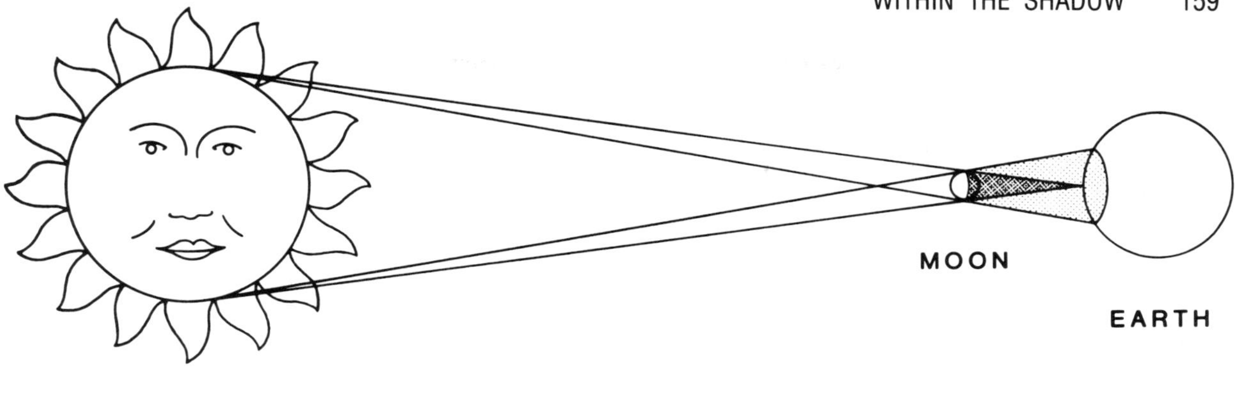

Accurate alignment of the sun, moon, and earth sends the tip of the deep central shadow cone of the moon to the earth. The tip of this shadow covers only a small area of the earth, and only there does the eclipse appear total. An outer shadow, shaped like a truncated cone, strikes a larger zone. Those who reside within it see a partial eclipse of the sun. (Griffith Observatory)

was bad news, but it was always consequential. This omen from R. Campbell Thompson's *The Reports of the Magicians and Astrologers of Nineveh and Babylon* lets us sample a little of the Mesopotamian meaning of the eclipsed moon.

> When the Moon is eclipsed in Siwan [one of the lunar months], there will be flood and the product of the waters of the land will be abundant. When in Siwan an eclipse of the morning watch happens, the temples of the land will be smitten. Samas [the sun god] will be hostile. When an eclipse happens in Siwan out of its time, an all-powerful king will die and Ramman [the storm and weather god] will inundate; a flood will come and Ramman will diminish the crops of the land; he that goes before the army will be slain.

During an eclipse, we either find ourselves in the shadow of the moon or watch the moon negotiate the shadow of the earth. In either case, we witness something unusual and spectacular. Solar eclipses demand our attention because they transform the ordinary world of daylight into an eerie, untimely night. Time seems to be suspended or even reversed when the sun disappears before its proper time at sunset and far from its proper place in the west. In an eclipse of the moon, the bright full moon goes dramatically dark and advertises the disruption of heaven. The eclipsed moon often

The sun, earth, and moon must be well aligned for a total lunar eclipse, too, but in this circumstance, the earth is between the sun and moon. The earth's shadow puts the moon in eclipse. (Griffith Observatory)

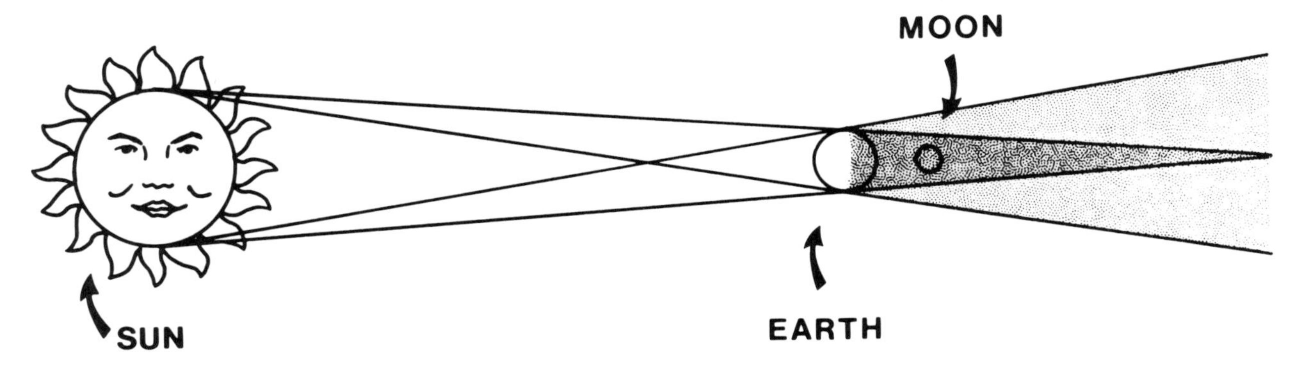

takes on a deep copper-red hue from sunlight scattered through the earth's atmosphere.

It's actually a coincidence that we ever see total eclipses of the sun. They occur, of course, when the moon cuts right across the sun's face. As the moon intrudes, the sun's light is blocked by the dark silhouette of the disk of the new moon. For their full effect, total solar eclipses depend on the fact that the disk of the moon looks as if it is exactly the same size as the disk of the sun. The moon's diameter is really about 400 times smaller than the diameter of the sun, but the sun is about 400 times farther away. By chance, the sun and moon appear to be identical in size. If the sun were larger or nearer, we might see it partially covered by the moon now and then. If the sun were smaller or farther, we might see it get easily lost behind our moon. But in neither case would we see the unique effects produced by a solar disk closely matched in size to the lunar disk.

Eclipses of the moon are a little different. They occur when a perfect alignment of the sun on one side of the earth with the moon on the other forces the full moon into the earth's shadow.

Photographed on the night of July 5–6, 1982, this total lunar eclipse sequence begins with the image on the right. The moon is slipping into the shadow of the earth, and it is completely enveloped in the central image. On the left, the moon is starting to emerge. The whole series represents nearly three hours of celestial motion. The northern half of the moon seems to disappear completely, and that irregular darkening of the moon was due to clouds of volcanic aerosols dispersed into the northern hemisphere by the El Chichón volcano in Chiapas, Mexico. (Photograph: Ronald E. Royer)

DARKNESS IN BROAD DAYLIGHT

Everything is out of the ordinary during an eclipse. Even when people know it is scheduled, they still get excited. It is easy to see why. A total solar eclipse is one of the grandest spectacles in nature. This eyewitness account of the performance the sun and moon staged over southern Europe on July 8, 1842, tells us what all the fuss is about.

> First the light dimmed as in a theater. The meager illumination which remained was of a bluish-black cast. Objects seemed smeared with the lees of wine, and our faces had a greenish hue. Oxen and asses stood still, cows stopped grazing, and even ants ceased their activity. Horses, on the other hand, continued plodding forward as reliably as locomotives. Many flowers closed, but the sensitive mimosa did not.
>
> A cry arose in the streets. At that moment, the Moon became surrounded with an aura, as large and glowing as a saint's halo. Pure white rays shot out, darting and flickering in all directions. We had been expecting an extraordinary sight, but we were unprepared for anything so magnificent. I confess that the weird and wonderful spectacle alarmed me. But in a moment all that we had seen was far surpassed. From the black rim of the Moon there suddenly shot forth three gigantic tongues of flame. They paused, motionless, like jagged mountain peaks in an alpine sunset. Each was different from the others, but all were many times larger than the white circlet of rays. It was as if the Sun, behind the Moon, were flaming up in monstrous volcanic explosions. (From Rudolf Thiel, *And There Was Light* [New York: Alfred A. Knopf, 1957], p. 272)

Many of the world's most prominent astronomers traveled to Italy to see the 1842 eclipse and to make the first modern scientific observations of a total solar eclipse. When the sun's disk went completely dark, the surprised and delighted crowds of onlookers in Milan cheered spontaneously. "Long live the astronomers!" in gratitude to the only ones they could credit for putting on such a great show.

Public enthusiasm for eclipses continues to be high. Nearly three thousand visitors assembled on the night of July 5, 1982, to watch the brilliant, round, white, and perfectly normal full moon go dark in total eclipse above the ramparts of Griffith Observatory. They could have all seen the event perfectly well from home, but for these visitors, the eclipse was more satisfying as a community event. At 10:33 p.m., Pacific Daylight Time, the moon entered the earth's shadow, in which everyone on the ground had already been standing since sunset. Darkness flowed slowly onto the white disk. By 11:38 p.m., the moon was completely engulfed and unevenly shaded. Its top half was completely black and so invisible. The bottom half was a deep copper red, tinted much darker than usual by the aerosols released into the atmosphere in late March by a volcano in Mexico. The huge crowd broke into cheers and applause for the gorgeous but eerie costume the fully eclipsed moon had donned.

DARK DISTURBANCE IN THE WORLD'S RHYTHM

The beauty of an eclipse was not always appreciated by those who saw in it a threat to the order of their everyday lives. Imagine what ancient people must have felt when they saw the sun darken without warning. The order of the sky was violated with an invasion of darkness. The irregular halo of the sun's pearly corona, the scarlet flames of solar prominences, and the jewels of sunlight splashing through mountain valleys on the moon's rim—Baily's beads—were only revealed during a total solar eclipse, and they turned the familiar image of the sun into a panic-stricken, wounded victim of an unseen celestial terrorist. It was a wild, uncanny event. To those who witnessed it, the foundation of the world began to slide toward the brink. The sun was under assault, and if the sun were menaced, the life and landscape below it were also at risk.

By tracing the word *eclipse* back to its Greek origin, *ekleipsis*, we learn that to the Greeks it meant "an omission" or "an abandonment." When either the sun or moon is eclipsed, it seems to disappear from the sky. It abandons us to the darkness. Such departures are more serious than the word *abandon* suggests, however, for eclipses were viewed as ruptures in the membrane of natural order. When the sun or moon blinked in eclipse, chaos winked from the sky, delighted with its opportunity for mischief or doom.

The Aztec also found eclipses to be upsetting. In Book Seven of Fray Bernardino de Sahagún's *Florentine Codex,* we can read what happened when a solar eclipse occurred over ancient Mexico. The sun

> . . . turned red; he became restless and troubled. He faltered and became very yellow. Then there were a tumult and disorder. All were disquieted, unnerved, frightened. There was a weeping. The common folk raised a cry, lifting their voices, making a great din, calling out, shrieking. There was shouting everywhere. People of light complexion were slain [as sacrifices]; captives were killed. All offered their blood; they drew straws through the lobes of their ears, which had been pierced. And in all the temples there was the singing of fitting chants; there was an uproar; there were war cries. It was thus said: "If the eclipse of the sun is complete, it will be dark forever! The demons of darkness will come down; they will eat men."

Later Hebrew commentaries on the Book of Genesis indicate that an eclipse occurred on the day the animals boarded Noah's Ark. Along with trembling in the foundations of the earth, lightning flashing from heaven, and the boom of thunder overhead, the darkening of the sun signaled the end of an era. Tupinamba Indians in northern Brazil see the possibility of world's end in every eclipse, and according to the *Völuspá,* one of the poems of the Norse *Edda,* the sun will turn black at the end of the world. Some people, like the Tang dynasty Chinese, saw eclipses as a sign of ongoing decay. These things didn't happen, they maintained, in the good old days, when the stronger moral fiber of the times inhibited such afflictions. To Mexico's Chamula Indians, Tzotzil Maya inhabitants of the Chiapas highlands, eclipses still demonstrate that the demons of chaos can gain the upper hand now and then, even in the realm of the representatives of order—the sun and the moon.

VICIOUS ATTACKS

Naturally, people tried to explain what was going on when an eclipse was seen, and conflict is frequently the underlying theme of their stories. Miwok Indians living near what are now San Francisco and Sacramento saw eclipses as signs of the continual struggle for power between the sun and the moon. Just north of Miwok territory, the Pomo name for a solar eclipse was "sun got bit bear," and the Pomo told a story about a bear out for a walk on the Milky Way. He met the sun, who would not get out of the way, and so they began to fight. When they fought, the sun was eclipsed. Then the bear continued and ran into the moon, the sun's sister. There was, of course, another fight and another eclipse. Some Arawaks of Guyana figured the sun and moon fought each other during an eclipse, and the Gê, Amazonian tribes of Brazil, also believe that eclipses are what result when the sun and the moon fight.

The Gê also say an eclipse occurs when the eye of the sun or the moon is pierced. A small boy inflicts the wound with an arrow. The damage bloodies and darkens the eye, but when a shaman removes the arrow, the eye is healed and the eclipse passes.

An eclipse over Babylon meant that seven devils had stopped the moon from carrying out his celestial duties and were holding him prisoner.

Mongol-speaking peoples in southern Siberia attribute eclipses to the monster Arakho. Apparently nourished by gnawing the hair off human beings, Arakho angered God with his questionable taste. To escape God's wrath, Arakho hid himself so well that no one could find him. The moon, however, knew his whereabouts and spilled the beans when God asked where Arakho was concealed. When God pulled Arakho out of his hideout, he cut him in two pieces. The top half of Arakho now constantly seeks—and occasionally enjoys—his revenge on the moon, which is eclipsed when caught in Arakho's spiteful grip.

Eclipses were often interpreted as attacks upon the sun or the moon by some monster like Arakho. The Balts were convinced that serpents, dragons, and witches occasionally tried to destroy the sun and moon. The Northern Altai of southern Siberia's Altai Mountains attributed lunar eclipses to a man-eating creature who resided in a star. Island Caribs, off the coast of Venezuela, blame Maboia, the devil, for eclipses. He sneaks up on the sun and the moon and cuts their hair, forces them to drink a child's blood, and tries to kill them. In New Zealand, the Maori described a solar eclipse as an attack by demons who then ate the sun. Despite this apparently fatal assault, the sun always recuperated.

SNACKING ON THE SUN, MUNCHING ON THE MOON

In Chinese, the earliest word for "eclipse" is *shih.* Inscribed in Shang dynasty oracle bones, it means "to eat" and reiterates an eclipse theme found all over the world: When an eclipse takes place, the sun or moon is devoured by one voracious beast or another. In the eyes of the Chinese, a dragon dined during an eclipse. The Armenians said the same thing, and the idea of dragon-induced eclipses persists in the astronomical terminology of today. As we shall see, the nodes of the moon's orbit—its intersections with the path of the sun—are important as far as eclipses are concerned. The time it takes the moon to circle back to the same node and the time it takes the sun to do the same thing are called the draconic month and draconic year, respectively. The Latin and Greek words, *draco* and *drakon,* both mean dragon. Archaic references to the ascending node as the "dragon's head" and the descending node as the "dragon's tail" clearly affiliate eclipses with a dragon, hungry for celestial wafers.

Eclipse glyphs in the lunar texts of the *Dresden Codex,* a Maya hieroglyphic divinatory almanac from Postclassic Yucatán, hint that the Maya also felt that something was eating the moon during an eclipse. In two cases, the eclipse glyph is suspended over the open maw of a toothy serpent. A post-Conquest book of Maya prophecy, *The Chilam Balam of Chumayel* (The Spokesman of the Jaguar of the city of Chumayel, in central Yucatán), calls a solar eclipse "the eating of the sun."

In the southernmost part of California, the Southern Diegueño Indians, also known as the Tipai or the Kumeyaay, called the first nick of an eclipse "nibbles the sun" or "nibbles the moon" and began a ceremony intended to mitigate the ill effects of the event.

Spirits of the dead who have taken the form of jaguars try to bolt down the potbellied moon, according to the Toba Indians of the Gran Chaco in Paraguay and Argentina. His wounds leave him bloodied, and he escapes only because people on the earth shout and make their dogs bark to scatter the jaguars.

Along the Trinity River in northern California, the Hupa Indians picture the moon as a man with twenty wives and a house full of "pets," mostly mountain lions and rattlesnakes. After a hunt, he carries the game he has killed back up to his pets and feeds them. When they are not satisfied with what he brings, they eat him instead, and he bleeds. The blood is the red color seen when the moon is eclipsed. During the eclipse, one of Moon's wives, the Frog, stands over Moon and knocks the animals away. The other wives collect Moon's blood together, and he recovers in time for his next sky journey.

An evil spirit in the form of a frog or a toad hungers for the sun during an eclipse as far as the Shan people of Annam (in what is now Viet Nam) are concerned. The Tatars of western Siberia and their Chuvash neighbors say the agent of eclipses is a vampire who tries to swallow the sun and the moon but spits them out again when they singe his tongue. Far to the east, the Buryats of southern Siberia explain that a beast they call Alkha ingests the sun and moon when they are eclipsed. Among the Yugoslavians, a Vukodlak—a cloud-chasing werewolf—stuffed himself with the sun or moon and caused eclipses. The Hungarians claimed a giant bird, the markaláb, or a snake dined on the light of the sun and the moon during their eclipses and then let them loose. Unspecified "animals" ate up the sun and the moon when the Finns witnessed eclipses. Galibi Indians of French Guiana believe the hungry culprit was a monster they could chase off with noise and arrows. In an interesting variation on the devouring theme, the Nootka on Vancouver Island maintained eclipses occurred when the sun or moon was swallowed by the "door of heaven," and their Kwakiutl neighbors on the southern coast of British Columbia got right to the point and called upon the "Mouth of Heaven" to vomit up the sun or moon when it was eclipsed.

THE HEAVENLY DOGS TAKE A BITE

According to the Vikings, a pair of ravenous wolves made meals out of the sun and the moon. *Grimnismal,* one of the poems of the *Elder Edda,* names them:

> *Skoll the wolf who shall scare the Moon*
> *Till he flies to the Wood-of-Woe:*
> *Hati the wolf, Hridvitnir's kin,*
> *Who shall pursue the Sun.*

Snorri Sturluson, the thirteenth-century Icelandic author, also mentions these two sky wolves: Skoll (Repulsion), he says, will eventually catch the sun, while Hati (Hatred), who runs ahead of the sun, will overtake the moon. When either orb is caught, an eclipse is seen, and people on earth rescue the imperiled sun or moon by making a racket to scare the wolf into dropping its treat. Snorri also tells us about Mánagarm (Moon's Dog), the son of a giantess from Ironwood (or the Wood-of-Woe), who will consume the moon and spill its blood all over the sky.

The Chiquitos Indians of Bolivia's eastern lowlands said that huge dogs constantly chased after the moon. When they caught her, their vicious bites shredded her flesh. Her blood poured out, dimmed her light, and reddened her face. Howling and wailing, the Indians scared off the savage dogs.

For all practical purposes, the story that was told in Bolivia's Gran Chaco was also told in Korea. There they said the king of Gamag Nara, the "Land of Darkness," tires of gloom and sends his horrible Fire Dogs into our world to steal the sun and the moon. On one of these raids, the worst of these creatures caught the sun in its jaws but could not hold it. The heat was too great. The king chewed out his pet and sent the second fiercest Fire Dog after the moon instead. Even though he caught it, the moon was so cold it froze the Fire Dog's teeth, and again the prize had to be abandoned. Despite these failures, the king persists, and every time the sun or moon is snatched by one of the Fire Dogs, there is an eclipse.

Some Chinese said that during an eclipse the sun was eaten by the Heavenly Dog, who also played an important part in an eclipse story told by the Kawa people, a non-Chinese minority of Yunnan Province, in southwestern China. Long ago, as the Kawa say, the moon managed to steal the herb of immortality, and the Heavenly Dog attempts every now and then to get the herb back from the moon. He bites pieces off of her, and she hides. The moon is eclipsed until the dog tires. Then she restores her shining light with the herb once more.

INTIMACIES BEHIND THE BLACK CURTAIN

Although violence and a taste for celestial appetizers activate many mythical explanations of eclipses, some eclipse stories are tales of love, not war. To Australian Aborigines, the moon and sun were husband and wife, and when they mated, there was a solar eclipse, with presumably a lot of action going on behind the curtain of darkness. The sexes are reversed according to the Tlingit Indians of British Columbia's most northern coasts. Sun is the husband, and Moon is his wife. After that, it's the same story. She makes a conjugal visit, and they enjoy each other's company, screened by the eclipse their coupling has created.

This idea of connubial bliss in an eclipse even finds its way into contemporary culture. After the *Fairsea* had successfully conducted a cruiseload of passengers to a total solar eclipse on October 12, 1978, in the northeast Pacific, a flag raised on the ship's mast portrayed the eclipse as a kiss between a lady sun and gentleman moon. Eclipse symbolism was also used in *Ladyhawke,* a 1985 motion picture with a medieval fantasy setting, to free two lovers, a lady and her knight, from a curse that supernaturally separates them. By day, the lady is a hawk, who travels on the wind or on her lover's arm. At night, her human form returns. But as she becomes a beautiful woman once more, the man turns into a wolf. He prowls the forest and protects her from danger, but they can never meet. Tormented every twilight and dawn, as they see each other slip from one shape into another, they grow less and less hopeful they can ever be liberated from the magic that keeps them apart. They can only escape this agonizing existence if the jealous bishop responsible for their unholy fate sees them together and human. That, of course, is impossible, for it would have to be day and night at the same time. That's where the eclipse comes in.

DISPELLING THE DARKNESS

Apart from a few benign interpretations, eclipses were usually seen as a threat. Understandably, people tried to fight back against the intrusion of chaos through ritual activity undertaken to liberate an imperiled sun or moon. The Kwakiutl, for example, burned boxes, blankets, and food in a big bonfire and made loud noises to send the big mouth in the sky to another diner. By screaming and throwing stones into the air, the Buryats figured they could make Alkha sufficiently unwelcome. South America's Tupinamba Indians also take action during an eclipse. While the men sing to the primordial tribal ancestor, the women wail. In their anguish, they and the children fall to the ground.

An old man of the Kawaiisu, an Indian tribe settled in southern California's Tehachapi Mountains would sing during an eclipse to cure the sun or moon of its illness, and south of what is now Los Angeles, the Luiseño Indians would sing songs prescribed for a sick moon in eclipse. At Mission San Luis Rey, people of this same tribe clapped and shouted during an eclipse to drive away the animal they said wanted to eat the sun or moon. California's Serrano Indians thought that spirits of the dead tried to eat the sun or moon during an eclipse. Everyone shouted together to stop the atrocity, watched the shamans and ceremonial assistants dance and sing, and avoided all food to starve the spirits out.

Mirrors stood for the moon in China, where banging on mirrors during an eclipse is a time-honored tradition. In Ch'ang-an (now the city of Xi an), the Tang dynasty (A.D. 618–906) capital in north central China, this activity was sustained throughout the city until the dragon coughed up the moon.

A legend connected with that inharmonious encounter of the sun and moon reported in the *Shu Ching*, and mentioned above, emphasizes the importance of proper conduct during an eclipse. The court astronomers in it are Hsi and Ho. While the eclipse was underway, the musicians, the officers, and the common people all got busy in an attempt to spare the sun the indignity it was suffering at the hands of the moon. Hsi and Ho, however, had been partying with a little too much rice wine. While everyone else was pitching in on behalf of the sun, Hsi and Ho were goofing.

> They were mere personators of the dead in their offices, heard nothing and knew nothing. Stupidly they went astray from their duties. . . .

And they were beheaded for neglecting their duties. Their example continues to exert a sobering influence on astronomers.

We have a Mesopotamian cuneiform tablet to fill us in on Babylonian methods for dealing with eclipses. Because the Babylonians could predict at least some eclipses, they had an edge in their interaction with the moon's transformation. The priests of the city of Uruk set up an altar beforehand, and when the eclipse started, they shouted their request that neither catastrophe, murder, rebellion, or eclipse should come calling on their city and its important temples, shrines, and palaces. Copper instruments—trumpets, harps, and drums—were carried outside from the temples and paraded on the moon's behalf.

Noise during an eclipse certainly seemed helpful, but many peoples were certain that firing projectiles into the sky would scare off the intruder. The Ojibway Indians of Minnesota and Ontario had a different idea, however. Judging that the sun in eclipse was risking death, they tried to rekindle him by firing arrows tipped with fiery coals at him.

It is easy to ridicule the various measures people have taken on behalf of a sun or moon suffering eclipse, but the record is quite clear: They never fail. The sun and moon inevitably return to their full brilliance and proper course. There may be a lesson here.

Because eclipses are departures from business-as-usual in the sky, people respond to them with unconventional behavior. This lunar eclipse over Tashkent, in what is now Uzbek S.S.R. in the Soviet Union, was witnessed on December 16, 1880. With the noise of cymbals, tambourines, and drums, they attempted to drive away whatever it was that was savaging the moon. (From l'Astronomie *by Camille Flammarion, 1887)*

Understanding how eclipses happen—and predicting when they will—becomes possible only after the movements of the sun and moon are mapped and monitored. It takes an accurate calendar and system of reference stars to do this, and the ancient Mesopotamians were among those who devised such a set of celestial markers. With those stars and constellations in "the path of the moon," the Mesopotamians entered a celestial sweepstakes that eventually rewarded them with reasonable success at lunar eclipse prediction and the ability to know when solar eclipses were either possible or out of the question.

NOT ALWAYS INCLINED TO ECLIPSE

It must have been realized thousands of years ago that an eclipse can only occur when the moon is either new or full. When the moon is lined up with the earth and sun, either in conjunction with the sun at new moon or opposite the sun at full moon, the moon is said to be in syzygy. This word is derived from the Greek *suzugia,* which means "a joining together." The three objects join together to put us within the biggest shadows we ever encounter. Of course, we don't see a solar eclipse at every new moon and a lunar eclipse every time the moon is full. Personal experience tells us eclipses are much less frequent than that, but why don't we worry our way through a solar eclipse and a lunar eclipse every month?

At new moon and full moon, the sun, earth, and moon are aligned, but not perfectly aligned. That's because the moon's orbit is tilted five degrees to the earth's orbit around the sun. To us here on the earth, the moon's path appears to be tilted five degrees to the sun's path. At new moon, the moon is a little above or below the sun, but rarely right in front of it, and it can't mask the sun's disk if it doesn't cross right in front of it. The inclination of the moon's orbit guarantees that most of the time it won't. Lunar eclipses are subject to the same limitations. The moon has to cross into the earth's shadow to be eclipsed. At some full moons, it passes above the shadow; other times, it slips below.

The moon must be new for any solar eclipse and full for any lunar eclipse; these are necessary conditions, but they aren't sufficient. Both sun and moon must be on both the sun's path and the moon's path. This happens every time they are both on a node, one of those two junctions between the moon's orbit and the ecliptic. When the sun and moon occupy the same node, we get hit with a solar eclipse. On opposite nodes, we endure a lunar eclipse. So if you want to know when eclipses will occur, you have to know where to find the sun, the moon, and the two nodes of the moon's orbit.

THE DECAPITATED DEMON OF ECLIPSES

The relationship between eclipses and the nodes of the moon's orbit was recognized early enough in ancient India to be included in the *Mahabharata,* the Sanskrit epic poem that is the source of many Hindu myths. According to this text, eclipses originated when the gods of India decided to produce

A five-degree tilt of the moon's orbit diminishes the accuracy of alignment and therefore the possibility of an eclipse at every full and new moon. Sometimes the moon is a little above the plane of the earth's orbit, and sometimes it is a little below it. If the full or new moon happens to occur when the moon is at one of its nodes, an eclipse is on the agenda. (Griffith Observatory)

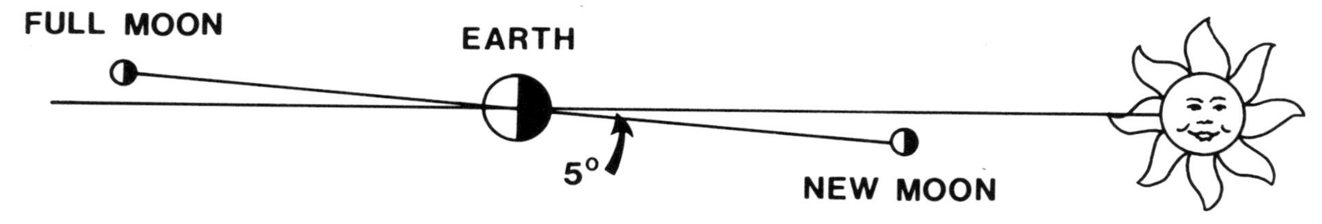

soma, the elixir of immortality. The Hindu gods enlisted the help of the demons to churn the cosmic ocean, and from this primordial chaotic soup, the ambrosia would emerge, as cheese curdles out of milk. Together, the gods and demons made a churning stick out of Mount Mandara, one of the massive peaks in the Himalayas, and began the work. Steadying the mountain-sized churning stick on the tortoise incarnation of the god Vishnu, they made it twist by pulling on the great serpent Vasuki coiled around it. Many things congealed out of this ordered movement: the moon, the sun, the goddess of fortune, the goddess of wine, the white horse of the sun, the magic tree, the magic cow, the great elephant Airavata, and other "gems."

When finally the soma appeared, Vishnu assumed the guise of an extraordinarily attractive woman, took all of the soma that was in the hands of the demons, and distributed it to the gods. By consuming it, they became immortal. The demons then woke from their enchantment and began to battle with the gods for a share of the elixir. In the midst of this confusion, one of the demons, Rahu, seized the opportunity to disguise himself as a god and steal a swig of soma. Just as he swallowed the precious liquid, he was spotted by the sun and moon, who recognized the imposter and reported his crime to Vishnu. Vishnu sliced off Rahu's head before the holy beverage of immortality could drop past his throat. His body collapsed on the spot and fell to earth, but his head had already been made immortal through contact with the soma and remained in heaven. There, out of hatred for the pair who informed on him, Rahu chases the sun and the moon. Now and then he catches one, and when he swallows, an eclipse occurs. Rahu may fail to swallow them completely, scared off by mortals below who bang with utensils on pots and pans. If, on the other hand, Rahu manages to consume either of them entirely, they always drop out of his severed throat and make good their getaway.

COUNTING ECLIPSES WITH THE MAYA CALENDAR

As a record of eclipses is accumulated, it becomes apparent that eclipses return in cyclical patterns.

Balinese Hindus call Rahu, the eclipse demon in the Mahabharata, *Kala Rau. In this scene from a traditional Balinese painting, Kala Rau is losing his head as he is discovered sipping the elixir of immortality. In anger, he circles the sky in pursuit of the two informers—the sun and moon. When he catches either of them, his severed immortal head swallows them, and an eclipse takes place. Then the ingested object drops out of Kala Rau's severed throat: The eclipse is over. (Drawing: Joseph Bieniasz, Griffith Observatory)*

In India, the two nodes of the moon's orbit are personified and named. The ascending node is said to be the head of Rahu and is called Rahu. Ketu, the descending node, is Rahu's tail. Rahu (left) and Ketu (right) are represented as planet gods on this lintel from the Temple of the Sun at Konarak. The lintel is now housed in a small building near the main temple, and Hindu priests make daily offerings to the planet gods on behalf of pilgrims and visitors.

Ancient skywatchers tackling the cyclical recurrence of eclipses realized that they occur only on certain days of the month and only in certain months. Eclipses, then, have "seasons."

In ancient Mexico and Central America, the Maya exploited the periodicity of eclipses to develop techniques for predicting them. Even though they did not picture the movement of the sun, the moon, and the nodes as we do, their systematic observations and records of celestial events enabled them to keep track of eclipses. Tables in the *Dresden Codex*, the Maya hieroglyphic divinatory almanac from Postclassic Yucatán, list groups of 177-day intervals with a 148-day period breaking the sequence every so often. These two spans of time are equivalent to six lunar months and five lunar months, respectively, and show that eclipse seasons were predicted from the cycle of lunar phases and not from the arrival of the sun on a node. About six synodic months are required to carry the moon from one eclipse season to the next, roughly half a year later. The Maya bundled these days together in a way that allowed them to anticipate the possibility of an eclipse. Apparently the table predicted subsequent eclipses from the date on which an eclipse was actually seen and used a 405-month period to do so. They wanted to count over enough lunar cycles to produce an eclipse-predicting interval defined by a whole number of days. How well they did this gives us a feeling for the accuracy of their techniques.

$$405 \times 29.530588 \text{ days} = 11{,}959.89 \text{ days}$$

The Maya called this 11,960 days and were off the mark by only 2 hours, 38 minutes over a 32-year-and-9-month period. They didn't express the length of the synodic month with the six-decimal precision specified here or observe the position of the moon with instruments capable of delivering that level of precision, but they did something just as good. They found, with a relatively small discrepancy, the total number of whole days equal to a total number of whole months.

PREDICTING ECLIPSES IN GREECE

Any account of the history of eclipse prediction must include the story of Thales of Miletus and his alleged forecast of the total solar eclipse that stopped a war. Thales, an Ionian Greek mathematician and astronomer, lived from 624 to 548 B.C. and earned superstar status as one of the Seven Wise Men of ancient Greece. In *The Histories*, the well-known and pioneering Greek writer and traveler Herodotus (484?–425? B.C.) discusses Thales and tells us that the Lydians and the Medes had fought each other for several years. Finally, after five years of intermittent warfare, the two armies stopped fighting in the midst of yet another battle when "day was suddenly changed to night." According to Herodotus, Thales had foretold this "change from daylight to darkness" by informing his fellow Ionians of the year in which it would happen. Modern calculations have identified a total solar eclipse which took place on May 28, 585 B.C. and was visible in northern Turkey, where the battle was fought. Apparently a total solar eclipse could still influence human affairs in the sixth century B.C. Suspending hostilities on a dime, the two startled armies immediately sued each other for peace.

Although this purported prediction by Thales has been the target of considerable skepticism, a review by science historian B. L. van der Waerden concludes that Thales could have done what Herodotus said he did. Some historians, especially D. R. Dicks, the author of *Early Greek Astronomy to Aristotle*, are convinced, however, that van der Waerden speculates far beyond what the actual evidence will allow. Thales could have made an educated guess of the year in which an eclipse would occur, and when his prediction was confirmed and remembered by his contemporaries, his stock rose, at least in Herodotus's book.

SHAKING DOWN THE LOCALS

The ability to predict eclipses continues to be associated in our own era with sophisticated scientific knowledge and modern technical skill. We can see that in Mark Twain's novel *A Connecticut Yankee in King Arthur's Court*, where knowing something about eclipses is a real asset for Hank Morgan, the Connecticut Yankee who is incongruously transported through time to sixth-century Camelot by a blow on the head. Morgan learns he has arrived on June 19, 528, but perceived by the court as a wizard, he is condemned to be burned at the stake at noon on the twenty-first. Remembering that there was an eclipse of the sun on June 21, 528, he contrives a plan to save himself from the flames by putting his knowledge of historical eclipses to work. He sends a warning from his cell that there will be dire consequences at high noon:

> . . . I will smother the whole world in the dead blackness of midnight; I will blot out the sun, and he will never shine again; the fruits of the earth shall rot for lack of light and warmth, and the peoples of the earth shall famish and die, to the last man.

This is obviously a very clever plan for a man caught out of time. All it takes is iron nerve and an uncanny recall of historical eclipses. This one, by the way, is entirely fictional, Twain's pure invention, but Hank Morgan is ready to put big juju on the Round Table. Then there's a serious snag. Morgan's incineration is moved up a day. Those Dark Age courtiers may not know much about eclipses, but they're not bad at preemptive strikes. Fortunately, Morgan's palace contact has confused the calendar, and the day the Yankee is led to his own cookout is June 21 after all. Morgan manages the eclipse with the finesse of a stage magician and secures an influential position in King Arthur's court, a beneficiary of the timely appearance of the shadow of the moon.

While awaiting execution in his dungeon cell, Hank Morgan had recalled another time when knowledge of an upcoming eclipse had proved advantageous:

> It came into my mind, in the nick of time how Columbus, or Cortez, or one of those people, played an eclipse as a saving trump once, on some savages, and I saw my chance.

He was lucky his memory of past eclipses was a lot less muddy than his recall of history. It was Christopher Columbus who used an eclipse of the moon on his fourth voyage to the New World to get himself out of a jam in Jamaica, where he was caught short of supplies.

Columbus had to know something about the sky, or he would never have gotten out of Spain's harbors. His astronomical tables told him there would be a total eclipse of the moon on the night of February 29, 1504 (according to our present Gregorian calendar). Stranded by his low provisions, Columbus faced real catastrophe when the Indians lost interest in trading food for exotic goods from across the sea and began to put their balance-of-payments deficit in order. He decided then to do a number on the Indians—the first in a long series of European fast ones. Taking full advantage of his knowledge of the upcoming eclipse, Columbus informed the Indians he would get rid of the moon for good if they failed to act as cooperative trade partners. At first, they were skeptical, but when the moon actually started to go, they changed their tune. Columbus knew high drama when he saw it, and he continued to kidnap the moon until just before the end of totality. Then he relented and interceded to bring back the moon. He was lucky, of course, to be beached on Jamaica. If he had been stuck in Maya territory, they might have been expecting the eclipse themselves.

You can't always count on manipulating the locals with arcane eclipse lore, however. A British astronomical journal, *The Observatory*, documented an effort by an English administrator to browbeat the chief of a tribe under his jurisdiction in the Sudan. Insisting on the chief's compliance with his instructions, the Englishman threatened to take a bite out of the moon the next night. Unimpressed by the warning, the chief informed the official that if he were referring to the expected lunar eclipse, it would be seen on the night after next.

MERCHANDISING THE DARKNESS

Today, the whole world is much better informed about eclipses. Demons like Rahu used to inject plenty of anxiety into them, but as their arrivals gradually were seen to follow a pattern, the demons once thought to menace the sun, moon, and cosmic order were domesticated. Now eclipses confirm instead of challenge our expectations of what takes place in the sky. So when the Hindu eclipse demon Rahu showed up again in his Indonesian identity, as Kala Rau, for the Great Ramadan Eclipse of June 11, 1983, over Java, he received a reception consistent with the times. It was one of the most commercialized total solar eclipses the world has experienced.

Kala Rau is the Balinese version of Rahu, and the Balinese, unlike most of the rest of the Indonesians, are Hindu and not Moslem. The eclipse in 1983, however, turned out to precede the most sacred month in the Islamic calendar, Ramadan. Like all Islamic months, Ramadan starts with the first crescent, right after new moon. The new moon before Ramadan, 1983, also produced a total solar eclipse. Although most of the Moslem inhabitants of Indonesia were preparing for Ramadan, that didn't stop them from appreciating the visit by a Hindu eclipse demon.

While preparing for the eclipse, you could stock up on at least two Indonesian beers specially packaged in commemorative eclipse cans with Kala Rau's portrait on them accompanied by the inscription *Gerhana Matahari Total*. That, for all practical purposes, means "total eclipse of the sun," but the literal translation of *matahari*, the word for "sun," is "beautiful, shining." It sounds like a good name for an alluring and exotic female spy.

Banners and posters proclaimed the coming of the eclipse, and some offered advice on what to do about it. Basically, the message was don't look at the totally eclipsed sun. Now this is rather strange, because thousands of people from all over the world had spent an awful lot of money traveling to Indonesia for five minutes and six seconds of totally eclipsed sun. There is a reason, of course, for this uninformed advice. Although the totally eclipsed sun is harmless and well worth viewing, the partial stages of a solar eclipse are dangerous to the eye. That's because looking straight at the sun can always damage the retina. People, perhaps, are tempted to look at the sun when it is partially eclipsed—and sometimes use completely unsafe darkened glass to do it. The part of the sun that is not blocked by the moon is just as bright as it is when the sun is whole. But people get confused by cautionary advice and conclude there is something about the eclipsed sun itself that makes its lack of light more dangerous than ordinary sunlight.

A portrait of Kala Rau, the Balinese eclipse demon, eyes the nearby solar disk on this commemorative can and contemplates, like the consumer of the beer, a tasty swallow ahead. (E. C. Krupp, private collection)

Besides eclipse-viewing advice, the Indonesians offered eclipse T-shirts, hats, Rubik's cubes, postage stamps, commemorative medallions, buttons, luggage stickers, the whole Gerhana Matahari Total schmear.

It was a gorgeous eclipse. I watched it from the grounds of Kwan Sin Bio Temple, a Chinese shrine in the village of Tuban on the north coast of Java. Two scaly green dragons threatened the ball of celestial flame over its front entrance, suitable emblems for celebrating an eclipse. About fifteen minutes before totality, Venus could be seen, brilliant in the darkening sky and far to the east of the nearly completely eclipsed sun. By then the landscape had taken on that eerie lighting so peculiar to eclipses. Everything is fully lit, but the light level is low. The shadows and colors are those of midday, but the intensity is off.

Just as the disk of the moon slid into place, the momentary diamond ring of sunlight marked the marriage. Then, with the sun in total eclipse, the corona became visible. Dramatic streamers stretched away from the blackened disk in four directions. Compared to the stark black disk of the moon, the sky almost seemed blue, and the black disk looked flat, a hole punched through the sky. Veteran eclipse-goers judged this eclipse to be one of the brightest they had seen. Around the horizon, the sky looked orange, not the deep red sometimes seen at the edge of the moon's shadow. The temperature dropped 13 degrees Fahrenheit.

The high point of the show was over when the sun slipped out of the moon's grasp. A second diamond ring announced Kala Rau's failure. Once more the sun had fallen from the ragged end of his throat.

11

Among the Nomads

If you had wanted to read stories about the planets in the 1940s and early 1950s, your next move was pretty obvious: You would have taken a hike to the newsstand, prowled through its rows and rows of pulps, and picked up a copy of the latest issue of *Planet Stories*. Its customary cover slogan promised "Strange Adventures on Other Worlds," and it delivered the goods. A typical sample covers the solar system—"Golden Amazons of Venus," "Black Amazon of Mars," "Lords of the Moon," "Cave Dwellers of Saturn." If those didn't work, the color illustration on the cover guaranteed a second look. There was always a high-contour woman, practically undressed for that era but wearing what readers today would consider a modest two-piece bathing suit, or wrapped perhaps in an ornamented bodystocking—a forecast of Spandex to come, or even dressed in something as at-home in a harem as on the heights of Olympus. No matter what she was wearing, she was a goddess, and she was in trouble. Threatened by a goggle-eyed, cobalt-blue alien octopoid whose tentacles waved luridly and snared her wrists, she hoped the tall, steely hero in the fishbowl space helmet would be able to rescue her. This was space opera—an interplanetary Western where ray guns replaced six-shooters. Spaceships, not horses, carried the good guys and the bad guys across the wide open spaces of other worlds.

Science fiction writer Leigh Brackett authored some of those fanciful visions of high adventure in outer space, and she called space opera "the folk tale, the hero tale" of our age and linked it with the tradition of legendary and mythological stories of the conflict between good and evil, light and darkness, order and chaos, the familiar and the unknown. In ancient myth and legend, however, the planets were not the scenes of slambang action, seductive intrigue, and heroic enterprise. They were the heroes and villains themselves. The shining, roving planets were gods—players in tales of quarrel, capture, love, and death. Powerful and divine, they moved at will among the fixed background stars. They were nomads in the night, and the Greek word for them, *planetai*, means "wanderers."

HOW THEY WANDER, WHAT THEY ARE

Besides the sun and moon, the ancients counted five other wanderers, or "planets," visible to the unaided eye. From about 100 B.C. our names for them have been the Roman names for five of the gods of classical Greece and Rome: Mercury, Venus, Mars, Jupiter, and Saturn. To the Greeks, those gods were Hermes, Aphrodite, Ares, Zeus, and Kronos.

Each god had his or her own character, power, and sphere of influence. For example, Kronos (or Cronus) was the father of the Olympian gods. He emasculated his father Ouranos, the starlit night sky, and took control of the cosmos until he was himself deposed by his own son Zeus.

Originally the divine agent of the shiny blue daytime sky, Zeus was the lord of atmospheric phenomena—rain, clouds, storm, lightning, and thunder. He gradually accumulated much broader power and was recognized as the king of the gods. Order, justice, wisdom, law, and sovereignty on earth all reposed in him. His son Ares had a taste for combat and slaughter and so was regarded as the god of war. A compromising affair with Aphrodite, the goddess of love and beauty, confirmed, however, that Ares liked to make love *and* war. His real adversary was wisdom and its goddess Athena.

Aphrodite's alliance with love and fertility can be traced, according to one account, to her birth from the sea's foam, pollinated by drops of blood spilled into the waves by the butchered genitals of Ouranos. Sensual and sexy, she played with passion, entertained romance, and showed off her voluptuous curves with provocative poses and minimal attire, in other words, a party girl.

Finally, Hermes presided over commerce, between the gods as their messenger and herald and among human beings as the patron of trade, travel, and communication. The root of his Roman name —Mercury—is found in the word *mercantile.* Crafty and swift, he flew to his destinations on winged sandals, wore a winged helmet and a cloak, and carried a caduceus, the staff of his office. Considered the inventor of music, astronomy, and weights and measures, he was associated with knowledge, writing, and records. As a psychopomp, he also conducted the souls of the dead to the banks of the Acheron, where they picked up the ferry to their subterranean reward.

These planets and their gods are colorful, and their stories are intriguing. Along with the sun and the moon, they made a celestial family whose relationships and interactions provided a steady stream of adventures in myth and in the sky.

We carried on the tradition of naming planets for the gods as new planets, unknown to the ancients, were discovered. Uranus, the seventh planet from the sun, was discovered by the English astronomer William Herschel in 1781. Its name is the late Latinized form of Ouranos, the primordial god of the starry sky. In 1846, a German astronomer at the Berlin Observatory, Johann Gottfried Galle, found the planet now known as Neptune, the Roman version of Poseidon, a brother of Zeus and the god of the sea. Pluto, the farthest known planet, was found by American astronomer Clyde Tombaugh in 1930 and is named for the ruler of the dark and remote realm of the dead.

The ancients, of course, couldn't say much about Uranus, Neptune, and Pluto, but even without telescopes, they could collect detailed information on the other planets, not as worlds in space but as divine lights making journeys through heaven. Their brightness, color, speed, route, and the duration of their trips all could be seen from below. Some of these things prompted people on earth to affiliate them with specific gods. The planet Mars, for example, is distinctly red, especially compared with the others. Red is the color of blood, and blood is shed in war. So naming Mars after the god of war makes sense. Also, when it comes to traveling through the sky, Mars is all over the map. It moves relatively quickly through the stars, migrates toward the east—as do all of the planets at least some of the time—but loops backward now and then, in a kind of temporary retreat to the west, before continuing forward to the east. It looks active and independent, an agitated red warrior.

Venus, after the sun and the moon, is the brightest object in the sky. She always shines beautifully, as either a morning or an evening star and fickle in the twilight, keeps the company of whatever celestial admirers may temporarily join her there.

Although not as bright as Venus at her best, Jupiter still can be conspicuous. It is the fourth brightest celestial object and, unlike Venus and Mercury, is not tethered to the twilight and the sun. It completes a round through the zodiac in nearly twelve years, one year for each sign, and so mimics what the sun does in twelve months.

To the unaided eye, the planets look like little more than extra-bright stars. Their special character is revealed in time, however, for they appear to move in systematic ways among the background stars. To the ancients, an assembly like this one—which includes Jupiter (near the top of the picture), Saturn (toward "7 o'clock" from Jupiter and fainter than Jupiter), Venus (below Saturn and brighter than Jupiter) in the company of the waning crescent moon—might have carried a special message about the kingdom or the king for the Babylonian skywatchers. This grouping actually occurred on the morning of November 4, 1980. We can presume the vestigial astrologers of our own day should have seen significance for a ruler in it, for it marked the dawn of Election Day. (Photograph: Paul Roques)

Mercury moves faster than any other planet. It travels first to one side of the sun—but never very far—and then doubles back to the other side, where it turns around again. A cosmic busybody, Mercury seems always to be informing the sun or the earth or anyone else on its trail, what gossip is worth repeating.

In sharp contrast with Mercury, Saturn takes a slow and dignified constitutional through the stars. His unhurried pace may be a sign of his advanced age or a symbol of his expulsion from the inner circle of power.

LIVING WITH THE PLANETS DAY-BY-DAY

We tell a planet story everytime we name the days of the week. There are seven days in a week, of course, one for each "planet" observed by the ancients. That means we count in the sun and the moon. There is no way to mistake Sunday's namesake, and *moon* is only slightly less obvious in *Monday*. It is also easy to hear the Saturn in Saturday, but after that, it looks like all we've managed to find is a long weekend. The other four days—Tuesday, Wednesday, Thursday, and Friday—don't sound much like planets. They are, however, and the connection is clearer in the names of the week's days in the Romance languages, or in Latin, from which they derive.

Sunday	dies Solis	day of the sun
Monday	dies Lunae	day of the moon
Tuesday	dies Martis	day of Mars
Wednesday	dies Mercurii	day of Mercury
Thursday	dies Jovis	day of Jupiter
Friday	dies Veneris	day of Venus
Saturday	dies Saturni	day of Saturn

In English, the other four days of the week are named after Norse gods. When the Roman seven-day week was introduced into the Germanic countries, what seemed like the best Norse equivalents were substituted. In this way Tyr (Tiw in Anglo-Saxon), the northern god associated with war, replaced Mars and provided Tiw's day *(Tiwes daeg)*. Although Odin (Woden in Anglo-Saxon) ruled the Norse gods of Asgard, his close connection with the souls of the dead forged an association with Mercury so that Mercury's day became Woden's day *(Wodnes daeg)*, or Wednesday. As master of the storm, Thor was identified with Jupiter. That made Jove's day, Thor's day *(Thunres daeg)*. Finally, Friday comes from Frig's day *(Frig daeg)*, for Frig, the consort of Woden. As a goddess of terrestrial and domestic fertility, she fulfilled some of the same functions as Venus.

Certainly the names of the days of the week match the Roman planetary gods, but this planetary week was not a Roman—or even a Greek—invention. Instead, it belongs to the Chaldaeans, who were the astrologers and astronomers of Mesopotamia from the sixth century B.C. on. They introduced the idea to the Hellenistic Mediterranean world in about the second century B.C. It had evolved out of their astrological notions, which emphasized the alleged influence of these seven celestial objects in the fortunes and failures of human enterprise. Detailed Hellenistic horoscopes mapped the positions of the planets against the signs of the zodiac and the "houses" of the sky, and from these charts, astrologers selected the most propitious course the celestial circumstances seemed to allow. Astrology then was what it is today, a system of magical thought with no scientific basis and no systematic evidence to validate it. Because the Chaldaeans so successfully seeded their astrological concepts in the Mediterranean mind, we now name the days of the week after Roman/Greek gods or their Norse substitutes.

GODS AND PLANETS IN ANCIENT BABYLON

Although the planets had technical names in Mesopotamian treatises and tables, they were closely identified with Babylonian gods at least as early as the Assyrian period (1244–626 B.C.). Use of the gods' insignia in astronomical contexts on Kassite (1600–1240 B.C.) boundary stones and on Akkadian (2371–2230 B.C.) cylinder seals implies that the tradition, or at least part of it, got started even earlier. Our knowledge of these Mesopotamian tutelary gods is understandably less complete than what we know about the Greek gods, but some known aspects of their powers verify that their planetary identities were carried on to Greece. Mars, for example, belonged to Nergal, the god of war, and Mercury was the planet of Nabu, the divine patron of scribes and the god of wisdom. In both instances, the Babylonian god of the planet foreshadowed the Greek. Ishtar was the Aphrodite of Babylon, a goddess who embraced fertility, desire, and love. Naturally, Venus was her star. Ninib's connection with the planet Saturn is certain but a little mystifying. Originally a powerful deity with a warrior's approach to problems and probably an incarnation of the victorious rising sun as well, Ninib became associated with hunting and kingship. As a former sun god, he may have been judged deserving of a star but not necessarily a bright, fast-moving planet. Banished to slow and modest Saturn, he reminds us of Kronos, who was also deposed and wound up with Saturn.

Jupiter, named for the king of the Olympian gods, personified the god Marduk in Babylon. Marduk's triumph over Tiamat, the monster of primordial chaos, qualified him for the big leagues, and kingship over the gods was his reward. The *Enûma elish*, or epic of creation, reports how the troubles with Tiamat began when in anger she declared war against the gods. Tiamat spawned her own army: the viper, the dragon, the female monster, the great lion, the mad dog, the scorpion man, the howling storm, the flying dragon, the bison, and more—eleven monsters altogether. She herself assumed the shape of a monstrous dragon.

Marduk agreed to take the dragon by the horns and slay her, but he had his price. He required the rest of the gods to recognize him as foremost among them and confer upon him sole authority to determine destinies and establish order.

With weapons and a snare, Marduk roared down upon Tiamat and her confederates. He was the thunderstorm, raining hard on Tiamat's parade. In a fierce and frightening holocaust, Tiamat was caught in his net. The four winds held down the corners on her. Shaking uncontrollably, she shrieked in defiance, and spitting venomous words, her maw gaped open. It was a furious living abyss in which a god could be swallowed and forever lost, but Marduk jammed her throat with an evil wind. Inflated like a frothy bubble, she was deformed and helpless. Marduk's arrow slashed her gullet, split her heart, and ripped her womb. Torn and slain before her allies, she persuaded them to surrender to Marduk, and he imprisoned them all.

By defeating chaos, Marduk earned the right to

Marduk was associated with the planet Jupiter, and he became the leader of the ancient Babylonian gods by vanquishing Tiamat, a monster who personified chaos. This relief from the palace of Assurnasirpal II at Nimrud is thought to represent an Assyrian version of the same story. Part of Marduk's affiliation with Jupiter is attributable to the planet's orderly behavior. (From The Dawn of Civilization *by Gaston Charles Maspero, 1894)*

establish world order, and the first step in the process was arranging the sky. He measured out the year and fitted it with a beginning and an end. He arranged the twelve months with their "three stars each." Then, entrusting the night to the moon, he appointed it "the ornament of the night to make known the days" by going through its phases. Marduk also directed the sun to complete the cycle from one New Year to the next and assigned to the sun the Eastern Gate, the twilights, and the day. A master of management and detail, Marduk knew what kind of cosmos he wanted, and he knew how to get it. The planet Jupiter was part of that cosmos, and the charter Marduk awarded the planet helps explain why Jupiter was so closely affiliated with him.

The *Enûma elish* informs us that Marduk set up a bureau for the planet Jupiter (Niburu) to establish the bands of heaven for the thirty-six stars and guarantee they would stay in place. Jupiter's own band is the ecliptic. With candidates like the sun and moon available, it may seem capricious to put Jupiter in charge of governing the sky, but Marduk needed someone to supervise the *night* sky. That puts the sun out of the running. Jupiter is the ideal candidate. It is bright, the next brightest object after Venus, whose motions are too promiscuous to be of any help here. Jupiter, on the other hand, follows a steady course through the stars, with one retrograde loop per year, and visits every constellation in the zodiac in a twelve-year cycle. It is, in a sense, a "night sun" whose travel seems to echo in twelve years what the sun does in twelve months. Finally, Jupiter clings closer to the ecliptic than any of the other planets. All this makes Jupiter an ideal regulator of the heavens and justifies Marduk's adoption of the planet as his personal star.

PLANETS AND GODS IN ANCIENT GREECE

Clear connections between the planet gods of Babylon and the planet gods of Greece verify the source of the scheme by which the planets were named. We also know that the Chaldaean tradition arrived fairly late in Greece, for earlier Greek names for the planets are quite different. Although Homer doesn't give us a complete list of them, he does mention *Eosphoros* ("dawn bearer") and *Hesperos* ("evening" or "evening star"), meaning the planet Venus in its morning and evening aspects. In the *Odyssey* (Book XIII, lines 93–94), Venus is called the "brightest of all the stars," and it "often ushers in the tender light of dawn." Homer compares Venus with the bright and moving point of the spear Achilles shook at Hector in the *Iliad* (Book XXII, lines 317–318). The references to brightness and motion make it certain Homer meant Venus.

The Chaldaean habit of allying the planets with gods starts in Plato's time (427–347 B.C.). The *Epinomis*, a dialogue attributed to him, identifies all five by name.

Mercury	star of Hermes
Venus	"Morning and Evening Star"
Mars	star of Ares
Jupiter	star of Zeus
Saturn	star of Saturn

Plato calls Venus the "Morning and Evening Star," but he says "its name is not known." He adds that it belongs to Aphrodite and that her name has been given to it, quite properly, by a "Syrian lawgiver."

Between 330 and 200 B.C., we encounter another set of names for the planets in Greek astrological and astronomical treatises:

Mercury	*Stilbon*	"twinkling star"
Venus	*Phosphoros*	"light bearer"
Mars	*Pyroeis*	"fiery star"
Jupiter	*Phaeton*	"luminous star"
Saturn	*Phainon*	"brilliant star" or "indicator"

These names probably owe something to an independent tradition in Greece and do not survive in our present system. Apparently they were invented after the time of Plato, for in his *Epinomis* he implies that prior to his time the planets' names were not known in Greece. He explains that a "barbarian"—not a Greek—first observed the behavior of the planets. That, he says, is where the names originated, and he adds that knowledge of the planets was accumulated by observers in Egypt

Antiochus I left this monumental planetary horoscope on the summit of Mount Nemrut, to accompany his tomb. The lion represents Leo, and a crescent moon is suspended from his neck. The three large stars above the lion's back are named on the slab, and the inscriptions confirm that they stand for the planets Jupiter (Phaeton, nearest the head), Mercury (Stilbon, in the middle), and Mars (Pyroeis, on the left). One interpretation of this grouping in Leo identifies the relief as commemoration of the coronation of Antiochus I on July 7, 62 B.C.*, but alternative dates have also been argued. (Photograph: Robin Rector Krupp)*

and Syria and passed along to the Greeks as well as to the rest of the world.

Certainly it is possible that skywatchers in "Egypt and Syria" were responsible for this relatively late transmission, but the tradition itself is much older and neither Syrian nor Egyptian. It was shared, perhaps, with Syria and Egypt, but it originated in Babylonia, where the planetary gods correspond to the Greek gods in a way they do not in ancient Egypt.

PLANETS AND GODS IN ANCIENT EGYPT

The planets carried several different titles in Egypt and can be identified at least as early as Dynasty XVIII (1570–1293 B.C.), during the New Kingdom. New names were added during the Graeco-Roman period (332 B.C.–A.D. 323). None of these names, though, suggests a connection with Babylonian or Greek and Roman gods.

Mercury had the same name in the earlier and later periods, but the meaning of the name is unknown. One New Kingdom text also refers to Mercury as "Set in the evening twilight, a god in the morning twilight." Identified as Set in the evening, it presumably carried evil connotations. The second half of the phrase indicates it was regarded as some other god when it was a morning star.

Venus, in the earlier period, was sometimes known as Osiris, and so death and resurrection, the primary themes of the Osiris myth, probably belonged to Venus, too. It is the planet that dies so often in the flames of the sun and is then reborn from them. This also sounds like the story of the phoenix, and the planet Venus was depicted as the *benu*, a mythological heronlike bird that was a kind of Egyptian phoenix. The bird's name is rooted in an Egyptian verb that means "to rise radiantly" or "to shine." According to the story of creation told at Heliopolis, the world began when the *benu* launched itself from the *benben*, the primordial mound of creation, to bring the sun into being. In other texts, the *benu* paves the way for the sun as Venus, the Morning Star. It is a herald in the darkness for the sun, the soul, and the existence of the entire world.

Venus was also called "the crosser" and "the star which crosses." This has prompted speculation that it, too, was recognized to be the same object in the morning and in the evening but that it was assigned separate identities for each aspect. Later we find it called "the morning star."

During the New Kingdom, Mars was called "Horus of the Horizon" and was said to be either the "eastern star of the sky" or the "western star of the sky." Further commentary added, "He travels backwards." This sounds like an explicit reference to the conspicuous retrograde motion the planet sometimes displays. Later, in the Greek and Roman period, Mars had another descriptive name: "Horus the Red."

All three superior planets—Mars, Jupiter, and Saturn—were depicted by the Egyptians as identical hawk-headed human figures. Only the hieroglyphics separated one from the other. Jupiter was called "Horus Who Bounds the Two Lands," "Star of the Southern Sky," "Horus Mystery of the Two Lands," and "Horus Who Illuminates the Two Lands." The "Two Lands" in every case are Upper and Lower Egypt, and all of these names are from New Kingdom sources. Later, Jupiter was known as the "Star of the Southern Sky," a name presumably preserved from the earlier era.

Finally, the New Kingdom inscriptions identify Saturn as "Horus the Bull of the Sky," which seems to be contracted to "Horus the Bull" in Graeco-Roman times.

Egyptian stories about the planets are practically nonexistent, but Mars and Venus are both mentioned in Spell 109 in the *Book of the Dead* and are

Egyptian portrayals of the planets are part of an astronomical tradition independent of Greece and Mesopotamia. From left to right, Mars, Saturn, and Jupiter parade as hawk-head incarnations of the god Horus across the ceiling of the burial chamber of the tomb of Sety I (ruled 1291–1278 B.C.) in the Valley of the Kings.

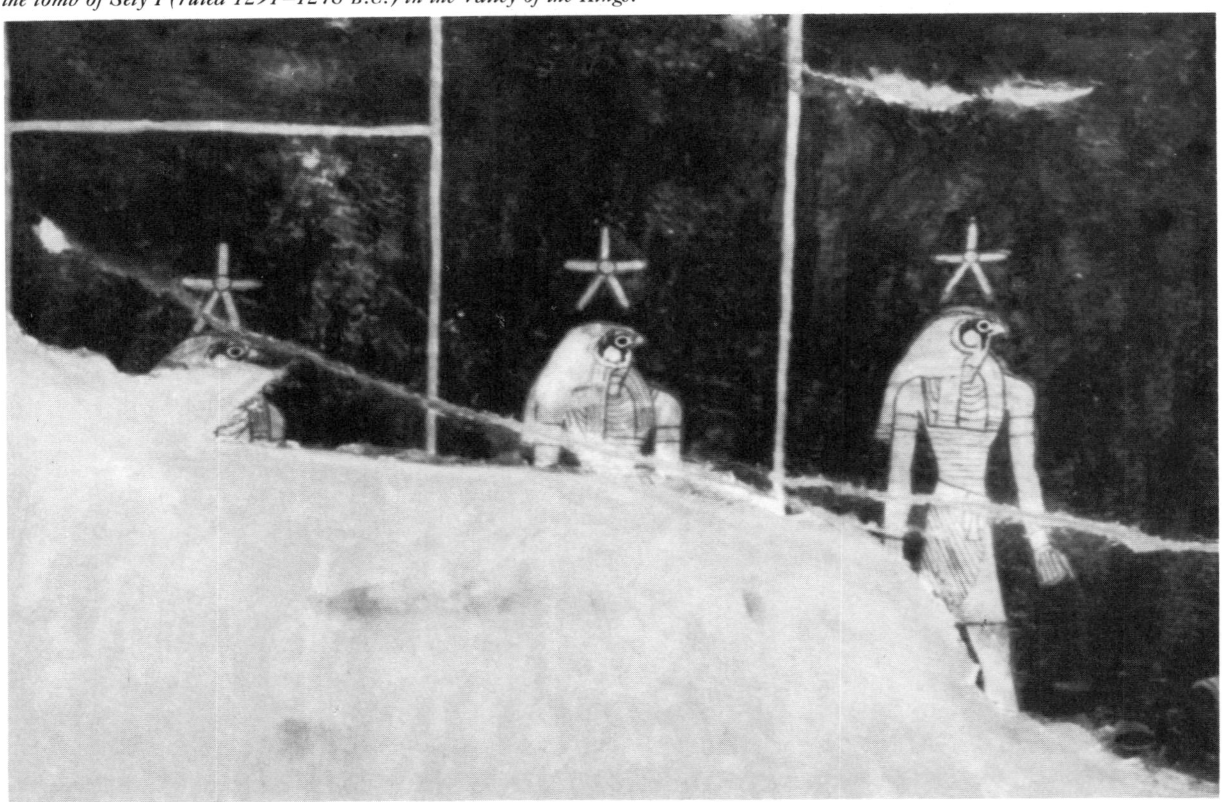

Nine planets are illustrated according to Hindu tradition above the entrance to the Rajarani Temple. From left to right, they are the sun, the moon, Mars, Mercury, Jupiter, Venus, Saturn, Rahu, and Ketu. Rahu and Ketu personify the invisible nodes of the moon's orbit.

counted among the "eastern spirits" the deceased encounters while traveling in the sun's company toward the deliverance of the dawn. More Egyptian planet lore must have existed, for we see representations of the planets in royal tombs, in temple ornaments, and in funereal papyri. What has survived is very sparse, but it is enough to confirm that the original concept of planet gods in Egypt differed greatly from what showed up in Babylon and Greece.

PLANETS AND GODS IN ANCIENT INDIA

An altogether different set of identities was assumed by the planets in India, and there their number was increased from seven—including the sun and moon—to nine. By name these planets were

sun	Ravi
moon	Soma
Mars	Mangala
Mercury	Budha
Jupiter	Brhaspati
Venus	Sukra
Saturn	Sani
ascending node	Rahu
descending node	Ketu

It is evident from this list that the two additional planets do not mean the ancient Hindus, without telescopes to detect them, knew about Uranus and Neptune. Instead, these extra planets are the ascending and descending nodes of the moon's orbit. Although invisible, they were judged influential enough to merit seats on the planetary council.

Such councils seem to preside on lintels over the entrances to Hindu temples built in the tenth and eleventh centuries A.D. in the city of Bhubaneshwar, about 300 miles southwest of Calcutta, near India's eastern coast. Parashurameshvara temple, also in Bhubaneshwar, is earlier. Constructed in the eighth century A.D., it portrays only eight planets on the lintel of its inner shrine. At that time, Ketu was not included.

In India the planets are associated with Hindu ritual. Because ritual sustains the gods, it must be properly performed. The nine planets therefore rise in succession to help bring ritual into conformance with sacred standards.

PLANET SPIRITS IN ANCIENT CHINA

Ancient Chinese ideas about the planets are also products of an independent tradition. Their names appear to have nothing in common with Mediterranean or Mesopotamian tradition.

Mars	*Ying huo*	"Fitful Glitterer" or "Sparkling Deluder"
Saturn	*Chen hsing*	"the Exorcist" or "Quelling Star"
Venus	*T'ai po*	"Grand White"
Mercury	*Ch'en hsing*	"the Hour-Star" or "Chronographic Star"
Jupiter	*Sui hsing*	"Year Star"

Jupiter was known as the "Year Star" because the Chinese counted out a twelve-year cycle based upon Jupiter's temporary "one-year" residence within one of twelve zones of the sky. Like the rest of Chinese astronomy, this program relied upon the north celestial pole and the equator for the frontiers of the divisions. The twelve zones were not twelve signs of the zodiac but twelve intervals along the celestial equator. By perhaps the eighth century A.D., a system of twelve talismanic animals was introduced into China, possibly from the Mongol-speaking peoples of north central Asia. One animal was assigned to each celestial domain. When Jupiter occupied that creature's territory, it was said to be the year of that animal. This tradition survives in the Orient today. In a dozen years, Jupiter makes a trip through the celestial zoo, one cage at a time: rat, ox, tiger, rabbit, dragon, snake, horse, sheep, monkey, cock, dog, and pig. Each animal confers a special character upon its year and appropriate personality traits on individuals born then.

By the fourth century B.C., each planet in China was associated with one of the Five Elements. These were not the same "elements" the ancient Greeks philosophers had in mind when they discussed the nature of the world. The Greeks counted four elements: Fire, Air, Water, and Earth—and they were talking about the essential and fundamental character of matter. The Chinese also included Water, Fire, and Earth in their list, but added Wood and Metal. To the Chinese, none of these Elements stood for a basic physical substance. Their Five Elements, or "Forces," were "perpetually active principles of Nature." Together they called them *Wu hsing*, and their proportion and influence at any given time defined the circumstances of that moment. The *Wu hsing* were said to emerge from the interaction of *yin* and *yang*, the two opposite but complementary and oscillating aspects of the cosmos. In Taoist thought, these are what keep the world in constant but balanced change.

Through a well-defined system of correspondences, the ancient Chinese linked the varieties of other aspects of nature together with the Five Planets and the Five Elements. There were five directions, five "seasons," five atmospheric conditions, five colors, five metals, five kinds of grain, and five tastes. Many other natural phenomena and types of human activity were also classified and organized according to this five-fold principle.

Jupiter, then, was the Wood Planet. Mars, not surprisingly, was the Fire Planet. Saturn, Venus, and Mercury were the Earth Planet, Metal Planet, and Water Planet, respectively. Together, all five were sometimes known as the Five Pacers, and each had its own rich set of prerogatives. Jupiter often concerned himself with sovereignty. Venus devoted attention to executions, sedition, and other unsettling matters. All of the Pacers heralded fortune or doom, as the vicissitudes of life in Imperial China took their inevitable turns.

PLANETS: THEIR TALISMANS AND ALLIES

The West had its own system of magical correspondences. It allied each planet with other symbolic components of the astrological tradition that gradually evolved out of Chaldaean and Hellenistic skylore. Unlike the Chinese system, in which five planets were paired with five elements, the European scheme incorporated all seven wanderers and affiliated each with one of the seven "primary" metals. We don't know how early these connections were made, but Greek philosophers and alchemists had adopted them by the fourth century B.C.

The sun's metal was gold. In color, this metal resembles the yellow sun, but its other properties enhance its solar character. Gold does not corrode, and this is a kind of immortality, akin to the immortality enjoyed by the sun through its daily and annual renewal. Gold and the sun are two of a kind: durable, pure, and perfect.

If the sun has an affinity for gold, the moon's taste runs to silver. Selene, the Greek goddess of the moon, rode a silver chariot and cast silver rays. So the moon gleams in silver, but it is not just the color that counts. Celestial objects shine, and so do metals. The ability to reflect light makes metal brilliant. Gold and silver seem to glow, and that, as much as color, links them with the luminous sun and moon.

Venus has her own precious metal—copper. Copper is harder than gold and more common than meteoritic iron, and all three were used in antiquity. Perhaps it's not surprising that such a significant and shiny substance would become associated with the third brightest object in the sky. Cyprus was a major source of copper in ancient times, and this island is also closely associated with Aphrodite. According to the Greeks, she was carried to the coast of Cyprus by the Zephyrs and welcomed there by the Seasons shortly after her birth from the sea's foam.

The astrological symbol for Venus is a cross topped by a circle. It is thought to represent a stylized mirror—a fitting emblem for a goddess of love and beauty—and copper mirrors were used in antiquity. But this fact only echoes, and does not explain, the fundamental relationship between the metal and the planet. It may be possible to trace it further, but to do so, we have to see the full set of correspondences. Planets and metals were just a part of it. Colors, precious stones, and emblematic animals also played a role.

PLANET	METAL	COLOR	STONE	ANIMAL
sun	gold	yellow	topaz	lion
moon	silver	white	crystal	dog
Venus	copper	green	emerald	dove
Mercury	quicksilver	grey	agate	swallow
Mars	iron	red	ruby	horse
Jupiter	tin	blue	sapphire	eagle
Saturn	lead	black	onyx	crocodile

These additional associations are completely consistent for the sun and moon. Topaz is a yellow stone and matches the color of the sun and gold. The tawny lion lords it over the other creatures as the king of beasts, and a bond between kingship and the sun was recognized widely in antiquity.

White is as appropriate for the moon as silver, and crystal—lucid and lustrous—shimmers like the moon. The dog is more obscure but may have something to do with canines baying at the moon.

Venus, according to the system, is bonded to the color green. Its stone, the green emerald, stresses the point. As Aphrodite, this planet meant love, fecundity, procreation, and life. Green is the color of vegetation, which symbolizes life and growth.

Planet symbols from left to right, top row: the sun, the moon, Mars; bottom row: Mercury, Jupiter, Venus, and the stylized sickle of Saturn. (From Symbols, Signs, and Signets *by Ernst Lehner, Dover, 1950)*

Copper's connection with Venus may also involve the color green. With time, copper develops a green patina. One of the most important copper ores is green, and copper burns blue-green in a flame.

Color, then, is a very tidy explanation for the link between Venus and copper, but the planet's identification with the goddess Aphrodite requires further commentary. Perhaps it is a product of what we see Venus do in the sky. Its two periods of visibility, as a morning star and as an evening star, each last about 263 days. This is close to the average duration of a human pregnancy—about 265 days. This near-equivalence may be the real reason why the planet Venus and the goddess of sexual love, new life, and growth are related. It may also explain in part why the Maya, the Aztec, and the other peoples of Mesoamerica contrived a ritual calendar cycle of 260 days. The dove, too, is an emblem of love and renewal of life in Graeco-Roman tradition.

Mercury is known for its speed and quixotic rush from one side of the sun to the other. Its talismanic metal is quicksilver, which is liquid at room temperature and seems to flow faster than water, with a life of its own. This metal is so closely identified with Mercury, planet and god, that it goes equally often by that name. Gray is a reasonable color for the metal, and aerial acrobatics may explain the connection with the swallow. This bird darts and turns like spilled quicksilver. Mercury's association with agate, on the other hand, remains enigmatic. Mercury's astrological symbol is the same as the symbol for Venus except that it is surmounted by a reclining crescent. This emblem is said to be a stylized caduceus, the herald's wand Hermes held.

The ruby logically belongs to Mars, the red planet, and color may also explain why the planet is named for the god of war: Armed combat spills blood. The astrological symbol for Mars—a circle with an arrow emerging up and to the right from the ring—stands for a shield and spear, weapons of the war god. Perhaps because the horse was an essential component of the war chariot, the animal of Mars was the horse. Iron also fits Mars, because it was a master of weapons, and the weapons of choice two thousand years ago were made of iron. Iron-rich earth is often red, and it is, in fact, iron oxide in the soil of Mars that makes the planet look red to us here on earth.

Jupiter's metal is tin, but it is hard to understand why. It may have some connection to Jupiter's role as a sky god. The Sumerian word for tin is "metal of heaven." Beyond this vague hint, we have little else to explain why tin was cast in this role. The color blue, the blue sapphire, and the high-flying eagle are all cleary linked with the vault of the sky. The planet's astrological symbol is a stylized lightning bolt, a favored missile of Zeus.

As the most distant of the planets visible to the unaided eye, Saturn moves more slowly than the rest. Its languid behavior makes it an appropriate bed partner for heavy, ponderous lead, the darkest of these metals. Dim Saturn was thought to be far from the center of light, the sun. Black fits, then, as the color of Saturn. Onyx, the gemstone of Saturn, is a black gem. Saturn's symbol is a stylized sickle, which recalls the god's agricultural associations as well as the emasculation of his father Uranus. The crocodile, however, is now behind our ken.

Despite the fact that we can't completely explain why certain planets were linked with certain metals, there is enough to see that the connections were not arbitrary. Instead they reflect the imagery our ancestors used to put sense and order into the world they experienced.

A review of ancient cosmologies and systems of organized knowledge quickly demonstrates that there are all kinds of ways to put together a picture of the universe, most of them dramatic departures from the schemes and metaphors of contemporary science. In general, we would say that our ideas about nature are more accurate than those of our ancestors, but being right isn't all that matters. Even incorrect perceptions of the cosmos have proved useful, because they, like our scientific understanding, gave their adherents balance, perspective, and a framework for action.

The system of correspondences between planets and metals is just a chapter in the laws of western magic. The purpose of those laws is establishment of the proper time and appropriate circumstances for a specific, symbolic act. That is also just what the Hindus said about the planets and ritual.

Magic is intended to manipulate the connections between things. At its heart, it is a quest for power, for power can make order out of chaos. Here on earth, only specialists like shamans or priests can acquire such power. It is a calling, a vocation. That power can be tapped by reproducing, in symbol and ceremony, the principles that govern the cosmos. Chinese astrology and Taoist magic do the same thing, but with a different set of rules, rules meaningful in the context of that culture. The sovereign principles such specialists seek are, in part, the correspondences that relate one thing to another. The ritual magician must spot these felicitous circumstances and insert himself or herself into them. It is a search for congruence, and it is why people watch the planets and tell stories about them. Those stories express relationships. Planets seem to be involved in celestial relationships because their free movement makes them interactive agents in heaven. They meet. They fly solo. They retreat. They disappear. They reappear. They lead full celestial lives.

Although the planets are free to move, their movements are confined to the ecliptic. Each moves at its own pace. Starting in the Hellenistic era in the second century B.C., the ancients listed them—not in the order of their distance from the sun, as we do—but according to the time each takes to complete its rounds. The traditional sequence for all seven wanderers, defined by the lengths of their trips, is

Saturn	29.46 years
Jupiter	11.86 years
Mars	686.98 days
sun	365.24 days
Venus	224.70 days
Mercury	87.97 days
moon	27.32 days

These are the times it takes each of these objects to return to the same spot with respect to the background stars and are called the sidereal (star) periods of the planets. To the Hellenistic astronomers, who thought the sky had depth, these intervals indicated the planets' relative distance from earth, with the moon closest and Saturn farthest.

In addition to their travels past the background stars, the planets pass through cycles of repeated configurations. Periods of visibility and invisibility and turning points in their motions run them through paces that really depend on their positions with respect to the sun and the earth. For example, Venus participates in a 584-day cycle that carries it through a stint as a morning star, visible before sunrise; a stint as evening star, visible after sunset; and two conjunctions with the sun, when the planet, lost in the sun's glare, is not seen.

The planet Venus follows five distinctive trajectories as a morning star in the eastern sky and five different paths through the western twilight as an evening star during an eight-year cycle that brings the same configuration of Venus back to the same date in the annual solar calendar. Each of these occurs in its own season in every eight-year cycle, and each involves an upward climb out of the glow of twilight and a descent back into the "fire" of the sun. Part of one of these sequences is shown here, as it might be seen looking west toward the Caracol and Castillo of Chichén Itzá in northern Yucatán. In mid-March, shortly before the vernal equinox, we catch Venus far from the sun. It lingers in the western darkness for several hours after the sun has set. A month later (April 15), Venus has dropped a little closer to the sun, but it is still rather high in the darkness. By mid-May, the planet has sunk noticeably closer to the sun, and in another month, just before the summer solstice, Venus is nearly lost from sight in the glare of the setting sun. (Drawing: Robin Rector Krupp)

MARCH 15

APRIL 15

MAY 15

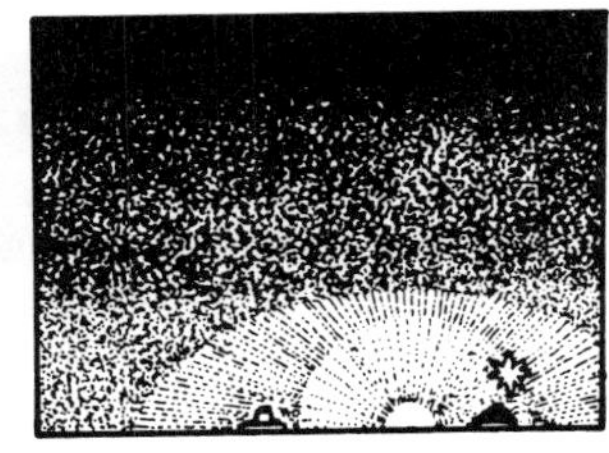
JUNE 15

To understand some of the stories that have been told about the planets and their affairs, we have to see exactly what it is they do.

THE WAY VENUS RUNS ON BOTH SIDES OF THE SUN

To describe what Venus does we can begin anywhere in its 584-day cycle. Imagine that it is on the far side of its orbit from us. It has accompanied the sun in the daytime for the last seven weeks or so, but now its orbital motion has carried it a little east of the sun. It rises after the sun and so is not seen during the day. (Actually Venus is so bright, it can even be seen during the daytime, but you have to know exactly where to look to spot it.) But after the sun sets in the west, Venus still lingers above the western horizon. Once it is far enough from the sun's dazzling light, we'll see it there, an evening star that announces the night that will follow. As the twilight grows darker, the earth's rotation carries Venus below the western horizon. Venus, however, also travels farther and farther east of the sun with each new day. It is a little higher each night, and it stays up a little longer. In time, though, after a long climb it reaches the end of its rope. Venus then doubles back, falling each night a little closer to earth, setting a little earlier, and slipping back toward the sun faster than it rose. Finally, it catches the sun again, and after about nine months (263 days on the average) as an evening star, it is in conjunction once more.

This time, however, Venus is on the near side of its orbit. From our perspective, it is traveling much more quickly, and so this conjunction doesn't last anywhere near as long. It can be as long as two weeks and as short as one day.

Now the orbital motion of Venus drives it west of the sun and out from behind the curtain of conjunction. To see it, we have to be watching it rise ahead of the sun. That means it's a morning star, visible in the east during the morning twilight before the sun inundates it with the glare of dawn. Each subsequent morning finds the planet a little farther west. We see it rise earlier and earlier. It climbs higher and higher before the sunlight touches it, and it scales the eastern sky faster than it floated to its greatest eastern excursion as an evening star. From here on, it's all downhill. Venus reverses direction and heads for the eastern horizon. By the time it reaches the bright neighborhood of the sun, another nine months have passed. It is back on the far side of its orbit from us and goes into conjunction once more. We don't see it again until it comes on stage for another performance as the evening star.

Mercury apes Venus but in much less time. What takes Venus 584 days (about one year and seven months) to accomplish, Mercury can do in 116 (about four months). Mercury's conjunctions are shorter, and at greatest elongation it is much closer to the sun, between 18 and 28 degrees, compared to 48 degrees for Venus. Because Mercury is a small planet with a fairly dark surface, it does not reflect much light. It is faint, then, in the first place, and its nearness to the sun makes it even harder to see.

Neither of these two planets ever stays up all night. Their inside orbits require each to alternate between appearances as morning star and evening star. In either guise, they must do what the sun does. If either is a morning star, it is only seen in the east, rising ahead of the sun. Other objects may occupy the western sky in the predawn hours, but Mercury and Venus cannot do so. In the same way, they have to put in their appearance in the west if they are evening stars. Other objects, of course, are rising in the east while the evening twilight fades, but not Mercury or Venus. They have to be in the west to follow the sun.

CATCHING VENUS: LOVE ON THE REBOUND

Venus turns up in one planet story after another and dominates many of the celestial soap operas. It flirts with love and death and runs through relationships as if there were no tomorrow. Its lore is so rich, the next chapter is entirely devoted to it, but here are a couple of stories that show how the cycles and movements of this planet have been incorporated into folklore and myth in widely separated parts of the world.

Tribes in the Nyanga River drainage of Gabon, in west central Africa, regard Venus as two different wives of the moon. One wife, Chekechani, is the Morning Star. She resides in the east and consorts with him in the dawn, but she isn't much of a homemaker. Because she doesn't bother to fix meals for him, the moon loses weight in her company. He gets fed up with the lack of food on the table and goes to see his other wife, Puikani. She lives in the west and meets her husband in the evening twilight, after the new moon. She is a much better cook, and once the moon gets to her apartment, he starts to put on weight again.

This story is, of course, just another tale of the moon's phases, but it verifies that these Africans notice Venus and are aware that it and the moon regularly meet. The moon goes through several monthly cycles while either the morning wife or evening wife are putting in an appearance, and so the story isn't a true, wife-by-wife description of what goes on in the sky. It emphasizes that the moon continues to wane in the dawn when it accompanies the Morning Star and that it waxes every time it meets the Evening Star after sunset.

Celestial relations get a little steamier in the Baltic *dainàs*, or folk songs. The Lithuanian serenades tell us how the Moon, who was the Sun's mate, left her for her daughter, the Morning Star. At new moon, the Moon was charmed by the Sun, his steady girl. But he withdrew from her embrace and tiptoed over to morning, where he put the make on Venus. The great sky god Perkúnas, not at all pleased with the Moon's betrayal of the Sun, knocks him back into the Sun's arms.

THE RAMPAGES OF MARS

Mars, Jupiter, and Saturn have much more freedom than Venus. Because their paths are outside the earth's orbit, they can move far from the sun. In fact, we sometimes see them on the opposite side of the sky from the sun (usually not at the same time). They are said, then, to be in opposition, and that is when they are brightest. A planet is brightest at opposition because we are closest to it, on the portion of our orbit between the planet and the sun. At opposition, these superior planets rise when the sun sets and set when the sun rises. That's when they stay up all night. They are highest at midnight.

The motion through the stars of Mars, Jupiter, and Saturn is usually eastward. The sun is also commuting eastward on the ecliptic, and because it appears to step through the zodiac more quickly than these outer planets, the sun often overtakes one of them, puts it into conjunction, and temporarily hides it from our view. The Sun then leaves Mars or Jupiter or Saturn in the dust. Before conjunction they were east of the sun, making them evening stars, visible after sunset. After conjunction, they are west of the sun, visible before sunrise, and therefore morning stars. Gradually the sun and any one of these planets—Mars, say—are far apart as they can be. For Mars, the time from one such opposition to the next is 780 days, or two years and almost two months. Because Jupiter and Saturn are substantially farther from us, their oppositions occur more often (every 399 to 378 days, respectively), taking a little more than one orbital circuit of the earth to get from one to the next.

If the earth were really the stationary hub of all this orbital activity, that would be the whole story. But it is the earth that is moving, not the sun, and the earth's own orbital motion makes the superior planets do something interesting. It makes them stop, reverse, stop again, and then continue along their normal eastward path. This happens because the earth, on an inside lane, overtakes them and makes them look as if they are going backward. Passing a slower train on an adjacent track does the same thing. That other train may be going forward, just as yours is, but if you are not conscious of your own motion when you pass, it can look as if the other train is going backward. This temporary "backward" trip to the west by a superior planet is called its retrograde motion, and it always takes place at opposition, when the planet is looking its best. It is most distinctive for Mars, which can backtrack a little more than 16 degrees, a distance in the sky equal in length to a chain of 32 full moons. About every two years Mars takes that turn for the west.

The cycle starts when Mars is on the far side of the sun and in conjunction. We first see it after about four months of hiding when the sun has

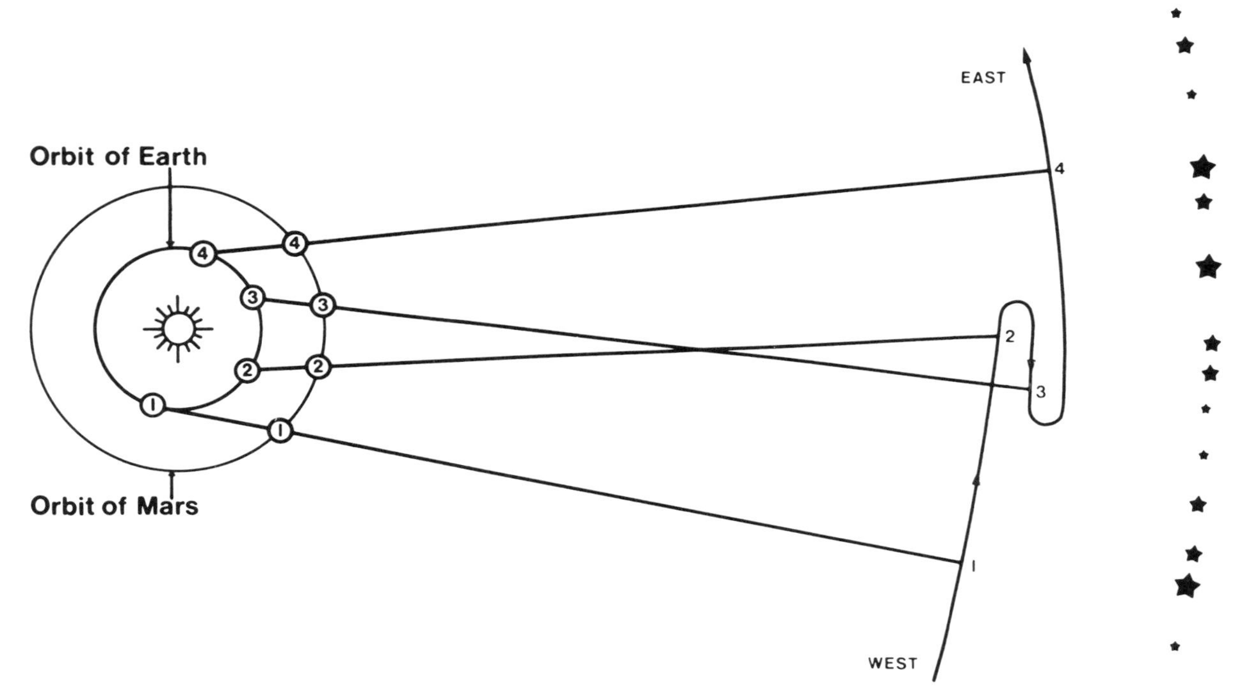

Mars almost seems to do as it pleases because its orbit is outside the earth's orbit but only about half-again as large. From our perspective on earth, Mars appears to move rather quickly against the background stars. Because the earth's orbit is smaller, the earth regularly overtakes Mars, and this makes it look as if Mars is temporarily moving in reverse. In this diagram, the earth has nearly caught Mars when both reach position 2. By the time each planet reaches its own position 3, the earth has passed Mars on the inside track. That makes Mars appear to travel in reverse, but eventually this retrograde motion ends. Mars then continues in the normal counter-clockwise direction. (Griffith Observatory)

moved far enough to the east of the planet to allow its designation as a morning star. Each morning the planet shows up a little earlier and a little higher in the sky. Compared to the sun, it is shifting to the west. At the same time, however, Mars is also moving among the background stars, through the zodiac, at a rate of about one sign every two months (57 days). So Mars is doing two things at once: It is rising earlier and earlier and so reaching a higher and higher point in the sky before the sun comes up; at the same time, it is gradually passing through the constellations of the zodiac and moving in the direction of the sunrise to do it. It is also getting brighter.

When it rises at midnight, it is on the meridian—due south and at the height of its arc for the night—at sunup. Eventually it rises past midnight and into the evening hours. We see it rise earlier in the night, and it stays visible until the dawn snuffs its light. But Mars is not an "evening star" until it sets a little after sunset. Long before that can occur, however, it must rise at sunset. At that point we see Mars, brilliant and red in the east, opposite the sun, which is going down in the west. The planet is highest at midnight, and it sets in the west as the sun rises in the east. All this time Mars is still moving eastward through the stars, but at opposition it shifts into reverse. Then, after several weeks, it pauses in this motion and gets back in gear toward the east. During this interval we have seen the planet "stop" twice with respect to the stars and turn a single loop through them. As the planet continues with its normal eastward migration, the sun is gradually approaching it from the west, making Mars rise after the sun, first in the afternoon, then at noon, and finally in the morning hours. Its time above the horizon at night grows less and less because it sets earlier and earlier. It is also growing fainter. When Mars is low in the western sky after sunset, it is a modest "evening star" headed for conjunction. It takes its last bow on one of those nights in the west and lays low in the sun's embrace until it restarts the cycle as a "morning star."

PLANETARY COURTSHIP AND THE MORNING STAR SACRIFICE

The motions of Mars—and Venus—fit the celestial journeys described in the Skidi Pawnee story of White Star Woman and the Morning Star. The Skidi Band of Pawnee Indians, who lived on the Great Plains of east central Nebraska, called Venus White Star Woman. To them, she was beautiful and powerful. Her home was in the west, and she was also known as Evening Star. Desirable, but without desire, she was courted by many stars and accepted the attentions of none. She opposed populating the earth with life and wanted to make no contribution of her own. Eventually, however, she was captured and overpowered by Morning Star, a great celestial warrior. Their child, a girl, was the first human being and a gift to the earth.

Much of what we know about White Star Woman and Morning Star was collected and recorded at the end of the last century by James Murie, the son of a Skidi mother and a Scottish father.

Murie begins his version of the story of Morning Star and Evening Star by associating the direction east with man and west with woman. Creation, he says, was conceived in the east and brought to fruition in the west. To make creation happen, Morning Star embarked on the quest for a wife, Evening Star. Even though the two prime-time players here are specifically associated with east and morning and west and evening, respectively, neither of them necessarily spent all of their time in the directions and hours assigned to them. East and west in this context are more like yin and yang, symbolic principles of nature. And if Morning Star must pursue his future wife until he catches her, he must migrate from the east. So according to Murie,

> Morning Star journeyed to the west. And ever as he journeyed, the Evening Star moved, came, and drew him towards her. (For men may see how the Evening Star moves nightly. One night she is low in the heavens, another night she is high in the heavens. Even so she moved and drew the Morning Star.)

So far, so good. Murie's very general description of Evening Star is accurate for Venus. Whoever Morning Star may be, he is headed west. That seems to mean he rises earlier each day and slowly takes his leave of the sun.

> Yet when the Evening Star beheld the Morning Star draw near, she placed in his path Hard Things to hinder his approach. Thus, even as the Morning Star first saw the Evening Star, she rose and looked on him and beckoned him. He started towards her, but the earth opened and waters swept down and in the waters was a serpent with a mouth wide open to devour.

Because he was a great warrior, Morning Star was able to defeat the serpent by hurling a ball of fire at it. Then he continued toward Evening Star, but she continued to put obstacles in his path, ten of them altogether. Morning Star was not going to be diverted by such distractions, however, and managed to eliminate them in succession. Finally he reached the lodge of Evening Star in the west, ready to consummate his marriage to her. Protected by four guardian stars of the four directions —Black Star, Yellow Star, White Star, and Red Star—the lady didn't make it easy for her admirer. Morning Star managed to force each of the World Quarter Stars to follow his orders, but he still had to follow the instructions of his intended bride. She insisted that he build special cradleboard for his future child. He also had to bring a mat for the child to lie upon. And he had to bring sweet, fresh water to Evening Star so she would be able to bathe the child. This water was the rain, and in the future it would refresh and cleanse the earth. Morning Star did all these things, and then Evening Star yielded to him. She gave birth later to a daughter, and when the girl was sent to earth, her mother gave her an ear of corn to be planted there. Her name was Standing Rain, and dropping through a cloud, this daughter of Morning Star and Evening Star reached the earth and met there a boy, the child of the Sun and Moon. United on earth, the two children of heaven parented the people who came to inhabit his land.

This is a celestial story of sex and violence, but the Skidi Pawnee brought the violence to earth by periodically sacrificing a young woman to the Morning Star. When one of the Skidi warriors perceived in a dream that the time for such a sacrifice would be approaching, he mobilized several other braves to conduct a raid on a neighboring tribe. The sole object of this enterprise was kidnapping the maiden who would become the next offering to the Morning Star. Now this was not just a fraternity-type panty raid by the Skidi Pawnee. A great deal of ritual preparation involved other members of the tribe, all of it orchestrated by the Morning Star priest. Special clothing and symbolic objects had to be fashioned for the man who had had the vision. He now impersonated the Morning Star, and the other warriors that accompanied him played the parts of various stars, including the four World Quarter Stars. Their stealthy trek to an enemy village took several days and mimicked Morning Star's own arduous campaign to the western lodge of Evening Star. At their camp, on the night before the capture was to take place, the raiding party performed a ritual reenactment of Morning Star's search for the woman in the west. Each dawn, they watched the Morning Star rise, and on the final morning the warriors positioned themselves around the targeted village according to the placement of the stars' positions in the heavens. The Morning Star impersonator stood in the east, and when the signal was given, he and the rest of the squad burst into the village, found an appropriate girl, and made their getaway.

In the Skidi Pawnee Morning Star Sacrifice, a woman from a neighboring tribe was captured in accordance with the configurations of the planets Mars and Venus. This miniature scene, on display in Chicago's Field Museum of Natural History, shows the ritual just before the captive is killed by an arrow fired by the Morning Star impersonator. The woman herself represents the Evening Star. One half of her body is painted red, the other half is black. The four rails beneath her are each painted with one of the colors in the Skidi Pawnee system of directional correspondences. The down-filled trench below stands for the "garden in the west" where the life of the world was conceived.

When Morning Star's team got back to the Skidi village, the girl was treated well but not informed of her inevitable doom. After what might be a lengthy period of captivity, the Morning Star priest judged that a celestial signal for the Morning Star ordained the start of a five-day ritual that would end with the young woman's death on scaffold built just for this sacrifice. Two upright posts supported four horizontal timbers on the lower part of the construction. Another crossbeam completed the frame at the top, and the whole multicolored rack was a little like a goalpost equipped with a rail fence to stop a ball carrier dead in the end zone. The vertical post on the north was painted black for the night and for Big Black Meteoritic Star, the World Quarter Star of the northeast, while the southern pole was red, the color of the dawn, the day, and Red Star, who presided over the southeast quarter of the sky. From bottom to top, the four lower crossbars were each painted a different color—black, red, white, and yellow—for each of the intercardinal world directions and each of the World Quarter stars. Four different trees were used—elm, box elder, cottonwood, and willow, each of which represented one of the world directions and colors, and lashings from the skin of each quarter's talismanic animal (bear, wolf, wildcat, and mountain lion) held their affiliated beam in place.

On the last night of the ceremony, the captive maiden was also painted, the entire right side of her body red, the left side black. Her hawk headdress represented the bird who carried messages to the Morning Star. As dawn neared, the ritual inside the lodge was ended, and the girl was ushered out to the scaffold and persuaded to climb. Standing on the fourth cross beam, her wrists were lashed to the log over her head and her ankles to the one that supported her feet. In climbing the framework, she had taken a step closer to Morning Star, and possibly the Skidi Pawnee saw her ascent in celestial terms, a passage through the four quarters of the sky to the Morning Star's embrace. With the black half of her body on the north and the red on the south, she faced the east, an earthly incarnation of Evening Star. Beneath her, the Pawnee had dug a small pit and lined it with down. It stood for Evening Star's "garden in the west," the place where she conceived Standing Rain, and the Pawnee word for it meant "bed." Everything was in place for Morning Star's arrival.

When the real Morning Star appeared, two men raced toward the girl from a hidden fire, east of the scaffold. Each gently touched one side of her body and loins with a torch. Then without warning, the impersonator of the Morning Star—probably the woman's actual captor—appeared in front of her and shot an arrow through her heart. As drops of blood fell to earth and fertilized Evening Star's garden, the captive maiden's spirit soared up to Tirawahat, the high god. He sent her to Morning Star, who dressed her in shining flint and placed her in the sky. She who had played the role of Evening Star had been sent back to Morning Star to reciprocate his gift of his daughter, Standing Rain, to the earth. The slain woman was now herself a star, a visible sign of the covenant between heaven and earth. Rain would come. Crops would grow. Bison would run. Victories would be won.

Fair enough, but who was Morning Star? He was really known as Great Star and was said to be the stars' most powerful warrior. His special color was red, and he was closely associated with fire, particularly the ability to ignite it. According to Von Del Chamberlain, the director of the Hansen Planetarium in Salt Lake City and the author of a detailed and comprehensive survey of the sky lore and cosmology of the Skidi Pawnee, *When Stars Came Down to Earth*, Morning Star, or Great Star, was the planet Mars.

The only red planet is Mars. It would, of course, seem to shift to the west with respect to the sun and eventually wind up in the western garden of Evening Star. If Venus were there at the same time, a conjunction of Mars and Venus would advertise their courtship. Chamberlain emphasizes that the five most reliably recorded Morning Star ceremonies—held in 1817, 1827, 1838, 1902, and 1906—all took place after Mars had completed its journey from the dawn to the western early-evening sky. A conjunction with Venus there does not seem to have been required, and not every passage of Mars prompted the quest. It is possible that several factors—some terrestrial and some celestial—combined to send the Skidi Pawnee on the errand they ran for Morning Star. A catastrophic smallpox epidemic had broken out in 1837. Mars had participated in several conjunctions on its route to the west. It had started the trip the middle of 1836, and by the fall of 1837 it was in the company of Venus, the Evening Star, in the western sky. After the close conjunction, both planets gradually moved out of the evening twilight. Venus was first to reappear—in the dawn, in March 1838. On April 22, 1838, the Skidi Pawnee sacrificed Haxti, a fifteen-year-old Oglala Sioux girl, in the traditional manner to the Morning Star.

Sometimes, however, the Pawnee seemed to confuse Jupiter and Mars. Jupiter, said to be the brother of Mars, on occasion fulfilled the role as Morning Star. One remark in the papers of Alice C. Fletcher, an American anthropologist who studied Skidi Pawnee ceremonialism in the 1890s, perhaps explains the ambiguity.

> We are told that there are two morning stars, brothers, the elder is red, and it is he to whom the human sacrifice is made; the younger is white, he is kind and does not share in these rites.

No matter what part Jupiter may have played in the Morning Star ritual, the red color of Mars, its commanding movement through the stars, and its obvious quest to the west make Mars the real Morning Star in the Pawnee myth. The Skidi Pawnee starred that planet and its brilliant partner Venus in a story about the origin of life on earth and in a ritual intended to maintain it. They saw meaning in the movement of those planets. With them, the Skidi Pawnee clarified the relationship between male and female, between heaven and earth, between life and death. Every time Morning Star went a few rounds in the sky, the Skidi Pawnee saw a cycle of world and tribal renewal in its capture of Evening Star.

12

By the Light of the Morning Star

ALTHOUGH the Skidi Pawnee apparently meant Mars when they spoke of the Morning Star, in most of the ancient world the Morning Star was the planet Venus. As the third brightest celestial object, Venus was a featured performer in many of the myths people told about the sky. The way it moved often prompted the ancient skywatchers—in the Old World and the New—to see a tale of a deity's descent from heaven into the underworld in the planet's sessions as a morning star and as an evening star and in its times of invisibility in conjunction with the sun. In several of these stories, divine pride accounts for the planet's disastrous plunge to earth.

THE AMAZON OF THE DAWN AND THE TWILIGHT TART

Alternating between evening twilight and dawn—and sometimes never putting in an appearance at all—the planet Venus behaved like Babylon's goddess Ishtar: capricious, willful, and heady. Ishtar was a member of the highest trinity of Mesopotamian sky gods, and her stories and symbols can be traced all the way back to Sumer, where she was known as Inanna.

Ishtar-Inanna was something more than the goddess of love and fertility represented by her Greek counterpart Aphrodite. Her interests extended beyond affairs of the heart to the judgments of fate, the vicissitudes of fortune, and the hazards of war. As the lover of Dumuzi (or Tammuz), Ishtar was also a kind of Mother Earth who incubated the earth's seasonal bounty and brought it to fruition.

As the morning star, Ishtar was a goddess of war who often traveled standing upright on the back of a lion. The physical power symbolized by her status as a warrior is actually related to her function as a force for fertility. Sumerian hymns to Inanna tell us she exercises this power through the rain. When she steps on the heavens, rain falls down. When she steps on earth, vegetation sprouts up. Thunder is heard when her lion roars.

False modesty never compromised Inanna's style. In another hymn, she boasts that she wears the heavens as a crown and the earth as her sandals. She is the master of kings and queens and while other gods are sparrows, she is a falcon. According to her, the gods of the earth and underworld "trundle along," but she is "a splendid wild cow."

Inanna may have been the master of the storm and a mistress of war in the morning, but she was a harlot at night. As the evening star, her interest switched to sex. Always erotic and sometimes lewd, she appeared in the twilight and kept the prostitutes company on their way to the bars to solicit business. From her upper-story apartment

The status of the planet Venus as one of three preeminent celestial objects in Mesopotamian tradition is confirmed by boundary stones like this one from the second millennium B.C. The eight-pointed star of Ishtar (Venus) is accompanied near the top of this stone by the crescent of Sin (the moon), in the middle, and the rayed disk of Shamash (the sun), on the right. The eight points of Ishtar's star may be a symbolic reference to the eight-year cycle of Venus. Other symbols on the monument refer to other gods, and some have celestial connotations. The inscription informs us that the stone deeds a parcel of land from King Eanna-sum-iddina to his subordinate Gula-eres in 1120 B.C. (Object 102485, British Museum, London)

window in the sky, she scanned the street below for customers.

Inanna knew what she had, and she flaunted it. When she leans against an apple tree, in a Sumerian poem "Inanna and the God of Wisdom," her vulva is said to be "wondrous to behold." She thinks so, too, and applauds her own anatomy. This heavenly vessel of hers is shipshape. In the hymn of her courtship with Dumuzi, she compares the unseeded furrow between her thighs to the "Boat of Heaven," the new crescent moon. As eager as that ready-to-ripen moon, Inanna is

Ishtar in her warrior goddess identity personified the planet Venus as the Morning Star. She grips the reins of her lion, and with one foot on its back, her brilliant rays are represented by the arrows quivered behind her back. Her horns tell us she is divine, and her wings allow her to travel through the sky. She emerges from the eastern horizon ahead of the sun. Its symbol is to the left of Ishtar, near the upper edge of this Akkadian cylinder seal impression. Such images were rolled onto clay envelopes to guarantee the integrity of contracts written on the clay tablets they enclosed. (Courtesy of The Oriental Institute of The University of Chicago)

primed for tilling. She asks, "Who will plough my high field? Who will plough my wet ground?"

A voluptuous, commanding Amazon, Inanna sent a welcome mixed message of power and sex to her admirers on earth. Unsatisfied, however, with keeping heaven on its toes, one day she decided to go to Hell.

DISROBING IN HELL

Inanna's descent into *kur,* the "great below," began when she turned her ear toward that underworld to gain an insight into its power and make its throne her own. Described as a house buried deep under the mountain of the world, the abyss where departed souls paid rent on the mortgages of their lives "separated the wicked and the good." Although no one escaped from *kur,* those who had conducted themselves honorably while alive had no reason to fear judgment there. Such reassurances probably offered scant solace to the souls on Hell's doorstep, however. The threshold of the "house of the setting sun" even made the dead pause, for

> *. . . the sill*
> *is a monster with jaws that gape*
> *and the jambs of the door are a sharp knife*
> *to slash down wicked men. The two rims*
> *of the river of hell are the rapier thrust*
> *of terror, a raging lion guards it.*

Despite the danger, Inanna had made up her mind to seize the throne of Hell from its queen, Ereshkigal.

Abandoning her temples on earth, Inanna prepared for this junket as if it were a business trip and put on her power clothes. In a fetching formal fall of hair, topped with a crown, she clasped a lapis lazuli choker around her neck and draped her formidable breast with two strands of lapis beads. Over her royal gown, a flounced and pleated garment, she fastened a jeweled breastplate. It must have been an effective, bosom-enhancing accessory, for it had its own name: "Come, Man, Come." Her eye shadow was a cosmetic advertiser's dream. It was called, "Let Him Come" and

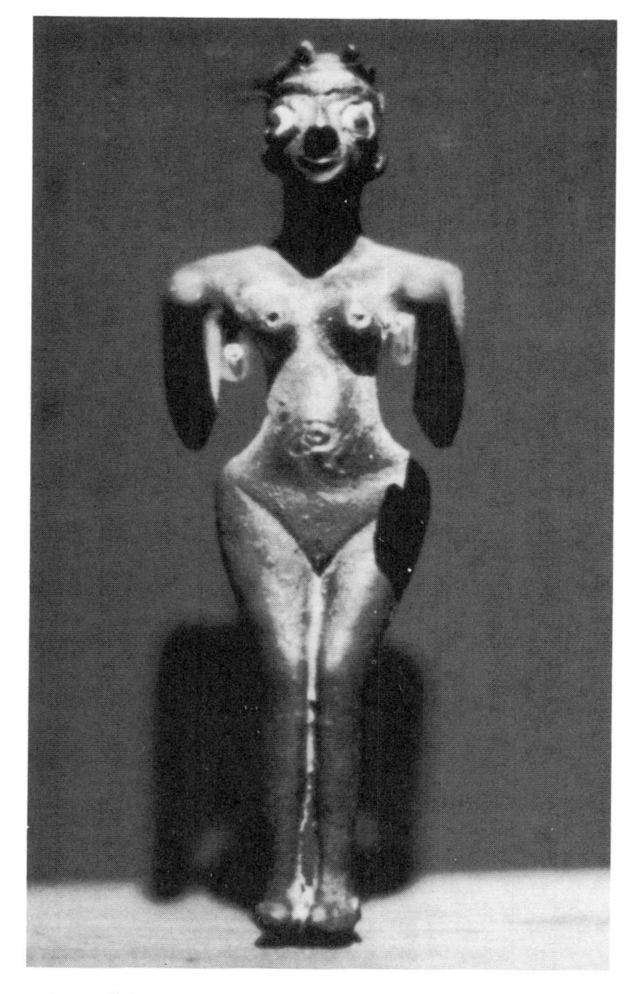

An explicit sexual message is conveyed by this small copper figure of a nude female goddess. She was found at Mari, a town on the Euphrates River in northwest Mesopotamia and was crafted about 2500 B.C. Although she carries no inscription to verify that she represents Inanna/Ishtar, the horns on her head indicate her divine character. She resembles similar representations of Ishtar and her other Middle Eastern counterparts known to have been associated with fertility, female sexuality, and the Evening Star aspect of the planet Venus. (Object in the Damascus Museum)

would probably be too steamy for commercial television. Dropping a gold bracelet over her wrist, she reached for her lapis measuring rod, another emblem of her earthly power, and headed for Hell, dressed to kill.

Sparkling like the Morning Star and accompanied by Ninshubur, her faithful advisor, Inanna knocked on the door of *kur*. As she waited for a response, she instructed Ninshubur to return home and wait three days and nights for her safe return. If she didn't show up after that, Ninshubur should get help from the gods—her father Enlil or Nanna the moon or the god Enki—and come back with reinforcements.

With Ninshubur packed safely back to the palace, Inanna shouts rudely to Neti, the gatekeeper of the underworld, and demands entry. A later Akkadian text of this myth says she also threatened to smash the door, shatter the bolt, break the doorpost, push the doors aside, and release the dead if she didn't get her way. Neti returns her insolence by yelling back, "Who are you?" She identifies herself, "Inanna, Queen of Heaven." She explains she has come down from the east to *kur* for the funeral of the Bull of Heaven, Ereshkigal's husband. This is the same seasonal celestial bull (Taurus) slain by the hero Gilgamesh and his friend Enkidu in Inanna's home town, Uruk.

Neti asks Inanna to cool her heels at the door while he goes to inform Queen Ereshkigal who has lighted on her doorstep. Ereshkigal is the true queen of the underworld and Inanna's older sister, and she rightly regards Inanna's descent into her realm as trespassing.

Neti gives Ereshkigal a jewel-by-jewel account of Inanna's costume. Biting her lip in anger, Ereshkigal pounds her fist against her own thigh and tells Neti to remove Inanna's jewels and clothing at each of the seven gates of the underworld according to prevailing policies and procedures. So Neti hustles back to Inanna, and when he permits her to enter the first gate, he also insists she turn over her crown. This unexpected demand catches her off guard. "What is this?" she asks. Neti cuts her off sharply. A guest, he says, has no business questioning the customs of the house. When in Hell, you must do as the souls that reside there. So Inanna turns over her tiara.

At the second gate, Neti insists on removing the lapis collar from her neck. Again she is astounded by his temerity and insists he explain. He gives her the same answer, uncouples the choker, and escorts her to the third gate. There she loses the beads from her breast. At the fourth gate, she relinquishes the jeweled corselet, "Come, Man, Come." Neti takes the gold bracelet at gate five and confiscates the lapis measuring rod at the next gate. Finally, at the seventh gate, he orders her to remove her robe. Now she is completely naked, stripped like the dead. Humbled before the queen of *kur*, she is judged by the Annunuki, the magistrates of Hell. From her throne, Ereshkigal stares with death in her eye at Inanna and announces the verdict: guilty.

Struck by Ereshkigal's wrath, Inanna, after a fatal striptease, died in the kingdom of the dead. She was nothing more than dead meat hung from a spike on the wall. Tainted and green, her corpse rotted in hell.

Three days passed with no sign of Inanna in her own territory, and so Ninshubur went to find help. Her father Enlil was out of patience. She had, he said, brought her fate upon herself. Her lust for power led her to the realm of the dead, and the rules require that those who go to its dark city stay there. Nanna had the same opinion. Enki, however, was deeply troubled about the loss of Inanna. He created two supernatural mourners to sneak into the underworld, find Ereshkigal, join her in her customary lamentations, and put themselves in her favor. These two spirits are really the symbolic essences of two functionaries in Mesopotamian funeral rites. They ingratiate themselves with Ereshkigal, and she promises them a gift. They ask for nothing more than the decayed carcass dangling from the wall. Ereshkigal tries to put them off by saying the corpse is Inanna's property. Politely insistent, the pair repeat their request. So Inanna's dead body is turned over to them. When they sprinkle it with the food of life and with the water of life, Inanna comes back to life. Escaping Hell, however, is not so easy. About to ascend out of the underworld, she is accosted by her judges, who remind her that no one gets out of death alive. She can't just pick up and go as she pleases. If, however, she provides an acceptable substitute, she will be permitted to leave

From here the story of Inanna carries on to the capture of her spouse, Dumuzi. When she sees him fat and sassy on the throne in Uruk while the rest of the world is mourning her death, she figures he can do her time for her in Ereshkigal's slaugh-

Ishtar, stripped of her fetching garments and vulnerable in the Underworld, seems to stand like a candidate for a firing squad between two administrators of Hell on this cylinder seal impression in the Hague Museum. (From The Dawn of Civilization *by Gaston Charles Maspero, 1894)*

terhouse. Accordingly, she delivers Dumuzi to the deputies of Hell.

As either a morning or evening star, Venus is a high-flying goddess whose trajectory inevitably snaps her back into the underworld. Infatuated with her own glory, she believes she can just push her sister aside and rule Hell the way she rules the sky. An ambitious regent of heaven, she hungers for power, and her hunger drives her to passion the way desire drives sex. She dies, however, as most celestial objects are said to die when they are lost from sight. Her descent to the land of the dead and subsequent rebirth is another example of the meaning the ancients saw in the behavior of celestial objects. In its seasonally tempered ascent and decline, Venus clarified the mystery of death and celebrated the power of sexual energy to repopulate the world with life.

PARADISE LOST AND PLANETARY PRIDE

As a soaring morning star that fell to Hell, the planet Venus was laminated onto the Judaeo-Christian concept of the Devil and provided him with another name—Lucifer. *Lucifer* means "light bringing" and originally was the Latin name for the planet Venus as the morning star that heralded the sun. Among the early Christians, Lucifer was the fallen archangel of God. Lucifer's pride prompted him to lead an insurrection against God and brought evil into the world.

Treated with grandeur in English heroic verse, the theme of the fallen angel is the foundation of John Milton's epic poem *Paradise Lost.* Published in 1667, it deals with one of the fundamental issues in Christian belief—original sin in the Garden of Eden. In the guise of a serpent marketing forbidden fruit, Satan seeded sin on earth, but to explain why Satan is engaged in this enterprise, Milton relied upon the story of Satan's fall from the side of God.

The infernal Serpent; he it was whose guile
Stirred up with envy and revenge, deceived
The mother of mankind, what time his pride
Had cast him out from Heaven, with all his host
Of rebel Angels, by whose aid aspiring
To set himself in glory above his peers,
He trusted to have equalled the Most High,
If he opposed, and with ambitious aim

Lucifer drops like a stone from the side of God's celestial throne. Headed toward Hell, he mimics the Morning Star descent of the planet Venus into the consuming fire of the rising sun. (Engraving by Gustave Doré, in John Milton's Paradise Lost, *1866)*

Against the throne and monarchy of God
Raised impious war in Heaven and battle proud,
With vain attempt. Him the Almighty Power
Hurled headlong flaming from the ethereal sky,
With hideous ruin and combustion down
To bottomless perdition . . .

In telling this story, Milton repeats doctrine formulated and refined by the early church fathers, including Saint Augustine, in the first few centuries of the Christian era. They, in turn, borrowed themes and images from the Old Testament and other Hebrew sources. It is there that we find a rebellious army of angels identified loosely with the stars. A passage in the Old Testament Book of Isaiah (chapter 14: verses 12–15) invokes the unambiguous image of Venus plummeting to Hell.

How art thou fallen from heaven, O Lucifer,
son of the morning! How art thou
cut down to the ground which didst weaken the nations!
For thou hast said in thine heart, I will ascend
into heaven, I will exalt my throne
above the stars of God; I will sit also
in the mount of congregation upon the sides of the north;
I will ascend above the heights of the clouds;
I will be like the most High.
Yet thou shalt be brought down to hell, to the sides of the pit.

Isaiah was not writing an astronomical treatise, but he did exploit the behavior of the morning star symbolically for his own purpose. Of course, he did not use the word *Lucifer*. In the Hebrew text Isaiah addressed the star as *Helel ben-shahar*, which means "bright morning star" or "bright son of the morning." Interpreters of Isaiah believe his message was metaphorical. He was really talking about the swollen pride of the king of Babylon, who would fall from his lofty perch in the same way Venus dropped from its celestial station into the fire of the sun and invisible death in the underworld. Regardless, Isaiah's treatment of Venus tells us that the planet was consumed by its sense of its own importance. In that sense, Venus is guilty of pride. It scales the eastern wall of heaven ahead of the sun, not as a herald but as a social climber, and it lingers arrogantly in the dawn long after respectable stars have withdrawn to their chambers. The vanity of Inanna provoked her desire to conquer Hell. Lucifer flirted with mutiny in heaven. In both cases, the planet took a fall.

Quetzalcóatl *means "Feathered Serpent," and this Aztec sculpture portrays him as a coiled plumed snake with a human face emerging from the serpent's mouth and with his coils clasped by human hands. In other aspects, Quetzalcóatl was also the god of the wind, Venus as the Morning Star, and Venus as the Evening Star. (Object in the Musée de l'Homme, Paris)*

FEATHERS IN HEAVEN SLITHER TO EARTH

Venus was observed systematically in ancient Mexico and identified with the god the Aztec called Quetzalcóatl, the "Feathered Serpent." When the Spanish conquerors of Mexico got to know a little about the people they had subjugated, they learned that the Quetzalcóatl was believed to have died and been resurrected. To the Spanish, this tradition seemed to mock the true savior. Like so many aspects of Aztec religion, the rituals dedicated to Quetzalcóatl and the beliefs surrounding his nature paralleled and contradicted the Christian doctrines of the conquerors. The Europeans came to believe that such deceitful similarity could only be the work of the Devil, who obviously had been locking up the religious franchises in the New World while Europe was building cathedrals. Furthermore, this god's transformation into the Morning Star made him sound like a Mesoamerican Lucifer. His story also involved a trip to the realm of the dead.

The Huastec priest engaged in ritual bloodletting on this relief from Huilotzintla, near Mexico's Gulf Coast, is dressed in the regalia of the god Quetzalcóatl in his role as Éhecatl, the divine power of the wind that animates the world with rain and is like the breath of life. The Huastec spoke a language related to Maya, and this kind of auto-sacrifice is known in Maya ritual, too. Quetzalcóatl engaged in similar self-mortification until an episode of intoxication compromised his credibility. (Object in National Museum of Anthropology, Mexico City)

Quetzalcóatl's jurisdiction was not narrowly defined. Rather his authority, like Inanna's, encompassed several important aspects of life. Fray Bernardino de Sahagún tells us in the *Florentine Codex* that Quetzalcóatl was also the wind, the guide and roadsweeper of the gods who brought rain. Linked closely with heaven's life-sustaining water, Quetzalcóatl engineered fertility in living things. Even his name suggests this. The iridescent green feathers of the quetzal bird symbolized vegetation. Because he has feathers, Quetzalcóatl is a sky serpent. Snakes are symbols of renewal and of the earth, for they crawl there. *Cóatl,* the word for "snake," also contains a root which means "water." The Feathered Serpent symbolizes, then, a lifegiving power drawn from water, earth, and sky.

Later, Quetzalcóatl runs into trouble with some sorcerers. They try to trick him into conducting human sacrifices, but he refuses. His own weakness of character, however, leads to his disgrace. The wizards persuade him to break his ritual fast with a few sips of an alcoholic beverage. Before he realizes what is happening to him, he grows tipsy and summons his sister. Together, they get so drunk, they fail to fulfill their official obligations. They quit taking purifying baths, they quit letting sacrificial blood, and they quit performing the rites they were sworn to conduct every dawn. When Quetzalcóatl comes back to his senses, he is deeply depressed. And the wizards don't let up. Their harassment eventually drives Quetzalcóatl away, and it was said that in the year One Reed he died.

Truly they say
that he went to die there
in the Land of the Black and Red Color.
They say in the year One Reed
he set himself on fire and burned himself;
they call it the burning place,
where Quetzalcóatl sacrificed himself.
And they say that when he was burned,
immediately his ashes rose up,
and all the exquisite birds came to see. . . .

When the pyre had ceased to burn,
Quetzalcóatl's heart came forth,
went up to heaven, and entered there.
And the ancient ones say
it was converted into the morning star.

The song makes Quetzalcóatl's relationship with the morning star plain, but there is more detail.

They said that Quetzalcóatl died
when the star became visible,
and henceforward they called him Lord of the Dawn.
They said that when he died
he was invisible for four days;
they said he wandered in the underworld
and for four days more he was bone.
Not until eight days were past did the great star appear
They said that Quetzalcóatl then ascended the throne as god.

What we have here in the *Anales de Cuauhtitlán* is an explicit description of Venus passing through inferior conjunction. Quetzalcóatl dies when he becomes "visible," that is, as an evening star. Sinking lower and lower to the western horizon and closer and closer to the setting sun, he is cremated in its fire and reappears eight days later, reconstituted from bone after being invisible. When we next see him, he is the Morning Star, rising in the east ahead of the sun.

The reappearance of Venus as a morning star drove the Aztec up the wall. In that guise they also recognized the planet as the god Tlahuizcalpantecuhtli, the "Dawn Lord." He was a hunter and a warrior, a fierce personification of military power, as was Inanna. His arrival in the dawn usually meant real trouble. In the *Anales de Cuauhtitlán*, we are told the old men always knew what Venus was doing and knew the name and number of the days on which it would be seen. On the day One Cipactli ("one alligator"), for example, Venus would rise heliacally and spear the old people by casting his bright rays at them. The Aztec felt the Morning Star's pernicious influence could bring disease into their homes, and so they covered up all of the openings of their houses at the time the Dawn Lord returned to the morning sky. They also killed captives in his honor and flicked their blood toward the planet with their fingers as an offering.

In Tlahuizcalpantecuhtli we also have a Venus that falls into the fires of the sun at the end of its time as a morning star. According to the *Anales de Cuauhtitlán*, the Dawn Lord once shot an arrow back toward the sun, from his position high above the sunrise, in an attempt to get the sun moving. He missed, however, and the sun fired a red-feathered spear back at him. Tlahuizcalpantecuhtli fell back to the underworld.

In Xolotl, the evening star aspect of Venus, the Aztec and their neighbors saw a western counterpart, a twin, to Quetzalcóatl. With some canine attributes, he operated on the downside of the cosmos and guided souls to the underworld. Xolotl also guarded the celestial ball court on the western horizon at the brink of the underworld.

VENUS TIPTOES THROUGH THE TABLES

Among the Maya of Yucatán, Venus was known by several names:

noh ek	"great star"
chac ek	"red star"
sastal ek	"bright star"
ah ahzah cab	"he that awakens the earth" [herald of the dawn]
xux ek	"wasp star"

and others. The highland Maya of Guatemala have names for Venus in their own langauge that means the same thing—"great star" and "daybringer." Most of these names make sense, although Venus would not usually be described as red. Probably red refers here to Venus as a morning star in the east, for the Maya associated red with that direction. Why it is a "wasp star" is less clear, but perhaps it refers to the planet's hostile intent as a morning star.

The Maya, like the Mesopotamians of the First Babylonian Dynasty (ca. 1829–1530 B.C.), knew that Venus completes five runs through its 584-day cycle in eight solar years. At the end of eight years, the same configuration of Venus is transported back to the same calendar date. This means, for example, that Venus will first appear as a morning star on the same date it appeared as morning star eight years before. Maya knowledge of the 584-day cycle of Venus and the equivalence between five such cycles and eight of the sun's tropical years is documented in the *Dresden Codex*. The long-term accuracy of the table is extraordinary. To see why, you have to appreciate what the Maya were trying to do. The *Dresden Codex* assigned aspects of Venus—emergence from inferior conjunction, for example—to dates in their 260-day ritual calendar. This calendar pairs 13 day-numbers and 20 day-names in all possible combinations to provide each day in a 260-day cycle with its own identity. The same kind of scheme was used by the Aztec and throughout Mesoamerica and apparently originated in the Preclassic era.

Kukulcán, the Maya version of Quetzalcóatl, makes several appearances in the Dresden Codex, *a Postclassic ritual almanac from northern Yucatán. Five different images of him that correspond to the five Venus cycles congruent with eight solar years are accompanied by numerical tables that refer unambiguously to intervals in the movement of Venus. This one, from page 47, portrays the heliacally rising Venus in the act of throwing a spear toward a victim down on earth. An element in his headdress, just to the left of his face and resembling the letter* m, *is a common Maya symbol for the planet Venus. (Akademische Druck-u, Verlagsanstalt, Graz, Austria)*

Armed with the relationship between the solar calendar and Venus cycles, the Maya didn't stop with eight years and five rounds with Venus. They wanted Venus to match up with their 260-day ritual calendar, too, and that meant going through more cycles. And they did, with an accuracy of ± 2 hours every 481 years.

These intricate calculations were not performed as part of a scientific investigation of the motion of Venus. Instead they were intended to help Maya prognosticators interpret omens based upon their calendar and the behavior of Venus. They installed their kings, sacrificed prisoners, and went to war by those omens. They were not watching a planet —but a god.

Each of the five pages in the Venus table also includes a picture of a Venus god acting like the Morning Star. Kukulcán's costume, regalia, and actual identity vary from one page to the next, but each time his action is the same. He fires spears at targets below, a different victim on each page. Something similar to this parade of the five victims of Venus was also known in central Mexico. Nearly a century ago, the renowned German scholar Eduard Seler recognized symbolic representations of the five cycles of Venus in the *Codex Borgia*, a pre-Conquest manuscript probably composed in the Mixteca-Puebla region of central Mexico. This elaborate and visually striking screenfold document is filled with material that relates to the gods, rituals, symbols, and calendar cycles of Mexico, and like the *Dresden Codex*, it indicates that the scribes who prepared it were watching Venus carefully. Although we can't yet tell why each of the five Venus cycles in the eight-year set had its own unique character, we do know that seasonal shifts in the planet's various aspects made it behave a little differently in each cycle. The way it rose and fell in the west, the details of its appearance in the east, and the time of year it was seen or not seen, all varied systematically in the eight years it took to go through five rounds of Venus.

HEROES IN HELL

We have another source of information in the *Popol Vuh* on what Venus meant to the Maya, and it links Venus once more to journeys to the underworld. Events in the *Popol Vuh* all turn upon one key story. Two brothers, twins and heroes, hike down the highway to Xibalba, the underworld realm of darkness and death, to prevail against the Lords of the Night who rule there. The hero twins are named Hunahpu (One Blowgunner, a day-name) and Xbalanque (meaning not clearly known), and they intend to beat the Lords on their own turf and succeed where their father and uncle had failed. Their father's name was One Hunahpu, and their uncle was Seven Hunahpu.

Sometime earlier, the father and uncle had been playing ball in their home court on the eastern horizon. Overlooking the precipice of Xibalba, it boasted an impressive view of the great abyss. The noise of their game, however, offended the Lords of Xibalba, particularly the two named One Came (one death) and Seven Came. Those two Lords sent owl messengers to invite the pair to play ball with them in the Xibalba stadium beneath the western horizon. Father and uncle walked down the road to Xibalba, the Milky Way's dark cleft in the constellation of Cygnus. There they were tricked and so failed tests that would have allowed them to survive in the underworld. They were killed and buried at the ball court's sacrificial altar. One Hunahpu's decapitated head was lodged in a calabash tree. Appearing as a skull to Blood Woman, the daughter of another Lord of Xibalba, One Hunahpu fertilized her by spitting into her hand. Impregnated with twins, she eventually gave birth to Hunahpu and Xbalanque, the young heroes who aim to settle accounts in the underworld.

The time comes for the twin heroes to play their own game in their elders' ball court, and the same thing happens to them. They make too much noise. The Lords of Xibalba get upset. A summons is sent. And Hunahpu and Xbalanque descend with their invitations. They, however, are ready with tricks of their own. Instead of being deceived by two dummies dressed up like Lords of Xibalba, they send a mosquito ahead of them to find out which Lords are real. Irritated by the mosquito's bites, each Lord reveals the true name of another, and in this way the twins learn some of the secrets of Xibalba. They are also too smart to fall for the practical but deadly jokes that brought their father and uncle to ruin. Instructed to keep cigars lit throughout the night, they fool the Lords of Xibalba by installing a firefly at the tip of each one. Ball games, nights of trial in dangerous hostels, and more tricks follow. Finally, in disguise, the twins perform stage magic for the Xibalbans, and the last illusion is the sacrificial death and revivification of Hunahpu. One Death and Seven Death are floored by this legerdemain and insist that Xbalanque do the same trick on them. He does, but this time it's the real thing, not sleight of hand. The twins then inform the rest of the Xibalbans that from now on the rules of hell are going to be different. They won't be getting any more offerings of human sacrifice, and only the guilty will be their prey.

The story starts when the hero twins are playing ball in the court on the eastern horizon. They are alive like the Morning Star, and they descend like the Morning Star. This does not mean they are the Morning Star. They just act like it. Probably the same could really be said for Inanna, Lucifer, and Quetzalcóatl. In any case, the tests of the hero twins in Xibalba take place during a period that mimics the time of superior conjunction. By the time they have finished the big ball game, the plot meshes with the period when Venus is an evening star. In this game, Hunahpu's head is the ball, and the imagery of death prevails in the symbolism of a surrogate head—a squash—that Hunahpu must don instead.

The short, inferior conjunction should follow the "evening star" chapter, but if it is expressed at all in the *Popol Vuh*, the symbolism is subtle. "Morning star" events should follow, and adventures of the hero twins on earth—their victory, for example, over Seven Macaw and his two sons—correspond to the morning star phase of Venus and the beginning of the cycle.

Archaeologist Michael Coe first recognized the connection between scenes on classic Maya funerary vases and episodes in the Popol Vuh. *Although Coe does not believe that the scene on this vase depicts the stage magic death of Hunahpu in Xibalba, it does appear to show a similar episode of sacrifice in the underworld. The figure on the left carries an ax and is dancing toward a baby Jaguar God of the Underworld. This god is reclining on an altar that is really the head of a monster. To the right we see a prancing, skull-headed Lord of Xibalba. (From* The Maya Scribe and His World *by Michael Coe, The Grolier Club; vase, The Metropolitan Museum of Art, New York, after rollout photograph by Justin Kerr)*

These parallels between the behavior of Venus and perils of the hero twins don't mean that the *Popol Vuh* is just an allegorical treatment of the 584-day cycle of Venus. It is also the story of the "sowing and dawning" of the cosmos, a creation myth that describes and validates the initial conditions of the human problem, the axioms of our existence. Sowing means both the planting of seeds and the setting of celestial objects. Whether Venus goes into the underworld or the twins go into the underworld, something is planted for a future season. *Dawning* means both the sprouting of a new season's young plants through the surface of the earth and the rising of celestial objects from the underworld. The season of dawn is the start of the present era, the time when the "sky-earth" in which we live begins to exist. This "sky-earth" comes into existence when Venus first rises in the predawn sky on a day named One Hunahpu (or One Ahau), for that is when the sun and moon also rise in the world's first dawn after the twins' victory in Xibalba.

13

Across Charted Territory

UNDER ideal conditions, the darkest, clearest nights permit us to view two to three thousand stars with our own eyes. That's only about half of the sky, however. The earth beneath our feet obstructs the rest of it. If we count every star in the entire sky, the total comes nearer to six or seven thousand stars. How many of those you can see over the course of a year depends on where you live. From where most of us live, we have a chance to see perhaps five thousand stars. That's not an unmanageable number, but it is more than most people are willing to catalogue or remember.

Even though the Greek astronomer Hipparchus could see that there were several thousand stars overhead, the catalogue he compiled in the second century B.C. contained only 850. His effort is nevertheless respectable. He listed accurate positions for all of these stars and presumably observed them with his own instruments to do so. About three centuries later, the Alexandrian astronomer Ptolemy enlarged Hipparchus's list to 1,028 stars. Even larger catalogues were compiled, and the invention of the telescope introduced us to even more stars, most of them much fainter than the unaided eye can detect. Astronomers now rely upon considerably more recent and more comprehensive catalogues. The Yale *Catalogue of Bright Stars* provides data for 9,110 stars, and *Uranometria 2000*, a recently published catalogue and atlas, takes advantage of computerized data handling to identify 332,556 stars as much as twenty-five times fainter than the limit of the unaided eye, along with 10,300 other objects. That's a lot of stars to be assigned a number, but there are a lot more that remain untagged. Our Milky Way is a spiral galaxy containing perhaps 400 billion (400,000,000,000) stars. In the entire visible universe there may be as many as 100 billion billion (100,000,000,000,000,000,000) stars.

Whether listing several hundred or several hundred thousand stars, astronomers have always found it helpful to organize and subdivide the sky into constellations. In antiquity, the constellations were delineated star by star. These constellations must have numbered about forty-five in Greece at the time of Eratosthenes (third century B.C.). Over the centuries, new ones were devised, especially for the part of the sky unseen by Europeans until the voyages of discovery. Some of these historically invented constellations have survived. Others have been abandoned, and in 1928 the official number was set at eighty-eight by the International Astronomical Union.

These modern constellations are based, in part, on the constellations of stars known to the ancients, especially the Greeks. Now, however, their borders are no longer established by the stars that

After the Big Dipper, Orion the Hunter is probably the most recognized pattern of stars. The three stars of its belt point upward from the center of this photograph. Close to the dawn glow on the eastern horizon, Sirius, the brightest star in the sky, has just risen. (Photograph: Tom Hoskinson)

give them their distinctive patterns, but by invisible property lines. Even though every corner is a 90-degree angle and every imaginary boundary conforms to the sky's basic coordinate grid, the boundaries are irregular and somewhat arbitrary. The constellations are gerrymandered pieces of celestial geography, but they do what they are supposed to do, what they have always done: They define celestial territory.

Systematic observation of the stars and constellations, over seasons and years, turned the night sky into familiar territory. The ancient Mesopotamians began charting the heavens with a set of thirty-six heliacally rising stars in the "paths" of the gods Enlil, Anu, and Ea. Later they also recognized eighteen stellar markers in the "path of the moon," and from those developed the constellations of the zodiac. The Mesopotamians, however, had no monopoly on imposing order over the star-strewn sky.

MOON WEDGES IN ANCIENT CHINA

Bronze Age Chinese skywatchers put their own system of moon markers to work during the Shang dynasty (1766–1122 B.C.). Names of the twenty-eight *hsiu*, or reference stars, show up in oracle bone inscriptions from this period and verify that the system was in place by 1400 B.C. Each marked the start of a zone in the sky that was something like a zone of longitude on the earth. A zone of longitude is like a wedge of an orange, with one end at the north pole and one end at the south pole. Slicing an orange in half across its wedges is like slicing the earth through the equator. All of these wedges of longitude are "hinged" to the earth's axis as the segments of an orange are attached to its core. Now imagine the celestial sphere as if it were an orange with twenty-eight segments. Let the navel stand for the north celestial pole. The opposite end of the pithy core represents the south celestial pole. Each segment of the orange represents the territory of one *hsiu*, and the zones themselves were also called *hsiu*.

Chinese astronomers could always tell where any particular *hsiu* star was located, whether it had set or not, by observing the position of a "partner" star they assigned to each *hsiu*. These partner stars could only work if they were always visible at night, and that meant they, unlike the *hsiu*, could never rise or set. Stars that never rise or set are called circumpolar stars, and the key star for each *hsiu* was carefully chosen so that a circumpolar star also shared its *hsiu* semicircle, or meridian. For this reason the *hsiu* were not all the same size.

Why the Chinese settled on twenty-eight stars to subdivide the sky into equatorial districts is not really clear. The moon requires 29.53 days to complete a cycle of its phases, from, say, one full moon

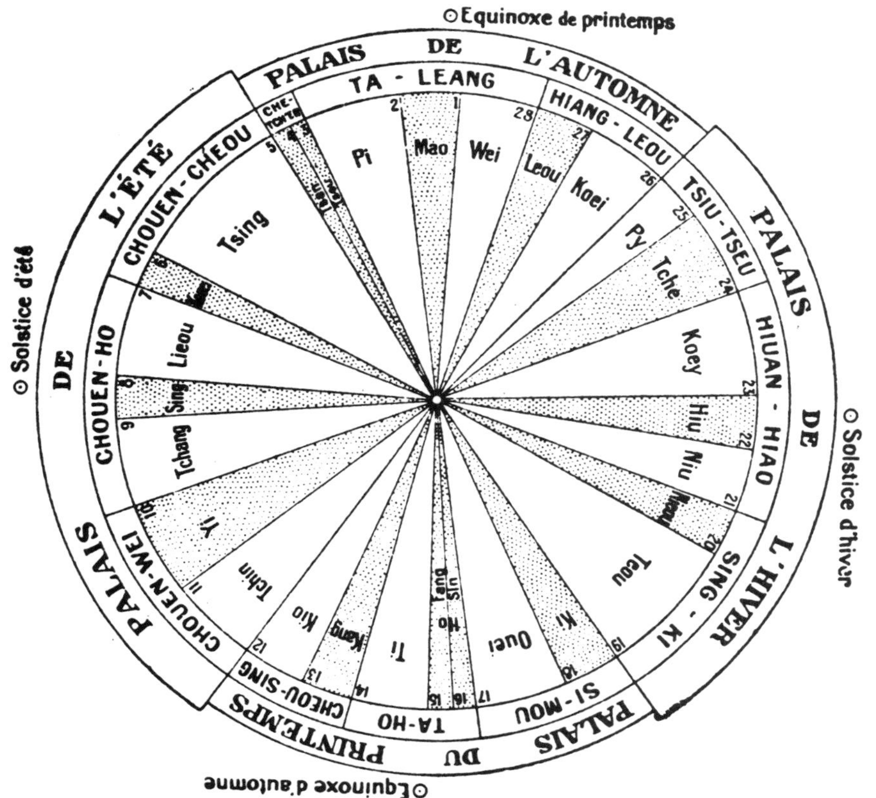

Bronze age Chinese astronomers organized the sky into twenty-eight segments marked by key stars, or hsiu. *These celestial zones were grouped into four seasonal "palaces." This diagram of the system looks a little like the cross section of an orange (From* Les Origines de l'Astronomie Chinoise *by Léopold de Saussure, 1930)*

to the next. It also travels once around the earth, and therefore through the *hsiu*, every 27.33 days. It is possible they wanted a number divisible by four so that the same number of *hsiu* could be allocated to each season and to four equatorial zones known as "palaces."

Grouping the 28 *hsiu* into four sets of seven, the ancient Chinese contrived four celestial "palaces," each assigned to a season and represented by a talismanic animal:

the Spring Palace and the Azure Dragon of the East
the Summer Palace and the Red Bird of the South
the Autumn Palace and the White Tiger of the West
the Winter Palace and the Black Tortoise of the North.

These Palaces each harbored a more or less central star that stood for that celestial realm and season. For the Spring Palace, the star was Antares, in Scorpius, and the Chinese called it the Fire Star. Alphard, the brightest star in Hydra, was known to the Chinese as the Bird Star, and it resided in the Summer Palace. The "star" of the Autumn Palace was the Pleiades. The meaning of its Chinese name is not known with certainty, but the word seems to include symbols for "sun" and "door." Finally, the Winter Palace is centered on Sadalsuud, or beta Aquarii. Its Chinese name, *Hsu*, means "emptiness" and may also have meant "the funeral mound."

BRIDAL SUITES IN THE MOON'S MANSIONS

A system of "lunar mansions" was also used in ancient India. The reference stars there were called *nakshatras*. The earliest use of the word *nakshatra* is found in the *Rig Veda*, but there it may have had a more general meaning. By 1000–800 B.C., however, the system of *nakshatras* was unquestionably in place, and at this time there were twenty-seven of them. Later Vedic texts, the *Atharva Veda* and several others, name all twenty-seven reference stars, and *Krttika*, the asterism we know as the Pleiades, always heads up the list. Later, the star Vega in Lyra the Harp was also included to bring the total to twenty-eight.

In its earliest form, the *nakshatra* system resembled the *hsiu*. Nine of the 28 *nakshatra* stars were also used by the Chinese to determine the *hsiu*, and 11 more are in the same constellations as other *hsiu* stars. No attempt was made, however, to ally the *nakshatras* with circumpolar stars as the Chinese partnered them with the *hsiu*. Both systems have stars in common—10 in the case of the *nakshatras*—with the set of 36 used by the Babylonians, and this fact has led some scholars to suppose that the Babylonian stars of Anu, Ea, and Enlil were the real roots of the Chinese and Indian astronomical reference schemes. No conclusive evidence confirms this idea at this time, however.

Eventually the 27 unequally spaced *nakshatra* stars were replaced by 27 uniform sections on the ecliptic. This is the same kind of transformation that occurred when the real zodiac constellations were ousted by uniform zodiac signs. It may have happened at about the same time in India as in Babylonia. The *Vedanga Jyautisa*, a Vedic text probably written in 600 B.C., refers only to *nakshatra* divisions in the sky and not to *nakshatra* stars.

Although the original purpose of the *nakshatras* is not stated in the Vedic literature, a number of indirect references suggests that they were at least used to keep track of the moon. They were called the "brides of the moon," and each night the moon would appear to enjoy the company of the next one in line. From one night to the next, on the average, the moon moves 13 degrees eastward along the ecliptic, and the average distance between *nakshatras* was 360 degrees (a full circle) divided by 27, or 13⅓ degrees. This was the length of each *nakshatra* arc once a system of uniformly spaced zones was adopted by the Hindu astronomers.

The 27 *nakshatra* stars may have been the moon's brides, but according to Hindu myth, all of the celestial ladies were not happy with their husband. Soma, the moon, was in love with only one of them—Rohini, the red star we call Aldebaran, in Taurus the Bull. Rohini was the second *naksha-*

tra, and its name means "the red doe." In the Puranas, the Sanskrit collections of Hindu myths assembled between A.D. 1 and 500, Aldebaran is called *Sura-vi*, "the celestial red cow." While Rohini's radiant beauty may have justified the single-minded devotion of the moon, 26 claims of matrimonial neglect forced the father of the brides to defend their interests with direct action.

The *nakshatras* were said to be the daughters of Ritual Skill, or Daksha. Daksha was one of the six *Adityas* that personified fundamental cosmic and human law. He conferred upon priests and magicians the skill they needed to communicate and negotiate with the gods. Angered by the moon's irresponsible behavior, Daksha cursed his son-in-law, and the curse inflicted a degenerative disease upon Soma. Because the moon had failed to keep his marital promise to every *nakshatra*, he would lose his vigor each month as soon as he reached his prime and decline in the two weeks after every full moon. That is what we see each month as the moon exchanges one wife for another night after night. Rohini only enjoys him at his best when the 29½-day cycle of phases and the 27⅓-day cycle through the stars felicitously combine to deliver her husband to her chambers with enough strength to satisfy her desire.

HOW THINGS GET STARTED

The formal systems of celestial zoning we find in ancient Mesopotamia, China, and India must have emerged after centuries of less elaborate skywatching. Perhaps at first only a few bright stars and highly distinctive patterns of stars were picked out of the five thousand or so available. The stars people first noticed and remembered were probably selected for their seasonal value and eyecatching appearance. Glowing landmarks, they transformed the wilderness of time into a well-tended field of events.

Much closer to our own time, we hear the same kind of celestial economy in the way the Cóchiti Indians of New Mexico explain the present state of the night sky. They see lots of stars up there, but they ignore most of them. A few are known by name and recognized because they mark seasons or perform some other useful function.

The Cóchiti live about forty miles north of Albuquerque and are Keresan-speaking Rio Grande Pueblos. Their account of the origin of the starry sky is a little like the tale of Pandora's box. Cóchiti's Pandora was a young girl named Kotcimanyako, or "Blue Feather." Long ago, after a great flood, she and the rest of her people had moved to the north. When the waters subsided, Uretsiti, the Mother of all Pueblo Indians, told everyone, including Blue Feather, to return south to the original homeland. After everyone had left but Blue Feather, Uretsiti entrusted to her a small white cotton bag. Tightly woven and tied well shut, the cotton bag was not to be opened.

Blue Feather slipped the bag onto her back and began her journey south. On the way, however, her curiosity about the bag sent her better judgment packing on its own trip. Blue Feather decided to take just a small peek into this mysterious bundle, and she carefully opened it, knot by knot. When she got to the last knot, it was impossible to keep a lid on the situation. The bag was swarming with stars, and they spilled out of the bag. Blue Feather tried to get them all back in the bag, but they scattered in every direction and scooted into the sky.

Blue Feather had let the stars escape and go where they pleased. She could only retrieve a few of them. Those she stuffed back into the bag, and when she reached Cóchiti, they were put where they belong. If she had not let the mob out of the bag, the sky would have been a tidier place. All of the stars would have been given names and been put in their proper places. Instead, only a few stars became familiar friends. The rest are strangers.

LOCAL ISSUES

People treat the sky like a menu and select patterns that are distinctive, recognizable, and useful. What they choose depends on where they live and what they need. For example, the Guajiro Indians of South America attach a great deal of importance to the bright, yellow-tinged star Arcturus, in the constellation we call Boötes the Herdsman. To the

Greeks, these stars were better known as the Bear Keeper or Bear Driver, and were associated with Ursa Major and Ursa Minor, the two celestial bears.

The Guajiro live on the arid Guajira peninsula of northern Colombia. They call Arcturus *Juyo'u*, the "eye of Juya," and Juya is the Master of the Rain in Guajiro myth. Arcturus is important to the Guajiro because it disappears in the west just after sunset in early October. That is usually when the heaviest rains of the primary rainy season fall. For that reason, early October is also the time of the New Year, and the Guajiro also use the word *juya* to mean "year." French anthropologist Michel Perrin, in his book *The Way of the Dead Indians*, explains that the word *juya* also means "rain" in general and specifically the rain of the primary wet season.

There are really four seasons in Guajiro territory. Wet and dry periods alternate. The longer rainy season continues from mid-September until mid-December. Then, a three-and-one-half-month dry spell arrives. Vegetation grows rapidly after the rain, but as this cooler period of drought continues, the growth slows. In the beginning of April, the second and slightly shorter rainy season starts. The plants start to grow again, and this season persists until mid-June. Then the year's fourth season, the second interval of dry weather, closes the year. Fiercely hot winds start to pick up. Nearly all of the plant life shrivels away. Expeditions in search of the remaining waterholes contend with dust and sand. Food and water become dangerously scarce.

The evening disappearance of the Pleiades on the western horizon coincides with the midpoint of the shorter period of rain. For this reason, the Guajiro see the Pleiades as a rainmaking rival of Arcturus. Known as *Iiwa*, the Pleiades compete with Juya for the favors of Pulowi, who is Juya's wife.

Juya and his wife Pulowi represent complementary oppositions in the world of the Guajiro. He belongs to the sky, while she is allied with the earth. He hunts the animals; she protects them. Pulowi is the mistress of wild plants and game and personifies the drought; her power is seen in the rainbow, which follows, and therefore opposes, the rain. They may be legitimately married, but they are not a fun couple. Their conflicting interests, however, are what mobilize the cycles of nature.

Not a particularly virtuous wife, Pulowi couples with her husband's rival Iiwa. Furious, Juya kills Pulowi's brother. Murder, however, has its price, even when committed in the heat of passion, and Juya is obliged to compensate his wife's family. He pays the debt with the lives of the Guajiro and the animals. Those who die of disease in the rainy season represent what he owes. To compensate for taking these lives, Juya gives rain—part of a reciprocal arrangement between the gods and the Guajiro.

A BIRD OF TIME, ON THE WING

The story the Guajiro tell about Arcturus, the Pleiades, and the rains still means something to them because it is an accurate reflection of what they see in the sky and what happens in their lives. Sometimes, however, people continue to recite stories about celestial objects that no longer perform as their story reports. Stars of the Southern Cross, for example, are said by the Warao, who live in Venezuela's Orinoco delta, to appear and disappear according to a very specific schedule. This constellation, however, does not behave as the Warao say, although in a different place and at a different time it did.

The Southern Cross is one of the most easily identified groups of stars in the southern sky. Its stars are fairly bright and arranged in a simple, cross-shaped pattern. Although it can be seen from parts of the northern hemisphere, you have to go pretty far south to catch it. Below the equator it is conspicuous, and peoples throughout Australia, South America, and Africa have made use of it. According to the Warao, the Southern Cross is Shiborori, a celestial bird with beautiful plumage. It flies, they say, to protect the young Warao children.

The Southern Cross actually looks like a bird, with the shorter beam of the cross forming a pair of outstretched wings. From Warao country, the

In the southern hemisphere, the Southern Cross is one of the best-known constellations, and the people who lived at southern latitudes recognized the pattern and made use of it. It looks like a small kite, just to the right of center in this photograph. The two bright stars that form a horizontal line in the left half of the picture are alpha and beta Centauri.

Southern Cross is visible most of the night in the months of April, May, and June. The Warao say it rises at about nine o'clock at night, crosses the meridian at midnight, and sets at about three in the morning. That is why an old woman of the village goes outside at about nine each night and calls to Shiborori. She is a cheerleader encouraging him on to the meridian.

Even though the Warao say Shiborori flies at the same time each night, the Southern Cross can't really do that. Like all the stars, the Southern Cross rises a little earlier each day. The times for the flight don't make any sense either. At 9 degrees north, the latitude of the Warao, the Southern Cross can't rise at 9:00 P.M., transit at midnight, and set at 3:00 A.M. Nevertheless, the Warao keep telling this story about the Southern Cross and keep watching it—and calling to it—because it still has important symbolic value to them. Even if the myth is not an accurate reflection of their sky and their lives, it is an accurate reflection of their highest concern—the survival of their children.

Warao traditions have been learned through decades of fieldwork by U.C.L.A. anthropologist Johannes Wilbert. Studying and living their way of life have convinced him the Warao have a tough job. Foraging a living out of the jungle and swamp is hard enough, but disease raises the infant mortality rate to 50 percent. The infants' fate in the next world is no more pleasant than their short lives on this one. The Warao believe that many, perhaps half, of the children become food for the spirits of the underworld.

That dark, unpleasant realm is located below the western horizon, between the points of winter and summer solstice sunset. It is ruled by Scarlet Macaw, a horrifying red-and-black supernatural bird that severs the heads of his prey. His house is actually on top of world mountain in the west. This mountain is bare: The earth on its slopes is spongy with the blood it has absorbed. Only a pallid, jaundiced light penetrates the house of Scarlet Macaw. His apartments are outfitted with hammocks made of dried blood. The furniture is framed with human bone and upholstered in human skin. The place reeks of rotting corpses. The floor is waxed with blood.

Warao children are destined to become meals for Scarlet Macaw and the dark gods and spirits allied with him, but some of the children will escape, protected by the humming wings of Shiborori in flight. The bird is pursued, say the Warao, by two hunters who want its gorgeous feathers. These hunters are alpha and beta Centauri, the two brilliant stars that point toward the Southern Cross and follow it across the sky. If the Shiborori should be snared by these stellar gods, the loss of its feathers would prevent it from flying and singing again. Without its protective song, all Warao children would die.

It is not just the behavior of Shiborori that fails to tally with reality. He also fails to resemble any bird indigenous to the Orinoco. To the Warao, Shiborori is a mythical turkeylike bird, and its plumage puts the feathers of a fan dancer to shame. The right wing is red with three rows of polka dots, and each row's dots have a different color: blue, green, and blue. The dots on the left wing are green, yellow, and green, and the wing itself is blue. Its legs look like a pair of unmatched socks. One is red, the other blue. Red eyes, red beak, and red feathers on its crest turn its head into a flame. Its body, tail, and head are mantled in green. Iridescent feathers frame in emerald the bone-colored badge on its breast.

The Indians do equate the black curassow, or *Crax alector*, with Shiborori and the Southern Cross. It is a turkeylike bird that lives in the Guyana region of South America and makes a chanting sound. A night singer, it is first heard in early April, at around midnight, but there the resemblance ends. No bird in the Orinoco has Shiborori's fabulous splendor.

Knowing that what the Warao say about the Southern Cross does not match what it really does, Professor Wilbert naturally wondered where, if anywhere, the Southern Cross would fly according to Warao itinerary. It was possible in the planetarium at U.C.L.A. to reconstruct where and when the Cross would have followed the Warao storyline. By shifting Shiborori's homeland to 23½ degrees north, the era to about A.D. 1500, and the flight date to the vernal equinox—about March 21 —Shiborori's flight plan matched the night schedule claimed by the Warao.

Confirmation of another detail in the myth added support to the idea that the planetarium reconstruction was on the right track. This detail involves the constellation we know as Cygnus the Swan and sometimes call the Northern Cross. To the Warao, the Northern Cross is also a celestial bird. They call it Akeuehebu and consider it to be Shiborori's twin. It makes a flight like Shiborori's to protect the Warao children. The Warao insist, however, the two birds must never see each other. If they do, neither would be able to sing, and the children would be endangered. When Shiborori sets, Akeuehebu must rise. According to the planetarium, the Northern Cross emerged from the east at about 2:00 A.M., close to the time the Southern Cross set. Later, at about 10:00 P.M. and not long after the Southern Cross rose, the Northern Cross set. The reconstruction in the planetarium showed that for all practical purposes, the two celestial birds did miss each other in the earlier time and more northern latitude.

The planetarium, then, pinned the Warao story to the proper latitude for sensible astronomy, but it created another problem. The Warao, after all, live much farther south. Why do they tell this story? Where did they get it?

Latitude 23½ degrees north cuts right across Mexico, and Wilbert believes there is ample reason to trace some of the Warao myths to Mesoamerica. Contact across the Caribbean was frequent and probably influential. There are also some other good reasons for thinking Shiborori groomed its feathers in Mexico. Johannes Wilbert points out a remark in Fray Diego Durán's sixteenth-century post-Conquest account of Aztec religious traditions. In the *Book of Gods and Rites* and *The Ancient*

Calendar, Durán describes "a beautiful bird with a bone piercing its body." This breast-piercing bone sounds a lot like Shiborori's "bone-colored badge." The Aztec called the bird Tozoztontli, and that was the name of the third twenty-day interval, or *veintena*, in the Aztec 365-day calendar. The primary festival of this period fell in early April, just as the rainy season was starting to get underway. Some of the ritual connected with this *veintena* was directed toward maintaining the health of young children and protecting them from black magic. These are also the main themes of the Warao Shiborori myth.

Even though no bird in Venezuela matches Shiborori, there is a bird from Mexico that seems to be Shiborori's dead ringer. It is the ocellated turkey, or *Agriocharis ocellata*. Its coloring is similar to what is described by the Warao, and it has a horny feature on its breast. Ocellated turkeys have polka dots on their wings and are native only to the tropical forests of Yucatán, Guatemala's Petén, and neighboring Honduras. This fact, too, suggests the myth of Shiborori originated farther north than the swamps of the Warao. In its homeland, it had something to do with the spring rains and the survival of young children. Somewhere in Mesoamerica someone charted the celestial territory and put a mythical ocellated turkey in the sky as a seasonal sign. Even though that bird keeps a different schedule in the Orinoco, the Warao still see power in the stars of the Southern Cross and urge it to keep flying.

HUNTING FOR ORION

We have already encountered an example of straightforward celestial metaphor in the account of Orion's run-in with the scorpion. When Orion embarked on his program to turn every wild animal into an endangered species. Gaia recruited Scorpius to stop the Hunter in his tracks. Killed by the scorpion's venom, Orion and his adversary were ordered to neutral celestial corners on opposite sides of the ring.

In another Orion story, told in the *Katasterismoi*, we find the Hunter punished with blindness and helplessly wandering after violating the hospitality and daughter of the king of Chios, an island off the Ionian coast. Hephaestus, the god of forge and fire, taking pity on the blind and fallen giant, assigned a young boy, Cedalion, to help and guide him. Lifting Cedalion up on his shoulders, Orion followed the boy's directions and headed east.

The ocellated turkey of Yucatán is the real counterpart to Shiborori, the supernatural celestial bird the Warao see in the stars of the Southern Cross. Its flamboyant plumage in many ways resembles the unlikely pattern the Warao attribute to Shiborori.

There, in communion with Helios the sun, Orion regained his sight.

The references to Orion, the sun, and the direction east alert us to the possibility of astronomical coding in the myth. Orion faces east when it is setting in the west, and it is only by going west that Orion can get to the east. The constellation's last evening appearance in the west marks the beginning of its period of conjunction with the sun. After this "communion" with sun, Orion rises in the dawn, and that heliacal rising in the east was, perhaps, what was meant by Orion "regaining his sight."

Orion's heliacal rising may also be symbolized in the *Odyssey*, where Homer alludes to the fact that the goddess Eos ("the Dawn") marries Orion. The Greek mythographer Apollodorus, writing his *Library* in the first century B.C., confirmed that the enamored Eos carries Orion off to the island of Delos, which was the birthplace of the god Apollo. Now, although Apollo was not the actual sun, he was very closely associated with the sun's symbolic meaning as an emblem of light, life, growth, order, reason, and artful endeavor. From Delos, Apollo's home turf, the sun was said to rise each day. It was after his arrival in Delos, according to Apollodorus, that Orion died.

Orion, then, is with the dawn in the place where the sun rises, but his death on Delos doesn't really make sense unless there is something else going on. Orion's transformation into stars and placement in the sky was one kind of death, but another may be intended here. The dawn does enjoy Orion's company from the time of his heliacal rising in summer until the first time he sets at dawn in winter. After that Orion sets before the dawn can join him. Those trips below the horizon that take place during the time Orion is never seen with Eos may signify his death, for as he descends, the Scorpion rises. Orion, then, remains "dead" until he can rejoin the dawn in the east. The Greeks didn't tell us if this is what they had in mind, but Orion, more than anything else, is a constellation in the sky. References to Delos, Helios, Eos, Scorpius, and the Pleiades in his myths strongly suggest they carried astronomical messages.

Orion the Hunter is one of those constellations singled out by nearly everyone. Along with the Big Dipper, it is also one of the patterns most contemporary observers can recognize. On modern maps of the sky, it contains two especially bright stars. Betelgeuse is red and marks the Hunter's right shoulder. His left knee is Rigel, which carries a hint of blue. Two other stars, Bellatrix and Saiph, form a rectangle with the first two, but the really distinctive attribute of Orion is the row of three closely spaced stars in the center of the rectangle. They are the "belt" of Orion, and they give him an hourglass figure. By name they are Mintaka, Alnilam, and Alnitak. What appear to be three fainter stars suspended at an angle from the Belt of Orion are known as his sword.

All of these names for stars in Orion are derived from Arabic names for the stars, most of which originated from ancient indigenous Arabic tradition. Because the Arabs preserved the Greek astronomical knowledge and later transmitted much of it to Europe, most of the stars' names are rooted in either traditional Arabic astronomy from before the eighth century A.D. or in the scientific Islamic Arabic astronomy that developed after that time. Orion itself was called *Al-jauza*, an Arabic name that implies they thought of the figure as female. Later, the Arabs called Orion *Al-jabbar*, "the Giant." The Sumerian name for the constellation Orion was *Siba-Zi-An-Na*, the "True Shepherd of the Sky," and he seems, in Mesopotamia, to have been the immortal spirit of the deceased Tammuz.

Nothing in the Arabic or Sumerian names sounds like the word *Orion*, and Assyriologist Robert Brown, author of *Primitive Constellations of the Greeks, Phoenicians, and Babylonians* (1899), believed that our familiar name for these stars originated with what he thought was the constellation's Akkadian title—*Uru-anna*, or "Light of Heaven." More likely, perhaps, is the suggestion the name comes from the Greek word for "warrior."

Not everyone, then, called Orion a hunter, but the Chinese saw in him the figure of a supreme general or warrior and knew him as *Shen*. They, too, saw a conflict between Orion and Scorpius, but for them, it was a quarrel between brothers. The stars of the Scorpion stood for the younger brother of General Shen.

Orion was also a warrior in India. The Hindus sometimes called him Skanda. As the general of a great celestial army, he rode a peacock. The *Mahabharata* says he let his arrows fly against the White Mountain, which is the Milky Way. Also known as Karttikeya, he became the supreme lord of battle in later Hindu texts. Largely through his leadership and strength, the earlier order of gods, the Asuras, was deposed.

Aborigines of Arnhem Land in northern Australia took a completely different approach to the stars of Orion. They imagined the Belt as three fishermen in a celestial canoe. These fishermen broke the fish and game regulations of their tribe and hooked a prohibited catch. Three men in a canoe were also seen in Orion by the Wasco Indians who lived along the Columbia River near The Dallas, Oregon. Their story also involves seasons and fishing.

To the Tewa Indians of New Mexico, Orion's Belt and Sword look like a sash hanging from a belt, and they call these stars Long Sash. He was an ancient Tewa warrior who guided ancestors back to the south along the Endless Trail, or Milky Way.

Confusion over the identities of the Aztec constellations makes it impossible to know what the Aztec saw in Orion, but a caption in an early draft of Fray Bernardino de Sahagún's *General History of the Things of New Spain*, identifies an illustration of two intersecting lines of stars as the constellation of the Fire Drill. It looks a little like the Belt and Sword of Orion, and its name refers to the two wooden sticks that are used to light a fire by friction. Other contradictions in the text, however, prevent us from confirming that Orion is where the Aztec positioned their Fire Drill.

Elsewhere in Mexico, the Maya may have believed the three stars in the Belt represented a turtle. A mural discovered at the Maya ruins of Bonampak, in eastern Chiapas, and painted at the end of the eighth century A.D., includes a picture of a turtle with three stars on the back of its shell in the upper—and celestial—zone of the wall. It

Three Maya star symbols form a "belt" down the back of a turtle painted near the vaulted ceiling of Room 2 of Structure 1 at Bonampak. Situated above a "sky band" with three other celestial images, it occupies the part of the room the Maya would have regarded as the sky. Many peoples recognized the three stars in a row in the Belt of Orion, and that is probably what is symbolized on the turtle's back in this commemorative mural. (Replica on display at National Museum of Anthropology, Mexico City)

probably represents the Belt of Orion. Despite additional evidence to support this idea, contradictory data keep us unsure.

Various Highland Maya names for Orion are reported—the Three Kings (los Tres Reyes), Three Stars, Three-in-Line, and the Three Marys (las Tres Marías). The Spanish names reveal clear European influence, however, and in the Middle Ages, the Belt stars were known as the Three Kings and the Three Marys.

Old Germanic names for Orion include the Rake, the Three Reapers, and the Plough, all of which indicate the constellation was used as a signal for harvest or ploughing. The Three Reapers are the Belt stars, and the Plough is probably formed by the Belt and the Sword. In the far north, these stars were specifically connected with the flax harvest, from which the thread for linen cloth was obtained. The Swedes, probably referring to the Belt and the Sword, sometimes called Orion a spindle and distaff. Another old Scandinavian name for Orion's Belt is just Frig's Distaff. Married to the high god Odin, Frig was really Mother Earth in disguise. Spinning thread was women's work in Norse society, and Mother Earth could spin with the best of them. Each day and night her spindle, the celestial axis, turned the sky to spin the thread of time. She had to pull the rough fibers from a distaff, and her distaff, Orion's Belt, was located close to the sky's equator, a seasonal and nightly signal of her progress with the thread.

In parts of Hungary, Orion was Kaszas the Reaper. He trimmed the star grass in the celestial farmyard of the high god and creator Ur Isten. This recalls the Norse tradition of Orion the Reaper mentioned above.

Some people saw the Hunted along with the Hunter in the stars of Orion. The Mongols, the Kirghiz, and the Siberian Buryats are convinced the three stars in the Belt are three stags. Pursued by an archer with a reputation for straight shooting, the stags saw sanctuary in the sky and rose into it just as the hunter was about to overtake them. The Teleuts, a pastoral, Turkic-speaking people of the Siberian Altai, told basically the same story. In most of these stories, other stars in Orion represent the hunters, their horses, their hounds, or their arrows.

In California's southeastern deserts, the Chemehuevi Indians saw the Belt as a line of three mountain sheep. It makes a lot of sense. In southern California, when these three stars rise in the east, they almost seem to climb straight up the sky, in single file and as sure-footed as the bighorns they are supposed to represent. Orion's Sword was an arrow shot at this celestial game by a pair of hunters below. Kumeyaay Indians of the southernmost part of California and northern Baja California pictured the Belt stars as an antelope, a deer, and a mountain sheep. Other Yuman tribes of southern Arizona's Gila River shared the same ideas about the stars in the Belt.

From the sample we've seen so far, we can feel certain that the stars in Orion were part of the territory people throughout the world inevitably charted in the sky. Their stories often tell us why. Orion is an ideal seasonal marker. Some of its stars are bright, and their arrangement, especially in the Belt, is distinctive.

CELESTIAL ARCHERY

The Belt of Orion could be three of anything, but sometimes the patterns formed by the stars really do look like something. In a story from India that is older than the Hindu tale of Skanda, Orion's Belt is an arrow. It was fired at the god Prajapati, who is represented by the other stars of Orion. The *Rig Veda* tells that Brahma Prajapati, the supreme god and creator, conceived a passion for his own daughter, the Dawn. To catch her, he changed himself into a buck deer. She transformed herself into Rohit, the celestial doe, and in that guise seemed willing—and able—to accept the buck's advances. As he approached her, however, the other gods observed this incestuous behavior and would not tolerate it. They enlisted the aid of a hunter, who fired the arrow at Prajapati and stopped him in his tracks. Rohit, the celestial deer, is actually the star *Rohini*, or Aldebaran, in Taurus the Bull, and the on-target hunter is the star Sirius.

This makes good celestial sense, for in one direction the Belt of Orion points toward Aldebaran, and in the other it points toward Sirius. The Southern Paiute, who lived in the Great Basin region of

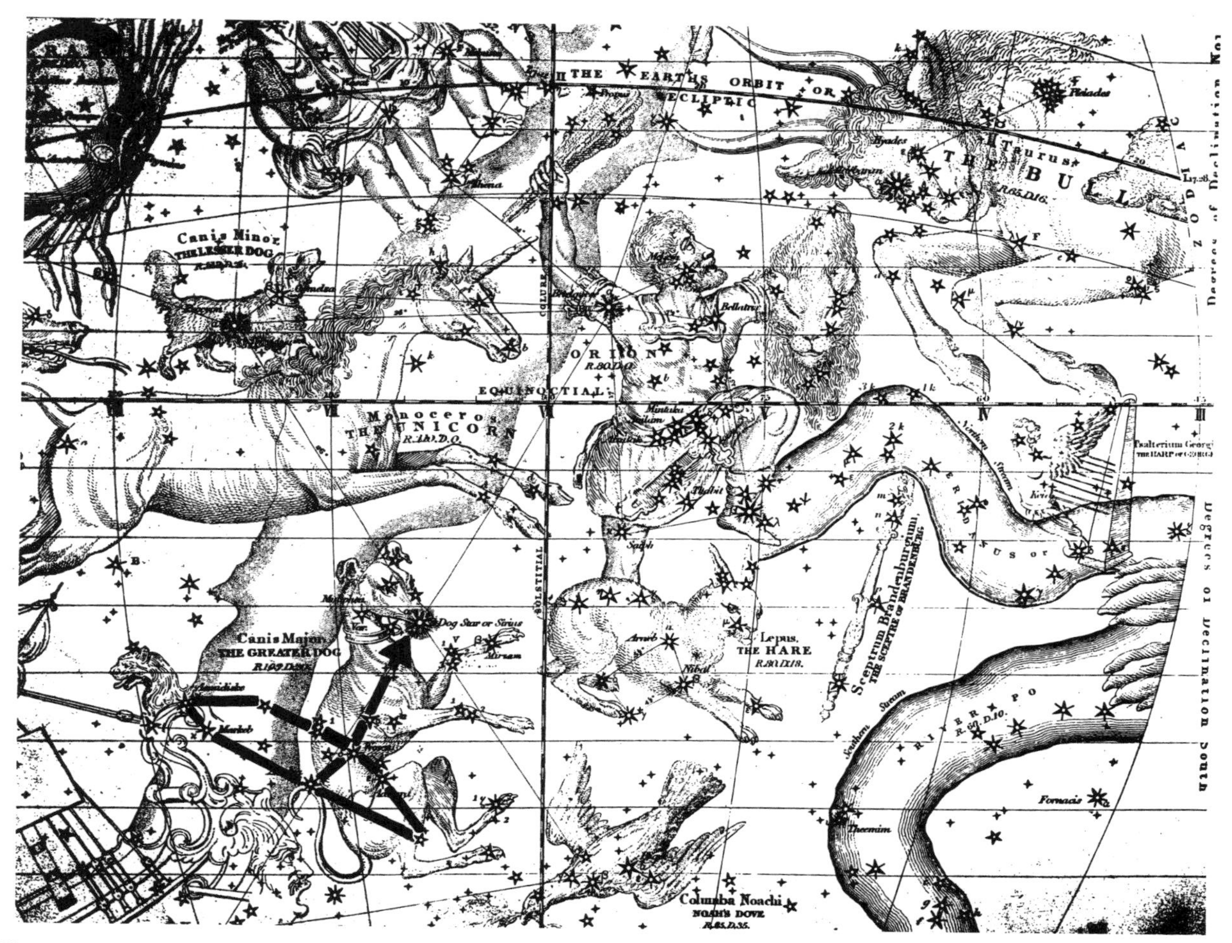

The Chinese and the Babylonians detected a bow and arrow among the stars of Canis Major the Greater Dog and other nearby constellations. Although their conceptions differed slightly, the skywatchers of both civilizations exploited the fact that the stars in that region really do resemble a bow with its arrow pointed toward or tipped by Sirius. Others, including the Hindus, made an arrow out of the Belt of Orion, not far away from the Celestial Bow. (Griffith Observatory collection, Atlas Designed to Illustrate the Geography of the Heavens *by Elijah H. Burritt, 1893)*

western North America, also described the Belt of Orion as an arrow and featured it in a story that seems to make the star Aldebaran the target.

Ancient Babylon, ancient China, and ancient Egypt all spotted a bow and arrow in this part of the sky, too. The famous Circular Zodiac relief from Dendera that is now in the Louvre represents the sky as the Egyptians saw it in the Graeco-Roman Period. Orion is striding forward as the god Osiris, and Sirius, as the goddess Isis in the form of a cow, follows in a boat behind him. Just behind Isis, the goddess Satis stands with her bow drawn and arrow aimed at Isis. Satis's name means something like "Lady Sharpshooter."

THE MISTRESS OF BEGINNINGS

In ancient Egypt, Sirius played the lead role in establishing the hours, mobilizing the calendar, and charting the sky. Sirius is the brightest star in the sky, and the Egyptians gave it credit for bringing the Nile to flood. They saw Sirius rising in the dawn at about the time of the Nile's inundation. The Nile's annual flood was important enough to prompt them to start the year with it. That is why they calibrated their calendar with Sirius and tied its first appearance in the dawn to the New Year. Egypt, it was said, is the gift of the Nile, but the Nile, as far as the Egyptians were concerned, was the gift of Sirius.

The goddess Hathor stands in front of a rocky island mass near Aswan. Ensconced in a serpent cave beneath the outcrop, Hapi, the god of the Nile, pours water from two offering vases. Aquatic plants crown his head. The vulture, the symbol of Upper Egypt, is perched upon the rocks, accompanied by a falcon. Hathor has the head of a cow and represents, in part, the principle of cosmic motherhood. To the right, and out of the picture stands Isis, the stellar goddess who coaxes the Nile out of this subterranean womb. This Graeco-Roman period relief is located on the north inside wall of Hadrian's Gateway at the Temple of Isis.

The original lunar calendar in Egypt normally included twelve cycles of lunar phases and fell about eleven days short of a solar and seasonal year. This means an annual event like the heliacal rising of Sirius would fall eleven days later in the lunar calendar each year. It was supposed to be seen in the month known as *Wep-renpet.* Although this name means "opener of the year," it was actually the year's last month. The "opener of the year" was Sirius, and it was supposed to make its first dawn showing in the month with that name. If the heliacal rising of Sirius took place during the last eleven days of that month, the Egyptians knew it would occur next in the first month of the following year. According to the rules of the calendar, that shouldn't happen. A predawn appearance of Sirius late in the last month of the year, therefore, told the Egyptians to add an extra month to the year. They did, and so they kept the heliacal rising of Sirius where it belonged.

When the Egyptians began using a 365-day solar calendar, they continued to keep the calendar by the moon as well and ran the two calendars separately. Choosing to start the solar calendar with the heliacal rising of Sirius, they reemphasized the star's importance to them. It was "a feminine sun which appears in heaven at the beginning of the year." Dendera's Temple of Hathor was primarily dedicated to New Year festivities, and these involved a ritual observation of the heliacal rising of Sirius from an open kiosk on the roof. Dendera texts inform us that it was a celebration of the birth of the new sun.

> Radiantly, above her father's forehead, the golden-one rises, and her mysterious form occupies the bow of his solar boat.

Sirius, then, was seen by the faithful on the roof at Dendera above the disk of the sun. She runs before the sun and so sits in the front of his boat.

> Her rays unite with the rays of the luminous god on that beautiful day of the birth of the sun disk on the morning of the new year's feast.

As the sun comes up, Sirius melts into the sunlight. Her rays "unite" with his. A procession, which is depicted in relief on the walls of the stairway to the roof, carried a statue of Isis to the open-roof kiosk. There, the statue was faced toward Sirius and the rising sun. Its light bathed the statue and dramatized in the sacred precinct of the temple what the texts said was going on in the sky.

ISIS ENTICES

The fertilizing power that fortified the flooding waters of the Nile belonged to the god Osiris, and Isis was his consort. As the star Sirius, Isis was "the mistress of the year's beginning" who "entices the Nile out of its source hole to provide life to living people." Another text from Dendera says she "causes the Nile to swell at the time when she shines at the beginning of the year." Osiris was the Nile River, and the handiwork of his talented wife stimulated him to rise and fertilize the land.

Osiris was also the moon, and he died in the last harvest's fallen grain, only to be reborn in the next season's growing crop. Osiris was never one thing; one of his titles made him "Lord of Everything." He personified a process, and that process seemed to make the world jump through its hoops. Osiris was born. He grew. He died. And he was reborn. Anything that goes through that cycle—and in the minds of the Egyptians nearly everything did—is Osiris: the land, the river, the moon, the sun, the plants, the animals, the power of the pharaoh, the souls of the dead . . . and the stars.

In the night sky, the Egyptians saw Osiris in the stars of Orion. They called him *Sah* in that form, and they said he was "fleet-footed" and "long-strided." Portraits of Osiris as the constellation Orion sometimes show him as a mummy in a celestial boat, but usually we see him striding forward, or at times standing in a boat. Wearing the crown of a pharaoh, he often has his arm raised and his face turned back. He is looking back at Sirius, either portrayed as a reclining cow in her own celestial boat or as the goddess Isis standing in one. In the sky, Orion does take the lead and so can be imagined to turn his head back toward his faithful wife.

Orion and Sirius performed in the sky the way Osiris and Isis performed in the myth. After Osiris was killed by Set and set adrift in the coffinlike chest on the Nile, he floated downstream and disappeared into the Mediterranean. Searching for him, Isis followed the same route. Eventually she retrieved his body and returned to Egypt. Although she was not able to bring him back to life, she managed to revive his sexual apparatus and conceive the child Horus by him. Later, the body

As the goddess Isis, Sirius sails in her celestial boat upon the ceiling (or "sky") of the burial chamber of the tomb of the pharaoh Sety I. Orion, depicted to the right as Osiris in his own boat, leads Sirius across this sepulchral heaven just as Orion precedes Sirius in the real sky.

of Osiris was found in the marshes by Set and cut up like the waning moon, but in this part of the story he is like the stars of Orion. He sails from east to west each night and dies on the western horizon. Sirius trails him and disappears there herself.

We know the Egyptians saw this process as a daily birth and death of the star. It is a consequence of the earth's rotation, and a passage in the *Papyrus Carlsberg 1*, a cosmological text written in the second century A.D., describes the appearance of a star in the east as birth from the goddess Nut, who represents the sky.

> . . . these stars travel outside her in the night when they shine and are seen. It is within her body that they travel in the day, when they do not shine and are not seen.

Although this document comes very late in Egyptian history, it repeats and paraphrases ideas that are much older, at least as old as the reign of the pharaoh Sety I (ruled 1291–1278 B.C.). The idea of stellar rebirth can, in fact, be traced all the way back to the Old Kingdom's Dynasty V (2498–2345 B.C.), when the *Pyramid Texts*, carved into the chamber walls inside the pyramid of Unis at Saqqara, declares the soul of the deceased pharaoh to be companion of Orion, who will "ascend from the east" with him, "renewed" and "rejuvenated."

LIVING AND DYING BY STARLIGHT

Sirius and Orion also live and die seasonally. Once they become lost for good in the light of the daytime sky—and are not seen at all during the night—they are said to have died. While still "alive," Orion gradually shifts from visibility in the morning sky to visibility in the evening. He sets sooner and sooner each night until he is last seen on the western horizon after sunset. His complete disappearance from the night sky, which follows, is now another kind of death. Sirius is not far behind him, and after a couple of weeks, she, too, dies in the glare of the sun. The pair return to Egypt when they rise heliacally seventy days later. Orion appears first, and Sirius follows a couple of weeks later. In ancient Egypt she showed up once more in the predawn sky, not long after the summer solstice and also close to the time the winter runoff from the distant mountains reached the Nile's First Cataract at Aswan. Beginning its annual flood, the Nile had been aroused by Sirius and was about to conceive her child, Horus. He was the new year, the revitalized land, the next crop of young grain, and the new sun that was born on the first day of the year.

Sirius burned with intense new life when she rose heliacally, but to do so, Sirius had to die. Stars were like souls. When they disappeared into the daytime sky, they were said to have died. Heliacal rising meant stellar resurrection. All of the stars that kept the calendar and told the time in ancient Egypt were selected because they followed the same pattern of Sirius, their leader. They were called decans, and except for a few extra ones needed to handle the last five days of the year, there were thirty-six of them. In 1955, historian of science Otto Neugebauer showed that by aping Sirius, the decans all had to come from a zone south of the celestial equator. Only stars in the zone he delineated could remain invisible for seventy days.

After dissolving into the evening twilight, Sirius remained unseen for seventy days. Every ten days, one of thirty-six decans would also slip, as Sirius had, into invisibility after making its last appearance on the western horizon after sunset. Referring to these decans, *Papyrus Carlsberg 1* tells us: "It happens that one dies and another lives every ten days." As one decan succumbs to the western twilight, another is born in the eastern dawn and takes the place of the deceased star. During the time the dead decan is gone, it is said to have gone into the earth and entered *Duat*, the same perilous netherworld the sun had to negotiate each night. There, in the "Embalming House" of the netherworld, the star takes the cure. Its impurities are shed. Untainted and new, it emerges seventy days later from the eastern horizon in the dawn to live once more.

Sirius, Orion, and all of the other decans went through a mummification that mimicked the way the dead were prepared for the next life. A commentary in the *Papyrus Carlsberg 1* equates the services for the dead stars to those for people who have died: "Their burials take place like those of men."

This logic worked both ways. Human souls had to walk in the steps of the stars. The deceased was said to be a star, and a proper mummification required seventy days to complete. Seventy days were required for mummification because that is how long Sirius and the other decans are each gone from the night sky in the Embalming House of the netherworld.

RISING FROM THE ASHES

In ritual and myth, each beginning is an echo of the first beginning and a reenactment of the time of creation. Because Sirius inaugurated the 365-day solar year in ancient Egypt and so "started time" over again each year, it also stood for the beginning of primordial time and the creation of the universe. For that reason the star was sometimes linked with the *benu*, the bird of creation, by the priests of Heliopolis. They also associated that bird with the planet Venus, which, like Sirius, made dramatic appearances in the predawn sky and heralded the coming of the sun.

According to Heliopolitan creation myth, the *benu* had alighted on the world's only place to stand, a primordial mound of earth that had emerged from the primeval waters. When the bird took wing, the newly created sun rose for the first time and brought light and life to the world. This is what Sirius does every year at heliacal rising. When it first takes wing in the morning twilight, it announces the New Year sun. That sunrise is like the first sunrise, and the start of the new year is like the beginning of time.

According to Herodotus, the *benu* was the same mythical bird the Greeks called the phoenix, which returned from the east after a 500-year absence. Although the Egyptian *benu* resembles the phoenix of Greece and Rome, there are some important differences between them. In fact, the ancient Egyptians never said the *benu* came back anywhere. They did, however, associate the creation myth with New Year's festivities at the heliacal rising of Sirius and the onset of the Nile flood. This implies a kind of periodic resurrection for the *benu*, but it never emerged from its own ashes with the flamboyance attributed by Greek and Roman authors to their phoenix.

Ancient Egypt's bird of primordial creation prepares to fly from its perch on a pyramid symbol. The pyramid stands for the first patch of ground to emerge from the embryonic waters of the world. In some contexts, this bird represented the star Sirius and its ability to re-create Egypt with an annual Nile flood. (Wall relief in room 26 of the temple of Ramesses III at Medinet Habu)

A magical bird that expires in a fire sounds, however, like a star that dies in the western twilight. Its return flight from the far east sounds like a star that rises heliacally in the east. These considerations and other similarities between the *benu* and the phoenix inspired some ancient writers to forge a link between them whether there should be one or not.

Tacitus, a Roman historian in the first and second centuries A.D., reported that the bird's journey to the east and back consumed 1,461 years. He probably picked this up from the Egyptians. They judged that it took the heliacal rising of Sirius 1,461 years to cycle completely through their 365-day calendar. This period of time is known as the Sothic cycle and is named after Sothis, the Greek version of the Egyptian name for Sirius.

By estimating the length of the year one-quarter of a day short, the Egyptians guaranteed that the heliacal rising of Sirius would arrive one-quarter of a day late the following year, one full day late in four years. This daily loss per year for 1,461 years (each 365 days long) adds up to 365¼ days, just about one true solar year. They knew the solar year lasted more than 365 days, but they chose to use that number and remained unperturbed by the drift of New Year's Day through the calendar and the seasons. It would, after all, wind up back where it started after 1,461 years.

Actually, small discrepancies affect these results. A true Sothic cycle lasts somewhat less than 1,461 years. No matter. The Egyptians thought it took 1,461 years to drive Sirius through time, and here it's the Egyptians who count. Tacitus took them at their word and didn't expect to see the phoenix again until the Sothic cycle had done its number.

DOG DAYS

A creation myth, funereal symbolism, an elaborate calendar, and a plan for subdividing the Egyptian sky with thirty-six stars were all direct consequences of the seasonal significance the Egyptians conferred upon Sirius. The Greeks and Romans also saw a seasonal signal in Sirius, and today we regard Sirius as part of the Greek constellation Canis Major the Greater Dog. For the Greeks and Romans, Canis Major was one of the two dogs that accompany Orion on his celestial safari. They called Sirius the Dog Star, and in the third century B.C., Roman farmers sacrificed a fawn-colored dog in May on behalf of the vitality and health of the fields, orchards, vines, and gardens. This was about the time Sirius was last seen in the west just after dark.

Sirius rose heliacally in July and was blamed for the sultry, stifling weather Rome then had to endure. Geminos of Rhodes, a Greek astronomer who lived in the first century B.C., had a little clearer insight into cause and effect.

> It is generally believed that Sirius produces the heat of the dog days; but this is an error, for the star merely marks a season of the year when the sun's heat is the greatest.

Notwithstanding the opinion of Geminos, this heat, and the disease and lethargy that accompanied it, prompted the ancients to call the forty days or so following the reappearance of Sirius the Dog Star in the morning sky—now often stated as the period from July 3 to August 11—the dog days.

MYSTERIOUS SIRIUS

Egypt wasn't the only place in Africa where the star Sirius played a role in primordial creation and returned each year to repaint the colors of the first dawn on the New Year's sky. The Dogon people, who live south of Tombouctou (Timbuktu) in the Republic of Mali, are well known to anthropologists for their elaborate indigenous cosmology. They believe Sirius, or rather an invisible companion they claim resides with Sirius, is one of the most important objects in the sky.

As early as 1931, the French ethnologist Marcel Griaule began collecting Dogon lore. In a few years he was joined by another French scholar, Germaine Dieterlen, who continued to study the Dogon after Griaule's death in 1956. According to Griaule and Dieterlen, the Dogon call Sirius *sigi tolo* and say that it has a very tiny, invisible partner. They call that second star *po tolo*, or "deep beginning," which seems to have something to do with creation.

The star *po tolo* is also closely associated with the fonio grain and is called the "star of the fonio." Fonio is a cereal *(Digitaria exilis)* native to Africa. The Dogon call it *po*, and it is the smallest grain they know. This tiniest of grains was the first of eight different seeds fabricated by the creator god Amma, and the star *po tolo* is just like it.

Because the fonio grain is very small and white, the star *po tolo* is very small and white. According to the story of creation, all things emerged from the star just as all things emerged from the primordial grain. The Dogon liken the star *po tolo* to the husk of a seed. Some of the blood of all the things that were created was left inside the star after they were released, and this makes *po tolo* very heavy. The Dogon say it is the heaviest of stars. It used to be located where the sun is now, but it moved away. All of the other stars moved out of the neighborhood except the sun, the Dogon claim.

Even though *po tolo* has moved, it is, the Dogon say, "at the center of the sky." They mean it is important and influential, and they think of it as a center in motion that governs the rest of the stars. Its influence makes them "stay in place." Certainly, then, it also regulates the behavior of Sirius. It is said to circle completely around Sirius every fifty years. As the companion of Sirius, *po tolo* still does the primordial seed's important work. Its spinning constantly seeds the world with new life in the form of infinitesimal grains.

Dogon notions about the stars and creation are complex. We have little time to explore here the intricate detail of their symbolic system of the world. But there seems to be a mystery here. The star Sirius really does have a small, dense, faint, and white companion—a white dwarf star—that revolves with Sirius around their common center of mass in fifty years. This companion remained unknown in modern astronomy until 1844, when the German mathematician and astronomer Friedrich Bessel deduced its existence from the tiny cyclical displacements he had observed in the position of Sirius for the last ten years. Now known as Sirius B, the orbital partner of Sirius was not actually seen in a telescope until January, 1862. By then the American telescope maker Alvan Clark observed it for the first time, with an 18½-inch refractor. He had the largest refracting telescope in the world at that time, and that's what it took to spot the Dog Star's faint friend.

Apparently informed about the vital statistics of the Sirius system, the Dogon preserve antique cer-

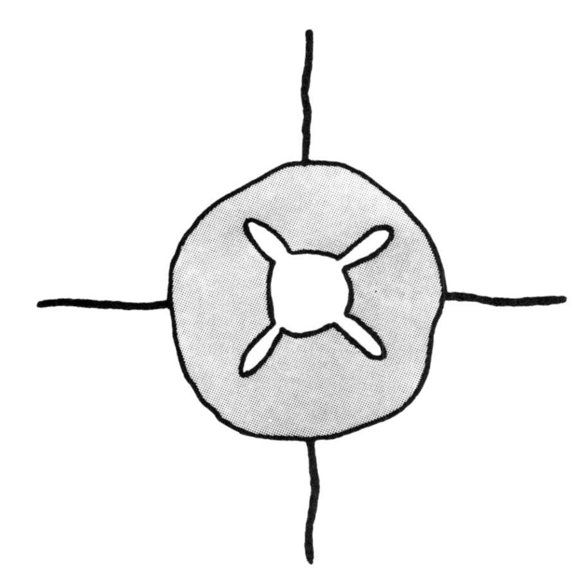

The Dogon drew a picture of Sirius (smaller four-pronged white image in the center) in conjunction with the larger red disk of the sun under an altar to symbolize the heliacal rising of Sirius. (Griffith Observatory, after The Pale Fox *by Marcel Griaule and Germaine Dieterlen, l'Institut d'Ethnologie, Paris, 1965)*

emonial masks used in a sixty-year ceremony said to be related to the stars in the Sirius system. The record of these masks confirms the ceremonies have been held since the thirteenth century. If the Dogon really know the story on Sirius, they've known it for about seven hundred years, long before anyone else invented the telescope or ever heard of Sirius B. Without telescopes how did the Dogon know about Sirius B?

One answer to that question was provided in 1976 in *The Sirius Mystery*, a book by Robert K. G. Temple, who sensationalized these Dogon astronomical traditions. Temple concluded that ancient astronauts, visiting earth from the Sirius system, delivered information about Sirius straight from the Sirians' mouths. Although most of those original reports were given to the Egyptians and others in antiquity, some of them were passed along by the ancients to the Dogon, who considerately preserved them for our mystification today.

Bypassing Temple's distorted interpretation of Dogon beliefs and byzantine handling of their origin, we still have to deal with what the Dogon have to say. Some have suggested that Dogon astronomy has been influenced by foreign visitors—from earth, not outer space. The Dogon, after all, have lived for centuries near a major West African trade node that hosted markets and trans-Saharan caravan traffic. They have been exposed to considerable foreign influence and other beliefs, especially Islam. We might accept this explanation were Dogon traditions not as demonstrably old as we know them to be.

Even if we rule out contamination of Dogon tradition by European travelers and traders, other explanations are still possible. Dr. Philip C. Steffey, an American astronomer, has examined the astronomical potential of Dogon symbols and myths rather carefully and suspects some of the stars mentioned by the Dogon may have been misidentified by the anthropologists. Planets and comets play an important and plausible part in his interpretations.

Also, Dogon astronomy and cosmology contain much that is at odds with the facts. For example, the Dogon say there is a third star in the Sirius system. They call it *emme ya tolo.* It is supposed to be larger and four times brighter than Sirius B. Despite a few reported telescopic sightings of a third star in the Sirius system, its existence has not been confirmed. Further painstaking analysis of the wiggling motion of Sirius allows no room for any more than two stars. If the Dogon say there are three stars in Sirius and we detect only two, perhaps it is time to review what else the Dogon said to find out what they really meant.

If we want to understand Dogon astronomical symbolism, we have to understand the seasonal agricultural cycle that governs Dogon life. Sirius, the rain, and the grain are all interrelated in Dogon territory. The Dogon occupy the rocky plateau and arid savanna in the Bandiagara cliff country in the Bend of the Niger River. They are farmers, and they try to grow enough sorghum, millet, and a few other commodities to get by. The soil, however, is poor, and so they must avoid heavy tillage. Otherwise, the thin topsoil will be lost. Drought and insects are also threats. It is hard to make a living in Dogon territory, and Mali is one of the world's poorest countries. The Dogon live on the edge of natural catastrophe, but they are persistent and resourceful.

Recognizing four seasons, the Dogon begin their year in mid-October, when the rains are finished and the millet is harvested. The first season is the "beginning of the dry season," and it is soon followed by the full dry season. That continues until about May. The last lunar month before the rains is the third season, and the fourth season begins in about mid-June when the rains start to fall. Usually the Dogon get rain each summer, but the rainy season is sandwiched by sizzling hot spells. The hot, dry month before the rains is important as a time for breaking the soil and planting. If this is done too soon before the rains, the hot wind will parch the seedlings and put the Dogon out of a crop.

Sirius disappears from the sky after its last appearance in the west in the early evening, at about this time in mid-June. When Sirius goes, the rains come. Sirius reappears in the morning sky in July, during the season of rain. It rises earlier each night but remains a morning star through much of the rainy season, Millet and sorghum are sown first, by mid-June. Rice follows a month later and is harvested a month before the millet.

The Dogon also plant *po* (or fonio) during this period, but it has to be handled differently. *Po* seed is scattered upon a field, not planted in furrows. When its stalks grow about calf-high, it is time to harvest the grain. This must be done as soon as the plants are ready, for the stalks are weak and will fall quickly. The grain will be lost on the ground. Planting the seed is easy but securing a harvest has its downside. The entire village has to mobilize when the *po* is ripe. The grain is threshed as soon as possible after it is cut. Harvesting at night allows the Dogon to take advantage of all hours for work, and the participation of the young is mandatory. Boys and girls are required to thresh the *po* to provide themselves—and the village as a whole—with future generative power.

Astronomical and agricultural associations in Dogon religious symbolism all suggest that the Dogon idea of an invisible companion of Sirius is a product of their view of the fonio kernel and of the seasonal cycle as it is linked with the broad theme of renewal of fertility. Normally the companion of Sirius can't be seen, but it does appear in the *po* crop. It dies that others might live and through death is itself returned to life. For the Dogon, too, the harvest is a necessary sacrifice. Ogotemmêli, an old, blind elder well-informed about esoteric Dogon religion, explained to Marcel Griaule that threshing *po* was like a sacrifice and that *po* was the victim. In that sense, *po* and menstrual blood were alike. Menstrual blood was a sacrifice that had to be made to the earth to end a period of temporary barrenness. Blood like this, the blood left in the womb after birth, was also said to be what made the star *po tolo* so heavy.

The parallel between the grain and the companion of Sirius also means that Sirius in some way participates in sacrifice. When the rains come, Sirius withdraws from the sky in a kind of sacrifice. Through such sacrifices—of the star, of the grain, or of the blood—fertility could be restored. By exploiting the seasonal behavior of Sirius in this baroque symbolic system, the Dogon, like so many others, contrived their own charted territory in the starry sky.

14

To the Lair of the Bear

In the 1987 motion picture *Dragnet,* Officer Joe Friday parks his convertible on a hilltop overlooking Los Angeles. He and the leading lady in the movie are starting to fall for each other. It's the kind of romantic moment that must happen tens of thousands of times, night after night, in southern California. She is overwhelmed by the beauty of the night sky and is moved to express what she feels: "Oh, Joe, look at the stars. There must be *dozens* of them." Not knowing any better, even a few stars go a long way with us today. From our light-saturated cities, the night sky resembles a desert. Because we see the stars so infrequently, we forget they are there. When, now and then, we encounter them again, unfogged by the genius of Thomas Edison, we are usually surprised by their number and by our inability to recognize most of them. There is one pattern in the sky, however, that most people still know—the Big Dipper.

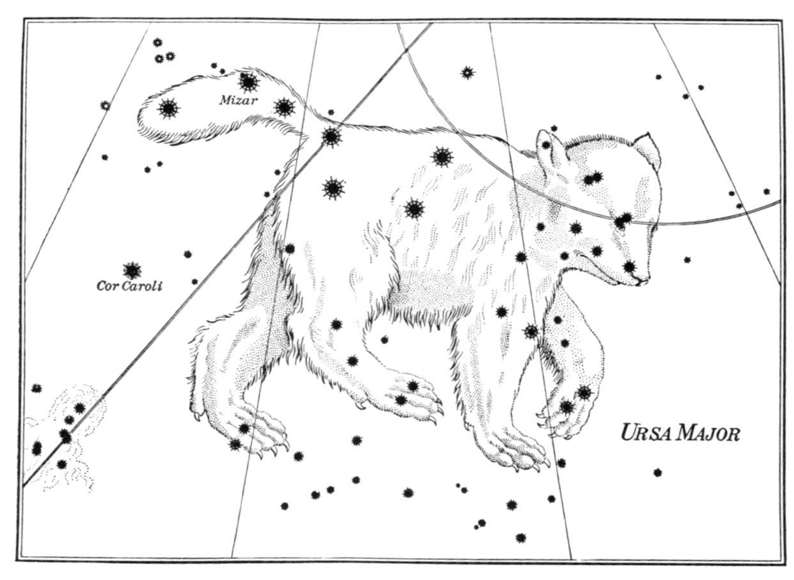

Greek star lore incorporated the seven bright stars of the Big Dipper into the constellation Ursa Major the Great Bear. The Dipper's handle corresponds to the Bear's unnaturally long tail, and the bowl of the Dipper establishes the Bear's back and hindquarters. (From The Stars in Song and Legend *by Jermain G. Porter, Ginn, 1902, after Johannes Bayer's* Uranometria *sky atlas, 1603)*

THE DIPPER IN DISGUISE

Although the Big Dipper is one of the best known patterns of stars, it is not one of the eighty-eight official constellations. Its seven stars belong to the constellation Ursa Major the Great Bear. It can be seen from almost all over the world, but it is especially conspicuous from the northern hemisphere. Its seven bright stars are arranged by chance in the shape of a cup with a long handle, or dipper, and because there is another similar, but smaller, pattern—The Little Dipper, or Ursa Minor the Small Bear—nearby, the bigger cup with the longer handle is known as the Big Dipper in the United States. It is one of the few patterns of stars that really look like what they are called.

Not everyone calls it a dipper, however. The British call it the Plough, and it's fair to say it also looks like a plough. In southern France, they call it the Saucepan, but the French are very thoughtful about food. In many European countries, it was pictured as a *wain*, or wagon. The Skidi Pawnee Indians of North America said it was a stretcher on which a sick man was carried. Among the ancient Maya, it was a mythological parrot named Seven Macaw. To many western Siberians, it was a stag. In ancient Egypt, these seven stars were portrayed as the thigh and leg of a bull, and they transported a celestial official around the sky over Imperial China.

Almost universally, the Big Dipper's primary stars—whatever they are said to portray—are numbered at seven. The stars of the Big Dipper are seven old men or seven thieves to Siberian Mongols, and the Kirghiz, a Turkic-speaking group of the southwest Siberian steppes, call them Seven Watchmen who guard stars that circle the sky's north pole. The Altaic Tatars, or Northern Altays, who live in the Altay Mountains of southern Siberia, identify them as seven "Khans," or tribal chiefs. In northern Caucasia, they were "seven brothers." The Buryats, a Mongol-speaking people of southern Siberia, call them seven blacksmiths or the seven skulls of the seven blacksmiths, or just seven old men.

Percy Bullchild, a Blackfoot Indian of Montana, retells the "history of the world" in his book *The Sun Came Down*, as the Blackfoot elders told it to him, and identified the stars of the Big Dipper as the seven sons of the sun and moon. In another Blackfoot story, the Big Dipper's seven stars are seven brothers pursued into the sky by a bear. To California's Chumash Indians, it appears they were seven boys transformed into seven geese. Hindu sky lore designates the Dipper's stars as the Seven Rishis, or Wise Men. Evidently the ancient Egyptians also imagined the Dipper had seven stars, for many of the pictures of the constellation of the Bull's Thigh—in temple reliefs, in royal tombs, and on coffins—show the leg of a bull surrounded by seven stars.

The seven stars of the Big Dipper ornament one of the smoke flaps on this replica of a traditional Plains Indian tipi erected for the opening of the Canadian Museum of Civilization in Ottawa.

Although the Egyptian idea that the Big Dipper

Astronomical symbols in the Ptolemaic-period Temple of Horus at Edfu represent the Big Dipper as a leg of beef. By topping it with the head of a bull, the Egyptians clarified its meaning, and seven stars surround it. The leg is held by a tether that apparently refers to the travels of the Dipper around the north celestial pole. (From the Description de l'Egypte, *collection E. C. Krupp)*

is the leg and thigh of a bull at first seems a little idiosyncratic, a closer look at the shape of that part of a bull's anatomy shows that it conforms reasonably well to the shape of the Dipper. The thigh is bulky on a bull, and the leg bends at the knee the way the Dipper's handle bends in the middle.

Further Egyptian allusion to the celestial seven appears in the wardrobe of the goddess Seshat. When shown on temple reliefs assisting the pharaoh in the ceremony known as the "Stretching of the Cord," she is outfitted with unusual headgear —a rod topped by a seven-pointed star. Over it hangs an upturned pair of bull or cow horns. Both the presence of the horns and the number seven remind us of the seven stars comprising the Bull's Thigh, which the Egyptians called *Meskhetiu*, and in fact, it is likely Seshat is wearing an emblem of the Dipper on her head.

Seshat was affiliated with writing and record keeping. Laying out the foundation plans of temples also falls in her jurisdiction, for that is the purpose of the Stretching of the Cord ceremony in which she was said to participate. The wall reliefs show her and the pharaoh stretching a loop of rope between two rods to establish the fundamental reference line for the temple plan, and accompanying texts tell us the stars of the Big Dipper were sighted by the pharaoh to determine the proper orientation for the line.

The spots on Seshat's striking skintight leopard-skin dress stand for the stars of the night sky, and in some portraits of her the spots are stars. Dressed in this garment, with a seven-pointed star on her head, Seshat presided over a ritual that locked the temple's plan to a reference line called out by the Big Dipper. By the time the pharaoh carried out the ritual, the line probably was already surveyed by his architects. His act was symbolic, but it mirrored a procedure that required the Dipper's services. Seshat was there in her celestial attire to make sure those seven stars performed.

People throughout the world have picked these seven stars out of the sky and told stories about them because they are so similar in brightness and so obvious in shape and because they do such helpful things. Most of the stories involve what we see the Big Dipper do. Always relatively close to the north pole of the sky, the Big Dipper circles around it and helps to locate it. In this way, it indicates the direction north. It's a compass.

A clear connection between the stars of the Big Dipper and the shape of the Egyptian constellation known as the Bull's Leg is evident in this detail from the lid of the coffin of Tefabi from Asyut, in Middle Egypt. Although the exact date of the coffin is unknown, it probably belongs to Dynasty X or XI, which together span the period from 2134 to 1991 B.C. *The female figure to the left of the Bull's Leg is Nut, the Sky Goddess. The couple on the right are Osiris (Orion) and his consort Isis (Sirius). (Drawing: Joseph Bieniasz, Griffith Observatory)*

Engaged in the Stretching of the Cord ceremony on a wall relief in the Temple of Harwer and Sobek at Kom Ombo, the Egyptian goddess Seshat pulls on a loop of rope with her wand. The symbol on her head includes the upturned horns of a cow, and the seven-pointed star inside them probably alludes to the seven stars of the Bull's Leg. (Photograph: Robin Rector Krupp)

Through the night the Big Dipper swings around the pole, and its changing position discloses the passage of time. Bernard Second, a Mescalero Apache who has collaborated with anthropologist Claire Farrer, uses the Big Dipper in timing songs for ceremonies at night and calls it his "Indian watch." So the Big Dipper is a clock, too.

The Big Dipper's orientation also varies with the seasons. By watching at the same time of night through the year—say, just after dark—you will see the Big Dipper in a different spot in each season. Each night the Big Dipper's starting block shifts a little along its path around the pole, and after a year, it is back to the same starting point at the same time of night. The Yi people of southwest China's Yunnan province still keep track of the year this way. That makes the Big Dipper a calendar.

HITCHING THE WAGON TO A STAR

In medieval Italy, the common name for the Big Dipper was *Carro,* which means "Wagon." Hungarian folklore identifies the Big Dipper as *Göncöl-szeker. Szeker* means "Cart," and Göncöl, the universe's first wagon driver, once spilled the load of straw that became the Milky Way. The Anglo-Saxons saw a celestial buckboard in the Big Dipper, and they called it *Irmines Wagen.* Irmin, it appears, was one of the old Germanic sky gods. The sky wagon was also known as *Wuotanes wagan,* or "Odin's Wagon," among Nordic peoples, and in the eighth century, Charlemagne (Charles the Great) became the vehicle's new owner. Teutonic peoples started calling the Big Dipper *Karlwagn* ("Karl's Wagon") then. "Karl" is just another form of Charles, and in England the Dipper became known as Carles Waen (Charles's Wagon) and Charles Wain.

In Cornwall, and perhaps other parts of Britain, the Big Dipper was named after King Arthur and known as Arthur's Wain. That is its name in Sir Walter Scott's *Lay of the Last Minstrel:*

Arthur's slow wain his course doth roll
In utter darkness, round the pole

A celestial vehicle also circled the pole over China. The Chinese portrayed it as a floating enclosed terrace in which a high celestial bureaucrat made his rounds.

Although the ancient Greeks called the Big Dipper a bear, they also said it was a wagon. The *Phaenomena* is our earliest Greek source (third century B.C.) for detailed information on the constellations, and its author, Aratus of Soli, said the Bears are also called the Wains. When Homer was composing the *Iliad* and the *Odyssey* in the eighth century B.C., the Greeks recognized only one bear in the sky, but in the *Iliad* (Book XVIII, line 487), they also named it the Wagon.

There is a fair chance that this idea of a celestial wagon originated in Mesopotamia. It is at least as old the seventh century B.C. We find the Sumerian name, *Ma-Gid-Da,* for the Big Dipper listed in the *mul-Apin* astronomical tablets, and this name is translated as "the Wagon." Because the name is Sumerian, the tradition could be far older than the seventh century B.C. Names of some of the stars catalogued in the *mul-Apin* as part of the celestial

The circling of the Big Dipper around the north celestial pole prompted the Chinese to think of it as a vehicle for the Emperor of Heaven or some other celestial bureaucrat. This representation of the chariot in the northern sky comes from the tomb shrine of Wu Liang in what is now Jining, Shandong province, in northeast China. Although some experts have identified the smaller star dot in the hand of the winged spirit at the end of the Dipper's handle as a star in Boötes, it is, in fact, Alcor, the optical binary companion of Mizar. (Drawing: Joseph Bieniasz, Griffith Observatory)

Wagon were studied by Oxford scholar E. Burrows. His analysis of the stars and a comparison with features of the Sumerian chariots excavated from the Royal Cemetery at Ur (2600–2450 B.C.) led him to conclude that the Big Dipper began as a four-wheeled chariot. The shafts of these Sumerian carts are known to have been curved in the same way the Dipper's handle is curved, and so the seven northern stars really did closely resemble the four-wheeled chariot known to exist in the middle of the third millennium B.C. This type of vehicle is thought to have transported royalty, particularly the king, from place to place and in processions of state. Many of these heavenly wagons were associated with kings or supreme sky gods, and that is because the unmoving pole, around which the Dipper turned, represented the world axis. The steady pole, the world axis, and the fundamental circular movement of the Dipper were all seen as signs of cosmic order and terrestrial authority.

In Mesopotamia, the stars of the Big Dipper were also said to be a chariot or cart, but the vehicle the Mesopotamians had in mind was probably a four-wheeled war wagon like that illustrated on the "Standard of Ur," a mosaic inlay that may have been part of a Sumerian musical instrument. (Object in the British Museum, London)

That eternal circling of the Wagon in heaven was the trademark of eternity itself. This idea is preserved in a German story in which the Big Dipper is called *Himmel Wagen* (Heaven Wagon). The wagon-driver is named Hans Dumken, from *dumeke,* which means "thumbkin" or "dwarf." He lived modestly in a small thatched cottage, alone except for his pig and his goat and the ox that pulled his wagon. He always wanted to travel and see the world, but because he had little money, he never had the opportunity. His kindness to a traveling stranger, who turned out to be Christ in disguise, was rewarded with the granting of his wish to be up with the stars so that he could see everything there is to see and travel all of the time. From then on, Hans Dumken has piloted his ox and wagon on an eternal trek through the northern sky. In the Heaven Wagon, he is the star Alcor. Alcor is a faint star, located very close to Mizar, the middle star of the Dipper's handle.

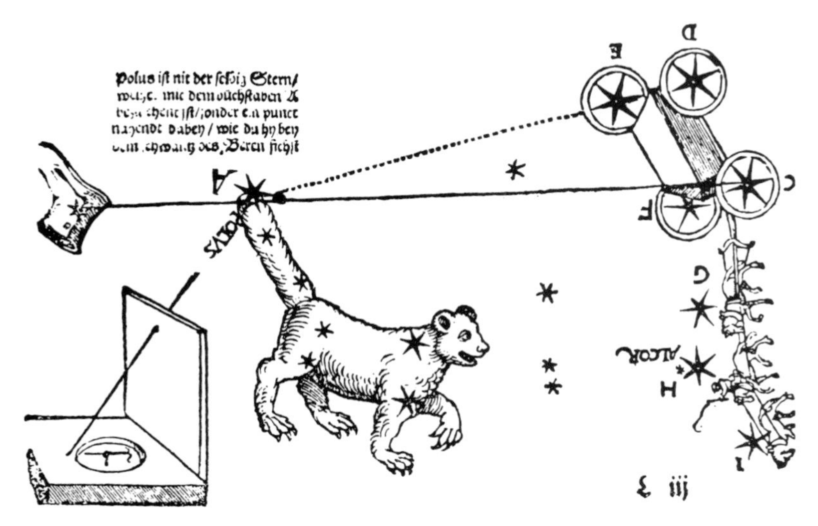

Many European peoples pictured the Big Dipper as a celestial wagon in this sixteenth-century German woodcut. The tip of the tail of Ursa Minor the Small Bear is identified as the pole of the sky. In the drawing of the wagon, the star Alcor is singled out with a caption and also shown as the rider on the middle horse. This rider is Hans Dumken, and he travels eternally around the pole. (From Ernst Zinner, The Stars Above Us, *Charles Scribner's Sons, 1957)*

HOLDING FORTH IN THE NORTH

Navaho traditions about the Big Dipper's unending turns around the north pole of the sky are explicit and concise. They make a single constellation out of it and the Pole Star and call it Whirling Male. Always circling close to the pole, the Big Dipper not only points the way north, but is a symbol of northern lands and northern skies. It appears on the flag of Alaska, the Union's most northern state, and in the nineteenth century, the Big Dipper was a symbol of the freedom runaway slaves could find in the North. Escaped slaves who rode the Underground Railroad "followed the Drinking Gourd" to the northern states.

Haftoreng, the general of the north in the divine army of the Zoroastrian god Ahura Mazda, was actually the stars of the Big Dipper. He guarded the gate to the underworld and prevented invasion by its demons and wizards. According to one of the Five Chinese Classics, the *Li ching* (Book of Rites), the black banner of Xuan Tian Shang Ti, the Spirit of the North and Lord of the Dark Heaven, symbolized the seven stars of the Dipper.

The word *septentrional* came to mean "north" because in Latin it means "seven plough oxen" *(septentriones)*, and the Romans pictured the Dipper's seven stars as a group of seven oxen ploughing circular furrows in the celestial fields around the north pole of the sky. In the thirteenth-century Dante was well-informed about the earth-centered Ptolemaic astronomy of his day. He called the stars of the Big Dipper "seven cold oxen" in *The Divine Comedy* and observed that the northern nations are all governed by them.

Our word for the far north is *arctic*, and it comes from the Greek word *arktos*, or "bear," and it is a product of Dipper lore among the Greeks. In the first half of the second century A.D., Ptolemy described the location of the northernmost lands as "towards the Bears." He meant, of course, Ursa Major and Ursa Minor, which were by then old neighbors in the northern sky. About four centuries earlier, Aratus had explained how the north pole is encompassed by two Bears which circle around it. One, he said, is known as Helice, and the other's name is Cynosura. Helice is Ursa Major the Great Bear, and Cynosura is Ursa Minor the Small Bear. The Achaeans (Greeks), he added, "judge where to direct the course of their ships" by Helice, while the Phoenicians "put their trust" in Cynosura. Although the Great Bear is "bright and easy to note, appearing large from earliest nightfall," the Small Bear is better. The "whole of it turns in a lesser circuit, and by it the men of Sidon steer the straightest course."

From this passage in the *Phaenomena*, we can tell that both Bears were known in Aratus's time and that Greek navigators preferred to steer their ships by the Big Dipper. The Phoenicians, on the other hand, used the Little Dipper, and for that reason the Greeks sometimes called it Phoenice. According to the Greek poet Callimachus (third century B.C.), Ursa Minor was unknown to the Greeks until Thales of Miletus suggested they give it a try in the sixth century B.C. That seems reasonable, for in the *Odyssey* (Book V, lines 273–277) Homer tells us Odysseus sailed eastward to Ithaca from Kalypso's island by keeping "the Bear," and the north, on his left. Homer only refers to one Bear, and he never calls it the Great Bear.

THE HIGH-FLYING BEAR

It is very clear the Greeks saw a bear in the Big Dipper, but explaining why they did is not so easy. Its stars look like a dipper and move like a wagon. There doesn't seem to be much resemblance to a bear. Regardless, the Greeks not only called it the Bear, they told a story about it. In that myth the nymph Callisto is transformed into a bear and transported to the sky to become the Great Bear.

As a member of the band of hunting companions in the tutelage of the goddess Artemis, Callisto was a dedicated virgin. She had the misfortune, however, to catch the eye of Zeus. The treatment of this story by the Roman poet Ovid (43 B.C.–A.D. 18?) in his *Metamorphoses* details how Zeus spotted her reclined in a private glade, "weary and unprotected and alone." Attracted to her refreshing unadorned style, he deliberated the prospect of ravishing her. He figured his wife Hera would never find out if he satisfied his desire right there on the spot. And even if Hera somehow did tumble onto his deception, it would be worth her reprimands.

Zeus knew Callisto's guard would be up if she saw him, and so he disguised himself as Artemis. When he got close enough to give her the kind of friendly kiss people share in greeting, he seized her and impressed far more passionate messages on her lips. She tried to fight him off but could not expect to prevail against the king of the gods. Later, after he had overpowered her, he disappeared into thin air, and she was left alone in the woods, disheveled and disoriented. Zeus proved that Callisto was not impregnable, and for a while the nymph tried to hide her condition from Artemis. Her pregnancy grew more obvious, however, and in the sorority of virgin huntresses, it was completely unacceptable. When "the moon's horns were filling out to complete their ninth circle" since Callisto's wrestling match with Zeus, Artemis and her party of nymphs decided to break from the hunt for a refreshing bath in a cool brook. All undressed for the swim except Callisto. Her excuses and delays prompted the rest of the nymphs to strip off her dress, and that's when her true circumstances were legibly registered in Artemis's ledger. By that time she was near term, and it is hard to understand how she managed to conceal for so long the little gift Zeus had left for her.

Banished from the company of the goddess of the hunt, Callisto gave birth to a boy she named Arcas. Not long after the birth of her son, she encountered an absolutely furious Hera, who had been biding her time and picked this moment for revenge. Hera then grabbed her by the hair and threw her to the ground, where, upon her back, Callisto began to feel fur bristling from her limbs and her nails curling into claws. Soon, the tempting mouth Zeus had kissed became the jaws of a bear, and Callisto, completely transformed and unable to speak, ran in panic to avoid the hunters and the hounds.

Years later, when the orphaned Arcas was a teenager, he was hunting in the forest. A bear, by chance, was just coming his way, and it was, of course, Callisto. Vague recognition of her long-lost son surfaced in her mind. Taking a step toward the boy, she frightened him. He had never known what had become of his mother, and the last shape he expected to see her in was the bear advancing toward him now. As he aimed his spear to kill what he was convinced was a dangerous bear intent on mauling him, Zeus intervened, held back the good throwing arm of Arcas, and swept them both up in a whirlwind to the sky, where the pair continue to whirl around the pole. That, according to some Greeks—and Romans, too—is how we managed to acquire a pair of Bears in the northern sky.

A no-frills version of the Callisto–Great Bear story is told in the *Katasterismoi*, the earliest collection of Greek star myths that still exists. In this account, Callisto gives birth to Arcas after she has been turned into a bear. Some goatherds trap the pair and put them in the custody of Callisto's father, Lycaon of Arcadia. Later, Callisto the She-Bear goes roaming and, unaware of the prohibition against bears in the sacred precinct of Zeus, tries to have a look around. Arcas and the Arcadians go hunting for bear, and as they pursue her, Zeus lifts her out of harm's way. In deference to their earlier relationship, he puts her in the sky.

Hyginus, the Latin author of *Poetica Astronomica*, blames Callisto's bear body on Artemis (Diana), who, as we already know, was not happy with the status of Callisto's virginity. The chapter Hyginus devotes to the Great Bear reports several alternative versions of the tale, but the same basic theme remains in all of them. A pure, young female is violated by the high god, transformed into a bear despite her innocence, and elevated to the sky when her life is threatened. There she remains, immortally circling the pole, never lost from sight. In what was then an accurate astronomical description, Hyginus added a comment, as Ovid did earlier, about the Bear's ability to pad around the pole and never dip so much as a paw in the ocean below. The reason this happens, he explained, has nothing to do with the Bear's effort. Rather, it is the work of Tethys, the wife of Ocean. She refuses the Bear admission to the sea because she had once been the nurse of Hera, and in Hera's bed, Callisto was a concubine.

THIS BEARS FURTHER SCRUTINY

The Greeks told a perfectly good story to account for the Bear, but there is no obvious reason why it had to be a bear. Max Müller, Sir James George Frazer, and Robert Graves all favored a more anthropological explanation that had little to do with the sky. The Callisto myth, in this view, was just an excuse to provide the Arcadians with a genuinely ursine ancestor. They traced their bloodlines to their king Arcas, from whom they acquired their name, which has some "bear" in its Greek. With the bears as their talisman, they fabricated a bear-parent for their king. This is plausible, but it doesn't explain why the Arcadians picked those stars to honor their ursine ancestor.

Müller conceded that the celestial Bear was probably always called a bear. The Greeks, he thought, just pasted their Callisto myth onto it. Müller also tried to explain the Bear linguistically. The Hindu name for the Big Dipper is *Saptarshi*. The word for "sage" or "wise man" in this title is *rshi*, but in Sanskrit the word was *riksha*. In one gender, *riksha* is supposed to have meant "star." It was a Sanskrit noun rooted in the verb "to shine" and in the adjective "bright." In another gender, though, *riksha* is said to mean "bear." It was argued, then, that the reason we call these stars the Bear is the confusion between the two meanings of the word *riksha* as time passed. Of course, we have no way to know if this clever idea is really true. We can, however, learn more about the meaning of the celestial Bear if we take a closer look at Callisto.

Callisto's name means "the most beautiful," and it is a title she shared with Artemis, who in some ancient stories also could appear as a she-bear. Callisto wore the same clothes as Artemis, did the same things as Artemis, and is at the core a double of Artemis. If we then say that Callisto the celestial Bear incorporates attributes of Artemis, we conclude that the Great Bear in the sky is, like Artemis, a spirit of wild nature. In the guise of the Bear, wild nature is renewed through celestial or seasonal death.

CIRCUMPOLAR BEAR

At higher northern latitudes, the stars of the Big Dipper never set. They circle around the pole without ever cutting below the horizon. They and all the other stars that do the same thing are said to be circumpolar. The Big Dipper was circumpolar in ancient Egypt, too, and the Egyptians had their own idea about what that meant. They watched those circumpolar stars take their turns around the north celestial pole and said they were "imperishable" or "undying" because they never suffered a term in the tomb below the earth's horizon.

What stars are circumpolar depends, of course, on latitude. Standing on the earth's north pole, the north pole of the sky is directly overhead, and all of the stars that are visible there are circumpolar. Stars below the sky's equator never come into view at the earth's north pole, and the paths of the stars that can be seen are horizontal, parallel to the horizon.

Scooting down to the earth's equator lets you see all of the stars at one time or another. The north and south poles of the sky are on the horizon, and nothing is circumpolar. Stars rise vertically and set vertically; the east mirrors the west.

If you are between the earth's north pole and its equator, the north pole of the sky is tilted away from the zenith. Its angle above the horizon is equal to your latitude. At the latitude of Greece, the north celestial pole is between 35 and 40 degrees above the horizon, and most of the Big Dipper, but not all of it, still circles it without disappearing. Two to three thousand years ago, precession put the Dipper even higher, so that it was circumpolar in the Mediterranean region. That is why neither Hyginus nor Ovid would permit the celestial Bear to touch down and why Homer in the *Odyssey* could say that the Bear is "never plunged in the wash of the Ocean."

BICKERING OVER THE DIPPER

Because it never set, the Egyptian version of the Bear—the Bull's Thigh—became a talisman of immortality. The Egyptians also gave the same name they assigned to the Bull's Thigh, *Meskhetiu*, to an instrument used in funerals, and it, too, stood for immortality. It looked something like an adze —a scraping tool with a chisel-like blade mounted on a handle. In profile, its shape resembles an upside-down Big Dipper. So in Egyptian symbolism, the thigh of a bull was interchangeable with the adze, and both could stand for the Big Dipper, all because of the obvious similarity in shape.

In pharaonic funerals, the *sem*-priest was in charge of the ceremony, and in paintings he is shown touching the mouth of the king's mummy with either a bull's leg or an adze. This funeral ritual was known as the Ceremony of the Opening of the Mouth. It was performed upon the mummified body of the deceased in order to restore its *ka*, the conscious personality with the power of speech and the senses. By touching the mouth and eyelids with a symbolic Big Dipper, the dead were reanimated.

Meskhetiu, the Egyptian name for the Big Dipper, identifies it as the Leg of Set, and it was one of the trophies in dispute in the story often called "The Contendings of Horus and Set." There, too, its distinctive celestial properties inspired the Egyptians to exploit it as a symbol of order, renewal, and kingship.

Horus was the son of Osiris, and Set was his adversary. Set's murder of Osiris and his challenge of Horus's legitimacy and right to rule all of Egypt were what ignited the quarrel. They inflicted serious trauma on each other in a series of battles. Set gouged out the left eye of Horus, but Set also suffered indignities. On one occasion his testicles were squeezed off. Horus may have worn the white hat here, but he was obviously capable of striking a low blow. Set's leg was also torn away. It's easy to see why Horus wanted Set's leg. As the symbol of the undying Dipper, it had the power to bring Osiris back to life.

J. Gwyn Griffiths, a Welsh classicist and Egyptologist responsible for *The Conflict of Horus and Seth* (1960), the most authoritative study of this myth, saw hints of a more cosmic scope—a clash between order and chaos—in its earliest form in the 4,500-year-old *Pyramid Texts*. As the son of Isis and Osiris, Horus was their agent and represented what they represented. A conflict between him and Set

The Egyptian constellation of the Bull's Leg symbolized immortality because its stars circled the pole without "dying," or setting below the horizon. A dipper-shaped instrument that carried the same name as this group of stars was used to reanimate the spirit of the deceased in a ceremony known as the Opening of the Mouth. (Detail from the tomb of the New Kingdom noble Inherkau at Deir el-Medina)

pitted cosmic order, perpetuation of life, and the fertile land against primordial chaos, mindless death, and the sterile desert. Chaos challenged order, and order was restored.

Although the death of Osiris is the fundamental reason for the war between Horus and Set, the contenders seem at times to feud over something outside the cult of Osiris. They are two primordial gods whose dispute, when resolved, puts Egypt and its kingship in order. Authorized succession of kings is an expression of ordered change, and in this story, the Big Dipper is an essential element in the process. It is one of the prizes at risk in a classic circuit through the story of the cycle of cosmic order.

PARROTING THE DIPPER

At more southern latitudes, the Bear does plunge below the horizon. That may be why the Aztec god Tezcatlipoca is missing a foot. Tezcatlipoca's name means "Smoking Mirror," and the symbol of a smoking mirror is what replaces his foot in some pre-Conquest pictures of him. His "smoking mirror" was probably an obsidian looking glass used by wizards to scry secret knowledge. Close affiliation with the night fit his reputation as a master of sorcery and divination. All-powerful, he was everywhere at once and knew everything. Tezcatlipoca had many attributes of the highest god. As the Black Tezcatlipoca, his domain was the north, the region of the dead.

An episode in the myth of creation recorded in *Historia de los Mexicanos por sus Pinturas*, an anonymously authored post-Conquest chronicle, tells us that Tezcatlipoca's constellation is the Great Bear and explains how he loses his foot. This story describes the conflicts between Tezcatlipoca and Quetzalcóatl at the end of each cosmic age. At the end of one of these ages, after Tezcatlipoca has had a run at ruling things as the sun, Quetzalcóatl whacks him with a club and swats him into the waters below the sky. There Tezcatlipoca transforms into a jaguar, the powerful nocturnal carnivore. Its spots, like the leopard's in Egypt, probably stand for the starry sky. The *Historia* continues by confirming that Tezcatlipoca's descent is mimicked by his stars, Ursa Major, which fall into the waters below. It is perhaps these dives taken by the Bear and Tezcatlipoca that inspired the notion that the god lost his foot in the jaws of the reptilian earth monster, but no primary source confirms that account.

A Big Dipper that drops below the horizon in Maya territory turns up as a parrot in the *Popol Vuh*. His name is Seven Macaw, and his wife, *Chimalmat* (Shield), is Ursa Minor, or the Little Dipper, perhaps enhanced by some of the stars of Draco the Dragon. Seven Macaw has a name that sounds like one of those 260 combinations of day-names and day-numbers that ran the *Tzolkin*, or ritual calendar, through its options, but "Macaw" is not a day-name in the Maya calendars. That makes it likely

In the Popol Vuh, *the Maya contrived a bird called Seven Macaw out of the seven stars of the Big Dipper. We see this baroque bird among the branches of the tree on the far left of this scene on a Late Classic funerary vase from Guatemala. One of the hero twins takes aim at this celestial parrot with his blowgun. When the dart strikes, the bird falls from the tree as the Big Dipper falls toward the northern horizon in its circuit about the north celestial pole. The tree, then, may be associated with the polar axis. (Drawing: Joseph Bieniasz, after a rollout photograph by Justin Kerr of a vase in the November Collection of Maya ceramics, Griffith Observatory)*

the bird was specified this way for other reasons. The "Macaw" part of the name is easy enough to understand. After all, the bird is a macaw. Because it also is the Big Dipper, the "Seven" very likely refers to the Big Dipper's familiar seven stars. The Dipper even looks like a macaw, if you equate the handle with the bird's long and ostentatious tail feathers.

In the darkness before the world's first dawn, Seven Macaw has actually been helpful to people. His light guides them in the long night, but he suffers from an inflated sense of self-importance and declares himself to be greater than the sun and moon. He gets away with this for a while because the sun and moon and other stars have not yet put in an appearance. Despite his grand claims, however, his realm of influence is restricted to the neighborhood around his own roost. Seven Macaw's sons—Zipacna, the maker of mountains, and Earthquake, the one who moves them—make equally tiresome claims. So the hero twins of the *Popol Vuh,* Hunahpu and Xbalanque, decide to knock that bird out of his tree. In this episode, Hunahpu and his brother play the part of the Morning Star warrior Venus.

Seven Macaw flies up into his tree each day to eat its fruit. After monitoring this habitual behavior, the twins hide below the tree and wait for the bird to find his perch. Hunahpu launches a dart from his blowgun at Seven Macaw. It splits his beak and knocks him over the top of the tree and down the other side.

Now, this is what the Big Dipper does. In a twenty-four-hour period it completely circles the pole. Once a day, it climbs to the top of its arc and then descends back to the horizon. At the latitude of highland Guatemala, 15 degrees north, it drops below the horizon and reemerges later.

Dennis Tedlock, an anthropologist and author of a new translation of the *Popol Vuh,* demonstrates that there is also a seasonal connotation to the story. Seven Macaw must be shot because he is offending Heart of the Sky, a god also known as Hurricane. Hurricane is supposed to bring the rains and flood in July, but Seven Macaw has been holding up their arrival by remaining in the sky. By July, however, the Big Dipper is gone. It rises at dawn, and that means it is up in the daytime and down at night. In either case, it is out of sight. The rains are free to fall.

SERVING SEASONS IN DIPPER-SIZED PORTIONS

Each night we can see the Big Dipper take part of its daily turn around the sky, but as the months and seasons go by, the change in the position of the sun along its annual path brings the Dipper up and down at different times of day. For that reason, the Big Dipper also takes an annual turn around the sky, but to see it, you have to look for the Dipper at the same time of night in different seasons. Seasonal changes in the Big Dipper were observed throughout the world. The Chinese, the Maya, the peoples of Siberia, and the North American Indians are among the many peoples who noticed what the Dipper was doing, made use of it in their calendars, and described it in their myths.

The Big Dipper's position changes from night to night, but the difference is too small to detect with the unaided eye. But after a month—or even less, depending on how carefully you watch—the difference is noticeable. Between seasons, the relocation is dramatic. It has to be. In a year, the Big Dipper, as seen at the same time of night, has to go completely around the north celestial pole. With four seasons in a year, we see it rotate one-fourth of a circle each season.

Begin, for example, in the middle of spring, around the end of the first week in May. In the early evening, the Big Dipper is directly above the north celestial pole, at the highest point of its circle. Its bowl opens toward the ground, and if it could carry water, it would be pouring out the spring showers. The handle extends to the right (east). This was probably the position of the Big Dipper a Kumeyaay Indian woman, Delfina Cuero, meant when she recalled an old tribal story about the Dipper and the olivella shells found on the beaches of southern California. She said they used to say, "When the dipper in the sky gets too full, it is dumped out. Then these small shells fall all around near the ocean." She added that the Indians used to "explain why the Dipper is lying differently in summer and winter."

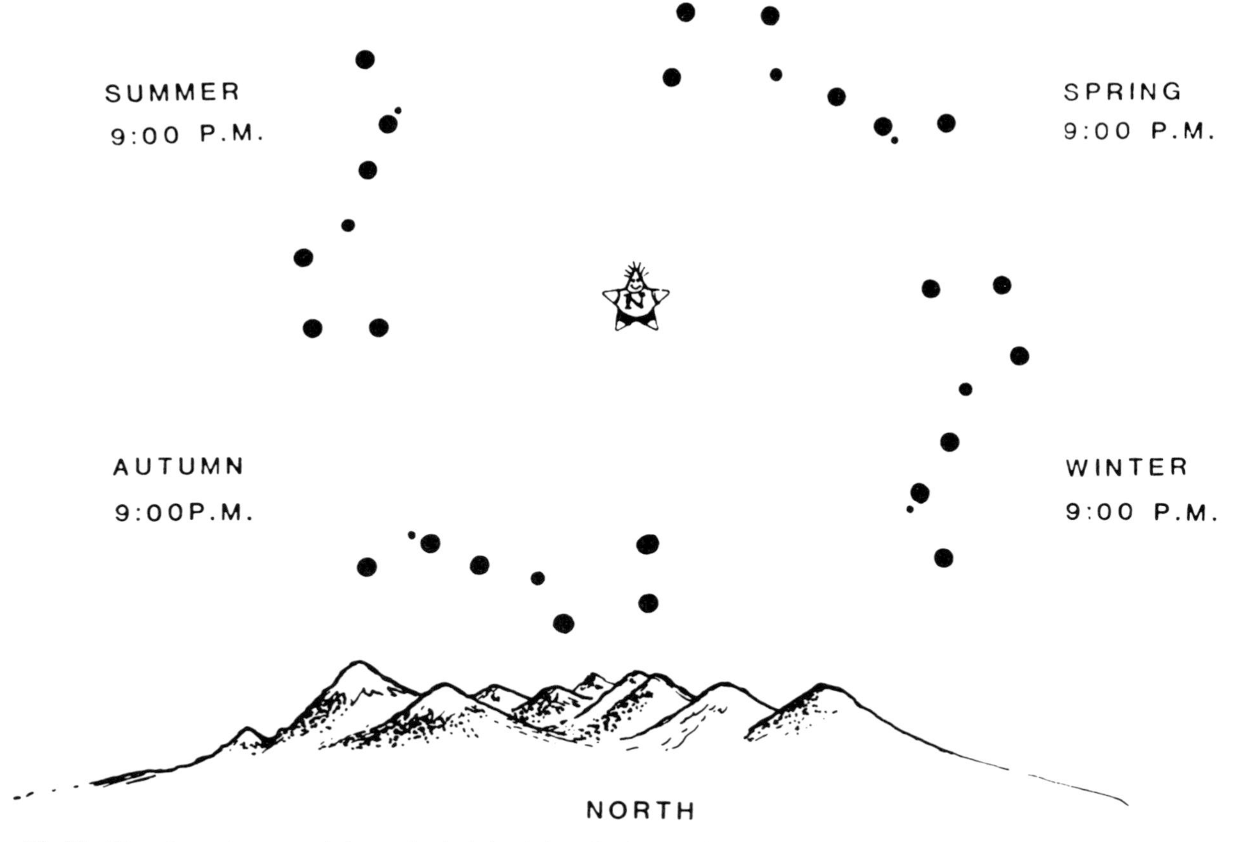

The Big Dipper's motion around the north celestial pole has also been used as a calendar and as an indicator of seasonal change. By observing at the start of each night at roughly the same time throughout the year, you will see the Dipper start its nocturnal circuit in a different position that is related to the season. (Drawing: Robin Rector Krupp)

Three months after the middle of spring, it is, of course, the middle of summer. At about the end of the first week in August, the early evening Dipper is to the left (west) of the north celestial pole. The bowl is lower than the handle and opens to the right (east). The handle extends upward and stands tall like the still-growing grain.

In another three months it is the middle of autumn, and by the end of the first week in November, the Big Dipper is low in the early evening sky. Part of it may even be below the horizon, depending on the latitude. With its bowl's bottom on the bottom, it is at the lowest turn of its arc. Grain from the harvest measured in this cup would stay put. The bowl is on the right (east), and the handle points left (west).

Just past the first week of February, it is the middle of winter, and the Big Dipper we see in the early evening has migrated again. Now it is on the right (east) side of the north celestial pole. The bowl is higher than the handle and opens to the left (west). The handle hangs down like a winter icicle.

Peoples in central and eastern Asia are known to have observed the Big Dipper in conjunction with the seasons just as described above. They said it is spring when the Bear's tail points east (right), summer when it points south (up), autumn when it points west (left), and winter when it points north (down). Because the Dipper looks as if it is attached at a right angle to an invisible armature coupled to the pole, its four seasonal orientations have been suggested as an explanation for the origin of the swastika, a very ancient symbol that conveys the idea of motion, direction, and cyclical change and has been linked with the passage of the year.

Everything the Dipper does involves change, but it tempers that change. It corrals change with time and direction, saddles it with seasons, and breaks it in with cosmic order. The Big Dipper is, therefore, a talisman of ordered change, and that is the theme on which its stories usually turn.

We can see this theme of ordered seasonal change translated into a tale of celestial pursuit told throughout much of northern Siberia, particularly by the Evenks and the other peoples of the Tunguso-Manchurian language group. For them, the stars of the Big Dipper were said to be Kheglen, the cosmic elk. Its calf was Ursa Minor the Small Bear, and both were hunted by Mangi, the mother-ancestor spirit. She took the form of a bear and appeared in the sky as Arcturus and other stars of Boötes the Herdsman or Bear-driver. The Pseudo-Eratosthenes said Boötes is the Bear-keeper and identified him as Callisto's son, Arcas. So in Siberia a bear runs down an elk, and in Greece a bear-keeper herds the bears, all because the star Arcturus chases the Great Bear's tail around the pole and through the seasons.

In practical terms, the elk was extremely important to the people of the Siberian Tungus. It was a walking convenience store that provided material for food, clothing, weapons, and tools. Because it lost and grew new antlers each season, it was an emblem of seasonal change and renewal. The Siberian elk spirit also certainly had something to do with the seasons. Sometimes it was said to take the sun away in its antlers and leave the world in darkness until it was hunted down. The sun was then returned, and so, we must guess, was spring. The Evenks say Mangi tracks Kheglen the elk all year until the end of winter. She catches and kills the elk, then, and as the elk dies, new life for the entire world emerges from its body.

BAITING THE SEASONAL BEAR

A Big Dipper story of seasonal pursuit was told in North America by the Micmac Indians of Canada's Maritime Provinces, with marvelously explicit detail. The Micmac didn't compare notes with the ancient Greeks, but for them and several other North American Indian tribes, the Big Dipper really was a bear, at least its bowl was. The four stars of the bowl indicated the bear's four paws, and looking at them this way, you can imagine the bulky body of the bear as the darkness inside those corners and even see the bear lumbering around the pole. The Micmac tale of the celestial bear was reported by Stansbury Hagar in 1900 in the *Journal of American Folklore* with an understanding of and care for the detail and self-consistency that make it a classic example of the well-tempered star story.

To appreciate the action, we have to meet the characters. Besides the bear, we have the hunters. There are seven of them—all birds native to Canada and known by Micmac as well as scientific names. The pot in which they intend to cook the bear is one of the props, and the bear's winter den also appears on stage. Three of the avian hunters are the three stars in the Dipper's handle. In order, from the bowl, they are the robin, the chickadee, and the moose bird. The moose bird is also known as the Canadian jay. Alcor, the faint star next to the chickadee, is the pot in this drama of birds and bear. If you continue to trace the arc of the Dipper's handle down to Arcturus, you pick up the other four hunters. They are all stars in the constellation Boötes:

the pigeon	gamma Boötis,
the blue jay	epsilon Boötis,
the owl	alpha Boötis, or Arcturus, and
the saw-whet	eta Boötis.

The saw-whet is a small owl with bright plumage on its head. The bear's den, located above the curved procession of starry bear hunters, is the circlet formed by the stars of Corona Borealis the Northern Crown with the addition of two stars from Boötes (delta and mu). The Northern Crown makes a fine, cozy den, and the stars from Boötes give it a short, tunnel-like entrance.

The stage is set, then, for the bear to awake from hibernation in late spring. When she does, she emerges from her den on the hillside and wanders down looking for something to eat. The chickadee in the tree above her is hungry, too, and has eyes sharp enough to see roast bear in his future. He is not, however, able to take the bear on by himself and recruits reinforcements. As a party of seven, they set off to catch the bear. With pot in hand, the chickadee flies between the robin and the moose bird so he doesn't take a wrong turn. Throughout the summer, they chase the bear, and through each summer night the bear swings down from the west, scuttles across the northern horizon, and runs up the east. In the fall, some of the bear hunters fall from file. The two owls are first to go. Heavier and more awkward than the others, they just can't keep up. It's time for them to head south anyway. The blue jay and the pigeon also lose the trail. Perhaps they, too, dream of more tropical climes at winter's onset. So now only three hunters remain, and in midautumn, in the early evening, they catch their bear as she tries to rise out of their reach. Erect on her hind feet, the bear decides to fight back. She is, however, a more vulnerable target. The robin fires an arrow into her breast and the impact knocks her onto her back. Desperate for dinner, the robin hops onto the slain bear, but he gets soaked with her blood. Needing to shake it off, he wings it over to a maple tree. With shimmies that would shame a belly dancer, he spatters nearly all of the bear's blood on all of the leaves of all the trees and makes them red in the fall. The maples are closest and therefore reddest. He's almost spotless, but a patch of red blood remains on his breast. The robin never does get it clean.

The moose bird had lost the trail in the last furlongs, but as the robin and the chickadee are slicing up the bear, building a fire, and putting the meat in the pot, he arrives for a share of the feast. He has taken his own sweet time, knowing that the other two birds were bound to bag their bear and would have to cook it anyway. This is the moose bird's habit. He always enters after the work is done and still insists on some of the haul. So the birds and the Indians also call him He-who-comes-in-at-the-last-moment.

The three birds enjoyed their bear banquet, but there is more to the story. During the course of the winter nights, the bear's skeleton floats up and over the pole on her back and down toward the earth. Although she is dead, her life-spirit searches out another bear, asleep, as if dead, for the winter in its den. There her spirit waits until spring. Then, resurrected from the tomb of winter's hibernation, she enters the world again to be hunted and killed again.

Because real bears hibernate, they are symbols of renewal and seasonal change. They represent the immortal spirit of life. We don't have a bear in the northern sky because the stars there look like a bear. We have a bear there because those stars act like a bear. The celestial bear prowls around the pole, performs a seasonal death, and returns to life when the world does. It is a celestial powerhouse, and the energy churned out by this revolving dynamo in the sky seems to activate the entire world. The celestial bear performs another one of those necessary sacrifices that ensure the perpetuation of life—this time in the northern sky. By dying, the celestial bear lives forever.

15

Beside the Seven Sisters

THE seven "planets" of the ancients and the seven stars of the Big Dipper already seem to have priority for the title of Heaven's Seven, but there is one more competitor—the Seven Sisters. The Seven Sisters are probably the best known seven in the sky, at least by name. Their real name is the Pleiades, and they comprise a group, or cluster, of stars—not the brightest stars in the sky, but bright enough to be seen without straining—in the constellation of Taurus the Bull. Since antiquity, they have been lodged in various parts of the Bull's anatomy by different authorities, but now we find them on the shoulder. They are lodged, as well, in our culture, and allusions to them show up in much of our literature, from the Bible to works by Rabelais, John Milton and John Keats. Alfred Lord Tennyson's appreciative eye is apparent in his treatment of them in *Locksley Hall*.

> *Many a night I saw the Pleiads, rising thro' the mellow shade, Like a swarm of fire-flies tangled in a silver braid.*

In describing them as a swarm, he echoed an impression held by many peoples around the world. The old Welsh name for them means "a close pack." The Hebrew word *kimah*, means "cluster" or "heap." The Andean Aymara speak of them as a "handful" or a "group," and one of the Peruvian Quechua names for them is "pile."

To the unaided eye, the Pleiades look like a small cluster of six stars. The group resembles a tiny pan or dipper, and this sometimes prompts those who are unfamiliar with the night sky to confuse it with the Little Dipper. The Pleiades are on the right side of this photograph. To their left can be seen the V-shaped pattern of the Hyades, a bigger cluster that forms the face of Taurus. The brightest star—at the top of the left leg of the V—is Aldebaran, the Bull's eye. (Photograph: Curtis Leseman)

SEVEN IN HEAVEN

Although *Seven Sisters* is a household name, only six stars can be seen without effort in the cluster by the unaided eye. Despite that, the Pleiades have been widely known as a gang of seven. The Anglo-Saxons and other Germanic peoples of northern Europe called them the Seven Stars. In a Dutch story, they are a baker's wife and six daughters. Romanians see in them a hen and six chicks. A story told by the Euahlayi tribe of Australian Aborigines of New South Wales cast them as seven sisters.

In the *mul-Apin* texts of ancient Mesopotamia, the name for the Pleiades is *MUL*, which literally means "star," but another Mesopotamian name for the cluster means "the Sevenfold One." Seven left eyes of seven Maori chiefs gaze upon New Zealand as the Pleiades in Maori tradition, and the Gê Indians of eastern Brazil regard the Pleiades as a group of seven brothers. Several North American Indian tribes, including the Navaho, the Natchez, the Arapaho, the Luiseño, the Karok, and the Monache, also counted seven Pleiades.

AMONG THE MISSING

Practically everyone had something to say about the Pleiades, and many were convinced that there ought to be seven of them. For example, the only surviving ancient collection of Greek constellation myths, the *Katasterismoi*, calls them "the seven-starred," but it adds that just six are visible. In one

Although only six Pleiades can be seen with ease, in European tradition they are usually known as the Seven Sisters. The Pleiades are shown as seven women in the Leiden Aratea, *a ninth-century Carolingian copy of the treatise on astronomy and weather written by the Greek poet Aratus in the third century B.C. This edition of Aratus's* Phaenomena *is one of the oldest surviving collections of representations of the constellations of the classical world. Not astronomically accurate, this family portrait of the Seven Sisters arranges them symmetrically. (Bibliotheek der Rijksuniversiteit te Leiden)*

This astronomical engraving from the Seleucid period (301–164 B.C.) in Mesopotamia is judged to include Taurus, the moon (in which a man appears to be fighting a lion), and seven Pleiades. The arrangement of the Pleiades looks more like the V-shaped Hyades than the Pleiades, however. (From The Mythology of All Races, *vol. 5,* Semitic, *by Stephen Herbert Langdon, 1931)*

passage its author says that the seventh is "extremely faint." Elsewhere he says the seventh Sister is "invisible."

With the tradition of a lost Sister so clearly in place in Greek star lore, it was natural for the Greeks to try to explain her absence. In the fourth century A.D., Theon the Younger of Alexandria said one of the Pleiades—Celaeno—disappeared when she was struck by lightning. Before that, the Pseudo-Eratosthenes, author of the *Katasterismoi,* identified the lost Sister as Merope, the one who married a mortal. After marrying Sisyphus, Merope realized she was the only one of the seven who had failed to wed a god. Shamed by her fall in status, she was said to have hidden her face. In the *Fabulae,* the Roman poet Hyginus explained that Merope's departure was a result of her transformation into a comet. Driven in shame from her sisters' company, she dropped her hair in grief, and it trailed behind her.

In North America, several different groups of Indians were convinced one of the Pleiades was missing. For example, the Iroquois of New York knew the Pleiades as seven brothers but said one of them had fallen back to earth. The Nez Perce of the northwest believed the Pleiades to be seven sisters. After one of them was ridiculed and embarrassed by the other six, she decided it would be better if she were not seen at all and covered herself with a veil she pulled from the sky. Although the Singing Maidens of southern Ontario's Wyandot Indians numbered seven, the Wyandot added that the seventh sister and the mortal husband who joined the Sisters in their basket in the sky always sit together in the back of the basket, in the shadows, and are not easily seen.

TAKING A DIM VIEW

If only six Pleiades are visible, why did so many people say the cluster had seven stars and why did some say that one is missing? Perhaps, it has been argued, one of the stars has gone dim. And if it can't be seen with the unaided eye, maybe telescopes can reveal its presence.

Modern astronomical studies of the Pleiades confirm that it contains far more than six or even seven stars; between three and five hundred stars populate this stellar cluster. These stars are fairly young as stars go, only about fifty million years old. Our own sun and the earth have been around for about a hundred times as long.

A long exposure of the Pleiades at the telescope also reveals veils of dust around the cluster's brightest stars, and reflection of the blue starlight by these cold clouds is what adds the romantic curtains we see wafting over the stars. It appears we may have some real options here for getting a faded Pleiad. Perhaps some time ago the dust thickened around one of them and reduced its light. Or if these stars are so young, one of them might still be experiencing growing pains. As a star ages, it breaks its adolescent dependence on gravity—the fast-food favorite of extremely young stars —and settles into a mature diet of the hydrogen in

Through a telescope, far more than seven Pleiades can be detected. Most of the stars in this photograph are, of course, background features against which several hundred members of the cluster are projected. The smoky haze enveloping some of the brighter stars is actually starlight reflected by clouds of interstellar dust. (Photograph: Paul Roques)

its core to provide the energy it needs to shine. In that process, the star's brightness will change.

Actually, stars like the brighter members of the Pleiades are more massive than the sun, and they consume their hydrogen sooner. They age more quickly, and that process also changes their brightness.

None of these explanations provides us with a paler Pleiad, however. The interstellar dust clouds in the Pleiades would need more than just a few thousand years to change markedly the brightness of one of these stars. Also, studies of stellar evolution tell us the brighter Pleiades are all well past adolescence, and when they start aging, they will turn into red giants and get brighter before they get fainter. Was there ever anything in the cluster that was brighter than it is now?

American astronomer Allan Sandage studied the cluster and concluded that it once did have two even brighter members that have since evolved into faint white dwarf stars. One of these has been located to confirm his prediction, at least in part.

But here again, time is not on the side of the legend of the lost Pleiad. Any star bright enough to be visible several thousand years ago has not had enough time to evolve into invisibility by now. Even fairly massive stars take more than ten million years to do it.

Those who wish to quench the light of one of the Pleiades astrophysically still have one more chance. Some stars are known to vary in brightness—both regularly and irregularly. If the seventh Pleiad were an erratically variable star, it might have been more conspicuous in the past.

Pleione, now the seventh-brightest star of the Pleiades, is an eruptive variable star. Every now and then it flings away an outer layer of gas into an expanding bubble. It did so in 1970, 1938, and 1888, and perhaps earlier, too. Observations of it in the 1930s by Harvard Observatory astronomer William A. Calder recorded a decline of about one-sixth magnitude over the three seasons of watching. Given its modest record of changing brightness, Pleione could have been at least a bit more conspicuous in the past. For that reason, the journal *Science* treated Calder's results as the "corroboration of a world-wide legend, rooted in ancient mythology, that once the six resplendent star 'sisters' of the Pleiades numbered seven."

The Japanese, who continue to acknowledge six stars in the Pleiades, have incorporated them in the emblem for Suburu automobiles. Their arrangement in the trademark corresponds well with the cluster's brightest stars. (Griffith Observatory)

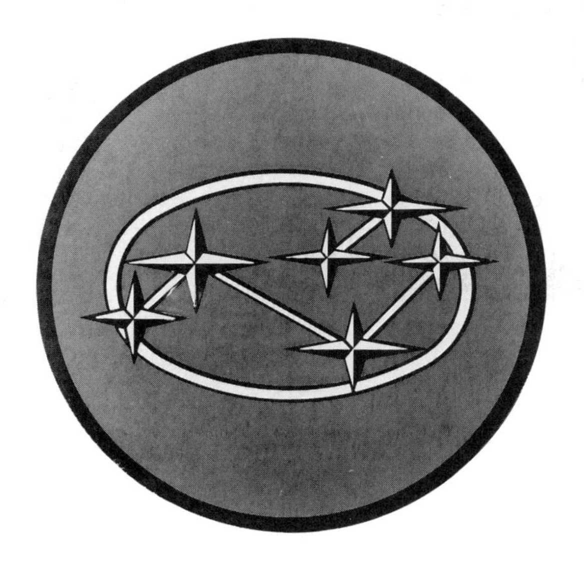

We seem to have a felicitous modern scientific confirmation of ancient wisdom, but the case is not quite so satisfyingly closed on the Pleiades. Here's why. The Greeks themselves did not really believe any of the stars had changed. Aratus, for example, after certifying that the Pleiades are seven "in the songs of men" but only six to the eye, had a few more words to say about the cluster's past: Not a star among them "has perished from the sky unmarked since the earliest memory of man." Aratus's poem was based upon the work of Eudoxus, renowned as an astronomer in the fourth century B.C., and it implies that those informed about astronomy in the third and fourth centuries before Christ believed that there had never really been a significantly brighter seventh member of the cluster.

IN THE EYES OF THE BEHOLDER

There is independent evidence to support the contention of Aratus that no more than six Pleiades have ever been conspicuous. In several traditions, the number of the Pleiades is said to be six without any comment about a missing member. A portrait of the Pleiades on the ceiling of a Tang dynasty tomb in Turpan, in northwestern China's Xinjiang Uygur Autonomous Region, as one of the twenty-eight Chinese *hsiu*, shows it in the familiar Chinese ball-and-stick style with six balls to indicate six stars. Northern Japan's caucasoid/australoid aborigines, the Ainu, see the Pleiades as six idle girls. The Japanese are unrelated to the Ainu, but they, too, see six stars in the cluster, which they call *Suburu*. That's why the emblem of the Suburu automobile sparkles with six, not seven, stars. Because there are only six obvious Pleiades, the Yurok Indians, who lived on the northern California coast and on the lower Klamath River, said the Pleiades are six women who live upriver, that is, at the end of the world, on the horizon. Another California group, the Tachi Yokuts, also numbered the Pleiades as six stars. To them, they were five girls and their husband.

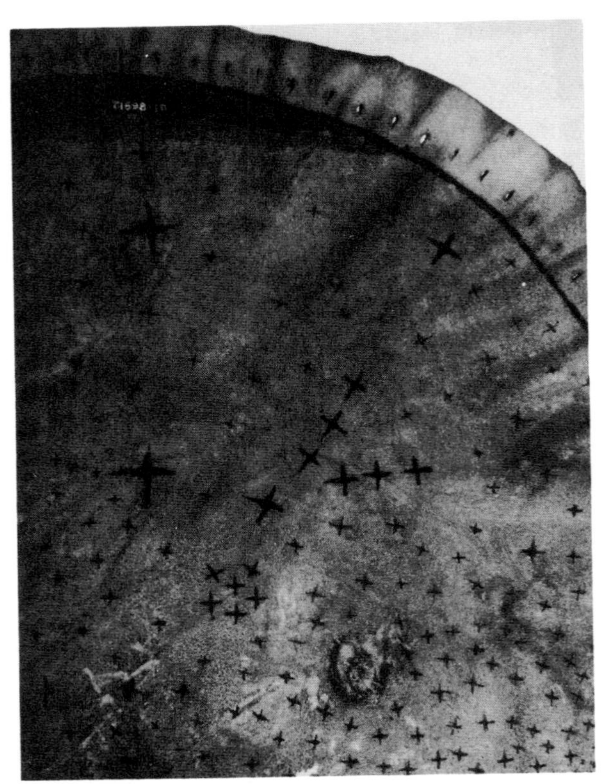

Six stars comprise a tight cluster on a ceremonial star chart from a Skidi Pawnee medicine bundle. They are the Pleiades, and despite their number on the map, they were called the Seven Stars. (Object in Field Museum of Natural History, Chicago)

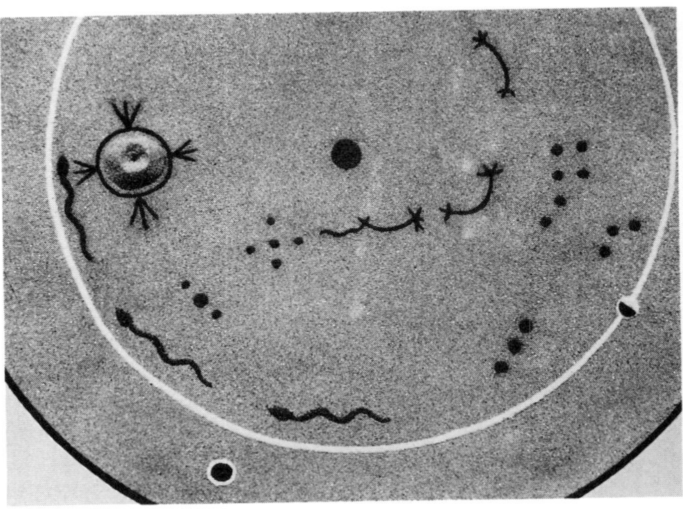

Southern California's Ipai Indians prepared symbolic ground drawings for the boys' initiation ceremony at puberty. Numerous celestial references were included in these drawings. In this replica, constructed by the Native American sand painter Paul Apodaca, we can recognize the six stars of the Pleiades in a dipperlike formation on the right. Indians reported that the bent line of three stars to the right of the Pleiades represented the Belt of Orion, and the row of three stars at the lower right were identified as the Shooting Constellation (possibly Sirius and stars in its vicinity). Another row of three on the lower left, with a brighter star in the middle, is Altair and the two stars in Aquila the Eagle that flank it. A cross-shaped pattern up and to the right of Altair looks a little like Cygnus the Swan, or the Northern Cross. It was called the Cross Star by the Indians. The large dot in the center of the white circle that stands for the horizon is the sun or perhaps some other important star.

Polynesian people of the Hervey Islands and others in the Cook Islands group southwest of Tahiti also see six Pleiades. At one time, they say, the Pleiades were a single star, and it was so flashy and overbearing, Tane-mahuta, the god of celestial light, decided out of jealousy to douse its glow. Everybody on the earth loved its radiance, but it submerged the rest of the stars with its light. Because it was no match even for the moon, people paid little attention to the other stars. Sirius, which in the absence of this perfect primeval Pleiades would have been honored as the brightest of stars, volunteered to help Tane punch out the star's lights. Aldebaran smelled blood and also joined the pack. Together they planned an ambush, but the star was ready for them and ran for cover in the light of the Milky Way, Tane's highway. Tane could see him, however, and fired Aldebaran at him like a shooting star. Struck by this missile, the great star broke into six pieces.

Those who counted seven Pleiades are, then, contradicted by some who counted six, and both are countered by others who came up with still different tallies. The Barasana Indians of the northwest Amazon region of Colombia speak of eight stars in the Pleiades. Quechua Indians in the smaller towns and villages in the Peruvian highlands provide various counts for the stars in the Pleiades: ten, thirteen, and sixteen all have been recorded. In the same area, the Indians also call them *las sieta cabrillas*, "the seven goats," but this name is clearly an import from Spain, where the Pleiades are known as *las sieta cabrillas*.

Attentive observers have always been able to see more than six stars in the Pleiades on clear, moonless nights. Michael Mästlin, a sixteenth-century professor of mathematics at the University of Tübingen, in West Germany, observed fourteen stars in the field of the Pleiades and mapped eleven of them in 1579, thirty years before the invention of the astronomical telescope. Hipparchus of Alexandria, one of history's greatest astronomers, said, around 130 B.C., it was possible to tally seven Pleiades under good conditions.

HOT ON HER TRAIL

Further scrutiny of Greek mythology seems to tell us where that missing seventh Sister has been hiding all along. According to the *Katasterismoi*, Electra was the missing sister. Hyginus also told us, in the *Poetica Astronomica*, that Electra is the displaced Pleiad and added that she migrated to the realm of the Great Bear. One of the ancient commentaries on the *Phaenomena* of Aratus is even more specific: Electra fled to the Great Bear, where she can be seen as the star Alcor, near Mizar, the middle star of the Big Dipper's handle.

In the Florentine Codex *a post-Conquest manuscript from Aztec Mexico, the Pleiades total nine stars. They are surrounded by a border of Aztec star symbols. (Griffith Observatory)*

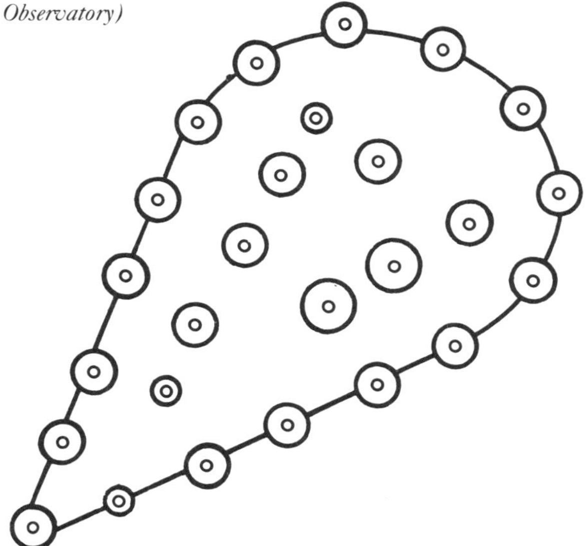

An Australian Aborigine bark painting from Arnhem Land provides thirteen Pleiades. They are women gathered in their camp, housed in a grass shelter. Below them are the Belt and Sword of Orion. The Belt stars are the women's husbands: three fishermen in a canoe. The fishermen broke the sun's law and caught kingfish they were forbidden to eat. For this transgression, they and their canoe were sucked into the sky. The stars in the Sword are said to be their campfire and their contraband fish. (Art Gallery of South Australia, Adelaide)

Other sources outside of Greece and Rome are just as specific about Alcor and confirm what is probably the real number of the unlisted Sister. The Altaic-speaking Mongols in Siberia called the stars of the Big Dipper "the seven old men" and said that they stole a star from the Pleiades, which numbered seven before the theft. Alcor, they added is the purloined star. The Kirghiz of central Asia have a similar tradition. A northern Caucasian story about the Big Dipper and an eighth star in its company being assaulted by the Pleiades also documents the connection. The Buryats, a Mongol group of southern Siberia, also give us the same lost and found Pleiad in the Dipper.

The Turkic-speaking Altaic Tatars tell a tale about the Big Dipper and the Pleiades that explicitly links them through the star Alcor. This story begins long ago when the world was very cold. A giant and powerful insect known as Metshin gorged on animals and people. He was not fussy about his food and made meals out of anybody—men, women, and children. This intolerable situation persisted until a khan, who resided in the Big Dipper, was informed by his horse that Metshin could be crushed to powder under a stamping foot. A cow overheard this and went looking for the gluttonous insect. When she found Metshin, she pulverized him under her foot and broke him into seven pieces. All escaped, however, through the cleft in her hoof. The khan managed to capture one of them, and it remains with him as the star Alcor. The other six escaped to the sky to become the Pleiades, and they continue to chase the Big Dipper today. As the Dipper drops down toward the northwest horizon, the Pleiades emerge in the northeast.

SEVEN BRIDES FOR SEVEN BROTHERS

Hindu myth from the *Mahabharata*, the post-Vedic epic composed about 500 B.C., also details a clear connection between the Big Dipper and the Pleiades. The stars of the Big Dipper were said to be seven *rishis*, or sages, and it was they who brought up the sun and caused it to shine. Their wives were the *Krttika*, or the seven Pleiades, and originally the couples were all happily married in the northern sky—the original seven brides for seven brothers. These virtuous ladies were doomed, however, to celestial estrangement.

Agni, the god of fire, emerged from the flames of a sacrificial offering performed by the Seven Rishis and was immediately infatuated with their wives. Unable to convince them to satisfy his erotic fantasies, he disappeared into the forest in an attempt to cool off. There he was spotted by Svaha, the star zeta Tauri, which marks the tip of one of the horns of Taurus the Bull. Svaha was as inflamed for Agni as he was hot for the Pleiades. Despite her fiery passion, Agni had his heart set on married women and declined her invitation.

Not quick to throw in the towel, Svaha pulled a fast one and disguised herself as six of the wives of the rishis, one after another. Each time Agni enjoyed a tumble with Svaha, he thought he was making time with one more of the rishis' wives. He was naturally delighted with his unexpected success with the attractive partners of the Big Dipper. The last of these ladies, Arundhati, was, however, so devoted to her husband, so praiseworthy, and so abstinent, Svaha's masquerades could not mimic her.

Svaha mated with Agni six times in a single day, the first day of the lunar calendar, and each time she bedded down with him, she collected his semen and tossed it into a mountain cave. There it conceived a child, a powerful boy named Skanda. News of his birth was accompanied by rumors that six of the rishis' wives were his mother. The wise men divorced their wives and sent them away. Taking up residence on the ecliptic, the spurned women became the six Pleiades. Arundhati, the one wife not included in Svaha's costume changes, remained with her husband as Alcor.

WHY SOME SAY SEVEN

Across Siberia, in India, and in the Mediterranean, we can find links between the Big Dipper and the Pleiades, and the Dipper's handle is clearly revealed as the haven of the misplaced Pleiad. Medieval Arabic tradition gives Alcor another name,

Early evening appearances of the Pleiades on the eastern horizon were followed by nights in which they climbed higher into the sky, and this seasonal ascent was described as a celestial dance by a variety of peoples. A nineteenth-century romanticized version of this performance, painted by Elihu Vedder, numbers the girls as seven. (From Star Lore of All Ages *by William Tyler Olcott, 1911)*

"the Lost One," and this may also be a reference to the same idea. The reason for this connection is straightforward. First, the Pleiades look like a dipper to the unaided eye, and when they are pointed out to people unfamiliar with the night sky, they are sometimes mistaken for the Little Dipper. This resemblance between the Big Dipper and the Pleiades may, then, have prompted the idea of an association. Both the Big Dipper and the Pleiades are also handy signals of seasonal change. The cycles of one can be related to those of the other. Because the Big Dipper has seven stars, it might have seemed fitting for the Pleiades to have seven, too. Because only six were visible, one had to be missing. The affiliation between the Big Dipper and the Pleiades made the Dipper a reasonable place to lodge the lost Sister.

Not everyone said there were seven Pleiades, and not everyone said the missing Sister was housed in the Big Dipper. But some people did. Although they did not explain why, similar shape and seasonal significance probably had a lot to do with it.

WHY WE WATCH THEM

It's not surprising that the Pleiades have garnered such international attention. Although not brilliant, they are reasonably bright stars. The brightest is third magnitude, and the next five are fourth magnitude. Scattered around the sky, they would all sooner or later wind up in one constellation or another, but individually they would not acquire the special status they enjoy as a set. Together, they turn our heads.

Location is also one of the things that makes the Pleiades important. They enrich the already opulent field of the winter sky, and as part of Taurus they also happen to reside very close to the ecliptic. By occupying a spot on the annual path of the sun, the Pleiades ally themselves with the yearly cycle of the seasons. Their conspicuous appearance combines with their placement in the sky and tempts people in any climate to see them as a herald of seasonal change.

The Pleiades started the year in ancient India. They were called *Krttika* and were described by Kalidasa, the fifth-century A.D. Indian poet and dramatist, to have the shape of a flame. The procession of lunar mansions, or *nakshatras,* began with them, and they occupied the front rank of those celestial markers because they established the New Year when, long ago, they marked the vernal equinox. By observing the full moon among the Pleiades, the equinox month was determined and given a related name—*Karttika.* The vernal equinox moon was itself known as the Child of the Pleiades.

SEASONED VETERANS

Seasonal connotations are evident in another of the Pleiades' names—the Sieve. They were known as such to the Letts, Lithuanians, Estonians, Finns, Hungarians, Koryaks, and the Chuvash. The term seems to have originated in the idea that cold weather and wintry draughts blow through this drain screen in the sky. Because these are well-known stars of winter, the idea makes symbolic sense.

In *Works and Days,* Hesiod recorded the seasonal meaning of the Pleiades for the Greeks of the eighth century B.C., saying that the time for harvest occurs when the Pleiades rise heliacally. In his era, in the Mediterranean, this occurred in early May. They reappeared then, in the dawn, after their forty-day interval of invisibility while in conjunction with the sun. Ploughing, of course, must precede the harvest, and Hesiod indicates that should have taken place when they went down. By this he means their first appearance on the western horizon at dawn. This would have occurred the previous fall, around the end of October or beginning of November, when ground was broken and seeds were sown to take advantage of the winter rains.

According to Hesiod, the Pleiades also herald the end of the balmy summer sailing season and the arrival of November's winter storms. Because the morning heliacal rising of the Pleiades in May opened the period of navigable sailing and their evening heliacal setting in November closed it, some scholars have derived the name "Pleiades" from the Greek verb *plein,* which means "to sail." This idea has not been accepted universally, however, and other explanations for the name have been suggested. They include the Greek word *pleoz,* for "full" or "crowded," and the word for doves, *peleiades.*

DOVES IN FLIGHT

There is good reason to think that the Greeks did sometimes speak of the Pleiades as a flock of doves, and that as doves they were seasonal heralds of rain and storm. Their place in Greek mythology is our route to this conclusion. To the ancient Greeks, the Pleiades were originally the daughters of Atlas, the giant condemned by Zeus to support the sky's vault upon his shoulders. Their mother was Pleione.

While traveling through Boeotia, northwest of Athens, the Pleiades and their mother had the misfortune to cross the path of Orion the hunter. Infatuated with the women, he started to chase them. It was a long race: Some say five years, some say seven. Zeus terminated this marathon by transforming Pleione and her attractive daughters into doves. Later they flew to the sky to become the stars we now see clustered together west of Orion and still outrunning him.

We know from Hesiod that the Greeks used the Pleiades as a seasonal signal of the beginning of the winter rains, and the myth of Orion's pursuit of the Pleiades equates them with doves. Other ancient Greek traditions, assembled and analyzed in 1932 by American scholar Alexander Krappe, showed a connection between the Pleiades, doves, storms, sky gods, and transcendental intoxicants.

The dove's role as the porter of ambrosia to Zeus—the sky god who sent the thunder, lightning, and rain—was a key element of Krappe's argument, and he assembled a convincing list of other birds that carried divine elixirs to Indo-European storm gods. The *Odyssey* tells us that wild doves must wing their way through the crashing rocks to bring the ambrosia to Zeus and that one is lost, like the missing Pleiad, on every passage. Also, the priestesses at Zeus's shrine at Dodona were called doves, and as "doves," these oracles could carry messages from Zeus. The dove was also the talismanic bird of Perun, the Slavic sky god of storm and thunder. Magical flight by birds to the sky with a precious liquid seemed, then, to demonstrate that wild doves—and therefore the Pleiades—are ancient seasonal symbols of thunder, lightning, and rain.

A CLUSTER IN THE CALENDAR

In North America, the Wyandot Indians, also known as the Hurons, said the Pleiades were seven celestial sisters—the Singing Maidens—who came to earth to sing and dance in the proper season—probably in the fall in the early evening. The sixteenth-century ancestors of the Narragansett, Eastern Algonquian Indians of Rhode Island, apparently informed the Italian explorer Giovanni da Verrazzano, when he sheltered in Narragansett Bay in 1524, that they timed the seeding of their crops in conjunction with "the rising of the Pleiades."

Although the seasonal cycle in the southern hemisphere is reversed from that in the northern, the Pleiades still count in the agricultural calendar. The Arawak of Guyana use their name for the Pleiades as their word for "year." In Paraguay, the year begins, as far as the Guarani Indians are concerned, when the Pleiades rise heliacally in May, and that is when they plant their seed. Gary Urton, an American anthropologist with an interest in indigenous skylore, discovered that the Quechua people in the highlands near Cuzco, Peru, pair the Pleiades off against the tail of Scorpius the Scorpion to keep the agricultural enterprise on track. They call the Pleiades *Collca,* which means "Storehouse," and that name is also shared by the hook of stars that constitutes the Scorpion's stinger. These two sets of stars are warehousing food, and that's why they are each called the "Storehouse."

Inhabitants of the small village of Mismanay, in the vicinity of the Inca ruins of Moray, observe the rising Pleiades in the predawn sky on June 24. In Peru, this is the dead of winter, and spring planting won't start before August. With the festival of San Juan (Saint John) just ended, the people of this area have celebrated in a Christian context the old winter solstice and New Year tradition of their Inca ancestors. The cluster's heliacal rising actually occurs before this time, on about June 3, but the agricultural year doesn't really start until the June solstice. As the new planting season gets under way, the people look at the Pleiades for signs of what to expect in the coming year. The more of them they can see, the better. If the Pleiades appear to be big and bright, the villagers anticipate a good harvest at season's end. On the other hand, they say the yield will be sparse if the stars in the "Storehouse" look small and smudged.

The cluster also semaphores a message about the proper time to plant. Further observations are made of the Pleiades in mid-August, and sowing usually begins in the second half of the month. Maize planting must also be coordinated with the waxing moon, and Urton believes the second *Collca,* the five stars in the curved tail of Scorpius, are also watched to pick the right time to plant.

Seasonal meaning for the Pleiades can also be found in a story told by Australian Aborigines—the Bunjellung tribe—in New South Wales. In this tale, the Pleiades exact seasonal retribution for a violation of sex and marriage taboos between tribal brothers and sisters. According to the Bunjellung tribe, the Pleiades are seven sisters, and all seven of the maidens had a reputation for cleverness and secret magical knowledge. One of them caught the eye of Karambil, a young man of their tribe. Amorous relations between them were forbidden by tribal kinship law and incest taboo, but Karambil grabbed her anyway and ran off with her.

The other six sisters were understandably upset

by this violation of community standards, and they decided to retrieve their kidnapped sister. They needed something to put pressure on Karambil, and so they embarked on a long journey to the west, where they found winter. Then they sent it back as frost, icicles, and cold to Karambil's camp, and they kept shipping cold weather back until Karambil promised to give up the girl. When he did, the six sisters traveled to the east to commandeer summer to disperse the ice.

The six sisters did in their story just what the Pleiades do in the sky. The Pleiades travel toward the west, where, in Australia, they disappear in the early evening in April or so. This is spring in the northern hemisphere, but it is autumn in Australia and headed for winter. When the Pleiades return to the east in the early evening sky in October, the Australia summer is underway.

HUNGRY IN HEAVEN

When people talk about the seasons and the Pleiades, they often talk about food, hunger, and magical departure from the earth. Parts of this pattern are very clear in the following story told by the Onondaga Indians of upstate New York. After reaching one of their favored hunting territories, a party of Onondaga decided to build their lodges by the lake and stock up on fish and game for the winter. By autumn, most of the work had been accomplished, and a group of the children decided to dance each day by the side of the lake. After their play turned into a habit, a strange old white-haired man, adorned with white feathers, appeared among them and warned them to stop. Not believing any harm could come from their dancing, they refused to take the old man seriously. After he had gone, they decided to enhance their enjoyment by having a picnic at their next dancing session. When they went home and asked their parents for food for the outing, however, they were refused.

Irritated but unwilling to give up the pleasure of dancing, they returned to the lake and continued their cotillion on empty stomachs. One day, as they danced hungry, they grew giddy and light-headed. Their bodies must have been light, too, for they started to rise into the air. Realizing that something uncanny was occurring, one of them warned the others not to look back to the ground lest danger befall them. A woman who saw them depart called them back, but they continued their ascent. She ran back to the winter camp and told the other adults what was happening. All of the parents then came out of the lodges loaded with food and called to their runaways. Even though they cried, the children would not come back. One did look back at the parents on the ground, and as he did so he was transformed into a meteor, or "falling star." The rest rose still higher until they found a place in the sky as the seven Pleiades. One of them sang all the way to the sky, and his singing made him fainter and fainter. By the time the seven reached their new home, the seventh child could scarcely be seen.

What happened to the children happens to the Pleiades. When they are first seen rising in the evening sky, just a little after dark, winter is around the corner. The Indians have to start relying on the food they've gathered from the autumn harvest and stored for the months to come. By this time, in late October or early November, the cluster appears in the east after twilight, reaches its high point at about midnnight, and sets at dawn. Each night the stars rise a little earlier and are a bit higher in the sky by the time we first see them after dusk. The light-headed stars are dancing higher and higher into the sky. Each night, through the season of scarcity and hunger, the Pleiades move farther from the surface of the earth. By mid-February, they are already crossing their highest arc—or culminating—in the early evening. On those winter nights, the Seven Onondaga Youths set at about midnight and are invisible until the next evening. We lose them entirely some time in May. They set in the early evening, just after the sun goes down, and remain unseen for the rest of the night. This game of hide-and-seek continues until June, when the cluster reappears for the first time in the predawn sky. By mid-August, the Pleiades are rising at midnight and culminating at dawn, and this is the time they have no place in the early evening sky.

Because the cluster's prime time is winter, it is easy to see why it is said to be a group of hungry children. When the Pleiades appear in the early evening and start to climb higher each night, they dance hungry into heaven, and it's no picnic. Winter makes the earth selfish. Their requests for food are denied, and starvation transforms them into stars. They depart the earth because food is withheld. For all practical purposes the seven brothers are dead. They may boogie in the night, but they are spirits now, not living people.

There is more to this story, however, than the obvious truth that winter means a dangerous encounter with starvation and death. To the Indians, the cluster not only signaled, but also ordained, the period when frost threatens neither newly planted seed nor nearly ripe crops.

To the Indians, the Pleiades seemed to run the farm. They informed the Onondaga when it was safe to grow food. Only a fool would fail to consult them when contemplating when to plant and when to harvest. Planting too early or harvesting too late could disastrously reduce the winter's food reserve. To make it through the winter, you have to plant enough seed. It is risky, however, to plant it all at once. A late frost could kill everything you've put into the ground. Holding some seed back is a sensible strategy, but if you wait too long, the first frost of fall will harvest the grain for you. Your larder will come up short. The cluster's connection with hungry children is not just winter's scarcity, then, but its power over the food supply.

SEEING THE JUNGLE WHEN IT'S WET WITH RAIN

Hunters and farmers in the temperate latitudes see a story of seasonal hunger in the Pleiades, and so do hunters and gatherers in the tropical jungles. For them, however, the seasons deliver rainy weather and dry, rather than frigid weather and warm. Rain in the jungle tropics makes game scarce. Food is in short supply, and disease returns to take human life. When the sky unloads its tears, children cry too. Illness and hunger threaten their survival, and the Pleiades advertise the danger.

The Barasana Indians of Colombia are one of those tropical tribes that monitors the seasons with the Pleiades and sees in those stars a story of seasonal hunger. Their territory is on the equator, along the Río Pirá-Paraná. They fish the river and hunt the forest in this northwest Amazon region and also engage in some slash-and-burn farming. They count eight stars in the Pleiades and call the cluster Star Thing. It is their most important constellation. Stephen Hugh-Jones, a British anthropologist who has done fieldwork among the Barasana, has described their use of the Pleiades to regulate their calendar. They think of these stars as a woman and equate them with the primordial Woman Shaman. She is the entire sky as well as the one who created the world. Her name is Romi Kumu, and as the Pleiades, she is the Star Woman who calls all the seasonal shots.

How, then, does Romi Kumu keep the seasons rolling? Each of her eight stars is actually a strip of wood, and with this kindling she reddens the dawn with fire. The flames and heat bring the dry season, but after the fires die down, the black ash lingers in the sky as the dark clouds of rain that take charge of the heavens. Imagining the eight starry firebrands with alternating stripes of red, for fire, and black, for ash, the Barasana see the Pleiades as the source of eternally oscillating seasonal change.

According to the Barasana, the Pleiades themselves produce the wet and dry weather. They kick off the year in November, when they first become visible in the eastern sky in the early evening. People on the earth are clearing brush at this time. This is the start of the dry season and the end of the rains. Even though the Star Woman is not in the sky at dawn, the fires she lights in the east at night scorch the dawn and give it a scarlet glow.

By January and February, the Pleiades are overhead right after dark. Summer is in full force. Now, after Romi Kumu has set fire to the sky, people burn the brush piles on the ground and prepare it for the summer manioc planting. The first intermittent rains arrive in March and are actually helpful. Even though they extinguish Romi Kumu's torches, they fertilize the crop abiding in the earth. A brief burst of edible plants, insects, fish, and frogs turns the world into a good provider.

By April, however, the Pleiades have slipped to the west and put in their last evening appearances there before disappearing with the sun. When they slip away, the real rains begin. As the rains get heavier, both the sun and the Pleiades make themselves scarce. One after another, the sources of food shut down, and the supplies diminish. This wet season of deprivation and hunger continues—except for a brief and capricious break in August—until November, when the Pleiades are seen once again in the early evening sky in the east. They then return with fire and dry air.

During the rainy season, life in the jungle can be pretty awful. Most social life is suspended. Disease seems to ride the raindrops. It was possible to hunt when the Pleiades were somewhere in the evening sky, but once the rains start, the game departs. Little else is available, but fortunately, the fruits of various wild palms mature at about this time. If their wild fruit develops enough nutritional value and if enough of the fruit ripens, it will get the Barasana through the season of want.

It is a poisonous time, and the stars now appearing in the Barasana evening sky sound pretty grim: the Poisonous Spider, the Scorpion, the Caterpillar Jaguar, and the Poisonous Snake. What we call Scorpius is the Caterpillar Jaguar. It carries some positive connotations despite keeping company with the depressing rains. During the rainy season, from June to August, caterpillars house themselves in pupae and prepare themselves for the transformation from larvae into butterflies and moths. Imprisoned in these self-imposed husks, they drop out of the trees. The Barasana collect them from the ground and add them to their diet. There is a brief break in the constant downpour, and a short dry season provides some temporary relief. As the caterpillar pupae fall like manna from heaven, the Caterpillar Jaguar constellation climbs higher and higher into heaven each evening. Caterpillar Jaguar seems to supply additional food and a change in the weather, but he is still associated with uncertainty and danger. The season of disease and hunger is far from calling it quits. Caterpillar Jaguar sinks below the western horizon in the early evening, in October, when the short and unpredictable Caterpillar Summer concludes. Heavy rains are possible at any time and return with force for one last opportunity to make life for the Barasana miserable.

When the rainy season comes, the Barasana say that Romi Kumu is menstruating, and the rain is her blood. Like a woman, the Pleiades are not accessible during the period the sky releases Romi Kumu's "blood." They are invisible in conjunction with the sun. Equating the Pleiades with Romi Kumu's vagina, the Barasana figure the flow of her blood is a threat to life. But once the process is complete, dry weather and new life return. The rain must fall and the blood must flow if the cycle of fertility is to start once more. For this reason, Romi Kumu's cyclical renewal makes the Pleiades a source of immortality.

The Barasana incorporate the theme of immortality into another aspect of their Pleiades lore. They think of the Pleiades as a gourd Romi Kumu once offered to the men on the earth. Because they would not eat the wax she kept inside of it, they lost their chance to live forever. But the animals that revitalize themselves by shedding their skin—snakes and caterpillars, for example—obviously did sample the delicacies of Romi Kumu's gourd and escaped death. Now each time they discard an old skin, they emerge renewed. Caterpillar Jaguar is the immortal, skin-shedding celestial counterpart to the Star Woman whose flowing blood guarantees a chance for new life.

Like the Quechua in the Peruvian highlands, the Barasana of the equatorial jungle see a kind of complementary opposition in the stars of the Pleiades and Scorpius. For the Barasana, too, these stars are on opposite sides of the sky. One is male, and the other is female. One of them is observed during one part of the year, while the other is seen in the second half. Both herald a dry season and disappear with the rain. Both deliver the groceries.

In *The Raw and the Cooked*, the first of a series of four books written to introduce his science of structural mythology, Lévi-Strauss assembled a collection of stories about the Pleiades. Many of them, drawn from several different marginal tribes of South America, have something to do with rain and hunger. In all of them, the message is the same: The absence of the Pleiades ordains the season when food is scarce. It is not, however, the message that interests Lévi-Strauss, but the way in which it is carried. In a structuralist approach to the myth, each storyline is indistinguishable from all others, and the true myth is the entire assembly of all of its versions. So the rainy season is more than just rain. It is the same thing as female menstruation, and because it is, everything connected with menstruation is also connected with the rain. Likewise, whatever is linked with the rain has a connection with menstruation. Both the rain and the menses are transitions from a fertile state, through a barren state, to another fertile state. Any other sterile condition that provides the same passage from one productive realm to another would belong to the myth. Human physiology, seasonal change, and the availability of food are three independent themes, but each one is part of the myth structured on barrenness and fruitfulness.

This makes a lot of sense, for a careful, worldwide review of stories about the Pleiades reveals not just one theme—say seasonal hunger and celestial withdrawal—but several. There are Pleiades stories about theft and pursuit, about the transformation through initiation from childhood to adulthood, about loss of innocence and participation in sex, about shamanic transformation, about the passage from life to death, about the curing of illness and recovery of health, and about the threat to cosmic order and its periodic renewal.

16

Along the Milky Way

MOST of us have very little chance to view the Milky Way in the night sky. It disappears in the glare of the artificial lights that now turn our nights into a honky-tonk parody of daytime. Even out in the country, far from the flood of city lights, moonlight often conceals the Milky Way in the celestial landscape. But when the sky is truly dark, the Milky Way stands revealed and admired: a pale, luminous ribbon dropped across the sky. Its subdued majesty makes it seem silent in comparison with the brighter stars. They sparkle and shout; the Milky Way glows.

The Milky Way seen on summer nights is captured with an all-sky fisheye lens. North is at the top of the picture. One end (upper left) of the Milky Way emerges from the northeast. The highest point of its arc crosses just a little below (and east) of the zenith (center). From there, the Milky Way is split by the Cygnus Rift, and a smaller stream of light branches to the right from the main current. The richer and wider part of the Milky Way that is in Sagittarius plunges below the horizon in the southwest (lower right). (Photograph: Mark J. Coco)

As a stream of divine milk, as a celestial river, or as a highway in the sky, the Milky Way linked the gods, the spirits, the dead, the shamans, and the souls of the living with the spiritual dimension of the sky. It might take special circumstances or knowledge to make use of this luminous umbilical with heaven, but it rolled overhead every night. Its nightly configuration also changed with the seasons. By reading its seasonal messages right, people turned the Milky Way into an activator of nature's cyclical renewal. When our ancestors saw it stretch from one horizon to another and cross through the highest point of the sky, they saw a seasonal and spiritual covenant, the night's pale promise. Like a spill of water turned to frost as it trickled across the window of heaven, the Milky Way looked as remote as the gods, as cold as the night. No one was fooled, however, by its frail light and casual appearance. They knew the Milky Way hid grandeur in its pockets and folds.

SPUTTERED MILK SKIES

If you could see the entire sky—what is below the earth as well as what is above it—you could tell that the Milky Way is a complete band of misty light. Tilted with respect to the celestial equator and to the ecliptic, it also encircles the sky. In the *Katasterismoi*, it is "one of the heavenly circles," and the Greeks called it *Galaxias kuklos*, or "Milky Circle." The Roman poet Hyginus used its Latin name *Via lactea*, or "Milky Way."

Not especially tidy, the Milky Way seems to slosh its way through the stars. The width of the band varies, as does its brightness. Fingers of its creamy white light bead and flow like milk. The Greeks and the Romans explained the Milky Way's irregular contours as a natural consequence of spilled milk, and according to the *Katasterismoi*, it was the milk of Hera, Queen of Heaven and wife of Zeus, whose milk splashed across the sky.

Sons fathered by Zeus on mortal women could not enjoy divine immortality unless they were nursed at the breast of Hera. Not overflowing with the milk of human kindness for her husband's illegitimate children, Hera shut down the dairy on them, including Heracles. His mother was Alcmene, another one of those exceptionally lovely

*The goddess Hera is taken by surprise, and milk from her misappropriated breasts splashes like sparks into the sky to form the Milky Way. (*The Origin of the Milky Way, *Tintoretto, National Gallery, London)*

earth girls for whom Zeus was forever falling. He deceived his way into her chambers by disguising himself as her husband, and he enjoyed himself so much, he prolonged the night for three days. Before long, Hera heard about this bedtime story and conceived a healthy hatred for Heracles, the child Zeus had conceived upon Alcmene. By the time Heracles was born, Hera had already worked up her fury enough to delay his birth a month, which pushed him out of line for the throne of Argos. That was a sign to the gods of Olympus that it was going to be hard to make an appointment for Heracles at Hera's nipples. While Hera slept, however, Hermes quietly placed baby Heracles upon her bosom. She woke with a start and shoved him away from her breast, but it was too late. Heracles had sampled Hera's hospitality. She had milk to spare, and as she pushed the child away, most of it spurted over heaven.

The Yakuts, who live primarily in northeastern Siberia, adopted a belief in a celestial goddess of birth and the nourishment of new life. Described as having "breasts as large as leather sacks," the Yakuts called her Kubai-Khotun. They prayed to this divine milkmaid under a variety of names, including Birthgiver, Birth-giving Mistress, and Birthgiving Nourishing Mother, and they said she would also dispense a milk-white elixir from heaven to the dying. The Yakut mother goddess in the sky also delivered souls to newborns by transporting them down from heaven at the time of birth.

WHAT THE MILKY WAY IS REALLY LIKE

The Pokomo people of East Africa say the Milky Way is smoke from the campfires of the "ancient people." According to the Bushmen of South Africa's Kalahari Desert, it is ashes thrown into the sky by a girl in ancient times. Many Siberian peoples liken the sky to a tent and refer to the Milky Way as the Seam of the Sky. The same idea is expressed in the Samoyed name for it—the Back of the Sky. Plato thought of the Milky Way as the seam that hemmed the two halves of heaven together, and Theophrastus, who also lived during the fourth century B.C. held a similar opinion.

These terms for the Milky Way are based on its appearance—its pallid, dusty light, the way it girdles the sky, or its riverlike path. Faintly emitting starlight from its rippled surface, it flows through the night. On the Adelaide Plains of South Australia, Aborigines believed the Milky Way to be a river, and the dark clouds of interstellar dust that seem to punch black holes into it were, they said, riverside lagoons. The Euahlayi tribes of New South Wales identified the Milky Way as floodwater spilled over the celestial grasslands from the modest stream that runs through them. The stars are campfires of the dead who live there, and the Milky Way's own tentative light is the smoke from those fires. Ancient Hindu tradition treated it as a heavenly version of the sacred Ganges.

In many parts of the world the Milky Way is said to be a path or road, and it does look like a luminous trail through the stars. The Milky Way was the trail left by an antelope in a race through the sky with a deer, according to the Yuman tribes of southern Arizona. The name California's Eastern Pomo Indians give to the Milky Way means "Bear Foot," and it refers to the path on which a bear in Pomo myth walks through the sky. The Dutch call the Milky Way *Vroneldenstraet*, or "Madame Hilda's Street." Hilda seems to be a Dutch version of Mother Earth, and in Germany she was known by several aliases, including the White Lady. The Milky Way was her own personal autobahn. For the ancient Greeks and Romans, the Milky Way was also heaven's Main Street.

The Greek name for the Milky Way gave us the word *galaxy*, which now means an immense system of stars, gas, and dust. That's what the Milky Way really is, but in antiquity it was hard to tell. The ancients didn't know that we are a part of that system and that the Milky Way is our view of it from the inside.

Even though the ancient Greeks were not quite certain just what the Milky Way was, many of them were willing to offer an opinion. At the end of the first century A.D., Aëtios wrote about the ideas of ancient philosphers; he said Parmenides considered the Milky Way to be a ring of luminous vapor exhaled by the wreath of celestial fire that encircles the earth. Anaxagoras and Democritus, in about 480 B.C., thought the light from the Milky

Way is really starlight. Somehow this light, emitted by stars out of the sun's reach, is reflected toward us in the circular band we see.

Aristotle, in the *Meteorologica*, a treatise dedicated to atmospheric phenomena, stated that the Milky Way was positioned below the moon and was therefore not celestial. Vapor released by the earth, he suggested, was ignited when it reached the sky.

Nobody seems to have gotten close to the truth until the first century A.D., when the Latin poet Manilius conjectured that the Milky Way's fluid light might be the merged beams of many "little" stars. That notion remained unverified until Galileo looked at the Milky Way with a telescope early in the seventeenth century. In 1610, he published what he saw in a small booklet entitled *Sidereus nuncius* (Messenger of the Stars):

> The galaxy is, in fact, nothing but a congeries of innumerable stars grouped together in clusters. Upon whatever part of it the telescope is directed, a vast crowd of stars is immediately presented to view.

What to our eyes is a majestic rolling wave of pallid light is really the combined light of tens of billions of stars, all too far for our eyes to resolve. But the telescope can separate them. Over the centuries we have counted them, mapped them, and analyzed their light to put together a picture of how they are packaged and where we fit into the picture.

The beltlike appearance of the Milky Way itself tells us something about the shape of the Galaxy: It is round and flat, a little like a compact disk or maybe a pair of frisbees stuck together (since it bulges in the middle). Because we can see the Milky Way completely around the sky, we know that we are inside the disk somewhere, viewing it edge-on. We see many more stars and clouds of stars when we look toward Sagittarius than we observe in the opposite direction, toward Orion. That means we are not in the middle of the disk but somewhere closer to the rim. The swollen ranks of stars we see in Sagittarius tell us that way lies the center of the Milky Way Galaxy. The stars thin out toward Orion because that way leads to the Galaxy's edge.

We now know that the Milky Way Galaxy is about 70,000 light years in diameter. Each light year is approximately six trillion miles, and that means a trip from one side of the Milky Way to the other spans about 420 quadrillion (420,000,000,000,000,000) miles. Our solar system is roughly two-thirds of the way out from the center. That puts us in the suburbs. Most of the Galaxy is concentrated in its central bulge. We are traveling around the center of the Galaxy at about 560,000 miles per hour, and so it takes us 182 million years or so to circumnavigate it. The earth has made that trip about 25 times since it was formed.

The starry disk of the Galaxy is about 3,000 light years thick. That means the diameter of the disk is about 23 times its thickness, roughly the proportion you would get if you shaved a quarter down to half of its actual thickness.

If we could fly far above the Milky Way Galaxy and look down at the disk, instead of through it as we do, we would see spiral arms winding toward the center. Bright and hot young stars and luminous hot gas make these arms visible. These arms, curling by us on inside and outside tracks, are partly what make the Milky Way look like a ragged, milky stream.

ALWAYS ON THE MOVE

Viewed from the earth, the Milky Way appears to roll and tumble in slow motion around the sky. Like some vast and quiet serpent, it lifts its great body up over the eastern horizon, rolls overhead, and eases itself down in the west. With its head apparently hidden below the southern horizon, this milky snake pushes forward through the night and uncoils more of its tail out of the north. As the night continues to turn, the sky serpent heads north and swings its tail through the south. The tip of its tail, like its head, remains out of sight. But when the trailing half of the Milky Way drops against the western horizon, the rest of it emerges again, billowing out of the east.

How much of the Milky Way we can see and when we can see it depend on latitude and season. In the northern temperate latitudes, for example,

If we could travel far outside the Milky Way Galaxy and view it from afar, it would probably look something like the Andromeda Galaxy, seen here. A little over two million light years from earth, the Andromeda Galaxy is thought to contain about half again as many stars as are in the Milky Way Galaxy. Although the Andromeda Galaxy is distant, it can be seen with the unaided eye. This telescopic photograph confirms that we view the Andromeda Galaxy at an oblique angle, which makes it look oval. The picture also reveals two small but conspicuous satellite galaxies in the neighborhood of this massive spiral. The numerous small dots of light are foreground stars in our own galaxy. (Photograph: Palomar Mountain Observatory, California)

the part of the Milky Way that starts in Lupus the Wolf, passes through Scorpius and Sagittarius, continues through Cygnus, and ends at Cepheus the King (the "front half" of the snake in previous paragraph), all rises just after dark in late June, about the time of the summer solstice. By dawn, it is stretched across the sky with one end in the northeast and the other in the west. This is the rich, summer Milky Way, and although the nights are shorter, this span of the sky's milky bridge is brighter than the part we see in winter.

Six months later, near the time of the winter solstice, we start the night the way we ended it in summer. Now more of the "tail of the snake" climbs out of the northeast and reveals the part of the Milky Way that begins with Cassiopeia, passes through Gemini and Orion, and ends in what used to be the old southern constellation of Argo the Ship. By sunrise, the winter half of the Milky Way is lounging on the western horizon.

Because the intersections of the Milky Way and the horizon appear to glide around the compass as the earth turns, the Milky Way seems to tumble over our heads. This happens because the Milky Way rings the sky at an angle to the celestial equator. In twenty-four hours, two different segments of the Milky Way cross the zenith. The sky does not remain dark during the day, however, and so we don't see the second crossing on the same day. It shows up at night in a different season. These effects make the Milky Way appear to sweep a different kind of path through the sky than is swept by the stars.

A portion of the Milky Way those in northern temperate latitudes never see passes through Carina the Ship's Keel, Crux the Southern Cross, Musca the Fly, Centaurus the Centaur, and Circinus the Drawing Compass. The farther south you travel, the more of it you see, and at the earth's equator, all portions of the Milky Way are visible at one time or another.

UP A MILKY RIVER

Seasonal significance was also seen in the dusky light of the Milky Way by the Quechua of Peru's highlands. Their name for the Milky Way, *Mayu*, just means "River," and they pictured it as a great celestial river, the seasonal source of the Vilcanota River on earth.

The Vilcanota runs from the southeast to the northwest, and to the Quechua, it is the terrestrial branch of the river in the sky. The Milky Way itself, they say, runs in two streams from an unseen, subterranean source in the north. The two

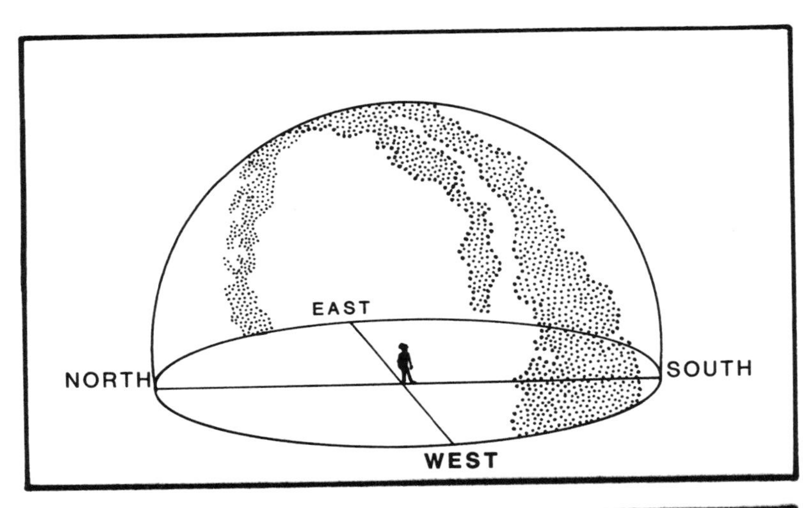

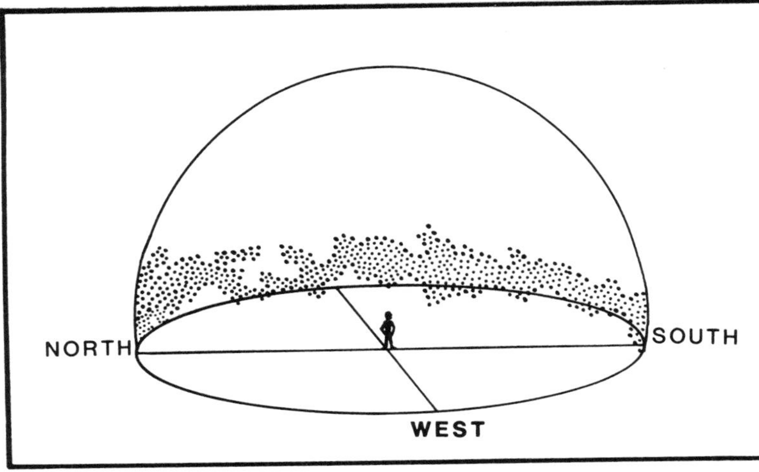

If you think of the sky as the inner surface of a domed ceiling like that in a planetarium theater and forget that it is really three-dimensional space, you can imagine looking at it from somewhere outside the dome, as represented in these two views of the Milky Way. In the upper picture is the summer Milky Way, mid-August at about midnight. One end of the ribbon rises out of the northeast, and the other hits the horizon in the southwest. Earlier on the same night, it crossed the sky at a different angle, and as the night progresses, it will continue to shift. If we wait twelve hours, the Milky Way will have the configuration shown in the lower picture. There, the part of it that was in the southwest is now in the north, and some of that section has disappeared. The part that was in the northeast is now in the arc that is bent toward the west, and a region we could not see before is now visible in the south. Twelve hours after midnight is noon, however, and in the daytime the Milky Way cannot be seen. Although it reaches this position every twenty-four hours at the north temperate latitude assumed here, to see it like this, you must observe it at night in a different season, at midnight in mid-February, for example. (Drawing: Robin Rector Krupp)

streams converge in the southern sky near the Southern Cross.

The Quechua also see two arcs of the Milky Way intersect at their zenith at one time or another. Although it is impossible to see both parts of it intersect each other there, they can imagine such an intersection as first one half of their Milky Way crosses overhead and then, twelve hours later, the other half does the same thing. The Quechua call this meeting of the arcs Cruz Calvario, and the point on the ground directly below it they call Crucero, which is also considered to be the center of Mismanay. In this village it is marked by the intersection of the two main footpaths through the village. There is a small chapel at the crossroads, and three crosses are stored in it.

A pair of diagonal lines is responsible, then, for the fundamental system of directions recognized by the people of Mismanay, and they link the seasonal limits of the sun, seasonal signals of chosen stars, the intercardinal orientation of the landscape, the organization of the village, and the circulation of water—all with the cross the Milky Way draws at the zenith.

In the early evening, at the June solstice, they see the Milky Way rise out of the northeast, the same place the sun rises in the morning. This is the middle of the dry season, and the sky provides no water. Six months later, in the middle of the rainy season, it is the December solstice. Now the sun rises from the southeast, and in the early evening the Milky Way climbs out of that same zone. Since the Milky Way advertises the seasons, it is also credited with supplying—and denying—the rain. This is completely congruent with the Quechua idea that the Milky Way is a river.

The system of orientation of the Quechua people of highland Peru is fundamentally intercardinal and based upon the points of sunrise and sunset at the solstices. These axes define a cross on the ground that is taken to be the center of the system, and that point is marked by the Crucero chapel. The zenith cross is formed by two configurations of the Milky Way that pass overhead. In one instance, the Milky Way starts in the southeast and ends in the northwest. In the other, it starts in the northeast and ends in the southwest. (Drawing: Joseph Bieniasz, Griffith Observatory)

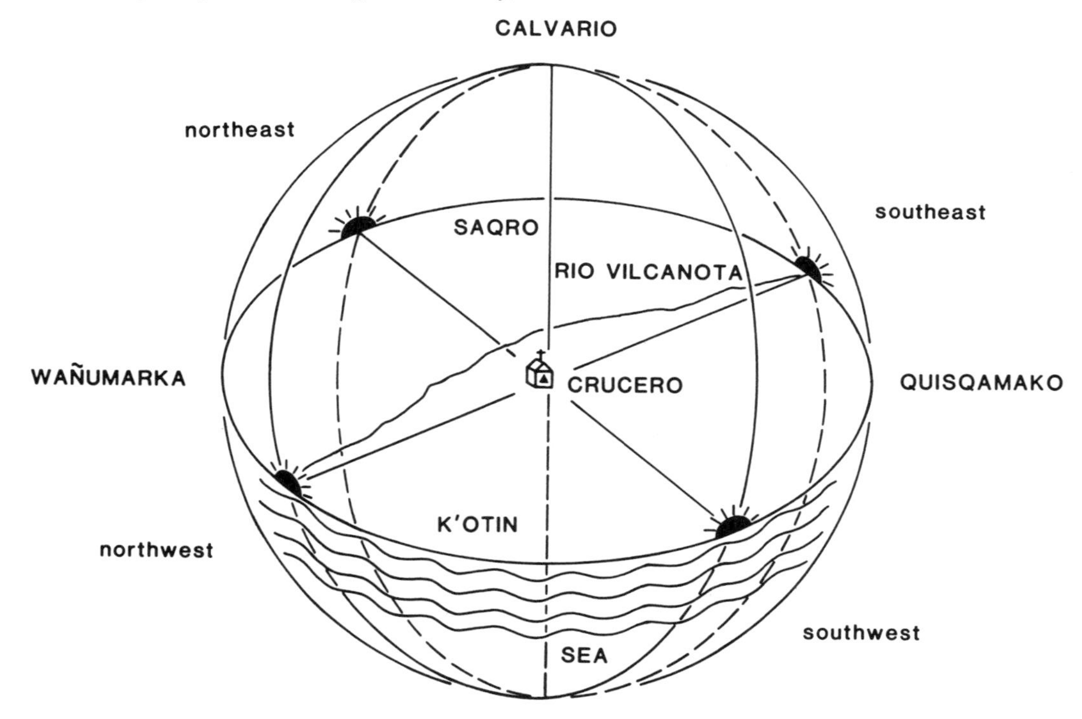

In the center of Mismanay, Peru, is this small chapel known as Crucero, inside of which is kept a cross. The two main thoroughfares of the village intersect here and are said to be intercardinal. The path on the right is Chaupin Calle, and the view is southeast. A small part of the northeast leg of the other path, Hatun Raki Calle, is visible to the left of Crucero.

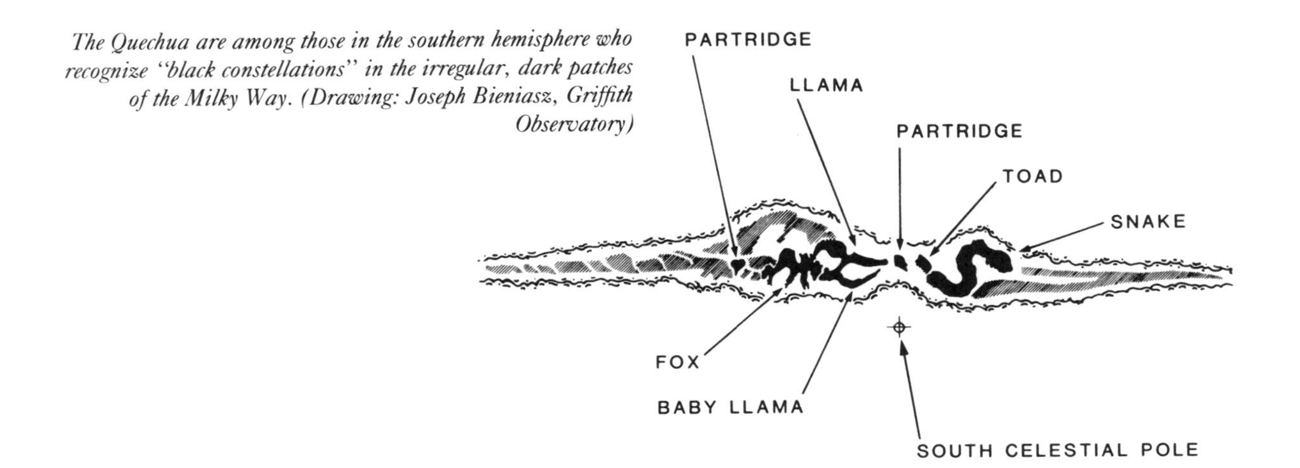

The Quechua are among those in the southern hemisphere who recognize "black constellations" in the irregular, dark patches of the Milky Way. (Drawing: Joseph Bieniasz, Griffith Observatory)

OBSCURED BY CLOUDS

The Milky Way also announces the seasons in the Peruvian highlands with the black clouds of interstellar dust that obscure parts of its southern half. The Quechua make "constellations" out of these inky nebulae, dark silhouettes in which they see the shapes of animals. There really aren't many of these black clouds blocking the view of more distant stars in the northern half of the Milky Way; in the southern sky, however, the obscuring clouds are abundant, and native peoples through the southern hemisphere—in Africa, South America, the South Pacific, Australia, and Indonesia—spot them and talk about them.

Probably the Coal Sack, right next to the Southern Cross, is the best-known dark cloud and the easiest to spot. This "dark hole" in the sky actually looks blacker than the average patch of background night sky because the interstellar dust so effectively subdues the light of more distant stars behind it. Anthropologist Gary Urton collected an unusually complete set of "dark cloud animal constellations" from the rural highland people in the vicinity of Cuzco, and to them the Coal Sack is a bird they call Yutu. It is a partridgelike bird better known by its Carib name, the *tinamou*. The Sky Tinamou is majestically lodged close to the "center" of the Milky Way, where its two streams meet and splash.

Another dark cloud, to the west of the Coal Sack but also near the Southern Cross, is said to be a toad. Yutu is constantly chasing the Toad, but the Toad is never caught. Urton also points out that the celestial toad rises high enough to be visible before sunup in early October, shortly after the terrestrial toads have emerged from hibernation in the earth for another season of toad romance.

Seasonal connotations also enhance the meaning of another dark constellation the Quechua manufacture from the Milky Way. They see a llama's head in the dark cloud just east of the Coal Sack, and they say its eyes are alpha and beta Centauri, the two bright stars that are embedded in the darkness of the head and point to the Southern Cross. In ancient Peru, sacrifices of black and multicolored llamas were scheduled for April and October, when the "eyes of the llama" are opposite the sun. The rest of the celestial Llama is visible in the black nebulae that continue down to the middle of Scorpius. The Llama's legs protrude clearly from the dark bulk of its body in the stars of Scorpius. Beneath it, the Quechua point out the dark image of a baby llama, which looks as if it is nursing.

Right behind the Llama, the dark Fox of the sky is lurking for a lunch of baby llama. The Fox is between the constellations of Scorpius and Sagittarius, and it is not too difficult to see a fox in the outline of the dark clouds there. The sun is in this zone in December, at the time of the southern hemisphere's summer solstice, and this is also the time of year when the people of Mismanay say that baby foxes are born. In fact, they single out December 25—Christmas—as the birthday of foxes. It is clear this is part of an older tradition that associated the young foxes with the December solstice. Foxes mate between the winter, or June solstice, and September, and the young foxes are delivered between October and the December, or "summer," solstice. In Mismanay, villagers point to the directional mountain in the northwest, the place where the June, or winter, solstice sun sets, and say that is where the foxes are born.

A CHINESE RENDEZVOUS AT THE RIVER OF HEAVEN

Living in the northern hemisphere, the ancient Chinese spotted no dark constellations in the Milky Way, but they did imagine it as a river in the sky. To the Chinese, the Milky Way was the Tian Ho, the "Heavenly Ho," a celestial counterpart of the Ho, or Yellow, River that flows through northern China into the Yellow Sea. The seventh-century Tang dynasty (A.D. 618–907) poet Hsieh Yen accurately observed the Milky Way is "frosty-white as shimmering silk" and "rotates on a slant" because it is hung across "the Purple Pole," which is the north pole of the sky.

The Chinese also saw a connection between the

Milky Way and the water that falls from Heaven to earth as rain. In winter, the dry season, the Chinese saw the thinner, more sparse, northern branch of the Milky Way, the piece that passes through Orion. Heavy rains, on the other hand, fall in summer, when the rich, turbulent, southern branch of the Milky Way takes over the sky. We have, then, in China what we also have in highland Peru—a seasonal link between the Milky Way and the rain.

One of China's best-known seasonal sky stories relates how the summer rain brings misfortune to a pair of starry lovers, the Weaver Maid and the Cowherd.

The story dates to about 700 B.C., during the first half of the Zhou dynasty, for a reference to the couple appears in one of the ancient poems of the *Shih ching,* or "Book of Songs":

In Heaven there is a River Han
Looking down upon us so bright.
By it sits the Weaving Lady astride her stool,
Seven times a day she rolls her sleeves.
But though seven times she rolls her sleeves
She never makes a wrap or skirt.
Bright shines that Draught Ox.
But can't be used for yoking to a cart.

The "Draught Ox" is the Cowherd star Altair, and the Weaving Lady that never weaves any clothes is Vega. The River Han in Heaven is the Milky Way.

The story of the Weaver Maid and the Cowherd begins by the banks of a peaceful stream on earth, where the Cowherd is quietly tending his water buffalo cow. The cow is wallowing with great satisfaction when seven maidens from Heaven drop down to the water for a frolic and a bath. All seven, the cow explains to the Cowherd, reside in a celestial palace where they spin the thread and weave the cloth worn by the gods in Heaven. Although all seven are lookers, the youngest, according to the buffalo, has the real corner on sex appeal. The buffalo cow advises him to steal the seventh sister's red robe. She is not be able to return to Heaven without her clothes and would have to ask the Cowherd to return them to her. Now, in old Chinese tradition, if a woman is seen wearing only a smile, she must become the wife of the man who catches her without her wardrobe. In this way, the Cowherd would acquire a lovely wife, and because she was a Celestial Immortal, immortality would be conferred upon him, too. The Cowherd has to agree it is a great way to pick up girls, and so he follows his water buffalo's instructions.

Things turn out just as the buffalo cow had planned. After the other six sisters finish their swim, dress, and fly back to the sky, the seventh sister is still looking for her missing clothes. Embarrassed but desperate, the Weaver Maid approaches the Cowherd for assistance. That naturally leads to marriage, and the Weaver Maid and Cowherd live happily together for three years. In that time, they have two children—a boy and a girl.

Nothing can separate the lovers . . . except Heaven itself. And concern in Heaven has grown. Because the Weaver Maid has left for earth and failed to return, her celestial loom is empty. She is the best weaver of the seven sisters, and for three years Heaven has been deprived of her work. The celestial garment industry is in a slump. As far as the gods in Heaven are concerned, the Weaver Maid's idleness cannot be tolerated. So one day, when her husband is out of the house, she is summoned back to her loom by her grandmother, the Queen Mother of the West.

Returning home to an empty house, the disconsolate Cowherd tries to comfort the kids. None of them knows what to do next. Then, the wise old water buffalo speaks. She is, she says, very near death, and when, in a few days, she dies, the Cowherd must remove her hide from her body and cloak himself with the two kids in it. In that form she will carry all three of them to the sky to be reunited with the Weaver Maid.

The Cowherd does exactly what his buffalo has told him to do, and just as before, the cow's plan works. Loading each child in a bucket and hoisting the carrier bar onto his shoulders, the Cowherd drapes his shoulders with the cowhide cape and floats to his wife's heavenly home.

Family reunions can be real celebrations, but this one doesn't last long. Within a matter of days the Queen Mother of the West realizes there is no new fruit of the loom coming out of the Weaver Maid's workshop. Fed up with the reunited lovers' lallygagging, she steps between the Weaver Maid and the Cowherd and draws a line through Heaven with her silver hairpin. That line cuts the sky in two and sends a celestial river, the Milky Way, between the pair.

Now, the Queen Mother of the West is a celestial goddess responsible for the world's rhythmic and seasonal change. She determines what is to happen to the universe, and as the master of its destiny, she controls the transformative powers of nature. Her presence in this story also tells us it has something to do with what happens in nature when the seasons turn. Whatever that something is, Altair and Vega are part of it.

Separated by the torrent in the Milky Way, Altair and Vega must live apart. The two children

The Chinese folktale about the Weaver Maid and Cowherd is really a Milky Way melodrama. Separated by the River of Heaven, they anticipate a seasonal rendezvous when the world's magpies manage to make a bridge of wings across the Milky Way. (From Outlines of Chinese Symbolism and Art Motives *by C.A.S. Williams, 1941)*

remain with their mother, and you can see them near her, two fainter stars a little closer to the river bank. With Vega, they form a small triangle. Starry echoes of the two kids can also be spotted on the other side of the Milky Way. There, two fainter stars in Aquila form a line with Altair between them and remind us here on the earth of the two children on their flight to heaven in buckets suspended over the shoulders of their dad.

The unhappy lovers pine for each other. Fortunately, the tears of the Weaver Maid, the Cowherd's own quiet grief, and the sadness of the children, have an effect on Heaven. The King of Heaven's heart is moved by their melancholy. In celestial court, he rules the pair will be permitted one annual meeting, and a few precious hours of intimacy, on the seventh night of the seventh month, if all of the world's magpies can build a bridge of wings over the raging waters for them. Fair weather will allow the birds to do it, but storm at that time of year could sabotage the magpies' campaign on behalf of family unity.

The seventh month of the year in the lunar calendar usually falls in the month of August, and it also corresponds fairly well to "Autumn Begins," one of the twenty-four subdivisions of the solar year in ancient China. Summer is passing, but it expires in the soft grandeur of sweet air, sunshine, and trilling insects. By this time the summer rains are usually over, and the dusty months of winter have not yet arrived. The night sky should be exquisite and clear. If so, the part of the Milky Way that includes Vega the Weaver Maid and Altair the Cowherd is visible most of the night. The mists and froth of Heaven's River seem to gleam as bright as ever in the transparent air. It's a quiet night on the celestial river.

If, however, there is a late summer storm, heavy rain on the seventh night of the seventh month washes out the one chance for the couple has each year to get reacquainted. In such a rain, Heaven itself cries for the pair's plight. Ever in sight of each other, on opposite sides of the river, the Weaver Maid and the Cowherd are then obliged to control their disappointment with the knowledge that another year will bring another opportunity.

An old Chinese holiday still commemorates the yearly attempt to bring the Weaver Maid and the Cowherd together. Always held on the seventh day of the seventh month, it is known as the Seven Sisters' Festival (or Maidens' Festival). Preparations for winter begin then. Women take warm clothing out of storage, repair it, and make sure it will be ready when needed. Cattle are transferred from the fields where they've grazed all summer back to the more protected quarters of the barn. It's no wonder people have their minds on weaver maids and cowherds.

LEAVING A TRAIL OF CELESTIAL STRAW

An entirely different tradition about the Milky Way treats it as a spilled load of straw rather than as running water, but the straw scattered in heaven, like the water in the celestial river, puts seasonal meaning in the Milky Way. These Milky Ways of straw are known from a wide geographic area that includes western Asia, northern Africa, the eastern Mediterranean, and parts of south central Europe. An Armenian tale from the seventh century A.D. has the earliest known reference to the celestial path of straw. According to this story, a native sun and fire god known as Vahagn got the better of Ba'al Shamin, a sky and sun god imported from Syria. They shared similar powers and competed for the same congregations. During one cold winter, Vahagn conducted a covert action in Ba'al Shamin's farmyard and stole a bundle of straw. Vahagn dropped pieces of straw as he made his getaway across the night sky, and we see those carelessly strewn stalks as what the Armenians used to call "the Straw-thief's Way."

There is also an old Persian story about a thief whose attempt to hide stolen straw in the sky was compromised by the Milky Way, the trail of it he dropped behind him. The Milky Way is known as "the straw-thief's track," and the same story was picked up by the Turks, many of the Balkan peoples, and by the Caucasian Tatars.

The Straw Path of Heaven is also known in Hungary, but it takes some interesting turns. Local names for the Milky Way include "Straw Road," "Straw-dropping Road," "Gipsy's Straw," "Gipsies' Way." Gipsies take a lot of heat from the people in the neighborhoods they visit, and so Gipsies are sometimes said to be the ones who stole the straw and dropped it in the sky.

Hungarians also blame the Milky Way on Gönköl, a wagon driver who spilled a load of hay upon the celestial highway he was traveling. Gönköl's wagon is the Big Dipper, and under Christian influence it became known as Saint Peter's Cart. Later stories about the Milky Way, therefore, blame Saint Peter for the spilled straw.

Additional data from Hungary help clarify why the Milky Way was said to be straw. A rural Hungarian informant explained, "Once, at harvest, the hay was scattered about, and the Milky Way is the sign thereof." Another Hungarian name for the Milky Way is "Harvesting Way." A seasonal configuration of the Milky Way probably once coincided with the period of harvest and planted the celestial Straw Path in the minds of ancient farmers.

FOOTPRINTS IN HEAVEN

Hunting peoples also see seasonal significance in the Milky Way. For example, the Udege people who live in the forests along the tributaries of Siberia's Amur River attribute the origin of the Milky Way to a long hike two brothers still make while hunting sable in the sky. The sable is a weasel-like predator, highly valued for its deep, dark fur. After the brothers killed all of the sables on the earth, the Master of the Mountains and Forest ordered them to go to the sky. The souls of sables hunted in Heaven descend to earth and repopulate the world. This is a story of world renewal, and in this story the path of the Milky Way confirms that the brothers are still hunting and that the world is seasonally renewed.

Yakuts in Siberia say the Milky Way is another kind of trail. They call it "the ski track of the son of God." In the Ostyak version the Milky Way ski track was left by a celestial hunter who was pursuing a fabulous six-legged, extra-sure-footed stag in the sky. After chasing the stag completely across the heavens—and leaving the ski track up there—the hunter runs the stag down to earth. He is unable to kill it, but he does manage to remove two of the stag's legs. The stag eludes him, however, and speeds off for the north as fast his remaining four legs can carry him. At the end of the line, the stag escapes the hunter by turning to stone. A celestial stag was placed in the sky to commemorate this great hunt. That sky stag is the bowl of the Big Dipper, a handy and familiar emblem of seasonal change.

PATHS OF CELESTIAL MIGRATION

When birds migrate, they fly a seasonal path through the sky. For that reason the Lapps and others have found a route for migrating birds in the pathlike Milky Way. Turi, the highland Lapp who wrote a book about his way of life, had two names for the Milky Way: Path of Birds (or Birds' Stairway) and Year Mark. A comment he added to one of his star maps may help explain what he meant by "Year Mark": "When the stars are many, then there comes the snow." In some way, the Milky Way puts in an appearance that coincides with the first snowfall and helps mark the year.

The language of the Finns and Estonians is related to the language of the Lapps, and their names for the Milky Way mean "the Birds' Road." Turkic- and Tatar-speaking Siberians name it the Birds' Way and the Wild Ducks' Way. Voguls and Ostyaks also sometimes call it the Ducks' Road or the Southern Birds' Road, and it allegedly guides the birds in nocturnal flight. Shortly after dark in the fall, we can see the Milky Way crossing overhead. It bridges the sky from the northeast horizon to the southwest and in that way, may be "guiding" the migrating birds south to their winter quarters. Behind all of these "bird path" names for the Milky Way, there is probably something connected with bird migration and winter.

The Estonians tell a migrating bird story that illustrates the Milky Way's role in this type of seasonal change. An eligible maiden named Lindu was the daughter of Uko, the same thunder and sky god known to the Finns as Ukko. Lindu was in charge of the birds and responsible for their successful travels in spring and fall. She was also pretty. Everybody wanted to marry her. Everybody tried to court her. One time the North Star pulled up to her house in a fancy carriage drawn by six brown horses. He brought her ten different presents, but she wasn't particularly attracted to him because he was boring. "You always stay in one place," she said.

Moon, too, showed more than a casual interest in Lindu and arrived in a sparkling silver coach pulled by ten silver horses and loaded down with twenty gifts. He, too, was rejected. If the North Star were too constant, Moon was too inconstant. And even though he couldn't keep a straight face from one day to the next, he ran around in the same circle month after month.

Lindu also wanted no part of Sun. He showed up in a golden phaeton powered by twenty tawny-red horses, and he had thirty surprises for Lindu. But she said he was just like the Moon, stuck in the same old rut.

Then Lindu's heart was struck by love. Northern Lights drove up in a diamond carriage, and he had a thousand white horses to pull it. When Lindu opened the door to welcome him, he brought an entire coachload of gifts into her townhouse. She was delighted with his presents and with him. He never did the same thing twice. He came and went as he pleased and wore very stylish and extravagant clothes for his public appearances. Lindu pledged her love and promised to marry him. Northern Lights was delighted but had to be back home in the far north by midnight. He asked her to plan the wedding, for he would soon return.

Northern Lights never did come back. Lindu waited anxiously for him throughout the winter and right through spring and summer. By fall her heart was broken. She sat by the river in her bridal dress, dropping her tears on the ground and oblivious to the birds that depended on her. Finally her father, Uko, realized something had to be done for his daughter and for the birds. He directed the winds to sweep her into the sky, where she lives today. She still wears that long white veil that never got to the wedding. It flutters from one side of the sky to the other as the Milky Way, and in it, Lindu, directs the seasonal flights of all of the birds once more. In the middle of the night she waves to Northern Lights, and he sometimes manages a brief visit with her in winter, when the Milky Way is off toward the north in the early evening.

NETTING THE SUN

Luiseño Indians in southern California also saw seasonal significance in the Milky Way, but for them the Milky Way was a kind of primordial trampoline that bounced the sun into the sky at the time of creation. It was also a net that kept the sun on course once it was launched into heaven. Before the sun got on board the ecliptic, the world created by Mother Earth and Father Sky was still dark. Those who lived there in the beginning—the First People—kept bumping into each other, and it was one of them who got the bright idea to make a sun. That was a great start, but they still had to get the sun into the sky. So the same man also wove a net out of red milkweed twine and asked everyone to meet him. They stretched the net out upon the ground, put the sun in the middle of it, and with magic chants and gestures bounced him into heaven. At first the sun went to the north, but everyone agreed that was not right. They put the sun back down on the net and tried again. This time the sun went south, but he came back again. So they made another attempt to launch him into the proper orbit. This time he went a little bit to the west and returned once more. Another snap of the net sent the sun into the sky again, and this time they got him in the east where he belonged. With a little more effort, their songs put Sun on a yearly course that never followed a straight line but carried him south and north in different seasons.

The sun must confine itself to the ecliptic, a path through the stars that does not coincide with the Milky Way. But the Milky Way and the ecliptic both circle the sky and therefore cross in two places. For the last several centuries, one of those places has been in the vicinity of Sagittarius and Scorpius. The other is between Gemini and Taurus. These zones are also where we find the sol-

The white netlike pattern on the ceiling of this painted natural rock shelter resembles the milkweed-fiber nets used by southern California Indians. In this view we are looking toward the back of the shelter. A boulder partly visible on the right forms the right wall of the shelter and helps support the painted ceiling stone. This site is located in what was once the territory of Luiseño, and one researcher has identified the net design as a symbolic picture of the Milky Way. Disks have been drawn at each end of the net and another on the left side. They are hard to see, however, because they are painted in red. If these disks represent the sun, they are located appropriately for the solstices, when the sun is "caught" in the Milky Way, and for the equinoxes, when the sun is to the side of the Milky Way.

stices. At winter solstice, the sun is in Gemini, and at summer solstice, Sagittarius is its home. The Luiseño story about getting Sun into the sky is really describing what the sun does every year. Once it is in the southern part of the net, in Sagittarius, it goes north. When it occupies a northern pocket in the net, it heads south. At other times of the year, it is "outside of the net," east or west of the Milky Way.

The symbolic meaning the Luiseño attached to the Milky Way inspired them to twine a net from milkweed fibers and use it in the boys' initiation ceremony to launch the youth into adulthood. This sacred net, or *wanawat,* was one of Mother Earth's first children, born from her in the earliest days of the universe, and it was also a symbol of the Milky Way.

The ceremony began with the digging of a large trench. About five feet long, it had the general outline of a human figure. The milkweed fiber net was somewhat similar in shape and was stretched across the trench. Three large, flat stones, collected from the seashore, were lodged at equal intervals in the webbing. Following some further instruction, each boy stepped into the net and crouched on one of the stones. With one foot on top of the other, he had to bounce his way from stone to stone and out of the net. If, by chance, he should slip into the net, the elders judged he was not long for this world. The Luiseño believed that at death our spirits retreat to the Milky Way and make their final home there. It is, they thought, a sign in the sky that tells us "we are only going to live here for a little while." Physical contact with the net, then, was probably like spiritual contact with the Milky Way.

GOING FOR THE HIGH SPIRITUAL GROUND

Tukano-speaking tribes of southeastern Colombia imagine the Milky Way to be a river in the sky. These peoples—among them the Desana and the Barasana—see an agitated, hazardous current overhead. The Desana say the Milky Way flows from east to west, and so its waters mimic the daily motion of the entire sky. They believe bundles of palm fibers float in this stormy stream and roll with its waves. Yellow-white in color, these fibers are associated with semen and turn the Milky Way into a flow of fertilizing plasma. The Milky Way also carries disease, waste, and rot. They rise from the dark, deadly part of the underworld below the western horizon, drift into the Milky Way's foaming waters, and float to its surface. It carries this dangerous residue through the rest of the underworld and into the sky.

According to the Desana, Creator Sun made the Milky Way at the beginning of the world, and now the Milky Way functions as a closed circuit whose fiber optics distribute the energy of Creator Sun. When an Indian kills game and eats it, he uses some of that energy. It is no free lunch, however. The Indian is obliged to repay this energy loan through appropriate ritual, which might involve the hunter's sexual abstinence.

The vitalizing, seminal power of the Milky Way energy bank is also at the disposal of powerful shamans, who are the real experts in spiritual enterprises. For them, celestial power is something of a cottage industry. They may go anywhere in the sky to acquire some of it, and they may get a little help from psychotropic plants to get them in the proper spirit for this kind of trip. In the Amazon, they undertake drug-induced visionary journeys to the Milky Way to acquire its power.

Tukano shamans believe power is needed to ensure the supply of game. Availability of game and fish is controlled by a supernatural spirit they call the Master of the Animals, and he can also dispatch dangerous creatures to scare or harm anyone guilty of bad behavior. The shamans are responsible for establishing good, professional relations with the Master of the Animals and know they need an edge in dealing with him. They get it from the Master of Snuff. The snuff is a narcotic powder ground from the inner layers of the bark of the *Virola* plant. Shamans take a hit and then travel to the Milky Way, to the home of the Master of Snuff. Together, the two of them pay a visit on the Master of the Animals and bargain for permission to hunt and to kill game.

Tukano tribes also journey to the sky on wings of *yajé*, a vision-inducing drink prepared from *Banisteriopsis caapi*, a jungle vine. The shaman orchestrates a *yajé* party in which an entire group of men take the narcotic drink. Such affairs strengthen social cohesion by providing the participants with special knowledge they can acquire only in a visionary trance. This knowledge reinforces their tribal beliefs.

Under yajé, the men first feel numb and a bit nauseated. Soon, however, this beverage creates a sensation of flight. Those who drink it begin to see colorful geometric patterns of light, and then they know they've reached the Milky Way. Not content to remain there, they try to glide farther into a visionary landscape of colorful scenes. These vistas include animals, monsters, and human beings. Guided by the shamans, the Indians believe they have entered the time of Creation and are witnessing the events described in their myths.

In the last stage of a yajé vision, tranquil scenes replace the mythological action. It is a soft, contemplative phase. Emerging from the entire experience, the participants say they are reborn. Some would say a hallucinogenic drug transformed their state of mind. But if their visions don't turn negative or frightening, such transformation gives them spiritual knowledge and experience powerful enough to create a feeling of growth.

HIGH ROAD OF THE GODS

In some cultures, the Milky Way was considered a turnpike, traveled by those who seek audiences with spirits and gods. The Euahlayi of Australia said the Milky Way leads to the home of celestial gods. Siberian Yakuts sometimes called the Milky Way "God's Footprints." In a Mesopotamian myth, Adapa had to take the Milky Way "road of heaven" for a mandatory audience with the high celestial god Anu. Pindar, the Greek lyric poet who lived in the fifth century B.C., described the Milky Way as the high road of the gods. Four centuries later, the Roman poet Ovid treated it the same way in the first book of *Metamorphoses* (lines 169–175):

> *Across the height of heaven there runs a road,*
> *Clear when the night is bare, the Milky Way,*
> *Famed for its sheen of white. Along this way*
> *Come the immortals to the royal halls*
> *Of the great Thunderer. . . .*

The Thunderer is, of course, Zeus, and the rest of the divine aristocracy are his neighbors.

By the fourteenth century, the English visionary William Langland was saying the Milky Way is a road to the Virgin Mary in his moralizing allegorical poem *The Vision Concerning Piers Plowman.* John Milton put the Milky Way road metaphor to spiritual use in *Paradise Lost:* He saw it as a "broad and ample road whose dust is gold," which led to God's eternal house.

GHOST RIDERS IN THE SKY

Thin, pale light makes the Milky Way look tattered and ghostly, and as a spectral road to the gods, it was sometimes a ghosts' road to Heaven and a home for human souls. The Osage, Plains Indians of southern Missouri and Kansas, believed the souls of the dead journey along the Milky Way and continue until they each find a star in which to reside. Among Skidi Pawnee spiritual leaders, the Milky Way was the trace traveled by departed souls on their way to the Southern Star, the final home of spirits. Ancient Hungarians sometimes called it the War Path and said that soldiers who died in battle marched upon it into Heaven. Both the Iroquoian and Algonquian tribes of the Eastern Woodlands saw the Milky Way as a path for the dead leading to the Village of Souls. American poet Henry Wadsworth Longfellow borrowed Eastern Woodland Milky Way themes and incorporated them into *The Song of Hiawatha,* where the Milky Way is "the broad, white road in heaven, Pathway of the ghosts, the shadows."

In Nicaragua, the Mosquito Indians say that dead souls find their way to the end of the Milky Way. When a Guajiro Indian dies, the soul crosses the Milky Way, "the way of the dead Indians," and continues on to the place where the silent spirits of the dead reside. The common Hindu name for the Milky Way is *Chhayapatha,* "the Path of the Shades of the Dead." The Pitris, or spirits of the dead, followed it to the subterranean realm of Yama, king of the dead.

New studies, over the last decade or so, of ancient Maya symbols and beliefs indicate that the Maya saw a map of the underworld in the night sky. To the Maya the underworld was Xibalba, a dark and dangerous realm of death and decay. It was ruled by death gods, and the souls of the dead had to journey through it. When the hero twins of the *Popol Vuh,* the Quiché Maya book of the world's creation, took a hike on the Black Road into underworld, they actually followed Milky Way.

The Milky Way is not really a "black road," but in the northern hemisphere part of it is black—the Great Rift in Cygnus, where we get a view of the corridor between two of the Galaxy's spiral arms. There is less action there—fewer bright stars and less star formation. This creates a dramatic split in the Milky Way that stretches from Cygnus all the way to Sagittarius. In this zone the Milky Way seems to branch into two paths. The northern trail is conspicuously fainter than the southern branch, and the dark gap that separates them is a huge cloud of interstellar dust that masks the light of more distant stars. The Quiché name for this dark feature of the Milky Way is the "Road of Xibalba," and they call the opposite, undivided part of the Milky Way the "White Road."

Southern California's Chumash Indians also saw a Y-intersection split the road of the dead at the Great Rift of the Milky Way. Astronomical analysis of the soul's pilgrimage to Shimilaqsha, the land of the dead, was undertaken by the late Travis Hudson and his coauthor Ernest Underhay in their book on Chumash astronomical traditions, *Crystals in the Sky.* In their view, a story told by María Solares, an Inezeño Chumash, maps the Milky

Way as it recites the itinerary of the trip of a dead woman to Shimilaqsha.

A young woman is mistakenly killed by the man who had recently married her. He keeps a vigil at her grave for three nights, and on the third night she rises from the earth and heads north. Perhaps disoriented or possibly visiting the places she had known during her life, she comes back, and her husband watches her set out for the east, the south, and the west. Each time she returns to her starting point, but she finally departs for the north again. This time she keeps on going, and he follows her. She realizes he is behind her and stops to persuade him to return. He can not go where she is bound. She is now a spirit, but he is still alive. Insistent, however, the young man convinces the soul of his dead wife to allow him to accompany her, at least for a short distance. They walk until dawn, and the light of day makes her as ghostly as mist. All you can see is her heels. At nightfall, she becomes completely visible again. She again directs him to turn back, but he refuses. They keep traveling like this and pass right by the Land of the Widows, which Hudson and Underhay believe to be the zone of the Milky Way around Cassiopeia.

Before reaching the land of the dead, a soul may have to negotiate a deep ravine on the other side of the Land of the Widows and contend with a road cut to shreds by the heavy traffic. Clashing rocks, giant eye-plucking ravens, and La Tonadora—the thundering white Scorpion Woman—all add spice to the trip.

After obstacles like this, the spirit wife and her living husband finally reach the gate of Shimilaqsha. It is like a bridge that spans the waters that separate the earth from the land of the dead. Souls must cross it successfully to reach their home, but it moves. And as spirits attempt to keep their footing, monsters rise out of the depths to scare the evil souls into falling off the overpass. Any soul not spiritually initiated and prepared for this ordeal is in serious trouble.

Once a spirit is safely across the bridge in Shimilaqsha, the road splits at a Y-junction. One branch leads to the land of the dead itself. The other is a driveway to the big crystal house of the old chief who governs Shimilaqsha. Equating this intersection with the Great Rift of the Milky Way, particularly as it is seen in early evening in winter descending into the west, Hudson and Underhay locate the bridge to the land of the dead near Cygnus the Swan and Aquila the Eagle.

Although recorded Chumash traditions do not prove that the Chumash saw a link between the souls of the dead and the Milky Way, many of their neighbors did, including the Luiseño, the Chemehuevi, and the Maidu.

Other stories about husbands who follow the spirits of their dead wives into the sky are known from the Gabrielino, the Miwok of Yosemite Valley, the Monache (or Western Mono), the Tachi (or Southern Valley) Yokuts, and the Shasta, who live in the far northwest part of California. The Shasta said the trail the woman walked to the other world was the Milky Way.

In the northeastern part of California, the meaning of the Milky Way to the Ajumawi Indians is just as clear. Floyd Buckskin, a member of the Ajumawi Tribal Council, has learned traditional lore from tribal elders, and he has recorded beliefs about the journey to the land of the dead. At death, the shadow, or spirit, departs from the body, and under normal circumstances it travels to the southwest along the Fall River Valley and continues to the Pit River. There it turns west and does not stop until it reaches the Pacific coast, where it turns north and flies to Mount Shasta for further directions. From there, the spirit picks up the Milky Way, which the Ajumawi know as the "Pathway of Spirits," and as it crosses over Mount Shasta, the shadow continues east to Hawisi the Creator at sunrise. On the summer solstice, the Milky Way rises out of the northeast before dawn and is said to be aligned with the sun and with the trail of the dead on the ground. Anyone who dies at this time of year is able to join the Creator with ease because all of the paths are so closely aligned.

OUR PATH TO HEAVEN

Scientific analysis of the Milky Way contrasts sharply with the ideas of the ancients, but the Milky Way is still seen as a primordial element of the universe, almost as old as the universe itself. And it conveys its true character to us in the same awe-inspiring way it revealed other identities to our ancestors. Even at its best, it shines with subtlety and restraint, but when we think about what it is, majestic visions are seeded in our brains. It is a vast hive of stars swarming to the tune of gravity. A halo of spherical star clusters, each with a membership of 100,000, surrounds the Milky Way. Countless glowing clouds of interstellar gas and dark lanes of dust, swirled into spiral arms, dwarf our whole solar system.

Old stars explode in the Galaxy with catastrophic fury. We see one light up like that only every few centuries, apparently without rhythm or pattern, but centuries are a twinkle in the eye of the Milky Way. Millions of years are moments on its time scale, and with that perspective, those dying stars look like the blinking lights on a Christmas tree. The Galaxy also gives birth to about twenty new stars each year, but they are lost among the 400 billion that are already here.

Something hidden is raging in the Galaxy's nucleus with 10 to 30 million times the fire power of our sun. Of course, we see none of this. The center of the Galaxy crosses through the sky, and to us it is a ghostly cloud that travels with the summer's stars.

The Milky Way's grandeur in time and space is camouflaged. It is just a silky stole of pale light draped over the shoulders of the dark. But once we realize what we are seeing, we feel its impact. Its soft light and slow turns shake us out of everyday complacency and put our imaginations back in touch with the cosmos. The Milky Way allowed our ancestors to reach into heaven. If we let ourselves lose it completely in the smudge and stain of artificial lights, our spirits are bound to be more confined to earth.

17

At the Top of the Sky

THE top of the sky told the ancients there was foundation and structure in their lives. Their stories about the high ground in heaven reflect what they saw going on there. It was an anchor that held the world in place. That anchor was sometimes but not necessarily straight overhead, at the zenith. Some saw it lodged at the north celestial pole, the stable, motionless hub of the night's orderly parade of stars. Some believed the sky was secured in other celestial zones, places dictated by other systematic motions of the sky. Wherever the sky was pinned into place, that "top" of the sky symbolized stability, order, and transcendental celestial power.

Each of those pivots in heaven was also linked to earth by a world axis. It might be described as a pole, a pillar, a tree, or a mountain. The world axis was a vital element of creation; it had strength and stamina, but it was threatened by chaos. Our ancestors spoke of all of these things in their myths of creation and world order.

This order was visible in the sky: sunrise on a key day in the calendar, the return of a particular slice of the moon, and the seasonal reappearance of an expected star in the predawn sky. The cyclical motions of the sun, moon, and stars apportioned time into manageable, meaningful bundles of hours, days, months, and years. Celestial phenomena provided a framework of time and space and so made an ordered cosmos out of the world. That's what the word *cosmos* means—"the ordered whole," and that order originated in heaven.

GETTING A CORNER ON CREATION

The old stories of creation and world order describe the processes we observe in the world around us in terms of gods and heroes. In them, fundamental order is established at the time of the world's creation, the era when the world as we know it was set up the way it is. A Chinese story about this event involves a divine primordial couple. Sometimes said to be brother and sister, as well as husband and wife, they worked hard to put the world in shape. Fu xi, the male partner, invented two calendar plants. One grew a leaf each day until it had fifteen fronds on its stem, then it would lose a leaf a day until the stem was bare. The other plant put on foliage more slowly—one leaf per month. After six leaves, and six months, the leaves fell at the same rate. Both of these plants set the pace for fundamental intervals of celestial time—the month and the year.

Nü kua was the female half of the couple. Her part in establishing world order was a little more demanding. A pair of primordial bruisers—the

You can't feel the earth turn, but you can see it happening by watching the sky spin around the north celestial pole. A time exposure targeted on the sky's north pole records trails of light left by northern stars as they follow their circumpolar paths. The north celestial pole is the only spot that does not move in the northern half of the sky. It seems to anchor the eternally turning heavens in place. Polaris, the North Star, is close to this center of motion, and in this picture, Polaris crowns the top of a tall tree. The ancients sometimes described the axis of daily rotation as a tree that connects heaven and earth. (Photograph: Michael P. McDermott)

monstrous god of water and the gigantic god of fire —had made a mess of the newborn earth, and Nü kua had to pick up after them. The two gods had gotten into a brawl, and after the water god, badly beaten in this bout, retreated in anger and shame, he slammed his head against Mount Buzhou, the "Imperfect Mountain," in the west. If the sky were the top of a card table, this was just like kicking in the hinge on one of its legs. The sky took a tumble, but Nü kua had no intention of leaving the cosmos crippled in the earth's northwest corner. She propped it back up with four solid columns, one for each corner of the world. For pillars she appropriated the four legs of a giant tortoise she killed for the sake of an ordered cosmos. Her work put the four cardinal directions back into place between the world's four corners.

This theme of world order is symbolically restated in many portraits of Nü kua and Fu xi. They are hybrid creatures. Their upper bodies are human, but from the waist down they are snakes. With their serpentine bodies entwined, they represent the creative combination of the *yin* and *yang*, the female and male principles of nature. Opposite but complementary, Nü kua and Fu xi wind around each other in turns that resemble the repetitious axial rotations of the sky, which itself creates the directional order of the landscape.

This Chinese account of the establishment of world order is not a scientific description. Instead it represents what seem to be underlying principles of world order—cyclical time, cardinal directions, and a stable marriage of earth and sky. In the Chinese view of the cosmos, the earth was square because it had four "corners," or directions, and the sky was round because the horizon encircles us with a canopy that looks domed.

Serpentine Nü kua and Fu xi, entwined with turns that mimic the rotations of the cosmic polar axis, establish world order with designers' tools. Nü kua holds a "compass," and Fu xi lifts a square and plumb line. Actually, these two instruments look like the Egyptian merkhet *(plumb line) and* bay *(forked sighting stick), which were used in measuring time. Other messengers of ordered time—the sun, the moon, and constellations of stars—accompany the pair, while at their waists, the earth is represented as a square apron with a world mountain in each corner. (Painting in local museum, Xinjiang Uygur Autonomous Region, Turpan, China)*

COSMIC BAKLAVA

Most ancient peoples recognized two realms in the cosmos—earth and sky—but also believed in a third kingdom, the underworld, where celestial objects sank when they set. It might be reached as well through caves in the earth. The standard, off-the-shelf model of the universe for many peoples came equipped then with three layers, which were often imagined as skewered on a world axis that punctured the earth at its center and speared heaven right through the celestial pole.

María Solares, an Inezeño Chumash, provides an example of this traditional, three-layered universe in her description of the Chumash cosmos: "There is this world in which we live, but there is

The axis of the world imposes vertical symmetry on the drum of a Siberian shaman. It connects an opening to the underworld, at the bottom, to the "hole" in the sky—the north celestial pole—near the top. A horizontal line separates the celestial realm, populated with star symbols, from the earth, with animals, plants, and a portrait of the shaman holding his drum. (From The Tree of Life *by Roger Cook, Thames Hudson, 1988)*

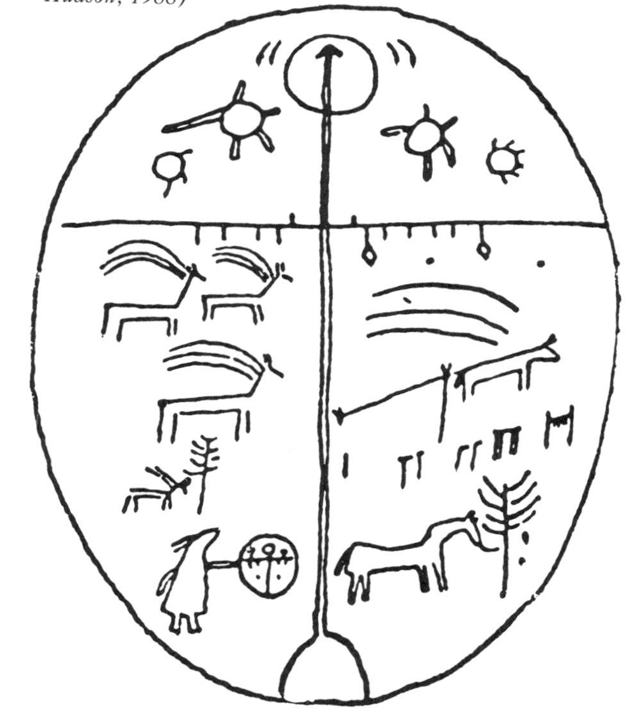

also one above us and one below us. . . . Here where we live is the center of the world. It is the biggest island. And there are two giant serpents that hold our world up from below."

We can find this same basic idea of three cosmic zones from Siberia to Mesoamerica, from Africa to Mediterranean Europe. In one variation or another, with nuances to suit local preferences, this concept shows up all over the world. People often added layers to the sky and to the underworld. The Dogon of west Africa's upper Niger say that the universe has fourteen terraces. The earth is seventh from the bottom and is in the middle of the cosmos. In Siberia, the Teleuts have sixteen heavens, and the Yakuts seem to believe in nine. The Chukchi have several levels in their heaven. Each floor of one is the roof of another, and all are centered on the polar axis.

Aristotle's stratified and earth-centered cosmos is symbolized in this diagram from an early sixteenth-century French edition of De sphaera. *This treatise on practical astronomy was written by John of Sacrobosco in the thirteenth century. The globe of the earth is surrounded by three rings that represent three concentric spheres of water, air, and fire. Beyond the zone of fire are seven more spheres representing each of the planets. The sphere of fixed stars is symbolized by the ring of zodiac symbols. This cosmos is contained inside one more sphere, the* primum mobile, *which Aristotle believed was responsible for imparting motion to all of the other spheres beneath it. (manuscript 67811, reproduced by permission of The Huntington Library, San Marino, California)*

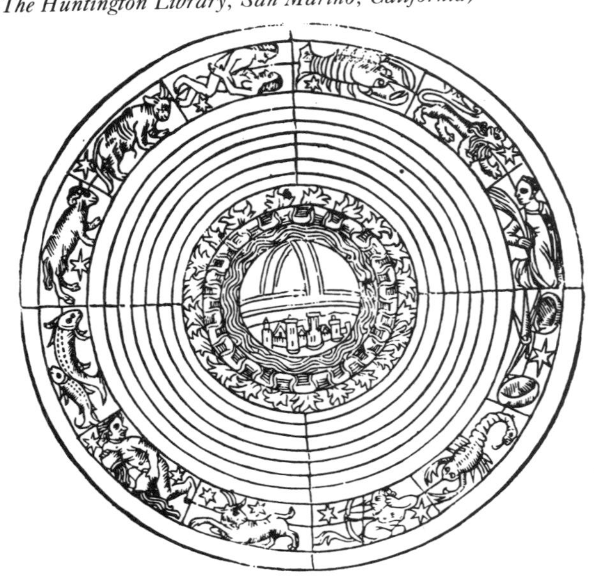

The Aztec and other ancient Nahua peoples of Mexico believed they lived in a vertically stratified cosmos with thirteen heavens and nine underworlds. The Aztec assigned specific personnel and power to each realm in their universe. The earth occupied the transitional zone between the celestial territories and the underworlds and was counted as level number one in both kingdoms. Above it were the zones of the moon and clouds, the star-skirted goddess of the Milky Way, the sun, and Huixtocíhuatl—the salt-fertility goddess. The next realm belonged to a single constellation —the Fire Drill—and emphasized its importance. Higher still were the green heaven of winds and storms and the blue heaven of dust. Thunder lived in layer number nine, and three more heavens— the white one, the yellow one, and the red one— lay above it. Above all of these, in the thirteenth and highest heaven, resided Ometeotl, the lofty, remote, and abstract creator of space, time, and the world.

Just below the earth was the place of water passage, where the dead crossed a river. Subsequent subterranean worlds included the place where the two mountains smashed together, the hill of obsidian knives, and the country of the icy, cutting wind. In the sixth underworld, banners were unfurled. Next the soul encountered a place where arrows were fired to kill. Layer eight was an infernal kennel where hearts were devoured by wild beasts. After passing through a narrow place of stones, the soul found its final home in Mictlán, the ninth underworld and the land of the dead.

Caught between heaven and hell, the Aztec earth was pictured as a mass of land immersed in a primordial, oceanic stream. The world's center was known as its "navel," and four cardinal directions divided the horizon and the land. Each direction had its own color, tree, bird, and tutelary god. Those supernatural sky-bearers stood out on the horizon in each cardinal direction to support the heavens, which merged there with the surrounding waters.

Our information about ancient Greek cosmology before the time of the early philosophers is largely limited to what we read in Homer and Hesiod, and what their mythological metaphors mean is not always clear. They did imagine an underworld of some kind, along with heavens above the earth, and a cosmic axis is implied by the image of the Titan Atlas, who, according to Hesiod (*Theogony*, lines 519–521),

> *. . . forced by hard necessity,*
> *holds the broad heaven up, propped on his head*
> *and tireless hands, at the last ends of Earth.*

Although the Greek cosmos was oriented by a cosmic axis, physical and mechanical ideas began to intrude upon the mythological universe of Homer's and Hesiod's time. The Greeks emphasized the concept of the sky's uniform circular motion and described the cosmic hierarchy as a set of concentric spheres. In Aristotle's picture of the universe, a spherical earth was surrounded by water, and air enveloped them both. Lighter than air, fire occupied the next level. It was the source of light, heat, and energy. Beyond the realm of fire circled the celestial objects: the moon, Mercury, Venus, the sun, Mars, Jupiter, Saturn, and the stars.

Later, Ptolemy imagined still higher spheres of influence. They were the source of certain long-term celestial cycles, and above them was the sphere of the Prime Mover, which put everything below into action. The entire system was cocooned by the immobile empyreal heaven, a transcendental realm identified in medieval cosmology with the celestial paradise—the angel-thronged home of God. In the early fourteenth century, Dante's *Divine Comedy* mapped out this universe in detail and gave it nine circles of Hell, thirteen floors of Purgatory, nine spheres of Heaven, an Empyrean with nine rings of angels, the throne of God at the summit of it all, and, of course, it also included the earth on which we stand.

NECESSITY'S SPINDLE

Even though the cosmic axis did not always coincide with the earth's axis of rotation, the sky's rotation about the pole is still its most fundamental motion. It is reasonable to guess that people noticed it at least tens of thousands of years ago, but because we have no writing from that time, we have no evidence of such awareness. We do find symbolic references to it in ancient myths people later told about the sky, and by the fourth century B.C., Aristotle's discussion is explicit. In his commentary *On the Heavens*, he says that circular motion is primary—or more fundamental than other motion—and adds that the stars are fastened to their own "circles."

Aristotle's teacher Plato also stressed the importance of the uniform daily rotation of the sky in his description of the system of world order, in the *Republic*. He called the axis of this motion the Spindle of Necessity. This spindle was part of the framework of the entire cosmos, and to see it without distortion, you would have to view it from beyond the earth. Plato put his earthbound mortal readers, then, in the company of "spirits" or "souls" who ascend from the earth and pass through all of heaven to a supernal realm above and beyond the sky. From there they saw a radiant ring—probably the Milky Way—that bonded the two halves of the sky together and crossed another celestial circle, probably the equator.

Ananke, the goddess who personifies the obligatory force that drives the universe, is enthroned at the highest point of Plato's world system. Her name means "Necessity," and the polar axis is her spindle. Her daughters, the three Fates, or *Moirai*, are stationed out there with her. Surrounding this cosmos at equal intervals, these three seated women assist Necessity, who spins the world's destiny.

NAILING DOWN THE NORTH

The motionless stability of the north celestial pole is also expressed in the names and myths attached to the star now seen closest to the earth's pole. Although we usually call it the North Star, it is also known as Polaris, Latin for "Of the Pole." The pole in this case is just the place where the earth's axis hits the sky. Latin acquired the word from Greek. In that language, *polos* means a "pivot" or an "axis" and is related to the verb "to turn."

Polaris is a Latin name, but it wasn't bestowed on the star until the Renaissance. At the time of Rome, there was no conspicuous star near the pole. It took the slow wobble of precession thousands of years to point the earth's axis toward Polaris and give us the North Star we have now. This means the pole isn't quite as motionless as it looks, but in the short run, from night to night, from year to year, its precessional movement goes unnoticed.

At the present time, Polaris is quite close to the north pole of the sky, but not right on it. It is about 44 arcminutes from the pole, or a distance close to one and a half times the diameter of the moon as it looks to us. This also means Polaris isn't completely motionless. It actually travels around the north pole in a tiny celestial circle. But for most of us—and most traditional peoples—it is close enough.

For that reason, the Samoyedic-speaking peoples of northwest Siberia say the heavens turn around what they call the "Nail of the Sky." Hammered into place at the north celestial pole, Polaris allows heaven to rotate but won't let it stray. It holds heaven steady and supports the sky. Lapps, like the Samoyeds, are part of the Uralic Language group, and they call the North Star the "Nail of the North." Finns and Estonians use the same name. The Paleoasiatic languages spoken by the Chukchi and the Koryak in Siberia's far northeast are unrelated to the Uralic group, but they, too, say Polaris is the "Nail Star." An old Norse term for Polaris, *Veraldar nagli*, means "World Nail," and in ancient Persia, the North Star was known as the "Nail in the Middle of the Sky."

THE CALIFORNIA LOTTERY

Anthropologist Travis Hudson became convinced that Polaris, the "Star That Never Moves," was the chief ally of the Chumash in their effort to survive nature's yearly gamble with the food supply. The star's role as the stable center of heaven was apparently what convinced the Chumash that Polaris had a balanced, stable universe and their best interests in mind. According to Hudson, Polaris was really *Snilemun*, or "Sky Coyote." Speaking of Sky Coyote, María Solares said, "He watches over us all the time from the sky." He was, she said, like a god to the ancestors. These descriptions would apply to unmoving Polaris. Hudson's interpretation is therefore probably right, but we don't know for certain. By the time Chumash Indians were asked about their traditions, the Spanish, the Mexicans, and the Anglos had nearly obliterated the old ways of life.

María Solares said a number of the Sky People would get together each night for a few rounds of *peón*, a gambling game in which opposing teams tried to guess which hand of a member on the other team was hiding a stick. Sky Coyote was the captain of one of the two great celestial teams who gambled in the sky throughout the entire year for the fate of living things on the earth. His opponent was Sun, a powerful supernatural being who controlled the seasons and brought death to the world as well as life.

To the Chumash Indians, the North Star seems to have been recognized as Sky Coyote, a powerful celestial supernatural being whose luck in winning an annual gambling sweepstakes rubbed off on the Indians in the form of a bountiful year. This Chumash portrait of a coyote was painted on a natural wall of rock on the Carrizo Plain.

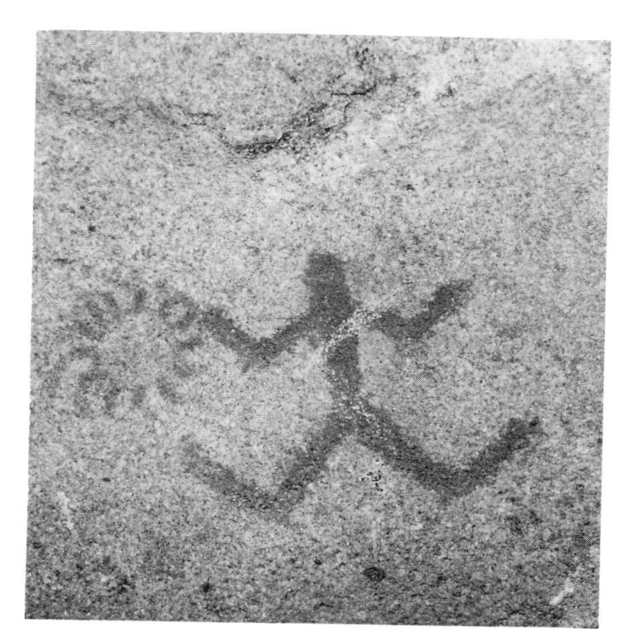

In order to improve their chances for acquiring ample food, the Chumash participated in seasonally significant ceremonies, presided over by shamans. This Chumash pictograph from the Carrizo Plain probably illustrates a shaman acquiring power from the sun or tempering its influence. (Photograph: Robin Rector Krupp)

If a player guessed right, Moon, a neutral observer, would award a counter stick to his team. The two sides would take turns guessing all night until one side won all of the counters. From night to night, moon umpired the games, kept score, and then tallied all of the statistics on the winter solstice to see which team had won the year's great game show in the sky.

The results of this year-round tournament made an impact on earth. Sky Coyote and his boys always wagered for the supplies in Sun's pantry. If the Pole Star's team took the match, it would be a rainy, abundant year on earth. As the scoreboard in the sky lit up with a final count in the North Star's favor, Sky Coyote would grab all of the goodies—geese, ducks, deer, acorns, chia seed, and other wild foods on which the Indians depended—and sweep them into the hole in the sky so that they would fall to earth for the benefit of those who lived there.

A first-place finish by Sun's team, on the other hand, meant real trouble for people on earth. When Sun wins the Big Spin at the winter solstice, he takes the jackpot in human lives. Winter and its solstice were dangerous, even in balmy southern California. At that time of year, there was little food to be found. The Indians had to depend on what they had harvested and stored. If it was enough, they would make it through the winter. If not, some would die. The world, as the Chumash saw it, walked a fine line. Nature's carefully tuned balance was symbolized in the cosmic gambling game, and sometimes their best interests slipped over the edge between life and death.

A THRONE AT THE POLE

The Chumash story of Sky Coyote's year-long contest with Sun illustrates the economic significance people might attribute to the motionless Pole Star and the balanced universe it was thought to uphold. Some people also put political meaning into the celestial pole. Its steady, unmoving station was a throne for Shang di, the "Sovereign on High," who ruled the cosmos from the summit of the celestial bureaucracy in Imperial China. Although he could not be seen, the north celestial pole symbolized his presence. All of the stellar functionaries circulated around him while, enthroned, he maintained the dynamic balance between *yin* and *yang* through judicious exercise of his power. His influence was felt in the cyclical patterns of time, in the seasonal transformation of nature, in the stability of the landscape, and in his endorsement of the incumbent emperor, the Son of Heaven.

Shang di's cosmic power sustained the emperor, and he acted as Heaven's agent. If the pole was the pivot of Heaven, the emperor was the hub of the world. As the terrestrial counterpart of the celestial sovereign, the emperor was obliged to exercise good judgment, maintain the calendar, and perform proper sacrifices at the appropriate season in temples designed to enhance the congruence of time and circumstances in every enterprise. It was his job to establish and maintain order, and he did so—through ritual, calendar corrections, and astronomical observations—with Heaven's mandate and the endorsement of the center of the sky. In

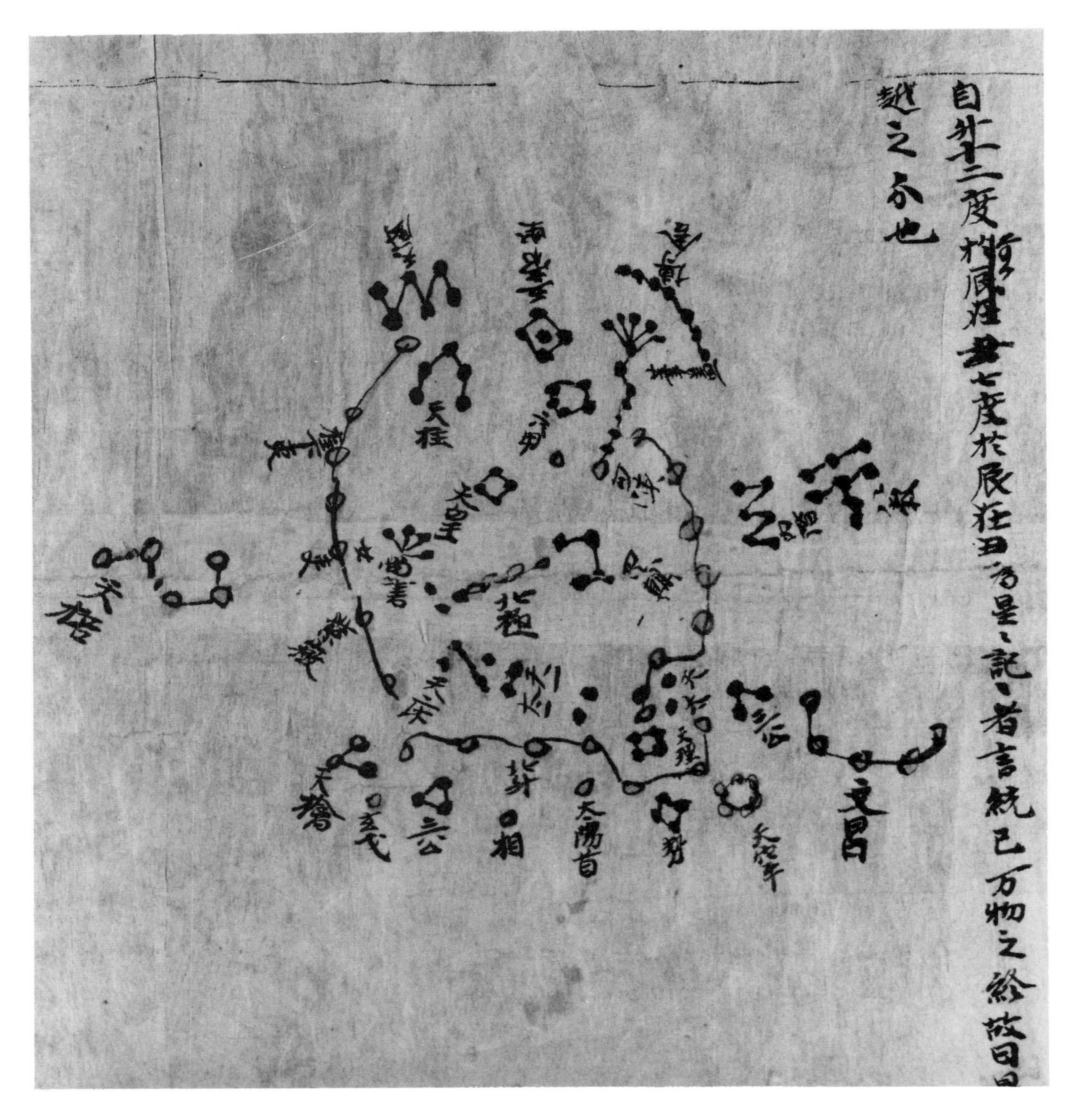

For the Chinese, the palace of the emperor was the terrestrial counterpart of the north polar zone of the sky. This star map from the Thousand Buddhas Grottos (Mogao Ku) near Dunhuang, in Gansu Province, was drawn in A.D. *940 and is part of the oldest surviving Chinese celestial chart. The manuscript was taken out of China by Sir M. Aurel Stein in 1909. Most of the figures on it are traditional Chinese constellations, which don't correspond to those more familiar in the West. It is, however, possible to recognize the seven bright stars of the Big Dipper in the dipperlike pattern near the bottom of the picture. The Chinese sometimes called this group the Northern Ladle or Grain Measure. (Stein ms. 3326, the British Library, British Museum, London)*

Imperial China, the earth mirrored the sky, and the two met in the capital. Because the Chinese capital—with its Imperial Palace—was the point of articulation between heaven and earth, the city was said to be the center of the world.

THE GUIDING LIGHT

Apart from helping maintain a balanced ecology for the Chumash and an effective government for the Chinese, the pole of the sky also provides immediate help as a compass for anyone trying to navigate through the night. The celestial pole not only creates the direction north, but with the help of Polaris it points out where it is. That, of course, is why we call Polaris the North Star, and that's what its name in Yucatec Maya—*Xaman Ek*—means, too. Its Navaho name, North Fire, is a little more colorful and no less informative.

Because the North Star signals the location of the sky's north pole and the direction north, it has guided travelers through the night. Chief Francisco Patencio of southern California's Palm Springs Indians wrote a book about the Cahuilla traditions of his people in 1943, and he said the North Star is "the most important star of all" and "guides all the world at night." Northern California's Pomo Indians regarded Polaris as the eye of Marumda, the creator. He watched people from his stationary post in the sky, and people kept track of his eye when traveling.

In the tenth century, Anglo-Saxon sailors called Polaris *Scip-steorra,* the "Ship Star," and used it to navigate. Related titles include Navigatoria and Steering Star. Speaking in Middle English between the twelfth and fifteenth centuries, people called Polaris the *Loode steere,* which meant the "Star That Shows the Way." Later this evolved into Lodestar. One of the old Greek names for the constellation Ursa Minor is Cynosura. In time, Polaris, at the tip of the Small Bear's tail, also acquired the name. Because the name *Cynosura* was also applied to Polaris, the English word cynosure has come to mean "something that strongly attracts attention" or "something serving for guidance or direction."

OMAHA INSURANCE

The ability of the celestial pole—and the North Star—to install direction into the landscape contributes to its status as a source of world order. The guidance and direction it provides in that capacity, along with the stability it generates as a steady linchpin of heaven, prompted Nebraska's Omaha Indians to turn the world axis into a symbol that enhanced social cohesion and tribal unity. Their myth of the Sacred Pole details how this happened, and Omaha ritual reveals how they kept that talisman of their society's equilibrium pointed at the celestial pole.

The story of the Sacred Pole of the Omaha was recorded in September, 1888, when Smoked Yellow, one of the last chiefs to whom it was entrusted, agreed to tell it to another member of his tribe, Joseph LaFlesche. Subsequently, the tale was reported in the Bureau of American Ethnology Report 27, *The Omaha Tribe,* by Alice C. Fletcher and Francis LaFlesche. This is the fundamental source on Omaha tradition.

The story begins during a time of tribal disruption, brought on by rivalries between Omaha chiefs. One of their sons left the village and went out on a hunt. While the council of chiefs was meeting to see if it could find a way to keep the tribe together and save it from extinction, the son of the chief lost his way in the forest. Disoriented and unable to return to camp, he pushed back the brush in an attempt to spot *Mika Em Thi Ashi,* the "Star That Does Not Walk." He hoped to use the Pole Star to guide his way home.

Instead he spotted a burning light in the distance. Thinking it was a campfire he headed for it. But when he got close enough to tell what it was, he discovered the light was produced by a burning tree. It was a supernatural tree, however, for despite the roaring flames around the trunk and throughout the branches, it remained unconsumed. When he touched it to test the heat of the fire, he discovered it was cool. Realizing he had encountered something extraordinary, he kept an eye on the flaming tree until dawn, and at sunrise, it returned to normal. The whole experience was so uncanny, he decided to stay another night and watch the fireworks once more. As the sky turned dark, the tree began to glow with the same magical fire.

By the next morning, the chief's son must have figured out how to find his way home, for he returned to his father and described what he had seen. They went back to the tree together to catch its next performance.

His father noticed that the Thunderbirds trailed fire when they arrived at the tree or departed. Each Thunderbird traveled from one of the world's Four Winds, and their flights and falling flame guaranteed four paths would be scorched to coincide with the four cardinal directions. Pilgrimages the forest animals had made to the tree had worn the four paths smooth with their footsteps.

Whenever the four Thunderbirds took their perches on the tree, it burst into the magical flames first seen by the chief's son, but the light from this fire could only be seen at night. The chief declared the tree to be a gift from Wakonda, the Great Spirit within all life and existence. Wakonda is the source of the world's rightness, or order. According to the Omaha, he "causes day to follow night without variation and summer to follow winter." They add, "We can depend on these regular changes and order our lives by them."

Concluding that the wonderful ever-burning tree would be just the emblem needed to keep the Omaha people from dispersing, the chief returned to the tribe to tell them about this miraculous tree. At his recommendation, a party of Omaha warriors went back to the tree in full regalia, ceremonially attacked the tree, cut it down, and brought it back to their village. There they reerected it and put responsibility for its keeping into the hands of a single clan, a clan designated to be the caretakers. Leadership of the tribe was invested in the keepers of the pole, and only through them could authority over the tribe be transferred. Through this tree, now transformed into the Sacred Pole, the threatened social order of the Omaha was strengthened. Problems, disagreements, and troubles were all to be brought, with presents and prayers, to the Sacred Pole.

Now, the Omaha really had a Sacred Pole that was handled ceremonially. It was the physical version of the tree mentioned in the myth, and we know it looks quite different from that tree because we have the pole; the Omaha turned it over to Harvard's Peabody Museum in 1888 for safekeeping. It is about eight feet long and about two inches thick, made of cottonwood, and very old. All bark has been removed, and both ends have been carved. The top has the shape of a small cone, while the bottom has a dull point. A scalp hangs from the end with the cone, and a small, hide-covered, twig basket containing cane feathers and down is bound near the middle. Fletcher and LaFlesche described a forked stick that accompanied the pole and was used to prop it up. They also said that "the pole was never placed upright but inclined forward at an angle of forty-five degrees."

Some modest detective work I undertook in 1982 revealed that at least some of the Sacred Pole's "supernatural authority" was really celestial authority. That remark about inclining it 45 degrees was the first clue. There is a simple astronomical rule that applies to the position of the north celestial pole. The altitude, or angle above the horizon, of the north celestial pole equals the latitude of the observer. Inclining the pole "about forty-five degrees," could make it point close to the north celestial pole if the latitude were right and if the Omaha aligned it to the north.

Omaha territory was centered at the confluence of the Missouri and Platte rivers, at about forty-three degrees north latitude. That's close enough for symbolism and the unaided eye. The angle is right, but how about the direction? Nothing explicit is said on this score by the Omaha, but there is additional information in Fletcher and LaFlesche's report. On ceremonial occasions, the Omaha arranged their camp according to a plan of clan organization based on their vision of the cosmos. Called *Hu'thuga*, this pattern specifies that a lodge for the Sacred Pole be built from reeds or poles from each family. More or less semi-circular, this lodge opened toward the center of the camp. The Sacred Pole leaned on its staff toward the center, and the sacred tent in which it was lodged was on the south side of the ceremonial camp. Taken together, all of these details tell us that the Omaha Sacred Pole faced north and therefore pointed at the north celestial pole. The Sacred

Pole was a cosmic emblem of the political authority of the chiefs, said to "hold the tribe together, without it the people might scatter." It was the "provider and protector of people," which corresponds exactly with the wonderful tree in the legend of the Sacred Pole.

SKY POLES

Many peoples, like the Omaha, have translated the concept of a world axis into a pole or pillar that props up the sky and underwrites world order. The pagan Lapps, for example, offered sacrifices to the Ruler of the World, the high god who made sure the sky didn't fall, at an outdoor shrine they called the "pillar of the world." Usually there was a split pole or close pair of stones at the site, symbolizing the celestial axis. One of these shrines, near Porsanger Fjord in northernmost Norway, was accompanied by an upright log. Planted in the earth next to the twin stone pillars, the log had an iron nail tapped into its top. In stood for the Sky Nail at the celestial pole.

The Lapp account of the world's end, told by Johan Turi in his book about about Lapp life and beliefs, documents the consequences of a collapsing pole. When on the last day of the world, Arcturus, the celestial hunter, shoots Polaris with his bow, the Big Dipper, "the heavens will fall, and then the earth will be crushed, and then the whole world will burst into flames, and everything will be ended."

In shouldering the sky, the celestial axis is also likened to the central pole that supports the canopy of a tent or to the pillar of a house that supports the roof. In Africa, for example, the Dogon center the cosmos on a world axis pillar they call Amma's House Post. Amma is the supreme god who created all worlds. Eskimos in Greenland equate the center pole of their tents to the pillar that holds up the sky.

When Nebraska's Omaha Indians erected the Sacred Pole, their talisman of tribal unity and community stabilizer, they aimed it toward the north celestial pole, the stable hub of the sky. An enclosure, open at one end, sanctified the location of the Sacred Pole, and symbolic figures of earth and grass were contrived on the ground. (Griffith Observatory)

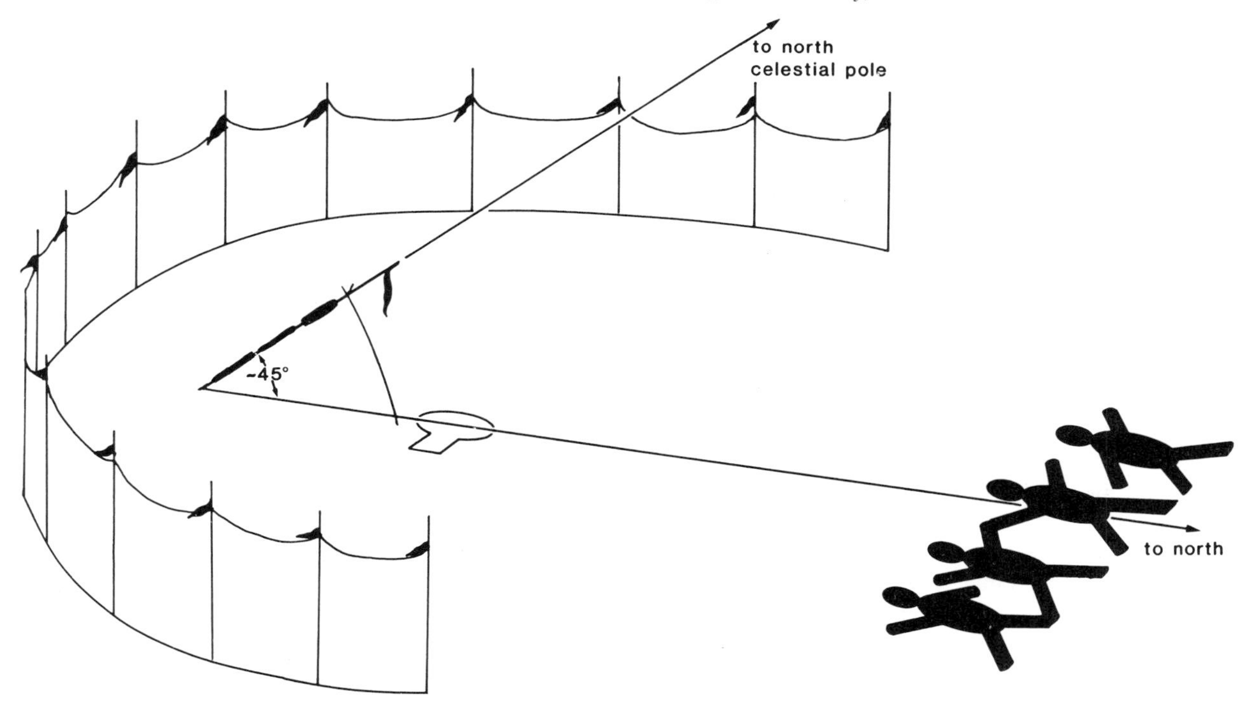

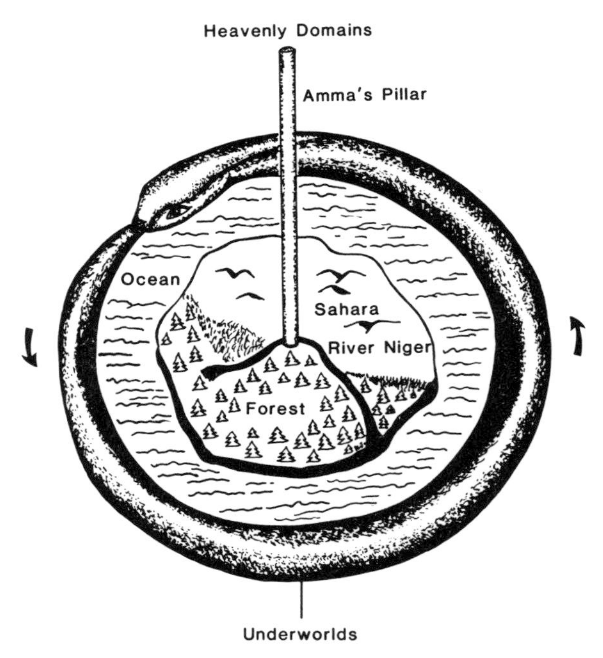

The world of the African Dogon is centered on a post that supports the sky, which they consider to be the roof of the house of their high god Amma. The earth is one of fourteen worlds that are skewered by Amma's pillar and that rotate around it. The Dogon think of the earth as a flat and roughly disk-shaped piece of land surrounded by the world's oceanic waters. Beyond the ocean, a great serpent swallows its own tail and perhaps symbolizes the cycles of time that renew the world as they are in turn consumed. (Griffith Observatory)

Any post responsible for holding up the sky must be strong and enduring. That is why the Kirghiz and West Siberian Tatars said the world axis is an iron pillar. It's an "iron tree" that matured with the growth of heaven and earth after the creation, according to the Yakuts.

All of these cosmic pillars had enough backbone to let heaven hold its head high. Residents on earth below, grateful that something in the universe was doing its job, knew the sky pole was valuable, and that is why so many other northern Asiatic peoples—the Mongols, the Buryats, the Kalmucks, the Northern Altay, and the Uygues, despite their different language groups—maintain it is a golden pillar. The Evenks (also known as the Oroqens in Inner Mongolia) call it a golden post. Copper represented value and wealth to the Kwakiutl Indians of the Pacific Northwest, and they pictured a copper pole connecting the earth with the sky and a realm below. It pierced the sky at the "door to the world above."

The axis of the cosmos in all of these systems was much more than a coat rack for the three kingdoms of the universe. It connected those subterranean and celestial realms with earth and transferred power from one zone to another. It also allowed those who knew how to negotiate that transcendental elevator to get off at any floor, at least in spirit. Mansi (or Vogul) shamans of northern Siberia climb seven stairs on their cosmic axis —one for each heaven—to reach the top of the sky. Northern Altay shamans also have to travel to the seventh heaven to reach the sky's summit and the home of the high god Bai Ülgän. In the Chukchi universe, there is a hole in every layer. All are aligned beneath the Pole Star, and a shaman can pass through these holes to get from one world to the next. The Pole Star itself is a hole in the vault of the sky, and through it a shaman can go to a higher heaven. The Siberian Koryak also said shamans could pass through the Pole Star on their way to audiences with celestial spirits. It was the central opening in heaven, the creator watched the earth through it. In shamanic traditions, then, the world axis had a spiritual dimension and facilitated a transcendental experience.

THE WORLD TREE

The Omaha Sacred Pole was certainly a pole, but it stood for a magical, symbolic tree. Because the Sacred Pole doubled as a world axis, the tree did, too. In some ways a tree is an even better image for a world axis. Anchored in the earth, it reaches straight for heaven. Its leafy branches broaden into a canopy that mimics the sky, while its roots dig deep into subterranean territory almost as good as the underworld.

Aboriginal tribes on the Great Australian Bight believe the sky is kept up by a giant tree named Warda. It had to be protected, in ritual and myth, to guarantee it would not let the sky drop on Australia. The Maya saw a world axis in the ceiba tree, which they called *yaxche*. This is the tree that supplies the kapok used to stuff pillows and life jackets. A gigantic ceiba is supposed to stand at the center of the world, where it connects heaven and the underworld with the earth. Souls of dead ancestors rise through the roots and ascend via its trunk and branches into the celestial realm. The sacred ceiba tree stands for the fifth world direction—up/down—and is the roost of Seven Macaw, the Big Dipper bird in the *Popol Vuh*. This connection with the Big Dipper supports the idea that the Maya world tree was aligned with the axis of the pole.

In ancient Norse cosmology, the world axis was symbolized by Yggdrasill, the towering transcendental ash that held the universe of nine worlds together. Its limbs umbrella the earth with a celestial awning, while its trunk passes through Midgard—our world—on its way to various levels in the underworld. Midgard is centered in this universe, and it is essentially a flat disk. A dwarf at each cardinal direction helps hold up the sky on the horizon, and Midgardsormr, the "World Serpent," curls around us in the surrounding world ocean.

Three separate roots from Yggdrasill reach into three different zones of the cosmos: one to the lowest world, which was a kingdom of clouds and death; one to the land of the Frost Giants, from which the cosmos originally was born; and one to Asgard, the home of the gods. Each root is watered by a sacred spring, or well. The one in Asgard is the headquarters and main sanctuary for the gods. They gallop there each day to hold council.

Asgard is the highest kingdom in Norse creation, and its ruler, Odin, is connected with Yggdrasill through the tree's own name. *Yggdrasill* means "horse of the terrible one." The "terrible one" is Odin, and the tree is his mount for supernatural journeys between the nine worlds. This world axis tree is, like others, a vehicle for spirit travel. It is probably connected with shamanic rituals, too, as a passage in the *Elder Edda* concerning Odin's search as a dead soul in the underworld for magical inscriptions of power implies.

A tall tree also turns the universe into a living system because the tree itself is alive. As an erect shaft, the tree may be equated with the male organ of procreation, and the world axis tree may then represent the male aspect of a system of world order that generates and sustains life. This, too, was part of Yggdrasill's character. The tree was said to support the existence of life, and it was directly associated with childbirth and the creation of life. It was an erect agent of life accompanied by nurturing female recesses: the wells that watered the stabilizing roots of the world tree.

A tree's ability to fruit and refoliate are processes of birth and renewal that reflect the female role in procreation. The female principle in the order of creation is therefore sometimes linked more directly with trees and the world axis. A Yakut myth, for example, describes the emergence of the Mother Goddess from the world axis tree.

IRMIN'S GERMANS

The world-ordering celestial power of the top of the sky was brought to earth by Chumash myth, the Imperial Chinese mandate of kingship, and the Omaha tribe's sacred pole. Shrines of the pagan Germanic peoples of Europe also directed this power of the celestial pole into the service of government and society. In the Irminsul, an immense wooden column that once stood at Eresberg, in West Germany, and in the great heathen temple at Uppsala, Sweden, the Norse concept of the world axis was duplicated in symbolic architecture and emblems.

According to Adam of Bremen, who wrote in the eleventh century, Uppsala's great shrine was an important place of assembly. It contained images of Thor, Odin, and Freya, and the Swedes met there, just as the gods gathered in judgment at the well of Urdr by the root of the World Ash. In fact, a huge evergreen grew beside the temple of Uppsala, to mimic Yggdrasill, and near its broad-reaching branches, a sacrificial well took the lives of drowned victims. Germanic people in other parts of Europe continued to gather for meetings at the base of a symbolic tree right into the thirteenth century.

The Irminsul, a Saxon version of Yggdrasill, was dedicated to the god Irmin. It fell in A.D. 772, when Charlemagne had his forces hack away at its

A monumental living evergreen tree grew beside the great temple of the pagan Swedes at Uppsala, and they drowned sacrificial victims in a nearby well. This sanctuary, therefore, acted as an earthly mirror for the divine realm, and the tree stood for the cosmic axis, the World Ash Tree. (From Historia de gentibus septentrionalibus *by Olaus Magnus, 1555)*

base. It took them three days to bring it down. The shrine, then, was already history in the ninth century, when the Irminsul's meaning was explained by Rudolf of Fulda. Writing in Latin in about A.D. 860, he said it was "a column of the universe that sustained all things." It symbolized, then, the world pillar and cosmic axis and represented the power of the god responsible for the world order. Charlemagne's motives for destroying the Irminsul are obvious. His act was not just a victory of Christianity over paganism. He brought down the entire world order of the Saxons and established his own authority.

We think we know what the Irminsul meant, but what about Irmin? Widukind, the author of a tenth-century Saxon chronicle, said Hermin, or Irmin, was the Saxon name for the god Mars, and he described a huge column they dedicated to Mars. This idea is strengthened by the name of the city in which the Irminsul was located; it is even now known as Marsberg. Earlier, the name of the place was Eresberg, which sounds like Aresberg, Ares being the Greek name for Mars.

Irmin is not really the Roman planetary god Mars, but the god the Romans interpreted as the Germanic counterpart to their god of war. The Norse counterpart of Mars was said to be Tyr, or Tiw as the Anglo-Saxons called him. Some scholars believe Tyr, under a name something like Tiwaz, was once the high god of the sky in northern Europe.

About forty or fifty towns in Lower Saxony had a mini-Irminsul dedicated to the god Ziu in the marketplace. These wooden or stone columns carried the image of a warrior and were called "pillars of Ziu." *Ziu* is the Germanic form of the Old English *Tiw*, and Tiw was also a warrior. That is probably why the Romans identified him with Mars.

There may be further evidence that Irmin, as Tiw (or Tyr), had a close connection with the north celestial pole that gave the Germanic world its axis and the Irminsul its meaning. Otto Siegfried Reuter, an amateur but disciplined German folklorist, collected and reconstructed old Germanic sky lore in *Germanische Himmelskunde* ("Germanic Sky Knowledge"), published in 1934. Reuter identified a relatively faint star, 32 (a.k.a. Σ 1694) Camelopardalis, in the celestial Giraffe, as the old Norse lodestar in use about A.D. 800 and said it was commonly known as Tyr. In the eighth and ninth centuries, Polaris was still not very close to the pole, but 32 Camelopardalis was. Tang dynasty astonomers in China were using it in the same era as their North Star.

According to Reuter, Tyr as the Pole Star guarded the celestial pole—and therefore the world axis and the entire world order—from the savage jaws of Fenrir the Great Wolf. Ancient prophecy had promised Fenrir would swallow Odin at the time of Ragnarök and world's end, and so when the wolf was still young but growing, the gods of Asgard decided to shackle him. Fenrir thought he could break any cuffs the gods could devise, but while testing one, he took Tyr's hand in his teeth to guarantee the gods would let him loose again. The gods had no intention of freeing the wolf, and when the fetter held, Tyr lost a hand.

Reuter said the ancient Teutonic peoples visualized Fenrir's jaws as a semicircle of stars in Andromeda, Pegasus, and Cygnus. The jaws are biting into the Milky Way right about where it splits, and the Milky Way is two streams of saliva flowing from the wolf's jaws. If Fenrir could put the bite on the sky's north pole, the divine throne of the high gods just beyond his teeth, the world would tumble. Tyr, however, protects the sky from the terror of the wolf, and the Milky Way dribbles across the night from the waiting celestial maw.

THE COSMIC MOUNTAIN

The same things that make a pole, pillar, and tree out of the world axis also turn it into a mountain, a cosmic mountain that supports the sky and is rooted in the earth. High mountains seem to touch

When Christianity replaced Teutonic paganism, the cross replaced the Irminsul as the chief symbol of world order. In this relief depicting the lowering of Christ's body, the cross is the central image and dominates the scene. The second-class status of the Irminsul is also documented, for it is the Irminsul, shown as a stylized tree bent over in defeat, that has allowed Nicodemus to reach the body of Christ and lower it to Joseph of Arimathea. Christ's mother, Mary, holds his head, while the sun and moon weep on high. The scene is carved on Germany's Externsteine rocks, the site of twelfth-century shrine grottos used by monks. Some believe that rock-carved chapels here earlier served as a pagan solar observatory. Below the foot of the cross and the base of the Irminsul —and out of the picture here—both are rooted in the same subterranean column, around which coils a serpent. (From Der Externsteiner Beneficiat Konrad Mügge *by K. Flaskamp, 1953)*

the sky and are said to be homes for celestial gods. Climbing a mountain carries a pilgrim closer to the sky and makes the idea of celestial ascent seem real. As the source of water and the stage of storms, the mountain sustains life. The springs on its flanks and pinnacles make it a reservoir of fertility. Lightning in remote peaks demonstrates that mountains transmit celestial power to earth.

A place where heaven connects with earth is sometimes called the "navel of the earth." The world is moored to the sky there as an unborn child is linked to its mother because a navel is a place where life and power are exchanged. Celestial order, power, and authority—conveyed by the world axis umbilical cord—converge upon a "navel" of the earth and make it the center of the world. The center is where heaven is in contact with earth, and that place is made sacred by the contact.

We find this concept in Mesopotamia, transformed into mountain-like ziggurats dedicated to celestial gods. In Mesopotamian cosmography, the world's central hill, like the world axis, connected earth and sky. The ancient Egyptians mimicked mountains with cardinally oriented pyramids that housed the bodies of their pharaohs and provided their souls with a mechanism for reaching heaven.

Mesoamerican pyramids were also artificial cosmic mountains, and they could symbolize the realms of heaven in the stack of platforms they pushed into the sky. They were places for communicating with the gods and were the sites of creation. Aztec destiny had its mythological origin on a cosmic mountain, and the Aztec transformed that mountain of birth into a pyramid—Templo Mayor—at the heart of their capital and the center of their world.

The world mountain navel and high sacred place were also known in ancient Israel. The Chumash of southern California believed that Mount Pinos, about an hour and half northwest of Los Angeles, in the Los Padres National Forest, was the center of their universe. The Greeks said their gods lived upon Mount Olympus, the highest mountain in Greece. Concealed in clouds and able to hoard snow all year round, it is, according to Homer

A stone cairn provides the summit of Chicoma Mountain in New Mexico with a Tewa miniature mountain shrine. The pile used to be larger, and a tall pole of spruce reached skyward from it. Perhaps once symbolizing the world axis or one of the directional trees said to support the sky at the world's corners, the pole is now long gone, but a small piece of wood lodged between the top stones seems to serve the same purpose.

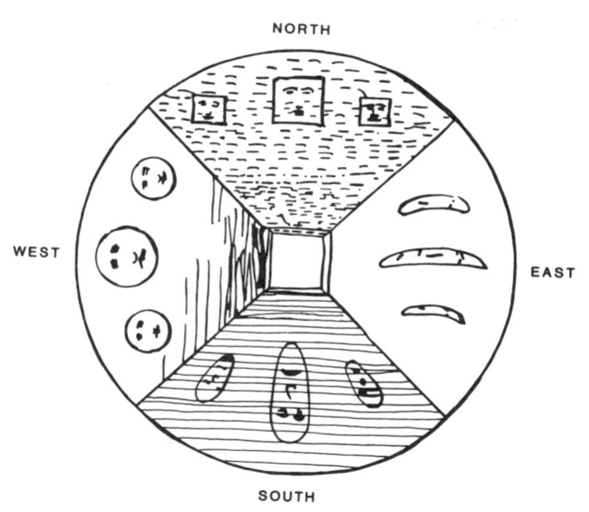

Kalmuck people in Siberia believe the world is centered on a great cosmic mountain they call Sumer. Its truncated summit is represented by the square in the middle of this picture they draw of their world. Each side of the pyramidal mountain emits a different color: south—blue, west—red, north—yellow, and east—white. Each region of earth that spreads out beneath one of the sides of the mountain reflects that side's light, and the people that live in those four "continents" differ in the shape of their faces. In the south, faces are oval; in the north, they are square. People living in the west have circular faces, and crescent heads are common in the east. The Kalmuck picture of the world is really a picture of world order and is, therefore, similar to a mandala. (Griffith Observatory, after The Mythology of All Races, *vol. 4,* Fionno-Ugric, Siberian *by Uno Holmberg)*

(*Odyssey,* Book VI, lines 42–43) "firm and unmoving forever." On Bali, all power is focused on the island's highest mountain, Gunung Agung (Great Mountain). World directions emerge from it and impose lines of supernatural force upon the landscape. Cosmic mountains also showed up in China, where they marked the world's corners and center.

Mount Meru is the true polar axis in Hindu belief. A vast mountain at the center of the flat world disk, Meru is the home of the gods, or *devas.* It is located at the world's north pole, and the Pole Star shines from its summit. The demons, or *asuras,* live at the south pole, and are as far from the gods as they can be. The Milky Way, a celestial Ganges, flows through heaven toward the top of the sky. The Pole Star welcomes the river, and it cascades to the top of Mount Meru. At the summit, its currents divide, and it flows down to the four quarters of the world. Other accounts of this mountainous world pillar place the whole disk, Mount Meru and all, upon the backs of four elephants—solid, strong, and cardinal upholders of the world. The elephants stand upon the shell of a great tortoise, and the tortoise rests upon the circled body of a great cobra.

A cosmic axis shows up in the key event in Buddha's life. His enlightenment, traditionally dated to 528 B.C., puts him under a symbol of cosmic order—the bodhi tree. The leafy canopy of this Tree of Knowledge symbolizes the divine character of the sky, and its trunk channels the awareness and compassion of enlightenment to the Buddha meditating at its base. Buddhist stupas are domed shrines built to honor Buddha, and they mimic the universe with symbols of its elements. The dome itself stands for the sky, and the umbrella pole that emerges from the top is the polar axis.

CLIMBING THE COSMIC MOUNTAIN

Javanese Buddhists saw a story of spiritual pilgrimage in the cosmic mountain and contrived an artificial mountain, a monumental representation of the spiritual and cardinally oriented mountain of the world, at Borobudur, about 25 miles northwest of Yogyakarta in central Java. The structure is about 1,100 years old.

Borobudur is acutally the world's largest stupa. It is also the largest ancient monument in the southern hemisphere. Built by the Sailendra dynasty in the ninth century A.D., it was abandoned soon after it was completed because the Buddhist kingdom in Java was overthrown by the Hindus. The structure is a small mountain, about 105 feet high and 403 feet long on a side, and it contains about two million cubic feet of stone. It is covered with 1,500 elegantly carved pictorial reliefs that depict scenes in the Buddha's life, his teachings, and other spiritual matters. There are also 1,212 ornamental niches and 400 more niches that originally contained a statue of the Buddha.

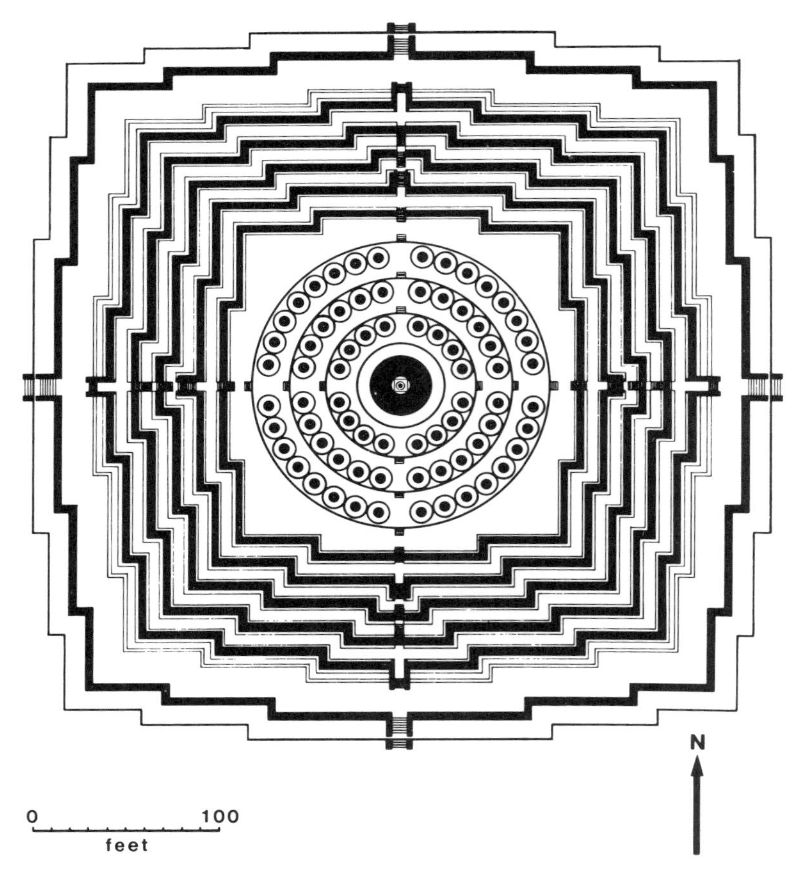

Borobudur, a medieval Buddhist shrine in central Java, was planned to resemble a mandala. It is cardinally oriented, and it transported pilgrims from their earthly existence to the celestial realm of spiritual enlightenment. (Drawing: Joseph Bieniasz, Griffith Observatory)

The plan of Borobodur is like a Buddhist mandala. It incorporates the circle that represents heaven and the square that stands for the earth. The four main stairways coincide with the cardinal directions. Ten terraces lead to the top, and each floor represents a stage in a person's life and spiritual development. Pilgrims walked its stairways and terraces, circling the center nine times, to reach the top.

A climb to the top of Borobudur was supposed to include a promenade around every level. The total ascent consumes three miles of turns and stairs, but through most of the journey you never see more than a few steps ahead of where you are. Finally, at the top, the terrace opens to the sky with a liberating effect. Spacious and architecturally abstract, the highest level is intended to awaken a change in consciousness, a kind of enlightenment. The physical message is clear, and it clarifies the spiritual message of the climb.

ERECT AT THE ZENITH

The persuasive world-orienting power of the polar axis is evident in the beliefs of many peoples of the world. Myths, rituals, shrines, and monumental architecture have been based upon the capacity of the polar axis to provide structure and stability to human society. It is, however, possible to contrive other world axes that perform the same function. The other most obvious "top" of the sky is the actual zenith, and a cosmic axis that rises vertically to the zenith is commonly encountered in the tropics. This is partly because the celestial pole (north or south) is low in the sky and less prominent. In addition, the sun in the tropics can pass through the zenith, and that is something it cannot do outside of them. For tropical peoples, solar zenith passage is usually an important station in the calendar, and that enhances the importance of the up-down zenith-nadir axis. In fact, it gives the world axis a seasonal meaning that directly involves the sun.

On the equator, the equinox sun rises due east and climbs to the zenith at noon. Because the equinox itself is a significant turning point in the seasons and the year, the zenith axis and the noontime equinox sun activate the story of creation told by the Desana and other Tukano tribes who live near the equator in Colombia.

According to the Desana, Sun Father was working his way west and upriver on a search for just the right place to create human beings. He was looking for the spot where the rod of his stick rattle would stand erect, without any tilt, and as he traveled, he stopped at likely locations and inserted the rattle into the ground. The rattle symbolized Sun Father's reproductive equipment, and so he was actually inseminating the earth with the rattle.

The rattle also stood for the vertical axis of the world, and in a properly organized cosmos it would stand upright at the world's center. Sun Father kept looking for the center, the place of creation, and when he finally found it, his creative apparatus stood upright. This was, the Desana say, his "yellow intention," and he penetrated the earth with the fertilizing power of his yellow light.

Performing this important act was also an exercise in timekeeping, for the Desana say Sun Father was "measuring the center of the day," which means high noon, when the sun is directly overhead. Its rays drive straight into the ground, and a vertical shaft casts no shadow.

The Barasana, another Tukano tribe, believe they know exactly where Sun Father's act of creation occurred—at the Rock of Nyí, a large triangular boulder near the Pira-Paraná River's Meyú Falls. Petroglyphs on the rock illustrate, in Barasana style, the intimacies of Sun Father and mother earth.

Just when Sun Father was able to get it up, around the time of the equinoxes, the rainy seasons begin in Tukano territory. With the rise in the level of the river, fish become scarce. They head upstream to spawn, and in fact, during these seasons the Desana believe the world is incubating for rebirth and the dry season's renewal of the food supply. Life has been quickened in earth's womb and is being seasoned there until it is ready to emerge.

The places in the sky where the sun's path

Ascending Borobudur means climbing a cosmic mountain. The main stairways coincide with the cardinal directions. Beyond this portal, the shrine provides the pilgrim with an unexpected vista of open sky and a pristine landscape of abstract symbols: ascending circles of bell-shaped stupas. The largest of these shrines, at the summit, is visible at the top of the stairs.

crosses the celestial equator and permits Sun Father to pass overhead at noon are, in Desana terms, a pair of copulating anacondas. Real anacondas move upriver at these times of year. Migrating through the seasonally deepened river to mate, they swim at night. As they travel west, on the trail of the sun, these serpentine symbols of seasonal creation raise nearly two-thirds of their huge dark bodies out of the river and slam back down on the waves like falling logs. These spooky thunderclaps of the anaconda water ballet can be heard in the Desana camps, where the Indians explain what's going on: The big snakes are just lifting their heads for a better look at the stars. They, too, are parties to a seasonal, world-ordering event defined by the erect axis of the equinox sun.

GOLD OF THE INCA

The mythical origin of the Inca empire, like the Desana story of the creation of the people, takes advantage of the zenith passage of the sun and the seasonal dimension the vertical axis has in tropical latitudes. In the case of the Desana, we have hunter/gatherers whose lives are framed by the arrival and departure of the rains. To the Inca, their agricultural production was the foundation of their empire. In his *Royal Commentaries of the Inca*, Garcilaso tells us that Manco Capac, the founder of the Inca dynasty, was looking for a place to center an empire. Manco was Sun's first son, and his father's instructions required him to find the place where a golden spike would sink straight into the earth.

Manco Capac and his sister and wife, Occlo Huaco, descended to earth at Lake Titicaca. They headed north from the great lake's Island of the Sun. That was also where the world first saw sunlight break through the darkness of the primordial flood. With the gold wand—about as long as man's arm and two fingers thick—the pair were looking for a place to call home. Each day on their journey they put the rod in the ground, but it failed to penetrate. At last they came to the valley of Cuzco. This time the golden pole slipped right into the earth and was gone for good.

Manco Capac and his wife knew that Cuzco must be the place. At that very spot, he built the *Coricancha*, or "Enclosure of Gold." It is now better known as the Temple of the Sun. This sanctuary became the heart of the Inca capital, whose name means the "Navel of the Earth." It marked the center of *Tahuantinsuyu*, the "Land of Four Quarters." Located at the confluence of two rivers, it was also like the "center" of the Milky Way, where Quechua today say its two streams meet, and like the "center" formed by the intersection of the directional axes of the village of Mismanay.

This story makes a world center out of imperial capital with a world axis aimed at the zenith and aligned by the sun. The erect rod, its golden color, and its vertical penetration of the earth all suggest the same kind of sun-inspired, fertilizing, order-establishing, creative event as described by the Desana in the northwest Amazon. At the latitude of Cuzco, however, about 13½ degrees south, the sun does not cross through the zenith on the equinoxes, as it does at the equator. If a zenith sun drove the gold rod into the ground in Cuzco, it had to do so in February or October. Also, the sun is at the opposite end of the axis, at the nadir, at midnight, in April and August. Is it likely that any of those dates would have any bearing on the founding of a world navel?

Those dates may be telling us that the real message of the myth is that agriculture cultivates empires. Anthropologist Gary Urton has emphasized the importance of the period between the middle of August and the end of October for agriculture in the Andean highlands near Cuzco and associates the saints' days that define this planting season to the midnight passage of the sun through the nadir in August and the noontime passage of the sun through the zenith in October. Crops planted too early before August may not survive winter frosts, and planting after the end of October may not give the crops enough time to mature before the next winter. Also, in an attempt to reconstruct the astronomical monuments and alignments of Inca Cuzco, astronomer Anthony F. Aveni and anthropologist Tom Zuidema have argued in favor of the significance of the sun's midnight nadir passage on August 18. This was said to be the time to plant in Cuzco. Taken together, these facts suggest that the procreative act of planting, in har-

In Cuzco, the center of the world was said to be the Coricancha, *which contains the "Tabernacle," the large, partly opened niche to the left of the door. The Tabernacle's stone blocks carry odd grooves that perhaps had something to do with the gold plates and settings of gems that were attached to it. Garcilaso de la Vega said that the Inca emperor took a seat in this gold-encrusted niche during the festivals of the sun (at the solstices), and Pedro de Cieza de León reported that the sun would hit the niche when it rose. (Photograph: Robin Rector Krupp)*

mony with the zenith and nadir sun, was part of the story behind the sun's gold rod that dropped straight into the earth, disappeared, and germinated an empire.

ENDING THE WORLD IN ANCIENT MEXICO

We depend on these cosmic axes to stabilize the world, but some people saw a potential for catastrophe in them too. Expectations of world doom were entertained by the Aztec every fifty-two years at the calendar completion ceremony they called the Bundling of the Years. On the appointed night, priests and other participants kept a vigil on top of Cerro de la Estrella, the "Hill of the Star," south of Mexico City, near Iztapalapa. They would watch the Pleiades climb to the top of the vertical world axis toward a midnight transit through the zenith. There, at the top of the sky, the Pleiades had the power to close down the universe.

At the time the years were bundled, the Aztec celebrated the New Fire Ceremony. All fires in the city—in temples and halls and homes—were extinguished. In a strange mutation of spring cleaning, the Aztec smashed used pottery, swept their houses clean, and threw hearth stones and votive

Four Aztec priests each transport a bundle of sticks to the altar of a temple whose fire has been ignited anew. These packages symbolize the 52 years that in their passing bring the 260-day calendar and 365-day calendar back to the same starting points. The New Fire was kindled on top of the Hill of the Star. Starting at the zenith, the cosmic axis reached down to earth and intersected it on that mountaintop. Its temple was charged with special meaning when the Pleiades made a midnight rendezvous with the zenith. (Griffith Observatory, after the Codex Borbonicus, *an early post-Conquest Aztec manuscript now in the Bibliotheque de l'assemblée nationale, Paris)*

statues into Lake Texcoco. They were either getting ready for the end of the world or the start of a new season.

A great procession of priests left Templo Mayor earlier in the day to make it to the top of the Hill of the Star in time for stargazing and proper ceremony. With them walked a sacrificial victim. If, at midnight, the Pleiades continued to cross high over the meridian—about 2 degrees off the hill's zenith—and stay on course to the west without the inconvenience of world cataclysm, the priest of Copulco would know that the Aztec had signed another fifty-two-year lease with the calendar. They could wait for the next complete set of zeroes to roll up on their cosmic odometer before worrying about another world's end.

The priest then took the fire drill he had brought with him and worked up a small flame on the breast of the victim. Once the fire was kindled, the victim's chest was abruptly opened with a flint knife. His heart, still beating, was removed and dropped in the fire. From these flames a bonfire was lit upon the temple altar, and it supplied the sparks that were carried back to every temple and neighborhood to provide new fire for every household.

The last New Fire Ceremony was celebrated at the Hill of the Star in 1507. Calculations tell us that a midnight zenith passage of the Pleiades had to occur on about November 14. Close to that date, on November 18–19, the sun passed through the nadir at midnight. So within the limits of the naked-eye observation, the sun was at the bottom of the axis—perhaps dead in the deepest and darkest underworld, or at least hiding its light under a bushel—when the Pleiades were up on top. If the Pleiades continued to the west without stopping the world, the sun might manage another sunrise. We don't know if the Aztec really thought this way about the axis and the Pleiades and the sun, but it is not so unreasonable. The sacrifice of the victim and the kindling of the New Fire both convey the idea of rebirth through necessary death. That victim's death was the end from which new beginnings emerged. Time was rekindled in the calendar, and somewhere, deep in the underworld, a rekindled sun had set out for the dawn.

TOPPLING THE POLE

Two historians of science, Giorgio de Santillana and Hertha von Dechend, wrote an entire book about cataclysms of the world axis and related ancient lore. Titled *Hamlet's Mill* and published in 1969, it attempted to show that much of the old mythology contained coded astronomical ideas and knowledge. Its authors believe a very ancient system was at work in the minds of our ancestors. The Big Dipper, the Milky Way, the Pleiades, the congruent cycles of the sun, moon, and planets, the world axis, and all of the other signs in the sky were linked in a vast novel of a tale—not just a collection of short stories—about the world order and what happens to it. In effect, de Santillana and von Dechend say that much of what we find buried in myth, legend, and fable is recognition of the shifting frame of the cosmos—what we call the earth's precession—long before the phenomenon was discovered and calculated by the Greek astronomer Hipparchus in the second century B.C.

Because of precession the world's steady axis is not so steady after all. Gravity's influence—from the sun and the moon—on the earth's equatorial bulge makes the earth wobble. This slow wobbling gradually swivels the polar axis around the sky. It points in one direction at one time and elsewhere later—gimbaling like a figure skater a little out of balance—and traces a circlet in the sky in twenty-six thousand years. Precession is what made 32 Camelopardalis, instead of Polaris, the Pole Star for the ancient Norse and the Tang dynasty Chinese and what gave the Greeks and Romans no pole star at all, despite what William Shakespeare anachronistically wrote in *Julius Caesar.* Asked to reconsider the banishment of one of the Roman nobles, Caesar says he will not be moved because he is "constant as the Northern Star." Hipparchus, of course, knew there was no North Star in his era and wrote, "at the pole there is no star at all."

JUST MILLING AROUND

In a number of ancient stories about the sky, the fundamental rotation around the celestial pole is likened to the uniform, circular grinding motion of a mill. Cleomedes, for example, authored a Greek manual of astronomy, probably in the first century B.C., and wrote, "The heavens there turn around in the way a millstone does." The Ostyaks of western Siberia have their own version of the heavenly mill. It grinds dependably and unattended. Its axis is a golden pole that extends all of the way to the sky. At its top is the golden cage where we find the Nail of the North.

Norse *Edda*s also seem to compare the spinning of the celestial axis—or the god that makes the sky turn—to the grinding of a mill in the name Mundilfari. It combines a word that means the sweep, or handle, of a mill *(möndull)* with the verb "to travel or move" *(fari).* Mundilfari was the father of Sol *(Sun)* and Mani *(Moon),* and his name tells us that the turning handle, or axis, of the sky is the parent of the sun and moon. By turning, the sky gives each of them life on the eastern horizon.

A magic mill, or something like it, is a central element in the *Kalevala,* a collection of traditional Finnish narrative poems or folk songs. It is called the *sampo,* and a careful review of its symbolic meaning indicates that it shares many traditional attributes of the world axis. Its name appears to be rooted in words that mean "pillar, post" and "prop, mainstay." It has a "lid of many colors" that goes around and around whenever the *sampo* is operating. So the lid spins on the *sampo* in the same way the multicolored sky spins around the polar axis. Somewhat enigmatic, the *sampo* is a one-of-a-kind magical object that is really three mills in one. It can grind out never-ending supplies of grain, money, and salt from three separate compartments. Prosperity, then, is its most important product, and that prosperity is like the stability and well-being that are guaranteed by the steady axis of the world. The *sampo* plants three roots—like the world axis tree Yggdrasill—to lock itself into position. One root goes to the sky. Another digs into the earth. The third plugs into a whirlpool, which probably represents the underworld.

We know Hamlet best from Shakespeare's play, but Shakespeare borrowed his Hamlet from the Danish historian Saxo Grammaticus. Saxo's "Hamlet" was Amleth, a legendary Danish hero. The story of Hamlet's mill is included in Saxo's *Danish History* and in Snorri Sturluson's *Poetic Diction,* where its owner is identified as Amlodhi. *Hamlet's Mill* traces these literary associations and concludes that Amlodhi, the original Icelandic owner of the mill, was really just an earlier incarnation of Hamlet. Both Snorri and Saxo say Amlodhi's mill is an

"Ocean-Mill" out beyond the earth or the horizon and that nine remote maidens run the machine. It is actually the eternally pounding sea that crushes stone islands to powder and salt.

Snorri also provided a longer account about another mill known as Grotti ("crusher"), which was owned by Frodi, the king of Denmark. Two female slaves—both of them Nordic amazons—were the only ones strong enough to drive the mill, and they ground out enough gold to make Frodi prosperous and powerful. He used these resources to impose the Peace of Frodi. During this golden time, all of the northern nations avoided war, and crime ceased to exist. It couldn't go on like that indefinitely, of course. Northern Europe is, after all, Viking territory. Frodi's kingdom was invaded by Mysing the Viking, who killed the king and stole the magic mill. He ordered Menja and Fenja to grind salt. They kept grinding until midnight, and by then there was a huge mountain of freshly ground salt on Mysing's ship. The women asked if that weren't quite enough, but he told them just to keep grinding. In a short time, the additional salt proved to be too much for the boat, and

> . . . the ship sank, and where the sea poured into the eye of the hand-mill was a whirlpool there afterwards in the ocean. It was then that the sea became salt.

In these stories, the magic mill is closely associated with the sea, and de Santillana and von Dechend believe the sea is the link to the precession and the displacement of the pole. The whirlpool in the sea, they say, is the symbol of disturbance of the pole. When the axis slips, there is churning trouble at sea. In *Hamlet's Mill* they contend that most treatments of this whirlpool share the same basic themes. First, the whirlpool is usually found near a tree (usually a Tree of Life), the axle of a mill, or some other symbol of the world axis. Second, it forms when something pulls the plug. By disturbing, removing, or relocating the tree or pole, water starts swirling down the drain. And third, the whirlpool is actually the entrance to the underworld and the realm of the dead.

Hamlet's Mill traces that stream that gurgles down the cosmic drainhole to an intriguing water reclamation plant—Tartarus and the Depths of the Sea. Tartarus, we know, is the deepest realm of the underworld, as far from heaven as you can get. *Hamlet's Mill* then attempts to extract a celestial location for Tartarus, in the southern half of the sky, from ancient star lore. Tartarus, the book convincingly argues, is the star Canopus, at one time in Argo the Ship. Now that the ancient constellation of the Ship has been dismantled by modern astronomy, Canopus is the brightest star in Carina the Keel. It is also the second brightest star in the sky. And—and this is important—it is located fairly close to the south pole of the solar system, or the south ecliptic pole. For that reason, Canopus is used to navigate spacecraft to other planets.

The north and south ecliptic poles are the centers of the polar circles of precession. Anything near them does not appear to precess. To the ancients, then, Canopus may have seemed exempt from precession. The south celestial pole traces its circle of precession around the neighborhood of this star, and the north celestial pole runs its ring around the north of the solar system, which also stays in one spot, guarded in the first coil of Draco, the Dragon. That means the track on which the sun, moon, and planets run—the ecliptic—remains in place compared to those two points, the north and south ecliptic poles. In a cosmos with a wobbling polar axis, the ecliptic axis is steady.

De Santillana and von Dechend rightly realized the ancients were deeply concerned about the sky because they believed the sky is the home of gods who establish the order of the world, enforce it, and defend it from threats of chaos. Their book is rich and interesting but not easy to read. Many different themes and an extraordinarily large and diverse collection of data fold over each other in its chapters like some origami nightmare. The arguments sometimes appear out of nowhere, end like a broken cable, and re-emerge unexpectedly elsewhere down the line. It is hard to prove that they are right. Much of *Hamlet's Mill* displays impressive learning, but many of the detailed examples —particularly those from the New World—don't hold up quite so well under close scrutiny. Many of the basic celestial themes the book identifies are genuine, however, and its authors make a good case for more astronomical symbolism in some myths than others have been willing to recognize. Their effort to unravel some new threads of archaic cosmology appears to be successful despite being diluted by unrelated, incomplete, or misinterpreted examples of what they are trying to prove.

GRABBING THOSE GOLDEN APPLES

De Santillana and von Dechend concluded that the ancients recognized the two ends of the ecliptic as divine and transcendent destinations of souls. The one in the north, in Draco, is affiliated with Paradise. The other, in the south, marks the subterranean kingdom of the dead. This cosmic axis allows us to extract the spiritual geography embedded in the constellations of Hercules and Draco and to understand just why the golden apples of Hesperides were so valuable.

The constellation Hercules depicts the hero kneeling with the Dragon underfoot, and that is the most important thing about this celestial configuration. When Heracles went to the land of the Hesperides to fetch the apples—one of his twelve labors—he found the garden was guarded by the dragon. Hera had assigned the dragon to this task because the Hesperides, the daughters of Atlas, were always sneaking into the garden and palming the apples. Heracles killed the dragon and appropriated the apples himself. The dragon was placed in the sky as a constellation.

Draco the Dragon was the guardian of Paradise, the realm of immortality. In the sky, he is still a guardian, only now he guards the celestial Paradise and the same tree that fruits apples of eternal life. That tree is a celestial axis, but it's not the polar axis. It is the rock-steady ecliptic axis, and in this context the apples that hang from its boughs are certainly the immortal stars and divine planets. Draco protects the unprecessing north ecliptic pole in its coils. That pole does not swivel through the sky because the Dragon is now immortal and eternally vigilant. If that weren't enough to lock him in place, the foot of Hercules the Strongman keeps his nose to the heavenly grindstone of the world mill.

Traditional drawings of the constellation Hercules often show him grasping a peculiar bough that bears serpent heads or apples or both. This woodcut from an edition of the myths of Hyginus published in Venice in 1485 clarifies the symbolism. Hercules is coming to terms with the dragon responsible for protecting the golden apples of Hesperides. This apple tree is another celestial axis, but it is not the polar axis. The reptile guards the ecliptic axis, whose north pole is among the stars of Draco the Dragon. Draco's coils keep the ecliptic pole in place, while the cosmos precesses around it, and in the later star maps, Hercules, with his foot on Draco's head, keeps the Dragon in line. (From Les Étoiles et les Curiosités du Ciel, *by Camille Flammarion, 1882)*

18

Down the Chimney

IN our time, we see the universe as a concoction of quasars, clusters of clusters of galaxies, undetected dark matter we believe must exist, and vast empty zones that make the whole works look like the inside of a torn and ragged sponge, laced with filaments and tunnels. Because the light from every distant object takes time—even traveling at 186,000 miles per second—to reach us, the farther out in space we look, the further back in time we see. Galaxies sail into the darkness, and each one sends us a postcard from a different moment of time. The universe we see around us is really a snapshot of all those different instants in the history of the cosmos.

The background of that picture is fogged by the faded glow of light emitted about three minutes after the Big Bang. That explosive expansion of a bubble of false vacuum occurred some time between ten and twenty billion years ago, and the physical laws that govern the behavior of subatomic particles now allow us to describe what the universe was doing from the first ten trillionths of a trillionth of a trillionth of a second (10^{-35} second) after its eruption into something we are able to discuss scientifically. We now speak of an inflationary universe that loomed out of absolutely nothing in strict accordance with physical law. As the cosmos cooled, the membrane of spacetime we presently inhabit congealed into what we now describe with the mathematical formulations of general relativity and quantum mechanics.

This testable scientific account of the cosmos is the most accurate description of nature yet devised, and there is no room in it for the world axis, cosmic trees, shamanic ascents to heaven, celestial signals of momentous events, or cyclical renewal of time and the earth. All of these themes are still embedded, however, in familiar traditions that still guide our behavior. Unrecognized, they reappear, for example, every December, when old sky stories and elements of ancient cosmological thinking show up again in time for Christmas in ornamented trees, Santa lore, and the tale of the Christmas Star. Usually crowned with an angel or a star, each Christmas tree reaches for heaven like a world axis, and every chimney Santa Claus descends is a channel between earth and sky.

DECORATING THE TREE

Although references to Christmas trees in England can be found as early as 1789, most people did not observe the tradition or even know about it then. The Christmas tree, as we know it today, did not become a popular custom in English-speaking lands until Prince Albert set up a tree in the palace for his wife, Queen Victoria, in 1840. Prince Albert

The traditional Christmas tree may not be able to document its descent from the world axis tree of pagan antiquity, but it reinforces many of the same messages: continuity of life, light in winter's darkness, and seasonal order. ("The Christmas Tree" by F. A. Chapman, reproduced from An Old-Fashioned Christmas in Illustration and Decoration, *ed. Clarence P. Hornung, Dover, 1970)*

was born in Germany, and his innovation for Victoria's holiday was rooted in German tradition.

Martin Luther, the sixteenth-century German theologian and leader of the Reformation, was said to have invented the Christmas tree after walking beneath the stars on a crisp and dark winter night. In an attempt to bring indoors the starry heaven from which Christ had come down to earth, Luther put the tree up for his children and lanterned it with more candles than they could count. Now, this story about Martin Luther is pure legend. The oldest known written reference to a decorated Christmas tree was actually written in 1605, in Strasburg, almost sixty years after Luther's death. It does confirm, however, that Christmas trees were cheering German households more than two centuries before Prince Albert made the custom fashionable in Britain. The story about Luther and the first genuine accounts of Christmas trees also demonstrate that someone in post-Reformation Germany was still thinking about sacred trees and the sky long after Europe's conversion to Christianity.

Scholars have tried, without real success, to

trace the Christmas tree back to earlier pagan sources. Lost somewhere in the pedigree of the Christmas tree, there probably reside the Yule log's midwinter rekindling of the sun and the world axis of the Irminsul, but we are unlikely to find them for lack of surviving evidence. What persists in the Christmas tree, however, are seasonal and cosmological themes that are well known from ancient times.

The Christmas tree, in part, celebrates the endurance of life through the world's seasonal crisis in winter and affirms the reestablishment of cosmic order. This role is consistent with the meaning of Christmas itself, for the early Fathers of the Church grafted the Christmas promise of mankind's spiritual rebirth onto the old story—the solstitial promise of nature's rebirth from winter's death—by selecting the winter solstice as the date of Christ's birth.

The Christmas tree is an evergreen, and because evergreens escape the seasonal death of deciduous trees, they have long been displayed at midwinter. Pine trees, stripped of boughs and bark, were placed outside of Swedish homes at Christmas time before the Christmas tree became popular. In Rome, at least as early as the second century A.D. and probably before, houses were ornamented with another evergreen—laurel—and with lights on the Kalends of January. As the first day of the first month of the year, the Kalends of January took place about six days after the winter solstice. It was an extremely important midwinter station in the Roman calendar. New consuls, the two chief magistrates of the Roman Republic, were elected at that time. People feasted and distributed presents. Although the Romans did not celebrate Christmas yet, they had Christmas spirit.

The Church didn't get around to deciding to celebrate Christ's birthday until the fourth century A.D. When the Church Fathers finally did sanction the custom, they had no birth certificate for Jesus —or other historical evidence—to tell them the true date of Christ's birth. Instead they picked the day that coincided with the winter solstice and with the Birthday of the Unconquered Sun, the chief state holiday of Rome.

Winter solstice themes also crept into Christmas through other traditions about special trees that bloomed or bore fruit at the holiday. Glastonbury, in southern England, had its miraculous thorn, allegedly brought to Britain by Joseph of Arimathea in the middle of the first century A.D. It was supposed to blossom on Christmas Eve. Seasonal rebirth at Christmas that involves a symbolic tree also shows up in an old Icelandic tale about a mountain ash that grew where two innocent victims had been executed. From their blood, the tree emerged, and despite the stormy winter winds, lights burned on every branch during the night before Christmas.

It is possible the Christmas tree has roots in more than one old tradition. The Paradise-play, performed on Christmas Eve in the Middle Ages, is another possible source of the idea. Based on the story of Adam and Eve, this folk drama was staged during the thirteenth century to affirm the distinction between the First Man and the New Savior, the Fall and the Redemption, the Old Testament and the New, and the old year and the new. Because the play took place in the Garden of Eden, the scene included the Tree of Knowledge. According to the German scholar W. Mannhardt, who in the nineteenth century wrote *Der Baumkultus der Germanen und ihrer Nachbarstämme*, a definitive story of the tree cult in ancient Germany, our medieval ancestors brought this tree on stage on so many Christmas Eves, it became a symbol of Christmas.

Even without a lineage clearly traceable to antiquity, the Christmas tree is an emblem of a stable and benevolent cosmos. Its yearly return to our homes is actually an acknowledgment of the cyclical renewal of world order. Even if modern cosmology makes no use of seasonal renewal and a world-ordering axis, something in many of us still longs for that promise of harmony and renewal Christmas makes each winter.

A RIGHT JOLLY OLD ELF

If a cosmic tree points the way to heaven for us every Christmas, Santa Claus undertakes the magical flight of a shaman. He is sometimes said to be responsible for erecting the Christmas tree sky pole himself. Descending vertically down the chimney, Santa returns by the same route back to the roof. Our chimneys, like the cosmic axis, carry him from one realm to another. There must also be something magical about the sack he carries, for it can supply toys to all of the children in the world. That bag is like the *sampo* or the magic mill that could grind out unlimited supplies of food, money, and salt and was associated with the World Axis. In addition, Santa Claus, like the gods and spirits, is immortal. His activity is focused on that critical night that was once the winter solstice.

In the pagan era, the winter solstice was the turning point of time and the birthday of the sun, the moment of new beginnings. All of nature was poised then to step over the border of the year. When it became the birthday of Christ, Christmas night became the hinge of the year. It commemorated the timeless moment when heaven through Christ came in contact with earth, and each year the anniversary of the event re-created once again the transcendent circumstances of that first Christmas. In Scandinavia, spirits of the dead could wander back home on Christmas night, and the old gods, storming down from the mountains, might carry off anyone so foolish to be abroad that night. Breton peasants were certain animals could speak at midnight on Christmas Eve. The Serbs maintained that the earth on that night was blended with Paradise.

Christmas night is supercharged with supernatural power. Santa Claus's vehicle is as strange as the night. In a sleigh pulled by eight flying reindeer, he is able to travel the entire world in one night and is never seen. He sees when you're sleeping, and he knows when you're awake. He knows if you've been bad or good. By the time "Santa Claus Is Coming to Town" became a popular Christmas hit, kids knew the score. Santa

The chimney is the means by which Santa Claus descends from the sky and returns to it again. In some respects, the behavior of Santa Claus mirrors the shamanic attributes of Odin. Whatever else Santa Claus may be, he is a seasonally significant spirit who arrives at about the time of winter solstice and whose workshop at the North Pole affiliates him with the world axis. (Thomas Nast, illustration for "A Visit from St. Nicholas," in An Old-Fashioned Christmas in Illustration and Decoration, *ed. Clarence P. Hornung, Dover, 1970)*

Claus is well informed about those who hope their address is on the itinerary of his one-night flight. You don't try to fool him.

Despite some intriguing parallels with characters in myth and legend, Santa Claus is not really the direct descendant of European gods and shamans. Neither was he fabricated solely out of the present era. He is actually a modern American amalgamation of several old stories that traveled to the New World with the people who told them. Once here, they were simmered together until Santa Claus as we know him was ready for the world.

Certainly one of the original inspirations for Santa was Saint Nicholas, the bishop of Myra, near what is now the village of Demre in Asia Minor. Born in Turkey to a wealthy family in perhaps A.D. 270, he eventually became known for his anonymous generosity and holiness. Christmas gift-giving doubtless has something to do with his reputation, and stories about him highlight his charity.

England's Father Christmas added the "right jolly old elf" component to Santa's image. By the fourteenth century, white-bearded Father Christmas was playing a Falstaffian role in *Saint George and the Dragon*, a Christmas mummer's play. With a wreath of holly round his head and a ready glass at his lips, he was more a celebrant of the holiday than one of its chief diplomats.

Washington Irving, however, was the real creator of the modern Santa Claus. In 1809, he wrote about the Dutch Christmas customs the Knickerbockers had transplanted to New York. Irving manipulated these traditions, fabricated many of his own additions to Santa Claus lore, and molded today's canonical Santa. Irving's *Father Knickerbocker's History of New York* contains the first references to Santa's habit of riding over rooftops, dropping down chimneys, putting a finger aside of his nose, and delivering the goods to children.

The Christmas Eve we now know so well was firmly in place by December 23, 1822, when, for the first time in print, Clement C. Moore's poem hung those stockings for Santa by the chimney with care. How those stockings originally got into the picture is hard to say, but in one of the old stories about Saint Nicholas of Myra, he left gold in the stocking a young woman had pinned to her fireplace. The tradition also probably owes something to the children in Holland, who left carrots and hay as a midnight snack for the white horse they said carried Sinterklaes—the Dutch version

The Santa Claus we know today owes much of his character to Sinterklaes, the Dutch version of Saint Nicholas, who wears a bishop's mitre and red robe. He is accompanied by Piet the Moor, who helps organize the presents. (Photograph: Bertus Bleiji)

of Saint Nicholas—to every home. After the horse enjoyed the goodies, Sinterklaes left goodies for the kids in their shoes.

Wearing the red robes and mitre of Bishop Nicholas, the Dutch Santa Claus also contributed part of Santa's present wardrobe. The fur trim and jingle bells, on the other hand, probably came from Germany, where the Santa Claus of record was sometimes accompanied by Knecht Ruprecht. A wild character with glowing eyes and a lolling, red tongue and dressed in furs, skins, and straw, Knecht Ruprecht looked more like a Norse god than a carrier of Christmas cheer. The rod in his hand was applied to backsides of children who couldn't say their prayers correctly. He would test them when he came to town, and if they had mastered their prayers, he would instead reward them with gingerbread, apples, and nuts.

Saint Nicholas and his pagan counterparts usually traveled by horse, but reindeer eventually entered the story in 1821. Santa's first reindeer pranced and pawed the rooftops than in an anonymously authored book, *The Children's Friend: A New Year's Present, to Little Ones from Five to Twelve.* One year later, Clement C. Moore multiplied the reindeer by eight and named them:

> Now, Dasher! now Dancer! now, Prancer and Vixen!
> On, Comet! on, Cupid! on, Donner and Blitzen!

What a role call! Selecting a name like *Vixen* and rhyming it with *Blitzen* took some kind of literary courage, no matter how critics disparage the verse. These are great names for reindeer, and the two in German invoke thunder *(donner)* and lightning *(blitzen)* in a proper bow to pagan Germanic sky worship.

The reindeer made sense. If Santa Claus had his headquarters at the north pole, northern creatures should provide the power for his sleigh. He was also a spirit of warmth in frigid winter's grip. Perhaps he needed animals accustomed to the season to make a nonstop flight around the world so close to the winter solstice. Domesticated like a horse but still exotic and a little wild, the reindeer were perfect. They imply Santa Claus really did call the North Pole—or at least some far northern latitude—home. When he settled there is not so certain.

Most people were never quite sure where Santa Claus lived. The Dutch had placed him in Spain. Charles W. Jones, Santa's most authoritative biographer, indicates that by the 1860s, Santa was believed to reside at the North Pole, but in the 1870s, Mark Twain put a return address on a letter from Santa to his daughter Susie, indicating Santa held court in the "Palace of St. Nicholas, In the Moon." Even for flying reindeer that trip would be tough. L. Frank Baum, author of *The Wonderful Wizard of Oz,* wrote his own biography of Santa in 1902, *The Life and Adventures of Santa Claus,* and wasted no time with that North Pole business. As far as Baum was concerned, Santa Claus could most often be found in the Forest of Burzee, on the southern border of Oz, or in the Laughing Valley of Hohaho just to the east. We do get a brief report in 1899, from *St. Nicholas,* a children's magazine, about Santa's arctic retreat: "Santa Claus lives somewhere near the North Pole, so he can't be interfered with." By 1931, Santa was unquestionably ensconced at the North Pole. Walt Disney's "Silly Symphony" movie short *Santa's Workshop* haloed the toy factory with aurora borealis and surrounded it with polar ice fields.

Santa Claus may not have started his career at the North Pole, but that is exactly where he belongs. He has his grip on the world axis. His yearly tour steadies us all. Remote, at the top of the world, in a land of eternal snow, he gathers the energy spent in the year's passing. Then he sails like a shaman, off into the solstitial darkness, to reallocate that power to the world's children. It's a promise in Christmas presents that through children and through seasonal renewal, the world is ever young.

STAR OF WONDER

The star that sometimes tops the Christmas tree Santa raises in the living room is an emblem of Christmas that goes back to the Book of Matthew in the New Testament. The Christmas Star was the sign in the sky that guided the three Magi, or Wise Men, west to Bethlehem to the Christ Child. Like Santa Claus, they were gift bearers, and tra-

A thirteenth-century mosaic within Istanbul's Kariye Camii (Church of Saint Savior) in Chora illustrates the Nativity with a Christmas Star that provides a transfusion of celestial light to the manger below. The star makes its own cosmic axis with earth and resembles what in a later age would be identified as a mesmerizing beam from a flying saucer.

dition makes their offerings of gold, frankincense, and myrrh the first Christmas presents.

The Star of Bethlehem turns this tale of the world's first Christmas into another old sky story. Its basic theme: Heaven heralds a great event. That is the real meaning of the Christmas Star, but if we believe there actually was something in the sky at the time Christ was born, it is reasonable to try to deduce what it was. That is not so easy, however. We know very little about the Christmas Star. Matthew was the only one who had anything at all to say about a special star visible at the time of Christ's birth, and he didn't say much; the star is mentioned only four times.

The Gospel of Saint Matthew was composed in the first century A.D., nearly two thousand years ago, and about fifteen centuries would pass before anyone asked if the Star of Bethlehem had a scientific explanation. Finally, the German astronomer and mathematician Johannes Kepler did just that in the seventeenth century. In our era, an extraordinary invention—the Zeiss Planetarium Projector—first put American audiences under a strikingly realistic image of the night sky during the 1930s, and by the 1940s, public planetaria had turned the scientific search for a Christmas star into a seasonal quest in a multimedia theater.

Able to duplicate the appearance of the night sky with impressive accuracy, the Zeiss Planetarium Projector can transport audiences to any location on earth and to any date thousands of years into the past or future in a matter of minutes. You would think we could just set the instrument for the date of Christ's birth and have a look at the sky to see if there is a Christmas Star available. It is not, however, so easy. For example, I am finishing this book in the year A.D. 1990 where A.D. stands for *anno Domini* and means "in the year of Our Lord," but it does no good to turn time backward to the night of December 24–25 one thousand nine hundred and ninety years ago, because we don't really know when Christ was born.

No one even numbered the years from the year of Christ's birth until A.D. 525, and then they didn't know it was A.D. 525 until Dionysius Exiguus worked out the details and converted Church authorities to his way of thinking. Dionysius was a monk. Scythian by birth, he worked in Rome, and by his reckoning, Christ would have been born in the year 1 B.C. Other ancient authorities differed on this point. Some put the first Christmas in 2 B.C. Others preferred 3 or 4 B.C. No one was sure.

There are, however, a few more clues that narrow down the date, and they are clues from heaven. Matthew said the star was seen by the Wise Men "in the days of Herod the king." Josephus, a Jewish historian writing in the first century A.D., mentioned that Herod had died after a lunar eclipse visible in Jericho but a few days before Passover.

There was a total lunar eclipse seen from Jericho on March 13, 4 B.C., and Passover that year fell twenty-nine days later on April 11. For various reasons, many Biblical scholars and historians have, therefore, accepted 4 B.C., as the year of Herod's death and concluded that Christ was born a few years earlier, perhaps in 7 or 6 B.C. In recent years, however, this view has been challenged, particularly by Ernest L. Martin, author of *The Birth of Christ Recalculated!* He argues that one lunar month was not enough time for all of the events reported to have taken place between Herod's death and the following Passover. Instead, he prefers the total lunar eclipse of January 9, 1 B.C. Although 7–6 B.C. has been the odds-on favorite for the era of Christ's birth, the period of 3–2 B.C. lately has been kicking hard on the door of historical acceptance. Our question here is whether we can find a satisfactory Christmas Star for either time.

The story of the Christmas Star suggests the appearance of a celestial object not seen before and not seen after the birth of Christ. Temporarily visible, it vanished from the sky not long after it made its debut. There are several possible candidates that come and go that way. Meteors show up suddenly and are gone in a flash, but even the occasional extraordinary bright fireball doesn't seem to conform to the description in Matthew. Meteors are too common and move too quickly to be considered for Christmas Star duty.

Comets, on the other hand, are a little more like it. A comet looks like a fuzzy star dragging a tail of fugitive light. Instead of whipping through heaven in a moment, like a meteor, or "shooting star," a comet lingers in the sky for days, weeks, or even months, nightly shifting its place with respect to the stars a little each night. We would need an independent account of a comet having been seen at the right time in order to verify this explanation. It so happens that the Chinese in that era were already keeping systematic records of comets and other celestial phenomena. We have, for example, Chinese reports of the passage of Halley's Comet in 12 B.C. along with three other comets in later years. None of them, however, appeared in the years in which Christ may have been born. There is no evidence to support the idea of a Christmas comet.

A star that explodes into temporary brilliance could also give us a Christmas Star. We now know that stars sometimes do explode, and when they do, some of them grow bright enough to be seen with the unaided eye. Such eruptive changes in stars are now understood fairly well, and they involve two stars, not one. Many of the stars in the Milky Way Galaxy are binary stars or systems with even more members in mutual orbit. As the stars burn the thermonuclear fuel in their cores, they age and evolve. One member of a binary may become a white dwarf before the other does. In this state, it is a highly condensed and exotic object with about as much material as is contained in the sun packed into a volume comparable to the earth. So far the two stars are still compatible, but when the second star expands into a red giant on its own way to becoming a white dwarf, it dumps some of its material onto its white dwarf neighbor. After accepting this transfer of mass for a while, the white dwarf finally no longer abides it. Physical conditions in the white dwarf won't allow the added mass to remain there, and it is exploded into space as an expanding shell. Where perhaps no star had been charted before, a new star, or *nova,* shines, at least temporarily. Ancient Chinese astronomers spotted such objects and kept records of them.

Even more spectacular explosions of stars are seen from time to time. These much rarer events are called supernovae, and when one takes place, the star is almost completely obliterated. Only five supernovae are known to have occurred in the Milky Way Galaxy in the last thousand years, and on February 23, 1987, a supernova visible to the unaided eye exploded out of the Large Magellanic Cloud, the nearest galaxy beyond our own. To our eyes, a nova may grow ten thousand times or more brighter than the star it once was, but a supernova can shine brighter than the rest of the galaxy in which it resides, at least for a short time. It may be hundreds of millions of times brighter than it was originally.

Chinese astronomical records do confirm sightings of two objects, in 5 and 4 B.C. respectively, that might have been novae, or even supernovae, rather than comets, as the accounts state, but the

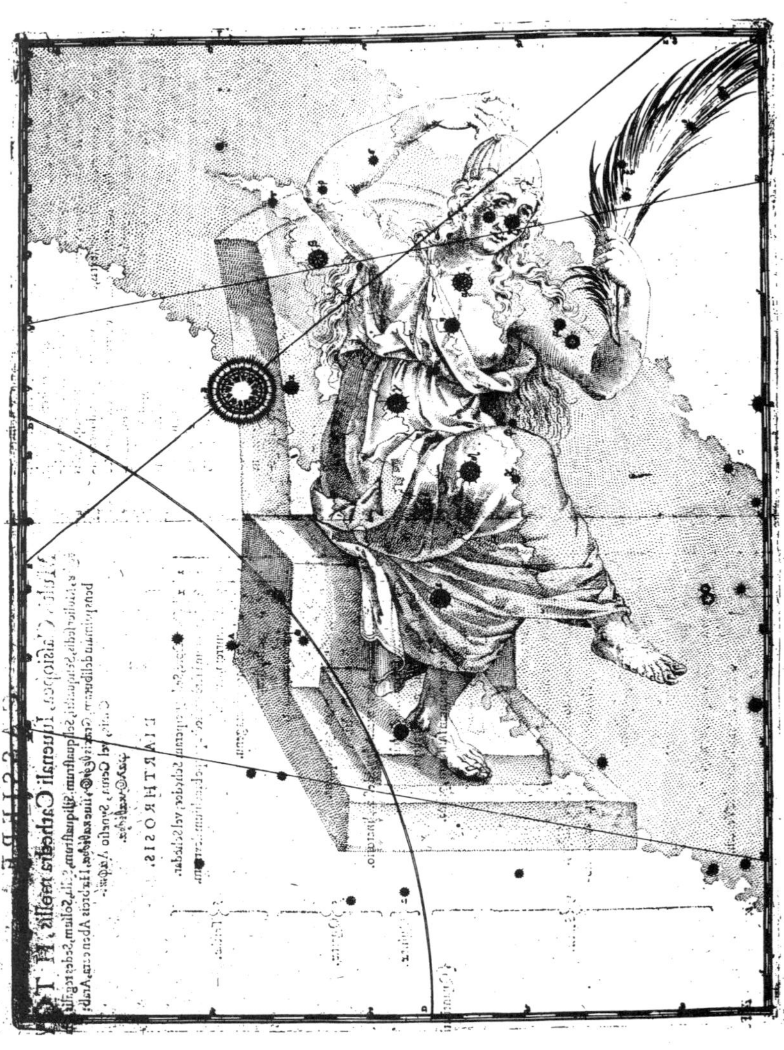

"New" stars a little like the Christmas star sometimes do appear. The bright star charted on the back of Queen Cassiopeia's throne in this map from Johann Bayer's Uranometria *was first observed and reported in 1572 by the Danish astronomer Tycho Brahe. By 1603, when Bayer published his collection of star maps, the new star had long since faded away, but it was still memorable enough to be included on the chart. Although Tycho speculated that this new—and temporary—resident of heaven might have condensed out of something up there among the stars, we now know he saw a supernova, the catastrophic, explosive finish of a massive evolved star. (Collection Griffith Observatory)*

dates don't really fit our best guesses for the time of the birth of Christ.

According to scripture, Herod didn't see the star himself. He had to ask the Wise Men about it. We would guess that an exploding star or comet might have been seen by Herod, but that doesn't seem to have been the case.

Tutored eyes did see the star, however, and they belonged to the Wise Men. These Wise Men, or Magi, are sometimes called the Three Kings, but the name given to them in the Bible implies they were specialists—men of knowledge. Originally the name *Magi* belonged to one of the tribes of Persia's Medes. Known to have played a dominant, controlling role in some elements of Iranian religion all through the Achaemenid dynasty to the end of the reign of Darius, its last king, in 330 B.C., the Magi developed their own Zoroastrian cult. Departing from orthodox tradition, they worshipped the evil god Ahriman along with the good god Ohrmazd.

Persian influence had already been felt in Babylonia and Assyria, and the presence of the Magi there is known by the sixth century B.C. The Magi came in contact with Babylonian astrology, and the term *magi* became synonomous for the Greeks with "Chaldaeans" and "astrologers" by the beginning of the fourth century B.C. It is reasonable, then, to think of the Wise Men as astrologers, and their Star of Bethlehem could have been something an astrologer would notice.

Astrology is concerned, above all else, with the positions of the sun, moon, and planets among well-defined zones in the sky, particularly the signs of the zodiac. At any moment, the placement of those celestial objects and the relations thought to exist between them establish the astrologically significant circumstances of the time and place in question. If the Magi were astrologers, it is possible they were, in some way meaningful to them, following configurations of the planets to Bethlehem.

In 1604, Kepler had been keeping tabs on Mars, Jupiter, and Saturn just before he discovered the last supernova seen in our Galaxy. Those planets were heading for a relatively close bunching, or massing, by the end of the year. When the supernova appeared in their neighborhood on October 10, Kepler wrongly concluded the planets might have created the new star. That thought inspired him to guess that something similar might explain the Christmas Star, and he started calculating the positions of planets for the era of the birth of Christ. He discovered that a similar massing of Mars, Jupiter, and Saturn had ornamented the sky in 6 B.C. and judged he had found his Christmas Star.

Now, when modern planetaria tell the story of the Star of Bethlehem, they usually turn back the sky to 7–6 B.C. and track an interesting set of conjunctions, or close meetings, between Jupiter and Saturn. Between May 27, 7 B.C., and December 1, 7 B.C., the combined movement of Jupiter, Saturn, and the earth, allowed Jupiter and Saturn to get together in the sky, separate, get together again, separate once more, and join up for a third and last time before finally departing each other's company. In 7 B.C., the first of the three conjunctions would have been seen in the morning sky, in the east, on May 27. The next one took place on October 5. On December 1, the pair of planets lingered together in the west after sunset. Conjunctions of Jupiter and Saturn occur about every twenty years, but "triple conjunctions" like the one in 7 B.C. are far less common. They do not recur with any definite cycle. Hundreds of years usually separate one from the next.

If you are looking for a Christmas Star, the triple conjunction of Jupiter and Saturn in 7 B.C. is pretty satisfying. It has symbolic meaning consistent with the vocation and expertise of the Magi, and it can be made to match the description in the Bible. But what about the date? Something happening in 7 B.C. might be fine had Herod died in 4 B.C. and Christ been born in 7 or 6 B.C. But the historic evidence may favor a later date. Was there an astrologically interesting Christmas Star in 3 or 2 B.C.? If so, was it as compelling as the triple conjunction that occurred four to five years earlier?

Between May 19, 3 B.C., and August 26, 3 B.C., there were nine major conjunctions between planets or between a planet and a bright star. All but one of them involved either Jupiter or Venus, the two brightest planets in the sky, and the most impressive of all was an extremely close conjunction between Jupiter and Venus on June 17, 2 B.C.

Throughout the most likely years of Christ's birth, there were a number of planetary conjunctions that might qualify as the Christmas star, but the conjunction of Venus and Jupiter on June 17, 2 B.C., was the best of the bunch. The two planets appeared to approach within half a minute of arc, enough to permit their disks to overlap. The two planets on that night in the ancient Near East would have merged into one "star" of unprecedented brilliance. If the Magi had been watching, the event might have sent them packing. (Griffith Observatory)

On that night, those two bright planets—not far from Regulus, the brightest star in Leo—were separated by no more than ½ arcminute. If the Magi were watching on that night, they saw Jupiter and Venus merge into a single and spectacular "star" in the western sky.

So even if Herod died in 1 B.C. and Christ were born in 3 or 2 B.C., we still get a great Christmas Star. Does this mean we've confirmed the story in Matthew? No, we really haven't. In antiquity, great events and the births of great men were thought to be signaled by celestial semaphores; the heavens themselves, for example, were supposed to have announced the delivery of Alexander the Great.

In astrological terms, there is always something significant going on overhead. If you give me a date for the birth of Christ, I'll give you a Christmas Star. One of them—the Jupiter-Venus conjunction of June 17, 2 B.C.—might look better than the others to our eyes, but a resourceful astrologer could manage without it if another birthday for Christ should seem more appropriate.

Whether there really was a Christmas Star in the sky at the first Christmas is not as intriguing as the desire we have to know about it. In a scientific age, we explore this symbolically rich detail in the Christmas story with the insight of modern astronomy. It is an engaging exercise, but we are just acting in accordance with the spirit of our own times and are trying to naturalize and rationalize an event that in its own time carried the symbolic and mythic meaning of a miracle. Finding a real Christmas Star doesn't confirm the Bible. And skepticism doesn't diminish the Star's real value. The story of the Christmas Star is a story of new life and a new world order. Those themes have always been valuable to us, and they are inevitably linked with the sky. The Christmas Star's real magic is its ability to put wonder and awe in the hearts of those who enjoy its tale, whatever beliefs may be quartered there as well.

19

Out of the Blue

OUR ancient ancestors, it seems, were always getting messages from the sky. Matthew's Christmas Star tells us that such messages had more meaning than any physical explanation we might find for the celestial couriers that delivered them. As far as Matthew was concerned, the whole event was miraculous—the Star, the virgin mother, the angels, the Wise Men, the birth of Christ—everything. Matthew would probably regard our attempt to understand the Star as a natural phenomenon as irrelevant. Heaven's intent was what mattered.

As long as the ancients were convinced there was a relationship between the stars and themselves, they were able to read reports and warnings in heaven. These dispatches helped clarify the destiny of kingdoms and people's affairs and offered guidance for prudent action. Effective because people believed in them, celestial notices resided comfortably within the mythic and symbolic landscape of antiquity.

Today, we don't exactly see a great fax machine in the sky, but we, too, sometimes act as if the sky were still delivering privileged information. Now, however, these bulletins are composed in modern jargon. Prophecies of earthquakes are tuned for superficially scientific ears with planetary alignments. The Harmonic Convergence heralds New Age spiritual development as "galactic evolution" with a solar system assist. And even as we were applying Newton's laws of motion and universal gravity to send a small squad of space probes on a scientific rendezvous with Halley's Comet, people on earth were creating a commercial space-age

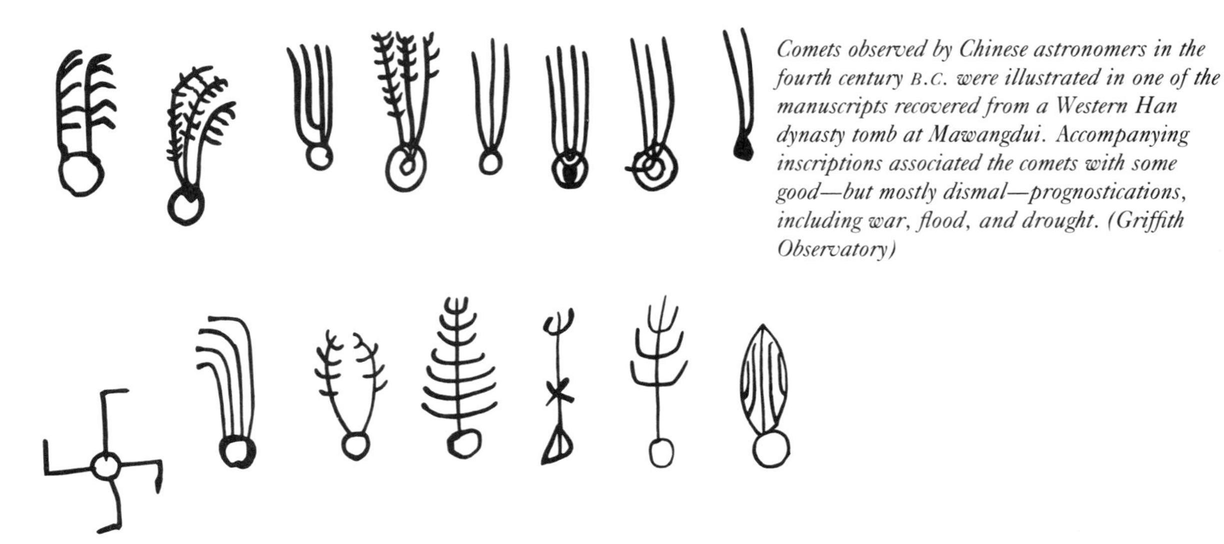

Comets observed by Chinese astronomers in the fourth century B.C. were illustrated in one of the manuscripts recovered from a Western Han dynasty tomb at Mawangdui. Accompanying inscriptions associated the comets with some good—but mostly dismal—prognostications, including war, flood, and drought. (Griffith Observatory)

product out of what was once a traditional herald of celestial doom.

THE WICKED MESSENGER

In the past, comets were generally considered bad company and were dreaded worldwide. People, for the most part, thought they could do without them. When a comet made an appearance in the night sky, wandering at will across the orderly celestial scene, people generally believed it meant trouble. As far as anybody could tell, comets did not follow any rules; they showed up unannounced, brazenly went where they pleased, and disappeared again after a brief but heady encounter with the sky.

In ancient Babylon, comets were associated with instability in government and the deaths of kings. In China, war, executions, deaths of generals and royalty, and imbalance in nature all rode in on the comet's tail. Comets were also bad news in Greece and Rome. In Homer's *Iliad,* the hero Achilles seems to refer to a comet's bad reputation and unpleasant disposition:

> *Like the red star that from his flaming hair*
> *Shakes down disease, pestilence and war.*

The English word *comet* comes from the Latin *comet* and the Greek *kometes,* which means "hairy."

The Comet of 1528 was so frightening that some people, according to Ambroise Paré, a sixteenth-century French surgeon, became sick, and others died of terror. It looked like a great curved arm threatening the earth with a sword. Hatchets, knives, other swords, and heads with beards and bristling hair accompanied it. (From On Monsters and Marvels *by Ambroise Paré, University of Chicago Press, 1983)*

The quotation from the *Iliad* unloads disaster from the comet's "burning" locks.

Comets never really shed their evil image in history. Halley's Comet—the best known comet today—was no exception. Our earliest record of it dates to the passage of 240 B.C. It arrived then in the seventh year of the reign of Emperor Qin, the first emperor of China, and the brief account of its appearance informs us that the comet's passage coincided with the wartime death of General Ao and also with the death of the Empress Dowager.

The earliest picture of Halley's Comet shows up in the *Nuremberg Chronicle,* which records the appearance of a comet in A.D. 684. and blames it for plagues, devastating storms, and a bad harvest.

Although the *Nuremberg Chronicle* portrait is the earliest illustration of Halley's Comet, it was not the work of an eyewitness. It was drawn, in fact, about a thousand years later with guidance from the original seventh-century account. The earliest eyewitness likeness of the comet appears in the Bayeux Tapestry, an eleventh-century Norman embroidery that depicts the Norman Conquest of Britain in 1066. The death of Harold, the Saxon king, at the Battle of Hastings was associated with the arrival of the comet six months earlier. Comet Halley returns about every seventy-six years, and every time it came back, it incited that old comet magic.

Others tried to understand the truth about comets. In the fifth century B.C., the Greek philosopher Anaxagoras thought comets were really conjunctions, or mergings, of planets. Aristotle, on the other hand, was convinced that comets were atmospheric phenomena and not components of the high sphere occupied by the stars. In his view, the realm beyond the moon was pure and unchanging; only heavenly bodies, traveling with uniform circular motion, could reside there.

In 1577, the Danish astronomer Tycho Brahe

The appearance of Halley's Comet in 1066 was documented on the Bayeux Tapestry and was associated with the subsequent death of Harold, the Saxon king, at the Battle of Hastings. Harold looks anxious on his throne while the comet seems to strafe the palace. This piece of political propaganda was embroidered by the Normans, who won the battle. The comet was actually a good omen for their king, William the Conqueror. (Drawing: Robin Rector Krupp, after the Bayeux Tapestry; in The Comet and You *by E. C. Krupp, Macmillan, 1985)*

finally proved that comets are not phenomena of the atmosphere but truly celestial objects, certainly farther away than the moon. His work didn't really convince all of the important parties, however. The well-respected German astronomer and mathematician Johannes Kepler, who inherited Tycho's observational data, decided that comets were atmospheric after all and that they traveled in straight lines, not celestial circles. Kepler's ideas about comets prevailed right through Halley's 1607 reappearance, which Kepler observed. It would take one more return of the comet to invalidate Kepler's and Aristotle's erroneous concepts.

MANY HAPPY COMET HALLEY RETURNS

English scientist Edmond Halley saw "his" comet in 1682. Using Newton's law of gravitation and measurements of the comet's path, he concluded that comets tour outer space in large, long elliptical orbits. Convinced that the comet he saw in 1682, the comet Kepler saw in 1607, and the comet Peter Apian saw in 1531 were all the same comet, he predicted that it would return again and hoped, if it did, that posterity would acknowledge that this discovery of a returning comet was first made by an Englishman. It did, and it has.

On a cold, Christmas night in 1758, just sixteen years after Halley's death, Johan Palitzch, a German farmer and amateur astronomer, spotted the comet from his fields with his telescope. Normally a comet is named for the first person to see it, but in this case, posterity has conferred upon it Halley's name.

Halley's Comet has returned three more times. On its last return, in 1985–86, it was greeted with considerable fanfare, and for the first time in human history, we engineered a closeup look of it as it zipped by the earth and the sun. Spacecraft from Japan, the Soviet Union, and Europe all converged on the comet for one week in March, 1986, and confirmed the scientific "dirty snowball" portrait of the comet nucleus theorized by American astronomer Fred L. Whipple.

We now know that comets are relatively small pieces of debris in the solar system. At the nucleus, or core, a comet is an irregular, mountain-sized bundle of various ices and dust. The nucleus of Halley's Comet turned out to be a bit larger, darker, hotter, and more irregular than we had earlier imagined. We have learned that it has the shape of a peanut, about nine miles long and five miles across. Its surface is extremely dark, darker than anything we have yet found in the solar system.

The nucleus of a comet orbits the sun on a long cigar-shaped loop and spends most of its time so far from the sun, it is too faint and too small to detect. When, however, this powdery popsicle gets close enough to the sun to respond to the heat, it starts to put on a fancier costume. Exposed surfaces of ice produce jets of gas and dust. Although the nucleus is mostly ice—water ice—its surface is hot during its closest approach to the sun, about 135 degrees Fahrenheit. The comet is a hot, furious gusher, a snowball in hell.

The volatiles evaporate off of the icy core, and the head of gas and dust grows to be many times larger than the earth, sometimes even as large as the sun. The tail can be millions of miles long, and the outer tenuous corona of hydrogen we never see from the earth may be as much as ten times the size of the sun. This sounds impressive, but it is all bluff and fancy. The huge size is misleading. There is practically nothing there. Comets are said to be "the nearest thing to nothing anything can be and still be something." A piece of comet tail as big as the Pacific Ocean, packed well, could fit into a suitcase.

BLUFF AND FANCY

Even before spacecraft reached Halley's Comet, its composition was already well documented. But that didn't stop some people from pulling the traditional comet lore out of the closet for Comet Halley's thirtieth recorded return. Halley's Comet showed it still had some of that old-time comet charisma. A paperback novel, *The Comet,* available on the racks of better liquor stores and convenience markets, readied us for another bout with disaster. The book's power blazes in blurbs across its cover: "A terrifyingly prophetic novel." Of the comet, it says, "When it returns—the sun goes out!" Inside we are asked, "What doom does the coming of Halley's Comet prophesy?" The answer follows on the next line: "A WORLD OF ETERNAL NIGHT!"

For those who preferred a "nonfiction" account of what Halley had in store for us, another book, *Comet Catastrophe*, published a warning about "the disturbing link between Halley's Comet and world disasters." In a modern application of comet power, airplane hijackings, train crashes, soccer riots, and the Bhopal chemical disaster were all linked with the comet's return.

A broadside was distributed to the windshields of needy cars parked at Los Angeles International Airport:

> STAR WARS
> HALLEY'S COMET IS COMING SOON!
> HERALD OF THE WORLD'S END
> COMET ATTACKS TYRANTS
> DOOMSDAY IS NEAR
>
> Today Satan's evil dominion is rampant: kings and rulers are rebelling against God; antichrists, false prophets and teachers are deceiving men; wickedness is increasing; crime skill (theft, robbery, kidnap, killing, adultery, debauchery, immoral living, drug abuse . . .) is more sophisticated than ever. Many countries are war-torn, the nations are engaged in a destructive arms race. The world is stricken with abnormal weather, sicknesses, and disasters (floods, fires, wind disasters, droughts, epidemics, earthquakes, hailstones, famines, epidemics, herpes, AIDS . . .) happen endlessly. Most awesomely, *Halley's Comet is coming in 1986!*"

The 1985–86 World Tour of Comet Halley was also a commercial success. People spent good money to travel all over the world to get a good view of it at its best. The comet was so well merchandised, one suspected it had hired its own agent. More books on comets were published that year than had seen print in the previous five decades. They shared the hypergolic Halley marketplace with Comet Halley buttons, cloisonné, pencils, posters, postcards, commemorative coins, coffee mugs, jigsaw puzzles, Matchbox cars, frisbees, comet pills, digital watches, backpacks, video cassettes, postage stamps, and popcorn. You could wear Comet Halley T-shirts, Comet Halley sweatshirts, Comet Halley bandanas, Comet Halley baseball hats, and Comet Halley rhinestone-studded ladies' outfits.

There is no doubt that comets still pack plenty of punch. Their power is still high octane, but apart from a few aberrant prognostications of comet doom, most of Comet Halley's last visit was a party that celebrated, through the comet's predictable return in the time of an average human lifespan, a feeling of connection with the past and the future. In that sense, the sky has once more fulfilled one of its traditional roles. It has given us a framework of experience that allows us to feel at home in the universe.

DOOM FROM THE SKY

Not everyone always hears cheering news in the celestial signals called out by the great quarterback in the sky. Astrological horoscopes, for example, can have their downsides. Astrology extracts rumors of both fair and foul futures from configurations of the sun, moon, and planets. It is a two-thousand-year-old system of magical thought that still commands widespread informal belief as well as what must be tens of millions of adherents.

Because astrology comes with its own kit of technical jargon and seems to deal systematically with celestial objects that are studied with care by modern science, many mistakenly believe the predictions are based on scientific fact. Even people who don't believe in astrology can be influenced by a mass epidemic of anxiety caused by pseudoscientific posturing. This is exactly what happened in the first half of 1988, when rumors of a major southern California earthquake, allegedly predicted by Nostradamus, a sixteenth-century French astrologer and physician, put tens of thousands, perhaps hundreds of thousands, of otherwise perfectly reasonable people on alert for disaster in May.

Halley's Comet met a commercialized reception on its 1985–86 return. A wide variety of Comet Halley merchandise was purchased from Griffith Observatory's Halley Galleria and from other discriminating museums and planetaria. (Photograph: Robert Webb, Griffith Observatory)

Michel de Nostredame, better known as Nostradamus (the Latinized version of his name), was the author of a series of annual almanacs, ten collections of prophecies sometimes known as *The Centuries,* and several miscellaneous publications. Read with the hindsight of history, his vague predictions, ripe with obscure symbolic references and composed in compact but enigmatic four-line verses, have been credited with anticipating developments long after the end of his own life in 1566. Among his supposedly triumphant forecasts are the rise of Napoleon and Adolf Hitler, the Russian Revolution, World War II, the atomic bomb, and the assassination of John Kennedy. Nostradamus probably would be as surprised as we are about these successes, for he was writing for a different audience in a different time. Today, however, modern interpreters are still in business without his endorsement but with new claims of catastrophes to come, including a World War III precipitated by Moamar Khadaffi. However reasonable that may sound, the news that drove southern Californians to contemplate their faults was the shrill alert sounded in *The Man Who Saw Tomorrow,* a 1981 film about Nostradamus and his predictions, narrated by Orson Welles. More recently released on video cassette, the film became a hot property in the video rental market. By late 1987, a slow but steady stream of inquiries about Nostradamus and an earthquake scheduled for May, 1988, prompted newspapers, radio stations, television news programs, and those who answer the telephone at Griffith Observatory to notice the development of a new piece of urban celestial folklore.

According to the film, Nostradamus predicted

the future with the unerring accuracy of an Exocet missile, and the time bomb he had dropped in the lap of California was about to explode. This is what Los Angeles was to expect:

> Fire from the center of the earth,
> The great Earthquake shall be in the month of May.
> Saturn, Capricorn, Jupiter, Mercury in Taurus,
> Venus, also Cancer, Mars in zero.

The film confirmed that the planets will move relentlessly into these positions by May, 1988, and when they do, the "New City"—Los Angeles or San Francisco—should expect a catastrophic earthquake. This was not good news.

Responding to the obvious concern of those calling for valid information, Griffith Observatory's John Mosley reviewed the claims of the film and the original text written by Nostradamus and made some interesting discoveries. First he checked on the locations of the planets. Although Nostradamus probably would have been using the astrological signs to position the planets, rather than the astronomical constellations, Mosley searched for a match between the verse and the actual placement of the planets, in either the signs or the constellations. It soon became obvious there was no match. Using the constellations, only two of five planets would be where Nostradamus put them. The signs boosted the agreement to three. "Mars in zero," sounds like nonsense no matter what system you use. It could mean Virgo, as some claim, but it is a fine example of the ambiguity that saturates the prophecies of Nostradamus. In any case, if there were going to be a May earthquake, it certainly didn't look like the planets were going to deliver it in 1988.

A close examination of what Nostradamus actually wrote in French revealed that the prophecy was incorrectly quoted in the first place. This is what it really says.

> A very mighty trembling in the month of May.
> Saturn in Capricorn, Jupiter and Mercury in Taurus:
> Venus also, Cancer, Mars in Nonnay,
> Hail will fall larger than an egg.

Well, fair enough. A "very mighty trembling" sounds more like an earthquake than "fire from the center of the earth" anyway, but the whole thing seems to finish with a hailstorm. Hail in Los Angeles is always remarkable, and egg-size hailstones are worthy of Hollywood. When they fall, you want to find shelter, but they're not the same grief as a major earthquake.

Despite energetic and successful attempts to get the true facts out through the mass media, concern about the great May earthquake continued to grow as May moved closer. Various self-proclaimed psychics began pinpointing the exact day and even the time of the coming quake. May was also the month of horoscopes in the White House, and First Lady Nancy Reagan had to endure that cover story on *Time* magazine, and throughout the news media, Joan Quigley, the San Francisco astrologer said to advise Mrs. Reagan, had predicted a major earthquake for May 5 and was understandably out of town that day.

Telephone inquiries and written requests for factual earthquake information from Griffith Observatory steadily grew during the early months of 1988. An automated recorded message machine on the observatory's public information hotline took about two thousand calls per week between April 19 and May 18, the period of greatest anxiety. Public use of the equipment was so great, the machine broke down.

Before long, nearly every day in May had been pegged for an earthquake by someone. With that kind of coverage, it is hard to be wrong. Every year Los Angeles sustains eleven thousand earthquakes strong enough to be detected on seismic instruments. That works out to an average of 30 per day, or about one every 48 minutes. Residents of Los Angeles are sublimely unaware of just about all of them, although now and then the earth shakes with enough violence to do serious damage. Earthquake scientists are also convinced that a great quake in Los Angeles is inevitable and that the chances are high it will occur in the next fifty years. That kind of assessment, based upon painstaking accumulation of data and analysis, is reasonable and important. The Cassandras who rode the Nostradamus bandwagon, however, should be ashamed of themselves. Their unsubstantiated claims put a lot of people in southern California unnecessarily on edge.

HARMONIC CONVERGENCE

Public reaction to the Nostradamus nonsense of 1988 tells us that celestial harbingers of doom can still call the tune if they seem to be outfitted for a new age with the trappings of modern science. Even though the promised planetary catastrophe failed to occur, the same story will return before long. Planets will align again and again, and our destiny will once more be delineated in a perennial misunderstanding of gravity's influence. In this kind of folklore—our own—the planets still have influence even without the quaint rules imposed by astrology. Our era's folklore still permits the sky to have influence, provided the folklore appears to share a border with our current understanding of the universe. When it does, we don't even recognize it as folklore. It is just another component of the belief system of our age and is for that reason understood to be true.

Planetary alignment played a similar part in the worldwide Harmonic Convergence observed a few years ago. Targeted for August 16–17, 1987, the *Los Angeles Times* called the Harmonic Convergence a "mysterious weekend event." It was, in fact, an exercise in mythic and symbolic interaction between human beings, the planet earth, and the universe at large. According to its organizers, the sky was still sending us important messages. Cosmological harmony, calendrical congruence, reanimation of the earth as a living organism, the concept of personal transaction with the universe, and many other traditional mythic themes concerned with life, existence, and our place in the cosmos, all merged in a modern expression of folk belief.

Our ancestors came to terms with these same issues by activating them in their symbols and myths. The same process was at work in the Harmonic Convergence, but it was a newly minted myth, plated in part with today's knowledge of the cosmos. It also borrowed ancient sky lore and tailored it with a heavy hand for an army of partici-

Casa Rinconada, a prehistoric Anasazi ceremonial chamber in New Mexico's Chaco Canyon, was identified by New Age enthusiasts as the earth's "west pole." Harmonic Convergence adherents converged on the canyon on August 16–17, 1987, in an as-yet-inconclusive attempt to recharge the global battery. (Photograph: Steve Northup, Time, *August 31, 1987)*

pants aligned with esoteric New Age spiritual beliefs.

The New Age movement is not really so new. Well-established themes borrowed from the earlier spiritual and occult movements, but repackaged with a flashy new eighties vocabulary and the glamor of the lifestyles of the rich and famous, are marketed as New Age novelties. Channelers provide outlets for phantom entities—either ancient, off-planet, or of unknown origin—unencumbered by the disclaimer "celebrity voice impersonated." Magic crystals, mystical healing, chromotherapy, astral travel, horoscopes, extrasensory perception, and—inevitably—UFOs are all possible but not required components of New Age belief. Within this chaotic spiritual salmagundi, the Harmonic Convergence provided a crucial, and cosmically ratified, reference point for the New Age calendar of cosmic order.

Intended to celebrate what participants felt marked the transformation of the spiritual condition of the planet, the Harmonic Convergence was, at its heart, a deliberately dispersed social gathering, unified by the electronic media and the practically instant communication possible only in our era of satellites, personal computers, and television. José Argüelles, author of *The Mayan Factor: Path Beyond Technology* and the principal missionary for the Harmonic Convergence, announced that the "galactic and planetary harmonic scale" would be calibrated in precise congruence on August 16 and 17. It would take, however, a global grid of spiritually enlightened persons functioning, as Argüelles said, as a "synchronized and unified bioelectromagnetic collective battery."

According to the New Age organizers, 144,000 self-appointed seekers were to converge upon a set of sacred sites, recognized "acupuncture points" in the spiritual circuitry of the earth, and establish a resonance between the planet and the greater cosmos. These spiritual nodes—thirteen evocative places on the planet said to be "generators" and "emanators" of energy to other sacred centers—included the north pole; Casa Rinconada in Chaco Canyon, New Mexico (the "west pole"); Lhasa's Potala Palace (the "east pole"); the prehistoric standing stones at Callanish in Scotland's Outer Hebrides; Palenque, an ancient Maya ceremonial center; Haleakala crater on Maui; the Inca citadel Machu Picchu; the Sphinx and pyramids at Giza; Ayer's Rock in the Australian Outback; the Dogon cliff shrines in Mali; Lake Titicaca's Island of the Sun; Easter Island; and the south pole. Although organizers of the Harmonic Convergence affirmed that everywhere is a "sacred site," helpful lists tabulated eighteen "Mystery Schools: Centers of Knowledge," nine "World Centers: Axis of the Universe," nine "Sacred Mountains: Abodes of the Gods," nine "Holy Waters: Sources of Life," nine "Emergence Places (Sipapus): Wombs of Origination," three "Cosmic Trees: Places of Enlightenment," three "Temples of Healing: Pivot Points for Peace," three "Halls of Records: Reservoirs of Remembrance," and plenty of second-string sacred sites for those who couldn't get to any of these. Getting to the Halls of Records was a special problem, since they included Atlantis, Shambala, and Paititi—all legendary kingdoms "presently hidden from the world."

What is going on here?

This has got to be an uptown, late-twentieth-century myth contrived by and for urbanized, technological pilgrims without orthodox religious portfolio. Certainly our ancestors didn't talk that way.

Just a superficial review of the themes in the list above demonstrates, however, the obvious: Our ancient ancestors did talk that way. The bulletins they saw posted on the sky came, as near as they could tell, from a realm of gods and so were supernatural in origin and power. Their brand of the supernatural has a hard time winning acceptance, however, in an age that has dispensed with it by transforming the sky. So people are instead always reappropriating the old stories their ancestors told and recostuming them in the spirit of their own times. Harmonic Convergence and the New Age are just the most recent examples of that process.

The Harmonic Convergence was, then, also a reflection of what historian of religion Mircea Eliade called our "nostalgia for Paradise." When Adam and Eve were turned out of the Garden of Eden, all of us were expelled. We always feel, however, as if we have just been expelled, and we long for a return. Harmonic Convergence literature frequently reminded readers that we have forgot-

Adam and Eve were evicted from Eden, and we still miss it. We express the loss of Paradise in our efforts to reassert a system of cosmic order in which we play a part. That gets harder and harder as the cosmos loses its center and its edge to the law of redshifts and General Relativity. (Gustave Doré, The Doré Bible Illustrations, *Dover, 1974)*

ten that the "earth is sacred ground." One purpose of the event was the reawakening of memory, a return to Paradise.

There is always good reason for thinking Paradise has evicted us. Something is always awry. That is what prompts New Age believers, and others, to put out the call for "rainbow warriors," spiritual guerrillas who will meditate the world into utopia. They wish for a new world order with "an end to the arms race, demilitarization, a redistribution of global wealth, an end to pollution, environmental harmony, and de-industrialization."

People always yearn for order and equilibrium and see the bright and shiny structure of their future in their stories of the primordial Golden Age. In that distant past, life, cosmic order, and meaning were created. Our ancestors described it in the tales they told about the origins of their ancestors and of the gods. We now use our ancestors as symbols in stories that reflect the hopes and anxieties of our own times.

According to Harmonic Convergers, August 16, 1987, was pinpointed in the prophecies of several traditional peoples, among them various tribes of North American Indians and the Zapotec people of Oaxaca, Mexico. They also claim it marks the completion of great cosmic cycles in the Maya, Aztec, and other ancient calendars. A promotional flyer from one New Age workshop organizer described the event as the "catalysing point in a 5000 year cycle, a time when the energies of the universe come together in such a way as to allow us to shift the prevailing planetary energy to one of harmony and unity."

Anyone who felt drawn to be a part of the grand shift was urged to learn as much as possible about it and to find a place to help anchor the energy on August 16. This key date also coincided with the anniversary of the death of Elvis Presley, but no one claimed the two circumstances were related (a religious cult is slowly and informally condensing around Elvis, however).

Described in its own advertising as "the most natural book of the twentieth century," *The Mayan Factor* supplied the "mathematical" foundation for the claim of pre-Columbian ancestry for Harmonic Convergence. Actually, it is unlikely that many enthusiasts really read the book and even less likely that those who did could make sense out of its arithmetical manipulations. It is filled with jargon, labored calculations, and unclear motives. Promotional descriptions say the book reveals "*scientific* information about prophecies, channels, and Native American wisdom." In fact, what it has to say about Aztec and Maya calendrical prophecy is completely inaccurate. For example, author Argüelles claims that August 16, 1987, closed the last of nine 52-year cycles that began when Cortés arrived in Mexico in 1519. In fact, the Aztec celebrated the last New Fire Ceremony and the end of a 52-year cycle in 1507, not 1519. That date according to Argüelles, also marked the end of the final 52-year cycle in a set of thirteen. They stood, he said, for the thirteen Aztec heavens, and the subsequent nine symbolized the nine Aztec underworlds. Compounding his error concerning the significance of the year 1519, Argüelles gives us a trim interpretation of Aztec calendrics and cosmology that is not supported by anything the Aztec had to say. In fact, their tradition contradicts it. Such details, however, did not deter Argüelles from informing us otherwise:

> According to ancient prophecy, August 16th marks the end of a cycle known as the fifth sun; August 17th is the beginning of a new cycle. . . . At the dawn of August 17th light a new fire and celebrate the beginning of a new cycle of life for the planet.

Borrowing the zero point, in 3113 B.C., of the Maya Long Count calendar, Argüelles also counts thirteen *baktuns*, each of which contains 144,000 days in the Maya vigesimal day count, to reach closure in the Maya calendar 1,872,000 days later on December 21, 2012 A.D. This 5125-year span of time corresponds, he insists, to the period in which the earth passes through a "galactic beam." In 2012 we are to exit the beam and enter a phase of "galactic synchronization." Harmonic convergence is just getting us ready for that.

Because the 3113 B.C. start-up date was specified in Maya terms as the completion of the last sequence of thirteen *baktun*s, it is possible they once thought of a cosmic age as lasting that long. The end of any *baktun* would have been important, and originally the end of thirteen of them may have been especially significant. But even though the Argüelles arithmetic in this case is okay, there is no reason to believe the Maya would have thought about these spans of time the way he does. The thirteenth *baktun* after the start of the present Long Count would not necessarily have signaled the end of the calendar. By the time the Maya started computing extraordinary and truly distant dates into the past and future, the idea of packages of thirteen *baktun*s appears to have evaporated. In fact, at the end of the Classic period, around A.D. 900, the Maya quit inscribing Long Count dates on monuments altogether.

Whether we are soon coming to a turning point in the Maya Long Count or not, between the reemergence of Nostradamus in the company of astrologically predicted earthquakes and Harmonic Convergence—the stalking horse for the New Age—we begin to get an inkling of what is at work. We are tuning up for the millennium. As we approach the year 2000, we anticipate that row of zeroes rolling up on our calendrical odometer and expect big changes to accompany the round numbers.

We bundle the years decimally. The Aztec preferred to package them in bouquets of fifty-two. The numbers may be different, but the story is the same. We tell ourselves the same tale of the passage of cyclical time, a story about beginnings, endings, and new beginnings, and that, at least in part, is what the New Age and Harmonic Convergence are all about.

Conventional astrology also wormed its way into the ancient Mesoamerican underpinnings of the Harmonic Convergence. The most concise description of the celestial ambience was provided by Joseph Jochmans, author of another New Age book, *Rolling Thunder: The Coming Earth Changes.* By his analysis,

> August 17, 1987, will witness a grand trine in fire signs—with Sun, Mercury, Venus and Mars in Leo; Jupiter in Aries, and Saturn and Uranus in Sagittarius; plus Scorpio Pluto sextile Neptune in Capricorn; Moon in Gemini opposing Uranus, sextile Jupiter.

What all this may mean is in the eyes of him who reads the horoscope. It was explained in the August 31, 1987, issue of *Time* as an August 16 alignment of three planets with the new moon. That's a little hard to figure since Harmonic Convergers saw a last quarter ("half") moon near the meridian at dawn on August 17. Even without the moon, it was a pretty loose alignment.

Most people who knew about the Harmonic Convergence didn't know that the closest grouping of planets in August, 1987, actually occurred on August 20–22. For most participants, what was happening was apparently not as important as the idea that "something was happening."

As the date of Harmonic Convergence moved closer, people naturally wanted to know what would be happening. Its organizers again came to our rescue with vital information: "Concerts, ritual, dance, and theater events, kite flyings, and outdoor extravaganzas of every sort will create a global fair woven together by electronic media into a unique celebration that promises to raise the human spirit through a single collective experience." Gary Trudeau's nationally syndicated newspaper comic strip "Doonesbury" laced his portrait of the Harmonic Convergence with battery acid. Acknowledging the uncertain outcome of the event, "Doonesbury" outlined the two most likely possibilities. On the one hand, the earth would experience "a new age of insight and understanding, of mass unification of divine and earth-plane selves." Alternatively, August 16 might just bring nuclear annihilation. Boopsie, the "Doonesbury" New Age channeler of Hunk-Ra, a 21,355-year-old warrior, concluded, "Either way there will probably be a crafts fair."

When the Harmonic Convergence finally arrived, thousands of pilgrims assembled at the selected power points. It is not clear that the required 144,000 all put in their appearances, but small crowds did gather at Sagaponack Beach on Long Island (certainly one of the prime spiritual synapses on the entire planet) and in Central Park, New York. Eight hundred arrived in time for the new dawn in Chaco Canyon, New Mexico. The biggest flock of the faithful converged upon northern California's Mount Shasta, a traditionally sacred peak to the Indians of this area. It was listed as one of the nine "Sacred Mountains" for converging harmonically. Some of those from Los Angeles who couldn't get away to Mount Shasta for the weekend converged instead before dawn in Griffith Park on the observatory's front lawn.

Meanwhile, strange things were happening. In the town of Mount Shasta, an "angel" mysteriously appeared on the screen of a television belonging to Diane Boettcher, a writer. The *Los Angeles Times* story about this and other Harmonic Convergence news from Mount Shasta also quoted Don Lovett, a Minnesota television repairman in town for the event. He called the image "an apparition materializing itself outside the rules of generated video signals." The November, 1987, issue of *Outside* magazine, on the other hand, reported the assessment of local television repairman Bob Wilson. The "angel" was the "result of a defective capacitor and a low-voltage power supply." Presumably no one discouraged the five thousand people who toured Ms. Boettcher's home for a view of the phenomenon. Mr. Wilson said, "There were people praying in front of the TV, and I didn't know how to break it to them."

To most people, believers or skeptics, the Harmonic Convergence is harmless and undeserving of criticism, but I am not so sure. If there really were something to these galactic beams, powerful cosmic energies, synchronizations with six other solar systems, and UFO-piloting space brothers that converged in the vision seen by José Argüelles while he was driving down Wilshire Boulevard in a rental car in 1983, I am not certain I want to trust the earth to 144,000 self-chosen "Sun Dance enlightened teachers." After all, what if they get it wrong? Then where are we? Fortunately, there is no basis in physical science to suggest we have anything to worry about. So is there any harm in Harmonic Convergers pursuing their fancies? I think so. By promoting the idea that more than a hundred thousand people should converge simultaneously on a selected group of ancient, prehistoric, or natural sites on a key date, they expose those vulnerable and delicate places, which are precious to us all, to use and abuse they really can't sustain, all in the name of sensitive and thoughtful stewardship of the earth.

Misappropriating the beliefs of our ancient ancestors is never likely to end because it does do something our psyches seem to need. But the superficial and false claims made about the past in the name of Harmonic Convergence dishonor the legitimate achievements and traditions of ancient and prehistoric peoples. We shall never know them intimately, but avoidable distortions of their beliefs and ways of life only muddy our own thinking and make us less able to evaluate objectively our lives and times.

As a belief, Harmonic Convergence contains an explicit streak of classical astrology, a mild dose of extraterrestrials, various references to ancient astronauts, and mysterious transcendental rays from outer space—"a new galactic type of energy" that was to resonate with the earth's own "frequency"—and brings us back into harmony with the cosmos. It is essentially a religious response to the chronic uncertainty of the times, and it is particularly well-adapted to the jargon of our age. It is another one of those tales told about the sky and intended to marry the cosmos with earth.

20

Beyond the Blue Horizon

We like to think that folktales are told by our less scientifically sophisticated neighbors and ancestors, not by us. They, after all, were superstitious. They believed in gods, spirits, monsters, imaginary realms, uncanny forces, and mysterious events. But we, too, operate according to belief, and the themes in our belief system suit the technology and the times. They include space travel, extraterrestrial life, UFOs, and contact with aliens—the real undocumented aliens, those not of this earth.

According to a 1986 National Science Foundation report, about 43 percent of the American people think it likely that at least some UFOs are "space vehicles from other civilizations." Those relatively neutral letters—U-F-O—stand for unidentified flying object, no more, no less, but they have become synonymous with another coined name, *flying saucer*. Unlike *UFO*, the words *flying saucer* are charged with special meaning. Flying saucers mean spaceships from other worlds.

Numerous books have been published on the subject of flying saucers and UFOs. Some are serious, scientific attempts to come to terms with the reported sightings of unidentified objects. Others speculate extravagantly about the nature and motives of the visitors without the slightest shred of interest in evidence and proof. And there are, as well, first-hand accounts by those who claim to

It is now the era of Groucho and E.T. Space travel and extraterrestrial life are so widely accepted, kids masquerade as aliens for Halloween. (Photograph: Robin Rector Krupp)

have made contact with the spaceships or their occupants. UFOs have been featured in specialty magazines and mainstream slicks, in science fiction, in motion pictures, and on television for decades. They now appear in toys, electronic games, comic books, and even food—Franco-American UFOs pasta and Sunshine Orbit Space Cookies.

Not everyone believes flying saucers and UFOs are real, however, and even those who are sure they exist can't make up their minds about what the strange flying disks and mysterious lights really are. Whether you believe in them or not, they are an accepted item in the vocabulary of the twentieth century.

THE COMING OF THE SAUCERS

Encounters with UFOs or with the occupants of flying saucers have been reported for more than four decades. The first use of the term was inspired at 3:00 P.M., Tuesday, June 24, 1947, when Kenneth Arnold, a forest fire and air rescue pilot and a fire-control equipment salesman from Boise, Idaho, was flying his own small plane in the vicinity of Mount Rainier, Washington. After sighting nine disks slicing in formation through the atmosphere at an altitude Arnold estimated as 10,000 feet and with a speed he clocked at 1,200 miles per hour, he reported his story to locals at the Yakima airport. His claims were met with disbelief. He flew on to Pendleton, and when he arrived, newsmen were waiting for him. He likened the flight of the disks to "a saucer skipping over water." The saucer imagery was exactly right for the newspapers. On June 26, the *Yakima Morning Herald* referred to the mystery objects as "saucer-shaped." Other alleged witnesses of similar events started coming out of the woodwork. The June 27 *Herald* used the terms "flying discs" and "sky widgets." By June 29, they were "flying saucers."

Arnold's story precipitated 850 more reports of flying saucer sightings in 1947. No one really knew what they were. Many explanations were offered, and most involved misidentification of atmospheric effects and astronomical phenomena. Interest seemed to be high in everyone except, as Arnold said, the U.S. Army and the FBI. High visibility hoaxes were also perpetrated. Then when Air National Guard Captain Thomas Mantell crashed his F-51 Mustang fighter on January 7, 1948, in an attempt to pursue a silver, cone-shaped flying saucer over Louisville, Kentucky, people began to worry about intentions of those who were piloting the disks. They seemed to be hostile.

We now know Captain Mantell died while trying to reach a high altitude Skyhook balloon. Flying too high, he probably lost consciousness from lack of oxygen, and his plane went into a dive. High-altitude balloons were secret devices in 1948, and for that reason the most likely explanation of Mantell's accident was not released.

Sightings of mysterious flying objects continued, and a Gallup Poll taken in 1950 demonstrated that most Americans were aware of the flying saucer phenomenon. Very few, however, believed the disks had anything to do with visitors from other planets. This was also the year that *True* magazine published an article by Major Donald E. Keyhoe, "The Flying Saucers Are Real." In it, Keyhoe, a retired Marine Corps officer, insisted that flying saucers are spacecraft navigated by extraterrestrials who have been monitoring the earth for nearly two centuries. Books on flying saucers, more articles in *True*, and stories in other major national magazines drop-kicked the extraterrestrial explanation of flying saucers into the American consciousness and scored big. By April 7, 1952, *Life* magazine was running "Have We Visitors from Outer Space?" and concluding that flying saucers might be from another planet.

THEY CAME FROM OUTER SPACE

Popular culture responded quickly to the reports of flying saucers. The movie *Killers from Space* (1950) brought a flying saucer attack from "another planet." In 1951 Hollywood managed to smash a flying saucer piloted by a murderous humanoid sentient vegetable into the icy Arctic, and *The Thing from Another World* inaugurated the cinematic invasion of earth. Although the creature is defeated, the film ends with a warning for the whole world: "Watch the skies, everywhere, keep looking, keep watching the skies." Dreamy flying saucers up to no good appeared in *Invaders from Mars* (1953) and executed interplanetary abductions in *This Island Earth* (1955).

In general, the saucers were hostile, but Klaatu, the alien who lands his flying saucer across the street from the White House in *The Day the Earth Stood Still* (1951), carries a coercive but helpful message to a world at grips with Cold War anxiety and the threat of thermonuclear war. Unfortunately, the trigger-happy people of earth shoot Klaatu—twice! He has to be brought back to life by alien science. Klaatu is a benevolent alien despot that has come to save us from ourselves, and we haven't exactly provided friendly hospitality. He explains that the earth's nations will have to give up their armies and weapons or Gort, the giant robot and interstellar policeman who has accompanied him, will destroy the planet. The earth will be burned to a cinder at the first sign of aggression spreading beyond its surface. If the earth threatens the security and peace of the federation of other worlds, its existence will not be permitted. Klaatu's instruction is simple: "Join us and live in peace, or pursue your present course and face obliteration. We shall be waiting for your answer. The decision rests with you."

Perhaps a more realistic attitude toward the earth and its inhabitants was taken by the alien chameleons who accidentally crashed their giant spacecraft in the Arizona desert in *It Came from Outer Space*. This was a 3-D movie, and the opening sequence with the spaceship's impact is spectacular and legendary. The aliens mean no harm and just need time and materials to repair their ship. They kidnap human hostages to protect themselves from exposure and attack from the locals, but it's a close thing. Astronomer John Putnam wants to go along with the program, but the sheriff wants to arrest the aliens. With the astronomer's help, the hostages are released and the visitors blast off. They didn't mean to come to earth. They didn't want to stay on earth. And they got away from earth as fast as they could. Putnam understands the limits of this first contact and says, "It wasn't the right time for us to meet just now. There'll be other nights, other stars for us to watch. They'll be back."

We can hear in all of these films a growing consciousness of the universe beyond our atmosphere. All of the aliens, however, are reflections of our own desires and concerns. Emphasizing the theme of alien invasion, they mirror the insecurity of the times in which invasion by, or at least conflict with, the Soviet Union and even Communist China seemed possible.

We also had a growing sense of future technology and the exploration of space. All of these ideas converged in the 1950s films about aliens and UFOs. Patrick Lucanio, the author of *Them or Us*, a careful study of the function and meaning of the alien invasion film of the 1950s, recognized most of these themes at work in those movies and concluded they were a natural and even useful consequence of what was going on in our minds. What seems to command our attention in any story we tell actually reveals our character. The flying saucers in these films had a job to do. Despite the exploitive packaging of these movies and the pulp thrills they provided, their story lines allowed us

An uncanny beam of light emitted by a craft from outer space abducts a small airplane and its occupants to another world in This Island Earth. *In many respects, this 1955 science-fiction film is a Sears catalog of UFO themes and images, and even its title emphasizes the impressions of some UFO advocates that we are isolated, helpless bystanders in a cosmos bustling with advanced and alien intelligences. (© 1955 Universal)*

to get used to a changing concept of ourselves and of the universe.

ABDUCTED BY ALIENS

The UFO story has persisted and evolved for more than forty years, and it continues to mutate with the changing times. In 1953, contactees like George Adamski, who claimed to have taken rides around the solar system with the saucer folk, were generally classified as lunatic fringe, while sightings of disks could be taken seriously. Most conservative saucer-investigation groups ignored the contactees, but this aspect of the UFO phenomenon would slowly grow in importance. By 1970, those who just claimed to have seen aliens were still often ridiculed, but not always. The late Dr. J. Allen Hynek, an astronomer, began consulting for the Air Force on UFOs in 1949 and made a place for sightings of extraterrestrials in the classification scheme he included in his 1972 book *The UFO Experience.* Such events were "close encounters of the third kind."

Encounters with aliens seemed to accelerate through the 1970s, and they began to turn into close encounters of the worst kind. Stories of abduction began to surface. Probably the most influential of these was the 1961 experience of Barney and Betty Hill. It was published in 1966 in John Fuller's book *The Interrupted Journey. The UFO Incident,* a television dramatization of the tale, increased public awareness as well as acceptance of the idea of alien abduction, and by 1987 New York artist and UFO investigator Budd Hopkins was prompting people to listen to the stories of numerous abduction victims and publishing some of them in a successful book, *Intruders.* Whitley Strieber, a successful author of horror novels, received a one million dollar advance from his publisher for his own "true story" of alien—but not necessarily extraterrestrial—abduction and assault. His book, *Communion,* was published in January 1987, jumped onto the national bestseller lists by March, and stayed there for months. It, like *Intruders,* was republished as a mass market paperback in 1988. Lots of people were now reading about aliens, abductions, and UFOs. Mass culture had moved those ideas firmly into the field of potential belief.

MAYBE THEY DIDN'T COME FROM OUTER SPACE

Most popular treatments of UFOs—both the allegedly factual accounts of sightings and strange experiences and the fictional use of them in entertainment and even advertising—start with the notion that their origin is extraterrestrial. If public acceptance for the extraterrestrial explanation of UFOs were not high, such advertising would not be possible. For example, one full-page photograph that appeared in several national magazines featured a strange disk hovering above the trees and rooftops of an ordinary neighborhood and accompanied it with one line of text: "We know why they're here." It wasn't a broadside on behalf of UFOs. It was a message from the National Federation of Coffee Growers of Colombia, and they wanted us to know their coffee is what the space visitors are really after.

Most believers in UFOs and many of those who claim to have had a personal experience with a strange craft and its unearthly occupants accept the idea that the ship and the aliens come from outer space. Critical reasoning and disciplined evaluation of the evidence, on the other hand, continually lead skeptics to conclude something else: Whatever UFOs may be, spacecraft from another world make no sense. Dr. George Abell, an American astronomer and a good-humored but unrelenting adversary of pseudoscience, used to say, before his death in 1983, that extraterrestrial visits were the least likely explanation of UFOs.

As recently as 1987, a grand afternoon stir was created in Los Angeles by the appearance of a shiny, starlike UFO that lingered for more than an hour over the Pacific and in full sight of anyone who cared to look. It was brighter than any planet and in the wrong place at the wrong time to be any familiar celestial object. Informed of the event by local media. I took the obvious step and looked at the mysterious object through the twelve-inch Zeiss refractor on the roof of Griffith Observatory. The telescope told the story in a second: It was a high altitude sounding balloon with a packet of instruments, picking up sunlight and reflecting it like a brilliant star. But for a while it was a superb UFO.

Most people are not used to observing the sky, are not familiar with what they might find there, and have a hard time describing precisely and objectively what they have seen. You don't have to collect UFO reports to find this out. It becomes obvious whenever we gather eyewitness data on a more normal phenomenon—the fall of a bright meteor. You could grade those reports with a bell curve. Most people more or less get most of the basics right. They know roughly when the meteor was seen, how long it lasted, which way it was headed, and how bright it was, at least in descriptive terms. A small number of people are remarkable eyewitnesses. They report more facts with more detail and more accuracy than anyone might imagine. And finally there are some witnesses who are just outrageous. They don't get anything right, not even the day on which they saw the meteor. We can assess these individual reports once we have a large enough sample to reveal the basic pattern of the event, but if we had to depend upon a single witness, we wouldn't know where we stand. That is the problem with UFO reports. Usually the information is anecdotal, supplied by one witness, and we can not always tell how reliable that witness is.

TOO MANY LANDINGS

Dr. Jacques Vallee, a scientist trained in astronomy and information science and an author of many books on UFOs, is the real-life inspiration for the intriguing and appealing UFO specialist LeCombe, who deciphered the mathematical language of light and tone "spoken" by the alien visitors in *Close Encounters of the Third Kind* (1977). Vallee is convinced UFOs are physically real, but he, too, says that alien spacecraft do not explain the UFO phenomenon. A fairly simple calculation by Dr. Vallee, however, shows that UFOs themselves contradict the idea of visiting spacecraft. It is possible, of course, to count all of the sightings we've collected, and Vallee decided to tabulate close encounters of the first kind—unambiguous sightings of near approaches and landings of UFOs. First he looked at the times at which they were seen and quickly discovered that most close encounters of the first kind take place after dark, peak at 10:30 P.M., and drop off after that except for a small increase again near dawn. If the pattern of these landings is random, however, we must be missing many of them. Most people are asleep through most of the night. So Vallee corrected the data for this effect, and he found that the vast majority of landings, seen or not, would have to be taking place between 1:00 and 3:00 A.M. From this information, it was possible to calculate the total number of landings that should have taken place in twenty years of UFO sightings. During those two decades, about 2,000 close encounters of the first kind had been reported worldwide. If we had seen them all, however, we would have records for 30,000. But this is still not the right answer. Our calculation started with the assumption that every observation is reported, and the experience of UFO investigators argues otherwise. Vallee estimates only one in ten is reported. That boosts the landings to 300,000. The original data, however, tell us that most reported UFOs are seen in remote areas, away from dwellings. If the people now concentrated in cities were redistributed evenly over the land of the earth, how many landings would be detected? Vallee estimates the final total at three million landings of UFOs around the world over two decades and then asks a simple question: What could possibly possess extraterrestrials to make three million earthfalls in twenty years? It doesn't make sense. Vallee then questions the original assumptions. Random visits by exploring extraterrestrials are out. Either the timing, location, and witnesses of landings are carefully selected in advance by the visiting aliens, or something else is going on. In either case, according to Vallee, the UFO events are staged. Who is mounting these little melodramas and why, he maintains, are the deepest mysteries of the UFO phenomenon. Maybe so, but there is some reason to think the invisible impresario of UFO performances is in our heads and not hiding out somewhere behind the wings or in the sky.

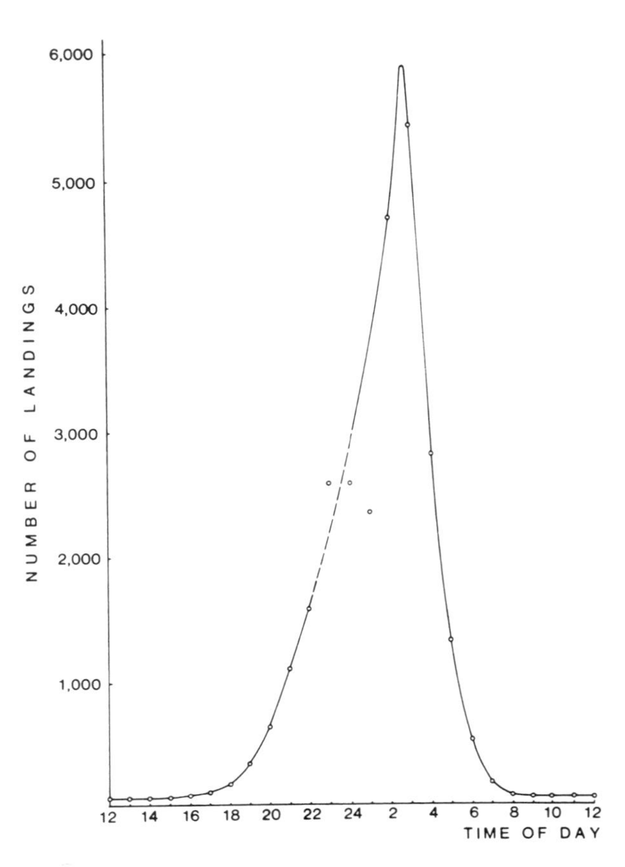

By examining the data statistically, Dr. Jacques Vallee learned that there have simply been too many UFO landings. This diagram shows that if UFO eyewitness reports are random, most UFO landings actually take place at about 2:00 A.M. Many of these go unseen, and a complete accounting for all of them over two decades of UFO reports implies that three million landings have actually occurred. That number makes no sense if we are dealing with visits by extraterrestrial spacecraft. (Griffith Observatory, after Jacques Vallee)

DINING WITH FAIRIES

In an earlier book, *Passport to Magonia,* Jacques Vallee explored the similarities between tales of contact with UFO occupants and tales of contact with the fairy folk. These are not the fairies of sweetness and light featured in "The Nutcracker Suite" segment of Walt Disney's *Fantasia.* They come instead from elsewhere, not this world, not this dimension, not this side of the highway. A stay with them takes a normal human being out of the familiar stream of time and cupboards him some place where time passes with a different beat. The secret commonwealth of the fairies is a world of peril. When it intrudes into our kingdom of cosmic order, we endure an encounter with chaos. Trafficking with the fairies is definitely risky and can mean death. Their territory is like the underworld and the land of dead. If you eat the food they serve in fairy land, you never return. It is more effective than the pomegranate seeds Hades persuaded Persephone to nibble.

Some traditional fairy tales, then, are perfectly good accounts of alien abduction in a UFO. "Connla and the Fairy Maiden" is one of them. Collected by the British folklorist Joseph Jacobs in the last century and published in *Celtic Fairy Tales,* it is an old Irish story that probably goes back to the seventh century and that contains elements of an even older tradition. In it, Connla, the son of a great warrior-king encounters "a maiden clad in strange attire" on a mountaintop. Although others accompanied Connla, including his father, only Connla could see the mysterious woman. They could hear her, however, and she said she came from the Plains of the Ever Living and that she lived inside the hills, within the earth, as fairies were believed to do. She announced her intention to take Connla away with her to the Plain of Pleasure. The others tried to prevent this with the help of Druidical magic, but they failed. The fairy tossed an apple into Connla's hands, and from then on he remained on the mountain, eating only the apple, and yearning for the fairy's return. After each bite, the apple was magically restored. After a month of nothing but bites of fairy apple, Connla left with the fairy for the magic land across the western sea in her round, crystal boat. The old king watched the shining craft glide over the sea into the setting sun, and Connla and the Fairy Maiden were never seen again.

This is not a story about interstellar spacecraft and extraterrestrials. It is about Paradise, death, and transcendence, and in its way it shares fundamental themes with the modern myth of UFOs.

UFO stories can also borrow symbols from the fairy tales. On April 18, 1961, Joe Simonton, a sixty-year-old Wisconsin chicken farmer, watched a shining flying saucer land in his yard at eleven in

Connla and the Fairy Maiden is an ancient Irish tale that could pass as an account of alien abduction. Connla is entranced by the Fairy Maiden and is about to depart with her in her "gleaming, straight-gliding crystal" round boat. (Drawing: John D. Batten, from Celtic Fairy Tales *by Joseph Jacobs, Dover, 1968)*

the morning. The hatch opened, and three short men emerged. Joe said they "resembled Italians." They made it known with gestures and objects that they needed some water, which Joe fetched for them. In the meantime, one of the men started frying something up on a flameless grill. Joe gave them the water, and they gave Joe a present. Then they climbed back into the ship. It rose into the air and shot out of sight to the south.

Now this is spectacular. Here we have an extraordinary story of an encounter with a UFO and its crew, and we also have physical evidence of the visit—the present the aliens gave to Joe Simonton. It wasn't a piece of strange metal not of this earth. It wasn't an unfamiliar object or an amazing new invention or a product of advanced extraterrestrial technology. The aliens gave Joe Simonton three little buckwheat pancakes. They were about the size of cookies. Joe ate one and said it tasted like cardboard. He turned the other two over to the U.S. Air Force, which asked the researchers at the Food and Drug Laboratory of the Department of Health, Education, and Welfare to analyze it. They did. Here is their breakdown: hydrogenated fat, starch, buckwheat hulls, soya bean hulls, and wheat bran. They were pancakes, perfectly ordinary terrestrial pancakes. It is not easy to know what to make of this, but food is a traditional gift from the fairy folk, and a fairy story told by the Celts of Brittany involved an unending supply of buckwheat cakes provided by the fairy folk. We seem to keep telling the same story in different ways.

NATURALIZING THE ALIENS

If UFO sightings are not really encounters with spacecraft from another world—and Joe Simonton's pancakes are enough to make us wonder about this—what are they? However compelling the mystery of UFOs may be, that is not what concerns me here. I am instead interested in the stories we tell about UFOs. They are sky stories, and their content and meaning reveal something about the way we use the sky symbolically today. In that sense, they are like the sky myths and sky lore of our ancestors. Flying saucers and visitors from outer space now take center stage in folktales we tell.

We are "folk," too. We are a "traditional" people with beliefs that touch us deeply and that have power over our behavior because they color the view we have of the world and of ourselves. We know we are entertained by movies, like *Close Encounters of the Third Kind* or *E.T.*, about aliens from outer space, but why are such stories so appealing to us? Millions of people wouldn't go to see *E.T.* and wouldn't respond to the merchandising of its main character into an icon of the times if the fable didn't have some power, didn't somehow get under our skin.

Other worlds, extraterrestrial visitors, and alien spacecraft have found a home in the symbolic imagery, in the storytelling, and in the belief system of our time because they let us make room for the miraculous in our lives. That is one of the things myth and folklore also did for our ancestors, but the things that made them marvel and wonder are consigned by us to the supernatural, a realm that enjoys less currency under the scrutiny of science. Our scientific approach to the universe has increased our knowledge of natural phenomena and at the same time banished the miracles that dazzled our ancestors. Some of their miracles were consequences of a cosmos in which celestial gods held court somewhere above the clouds and had the power to settle accounts here on earth. Our Big Bang relativistic universe, on the other hand, has no place for sky gods and supernatural journeys to other worlds. But a capacity for the miraculous—along with the rational—seems to be standard equipment in the human psyche.

To accommodate the miraculous in a scientific age, we have disguised it with the livery of science. Evidence may not endorse the existence of extraterrestrial intelligent life, but we have devised a picture of the universe that leaves room for it. In doing so, we have made a place at the table for uncanny beings and their mysterious means of transportation by making them conform to at least the soft edges of our world view. Because we can imagine a universe populated with perhaps a hundred billion galaxies, with hundreds of billions of stars in many of those galaxies, and with anybody's guess how many planets may orbit at least some of those stars, we can deal with the mysterious encounters people report as if they were part of physical reality. In addition, our own experience with high technology and space travel and our ability to extrapolate that experience permits us to turn strange aerial phenomena into machinery.

This story we are now telling about aliens and UFOs is an old tale recharged with images borrowed from genuine science to let mystery reemerge. Rationalizing shining celestial disks and strange beings, this myth inspires religious awe and spiritual adventure with the language of interstellar travel. We have made the miraculous compatible with a universe in which we believe.

The supernatural was an essential ingredient of the ancients' world view, but it conflicts with ours. By turning the supernatural into part of the natural, physical universe, we allow it to mix in rational company. Our UFO folklore resolves the conflict between the miraculous and the ordinary. That doesn't domesticate or diminish the miraculous. It lets us acknowledge its existence.

Our need to recognize the miraculous is a religious response and is probably, at the core, a product of our perception that the cosmos doesn't really make sense, at least not all of it. It contains a mystery, and the mystery is not trivial. The mystery is not how it came to be or why it is the way it is. Modern cosmology and particle physics are apparently addressing those questions with some success. It is a profound mystery to us, however, that anything exists at all, and our imaginations have managed to deal with that uncanny mystery by recognizing the miraculous.

Although the idea of visitors and spacecraft from other worlds fits into our picture of the physical universe as a possible, if unproved, option, UFOs also have a symbolic and religious dimension, and that is what their traffic with the miraculous is all about. There are many ways to look at it, however, and some are less sophisticated than others. Barry H. Downing, a Presbyterian minister and the author of *The Bible and Flying Saucers,* takes a divine brute-force approach to the problem. He spelled out his perception of the religious function of UFOs in comments published in the March, 1988, issue of *Omni* magazine. "It's really an issue of faith," he said. "God has to stay out of sight so he can test faith, yet remain close enough to keep us from despair. UFOs have taken up this role in our culture. They stay out of sight enough so we have to go on faith, but they give us a sign that someone who cares is watching over us. They're like shepherds watching over their sheep." Without refinement or subtlety, Downing has equated UFOs with the miraculous in a conventional Christian setting.

In our times, thermonuclear weapons understandably represent the ultimate intrusion of chaos and operate as a symbol of the apocalyptic conclusion of the era in which we live. The image of global annihilation these warheads inspire has prompted some people to see the alleged visitors from outer space as guardian angels who can deliver us from our own folly. These wise and benevolent New Age aliens who are waiting to deliver to us a cosmic love letter of revelation and hope are actually rooted in Klaatu and Gort, the alien strong-arms with the carrot and the stick in *The Day the Earth Stood Still.*

There is, then, a component of the population that approaches the concept of extraterrestrial visitors from a religious point of view. For others, the connection between UFOs and the sacred is spelled out as a morality play with lessons about environmental and thermonuclear catastrophe, but even for them, the real message of extraterrestrial spacecraft is celestial transcendence. We have replaced the old sky gods with new ones better adapted to the universe we believe ourselves to inhabit.

This is not really a surprising development. The imagery with which we describe the universe and define our place in it has always originated from the contacts we make in everyday life. Our ancestors maintained a daily intimacy with the sky, and the sky put a frame around their lives. We don't have that same kind of daily interaction with heaven, but the high technology that frames our lives puts us in regular contact with our own vision of outer space. It is no wonder we are also preoccupied with the possibility of visitors and spaceships from other worlds.

FUELED BY THE SACRED

Despite the references to space travel and inhabitants of other worlds, many of the stories about UFOs tell us they are really encounters with the sacred. Now by *sacred,* I do not mean the relatively secure and conservative environment of a church or temple or the familiar doctrines of rational, intellectualized religion. I mean *sacred* in its true sense, the sense we sometimes have of the mystery of existence and the power of the world's order. Direct experience of the sacred in these terms is the revelation that is at the core of our belief systems. It is something that emerges in the ecstasy of religious myths. It is the realm of power and peril entered by the shaman, the technician of the sacred, who seeks cures, quests for knowledge, and guides the dead. To experience the sacred is to cross a border always in sight but only sometimes approached, like the divider line on a two-lane highway, which is always there but crossed only now and then and with deliberation. When we do cross into the realm of the sacred, we suddenly find ourselves immersed in different, and sometimes threatening, territory.

It is hard to put a finger on exactly what the power of the sacred really is, but we distinguish it from the rest of our experience by recognizing that it is wholly, uniquely different. Its realm and its experiences are sometimes called *numinous*, that which, according to the dictionary, "surpasses comprehension or understanding." Mircea Eliade, in *The Sacred and the Profane*, explains that it is what is "induced by the revelation of an aspect of divine power." It is the other side of the highway. Traditions of peoples worldwide and accounts from ages before our own turn this transcendent sense of the sacred into a key component of religion. The person who confronts it, by any route, is usually overwhelmed. Emotions are charged. Feelings of awe, majesty, terror, spiritual liberation, and personality integration are all potentially a part of it. If the intensity of the confrontation does not shatter the personality, the person comes away from the experience transformed, renewed, and reintegrated. This may sound rather lofty and sublime, and more than a little muddy, but people do describe these experiences. That is what counts here, how these experiences feel, not what they really are. Traditional belief systems incorporate several different mechanisms for activating them and giving them meaning, including the shaman's mystic vocation, the spiritual and cultural initiation of the youth, the ceremonial consumption of hallucinogenic drugs, and the rites of birth and death.

Close encounters with UFOs are the sacred re-emerging by hook or by crook in a secularized world. We smile generously at the folk beliefs of simpler societies and run into the same old cosmic mysteries that commanded their attention. In our case, we dress them up so they are presentable for the space age.

According to Allan Hendry, author of *The UFO Handbook*, eyewitness accounts of close encounters with UFOs are human interest stories, not news, and for that reason we rarely find them in the headlines or even in the newspaper at all. Most people are not aware of the character of these tales, unless one of them is dramatized for television or included in a bestseller. But while the rest of us are recreationally pondering the possibility of extraterrestial life and visitors from outer space, a few of us are experiencing something overwhelming. It threatens personal identity. It can mean transformation and change. In the end, it may sparkle and glow. But getting there can be terrifying.

GETTING TAKEN FOR A RIDE

Barney and Betty Hill's UFO abduction was one of those stories of terror and transformation. It began on the night of September 19, 1961, when they were driving home through the Connecticut River Valley. They were returning from Canada to Portsmouth, New Hampshire. At about ten-thirty, Betty Hill noticed a strange light in the sky. She also observed that the moon was bright and that there was a star below and to the left of the moon. The strange light was starlike, but bigger and brighter than the one below it, and it was the beginning of a tale of chase, fear, capture, abduction, release, and amnesia. Whatever happened to Betty and Barney Hill was so traumatic, they repressed their memories of it. It didn't reappear until two years later, when they entered the care of Benjamin Simon, M.D., a Boston psychiatrist, for treatment of chronic anxiety. Extracted under hypnosis, the story contains many interesting and oddly contradictory details of their experience with aliens aboard a flying saucer. Both clearly believed they had been kidnapped and medically examined by the aliens, but their accounts of the exchanges between them and the presumed extraterrestrials don't really sound like what we would expect from an unprecedented and perhaps experimental encounter between two independent intelligent species. Their conversations sound like something else. Dreamlike, contemplative, and absurd, their story puzzles the reader with what seem to be non sequiturs and childlike reactions on the part of both the humans and the aliens.

Betty is terrified by a huge needle one of the aliens intends to insert in her navel and cries in pain when he carries out the procedure. He tells her it is a pregnancy test. Certainly it's an unsettling incident, but the experience has comic moments too. One of the aliens runs into the room where Betty is held, and he's carrying Barney's false teeth. Excited over Barney's dentures, they try to remove Betty's teeth, but hers are real and won't come out. The aliens are puzzled by this and ask her why. Betty laughs about the confusion and tries to explain how it is that people may lose their teeth and need dentures. The alien asks her if it happens to many people, and she says it happens to almost everyone as he or she gets older. Now this really confuses the visitors. They don't know what "older" means, and although Betty tries to explain it, they just don't get it. Betty is now educating the aliens. It becomes clear they don't have the slightest understanding of age, lifespan, or even a year. So Betty tries to explain to this visitor from outer space, who must have traveled light years between the stars and crisscrossed the solar system to reach the earth, how the earth is in orbit around the sun and how it goes completely around in a year, and he never does figure it out.

Betty tells the alien she knows he isn't from the earth, and she asks him where he comes from. He asks her if she knows anything about the universe. She says a little, not much, and then the alien walks across the room to the head of a table, does something that opens the metal wall, and pulls out a map. The map has lines and dots on it. Betty asks him what they mean, and he says the heavy lines are trade routes, thinner lines mark the routes to places they visit occasionally, and broken lines represent expeditions. Betty asks him where his home port is, and he says, "Where are you on the map?" Betty looks, laughs, and says, "I don't know." So the alien replies, "If you don't know where you are, then there isn't any point of my telling you where I am from." This sounds like one of those exchanges between Alice and the Caterpillar in Wonderland. After the alien puts the map back into the wall, Betty feels stupid because she doesn't know where the earth is located.

Much of this story displays a sense of disorientation in space and time, something that often occurs in encounters like this. Attempts have been made to match the pattern on the star map, which Betty remembered and later drew, with real stars, but that makes as little sense as Betty's own story. Alleged similarities to the real geography of outer space depend on assumptions and arbitrary interpretations. It also makes no sense that there would be a map in the first place, unless it was for classroom demonstrations. We don't navigate our spacecraft by maps, and it's even less likely interstellar visitors would find them useful.

The story of Barney and Betty Hill is much longer and more complex and more enigmatic than what has been detailed here. Much of it can't be explained, but we think we know what that light was that seemed to trigger the whole affair. Astronomical analysis of the sky on the night of September 19–20, 1961, was done for the first time in the fall of 1975 by Dr. Edward K. L. Upton at Griffith Observatory and independently by Robert Sheaffer, a skeptical UFO investigator, professional sci-

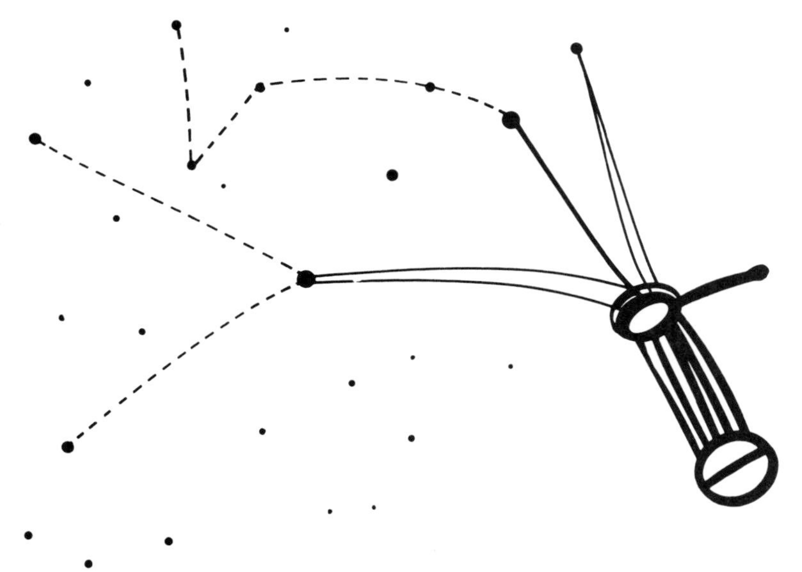

Under hypnosis, Betty Hill described an ordeal she believed took place aboard an alien spacecraft. While in the flying saucer, she saw a star map on the cabin wall. Her abductors told her that the heavy lines indicated routes that were traveled frequently. Thinner lines meant less interstellar traffic, and the broken lines led to the boondocks. Later, Betty Hill drew the chart from memory, and her map is shown here. One attempt to match the map with stars in the solar neighborhood identified the two big stars (apparently connected by a regular shuttle service) in the lower right corner of the map as the two components of a binary star we call zeta Reticuli. (Griffith Observatory, after Betty Hill, in The Interrupted Journey *by John G. Fuller, Berkeley, 1980)*

ence writer, and the author of *The UFO Verdict, Examining the Evidence.* The moon was out that night, and there was a bright "star" below it, the planet Saturn. Right where Betty said she saw the spacecraft, there was another bright object, but it was really an even brighter planet—Jupiter. Driving the car through the irregular countryside, it would have been easy to get the impression that Jupiter had suddenly appeared and was moving, even following, the car.

Even with a good astronomical explanation available for some of the factors in the Hill abduction, there is still plenty in the story to mystify us. There is a curious psychology in it, and it also has a religious component. When Barney was returned to the car with Betty after the abduction, he chuckled and mused to himself about the experience. After acknowledging that he never believed in flying saucers before, he now wondered where they came from. And he wished he had gone with them. Such a trip, he says, might prove the existence of God, and he added, "Isn't that funny? To look for the existence of God on another planet?" He had seen and been a part of something different from anything he had ever encountered.

TUNNEL VISION

Some of those who believe they've had a UFO encounter describe an ordeal that parallels what other people—survivors of serious physical trauma—describe as the "near-death" experience. In this, too, space and time take on different meaning, and the person may have a sense of floating outside the body, perhaps in a big room. Floating and roomy interiors of spacecraft are frequent components of abduction accounts. The sensation of a bright light and geometric patterns may initiate the encounter. In a near-death experience, the person sometimes feels as if he or she is moving down a long tunnel, corridor, or tube, and approaching a border or door. There, the individual may encounter a shining being. Telepathically communicating with the spirit of light, the person near death reviews all of the events of his or her life and is subjected to an examination of the moral character of that life. There is a feeling of physical paralysis, and a message, usually having something to do with some right action to be taken by someone, perhaps by many people, in life. The person having this vision then returns to life and consciousness. In the aftermath, a significant personality change may take place, and the person continues to recall the inexpressible nature of the unusual experience and is guided and comforted by the memory and knowledge of what has been felt. Many of these features occur in UFO abductions, in drug-induced hallucinations, in mystical experiences, in shaman's trances, in other intense experiences of the mind, and in various traditional accounts of the soul's journey after death. It is a serious adventure. The person who has it feels as if he or she has dealt with something fundamental, profound, extraordinary, and ineffable. This is the awe, the special kind of fear and wonder, induced by what is felt to be a revelation of the sacred.

ALTERED STATES

In the years that followed the publicizing of the Betty and Barney Hill case, more and more abductions were reported. Often the victims felt a transcendental change in personality and attitude. Some longed for additional contact with the aliens. Others never wanted to encounter anything like it again. Reviewing these accounts tells us much more about the altered state of mind experienced by the witnesses than about the facts of the encounters. In his commentaries about these experiences, Jacques Vallee confirms that abductees' lives are often substantively altered. "They develop," he writes, "unusual talents with which they may find it difficult to cope." The experience has a mystical or religious quality to it. It may involve an odd twist in the perception of time and space. A witness may be unable to account for his or her whereabouts for a certain period of time or may have a sense that the environment in which the encounter took place was unfamiliar or permitted things that are impossible in ordinary, three-dimensional space. Later, the witness may be preoccupied with the concepts of time and space. The experience is related to religious revelation, and the witness may respond like a prophet.

Whitley Strieber's scary story of his own experiences with what he called the "visitors" contains many apparently supernatural elements and does not really involve UFOs and extraterrestrials. Much of what he had to say, however, parallels or matches what turns up in other accounts of UFO abduction. Budd Hopkins has been counseling those with similar experiences for a number of years now, subjecting to hypnosis those who are willing, in order to obtain the details about what happened to them and ostensibly to help them cope with it. He has also been publishing books on what he believes he's learned. The human race, he concludes, is the subject of an extraordinary genetic experiment conducted by alien intruders. Why they are doing this is not clear. Strieber doesn't necessarily look at the problem the same way, but at one point he expresses the visitors' motives in related terms: "In some sense, their emergence into human consciousness seemed to me to represent life—or the universe itself—engaged in some deep act of creation."

There is no doubt that Strieber is a visionary. That may be something he shares with all of those who have encounters with aliens. Dr. Robert A. Baker, professor of psychology at the University of Kentucky, reviewed the experiences and behavior of those who believe they have been abducted in an article in the winter 1987–88 issue of *The Skeptical Inquirer* and demonstrated that what is reported is not as exotic or unknown as we might think. Baker shows that such experiences are consistent with hypnogogic and hypnopompic hallucinations and regards Strieber's description as a "classic, textbook" account of the latter. He emphasizes that perfectly sane and rational people have such hallucinations. They mean neither psychosis nor the presence of aliens. Hypnotic regression, he adds, is no certain path to real events. Confabulation, creation of false memories, inadvertent cueing by others—Hopkins, for instance, who may be present with their own agenda, and the unique psychological needs of the subject, whether an abduction has really occurred or not—all will modify the story and fabricate details that the subject will believe to be true. Getting to the bottom of alien abductions with hypnosis is not so easy. Philip J. Klass, for thirty-five years a senior editor for *Aviation Week and Space Technology* and the author of three skeptical books on UFOs, also concludes that hypnotic regression carries real risks. In a fourth book, *UFO Abductions: A Dangerous Game,* Klass examines the current abduction vogue critically, judges it is less than it seems, and argues that hypnotic regression, improperly performed, can damage a person emotionally.

MAKING THEM UP AS WE GO ALONG

By 1977, it was obvious that people who suffered unauthorized detention with the saucer people often told interesting stories under hypnosis. Dr. Alvin H. Lawson, professor of English at California State University, Long Beach, and his associate John De Herrera decided then to find out what would be said under hypnosis about imagined abductions by people who had never had a UFO experience. With the help of Dr. William C. McCall, a physician who had already used hypnosis with many "genuine" UFO victims, Lawson and De Herrera pulled abduction stories out of sixteen individuals—ten women and six men—who had volunteered for the experiment. Individuals with extensive knowledge of the UFO literature were excluded from the test. One at a time, Dr. McCall put them into a hypnotic trance and then guided them into thinking about seeing a UFO. They were to imagine boarding the vehicle and receiving a physical examination. After that, it was up to them to fill in the blanks. To everyone's surprise, the imaginary abductions were saturated with intricate detail and sounded a lot like the abductions other witnesses reported as real.

There were some differences between the two kinds of witnesses. For one group, participation was voluntary and so essentially benign. For "real" UFO victims, however, attendance had been compulsory, and that made some victims feel vulnerable and powerless. Unlike those who truly believed they had been abducted by aliens, Lawson's subjects maintained control of their emotions during the encounter, experienced no "missing time," showed no physical or physiological signs of their involvement with aliens, suffered no amnesia, were subjected to few dreams or nightmares after the experience, had no disturbing psychic or emotional effects in the weeks that followed, and had no conscious memory of the abduction. Despite these differences, the visions themselves are strikingly similar. What these experiences seem to share is the imagery adopted by the brain to cope with a certain type of perception while in a trance-like state. The difference between "real" abductions and imaginary abductions is the trauma and its subsequent impact. Something affected the "real" abductees much more strongly.

Lawson also learned that most UFO folk could be categorized into six distinct types. They include aliens that look basically human and often wear a one-piece "flight suit." Humanoids, on the other hand, resemble human beings but have disproportionately large heads and something of a fetal appearance. Their large eyes, small size, frail appearance, stiff movement, and considerable strength make them hard to miss. Animal-like aliens, often equipped with claws, a tail, pointed ears, big teeth, feral eyes, a strong odor, fur or scales, and hostile intentions, also are encountered. Robots, based on a bipedal human design, lurch along awkwardly or float easily from place to place. Lawson calls the fifth type of alien "exotic." It has very bizarre and grotesque features and sometimes combines otherwise contradictory elements. The last type of alien is like an apparition. It can materialize and dematerialize spontaneously and seems supernatural rather than physical. These aliens all seem to do business with the same tailor. Usually dressed in a seamless, one-piece suit, their bodies are covered and their heads are exposed. They wear belts and usually have some kind of conspicuous buckle, medallion, cross-straps, or light on the abdomen. Intrigued by this pattern, Lawson discovered he could find the same types of beings in lots of other places besides UFO reports. Greek mythology, Christian belief, the lore of demonology, Celtic folklore, Shakespeare's plays, Lewis Carroll's *Alice in Wonderland,* L. Frank Baum's Oz books, science fiction stories, superhero comics, and characters on the boxes of breakfast cereals—all fit the pattern. Lawson's study leaves the impression that those who witness aliens are really drawing upon a cultural fund of images of considerable antiquity and still in active use.

We also see the process at work in Walt Disney's *Pinocchio* when the Blue Fairy begins her descent from the sky as a star and arrives in Gepeto's room as a luminous sphere from which she materializes. Something very similar happens in MGM's *The Wizard of Oz.* Glinda, the Good Witch of the North, suddenly appears as a distant glistening, iridescent bubble in the daytime sky and quickly covers the distance to the Munchkin village, where she hovers into a landing on the Yellow Brick Road and condenses into her human form.

From Lawson's experiments and from intense spiritual experiences, we may guess that the brain responds in a similar way to a number of parallel stimuli that put it in a similar state. Amazonian shamans, victims of severe trauma, consumers of hallucinogenic drugs, and UFO abductees may encounter many of the same marvels: an initial bright, pulsating light, images of tunnels or tubes, varied and intense colors, rotating or spiraling images, luminous geometric patterns, erratic movement of images, large, domed rooms, the sensation of weightlessness, communication with supernatural beings, personal participation, and a feeling of mystery and significance.

These experiences are linked with some known brain behavior. For example, the spontaneous triggering of phosphenes on the retina without visual stimuli produces some of the effects of color and pattern. Hallucinogenic drugs, shamanic trances, and hard knocks on the head can induce phosphenes. Visionary images, possibly related to mismanaged memory retrieval, are also generated by drugs and shamanic techniques. The absence of sensory information may allow the memory to release intrusive, hallucinatory images. Stress can also knock out the mechanisms that normally inhibit such visions. All of this must have a neurochemical foundation, and ongoing research in these areas is targeting it.

Alvin Lawson speculated that the long tunnels, shining lights, big rooms, medical examinations, and other aspects of the near-death experience and the alien abduction were really muddled memories of birth. Astronomer Carl Sagan has promoted the

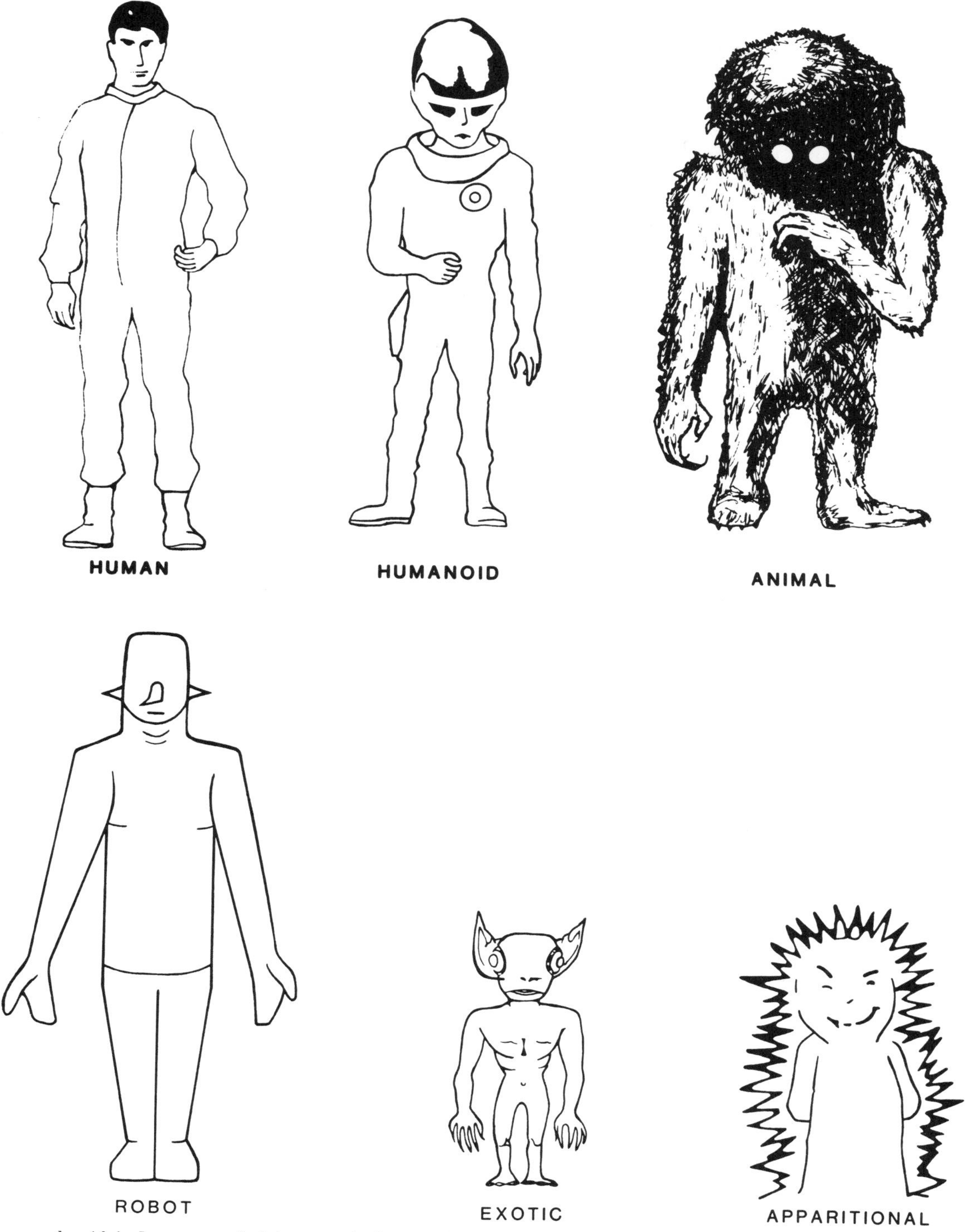

UFO researcher Alvin Lawson studied the types of aliens people reportedly encounter and discovered that they fall into six basic categories. (Griffith Observatory)

same explanation of near-death experiences. Birth is, of course, the first great transition in all of our lives and perhaps teaches us some of the first lessons we learn about the world. Symbolically, it means transformation, renewal, and new life. This is a very appealing and persuasive way of looking at these otherwise mystifying reports, but Dr. Susan Blackmore, a researcher in the brain and perception laboratory at the medical school of the University of Bristol, soundly demolished the notion of hallucinatory reruns of birth in the May 5, 1988, issue of *New Scientist,* in her article "Visions from the Dying Brain." For example, the birth canal would not look like a tunnel, even if the fetus could look at it. Also, those born by Caesarean section should not experience tunnel visions and out-of-body experiences if they are related to memories of natural birth. But Blackmore's surveys show that they do.

Dispensing with birth memory, she tries to provide an alternative explanation for the brain's behavior. The tunnel effect is always with us and is probably a natural consequence of the structure of activity in the brain's visual cortex. It maps onto the retina in rings, tunnels, and spirals. The dying brain, she says, experiences massive disinhibition. It relies on memory for sensory input to stabilize itself and orient it. Those who research drug-induced hallucinations see the same process at work there. These apparently common and universal visions depend on just what is happening in the brain's failing information processing system. It constructs a model of the world, and consciousness depends on it. If the model is traumatically affected because the brain is denied information from all of the usual sources, consciousness is going to be altered, and it is. Dr. Blackmore does not, however, dismiss these experiences as "just hallucinations." They are, she insists, "life-transforming and important hallucinations and ones we would do well to try to understand."

If this is what we are seeing in UFO abductions, it is because something has caused a breakdown in mental functions. Maybe it is just a random slip of the gears. Jacques Vallee thinks it is part of a conditioning program proctored by unseen agents for purposes beyond our ken. Whatever it is, it's like an EMP strike, an electromagnetic pulse generated by a high-altitude nuclear blast that burns out all of the electrical and electronic circuits in a nation's power supply system and communications network. Only in this case, it's the brain's circuits that blow. It's a temporary breakdown, however, and as the old standby neural circuitry kicks back into action, it helps the scrambled consciousness reassemble a more familiar map of reality. But the memory of the fireworks lingers.

What triggers the experience in the first place? Some life-threatening crisis or deep personal anxiety, according to psychologist C. J. Jung, is the culprit. He also had some ideas about those shining celestial disks or spheres that seem to have so much to do with these affairs. His book, *Flying Saucers (A Modern Myth of Things Seen in the Sky),* interprets the round and luminous objects as symbols of order, equilibrium, orientation, and integration in a modern world defined by technology. Like the mandalas of traditional people, UFOs appear in times of disorientation and when chaos intrudes. When needed, they facilitate the restoration of order. If that is the case, they seem to be modeled on those shining disks and celestial lights that really do establish order and balance—the sun, the moon, and the stars. Although a UFO experience may seem to begin with the sighting of such an object, it may actually be something added to the story after the brain has gotten its act back together and is reconstructing what went on when the lights went out. It is also possible some other neurophysiological process stimulates the vision of a flying disk and does it much more frequently and easily than it fabricates aliens.

While popular culture and public commentary continue to portray the UFO experience as a consequence of the same kind of exploration of the cosmos in which we are ourselves engaged, the real story about UFOs seems to have very little to do with spacecraft, hardware, and traffic between the stars. Instead we are dealing with the human psyche, with myth, and with belief. It is reasonable, however, to find myth and belief speaking in the vocabulary of the space age. And that, it seems, is exactly what they're doing.

UFOs may well be symbols of personal psychic integration, but we are also using them culturally. They deliver the sacred in a secular age. We make it harder and harder for ourselves to experience awe. All our ancestors had to do was look at the sky. Awe must reemerge, however, and it seems to be piloting a UFO.

BEYOND THE BLUE HORIZON

Once upon a time, dreams and visions delivered awe from heaven to our doorstep. Angels paraded down Jacob's ladder and back again to the sky. Elijah, at the end of his life and work, rode to heaven on the whirlwind, in a chariot of fire. Contact with the sky meant commerce with the divine, and it was not a business for the fainthearted. The interview was purchased with terror, and exaltation was returned as change. According to Genesis (chapter 28, verse 17), Jacob's stone pillow, the site of his dream, was "dreadful." It was also the "House of God" and the "Gate of Heaven."

Today we no longer see the universe as a sacred landscape, but awe and wonder can regularly be read between the lines in any astronomy textbook. In astronomy we contemplate the biggest, the farthest, the oldest, the brightest, and the strangest things we know. Telescopes and space rockets are real Jacob's ladders to heaven. They carry us beyond the earth to a sky that once passed as the kingdom of gods alone.

Unlike Jacob, however, most astronomers and astronauts do not enjoy an audience with angels while gathering photons from the far corners of the cosmos or while soaring in orbit. But this age—our childhood in outer space—has supplied us with many people who report encounters with aliens and shining disks. And even if flying saucers pass beyond the pale of at least part of our credulity, the concept of life on other worlds is cocktail conversation. Most people I talk to want there to be intelligent life in outer space. When we wish upon that star, we are really saying we just want to feel at home in the universe, and we would feel more at home if we were not alone.

Science fiction now makes the bestseller lists, and E.T., close encounters, and mysterious black monoliths are big at the box office. These are some of the stories we tell ourselves about the sky today, and they have something in common with the celestial visions of our ancestors. They show that the sky still moves us.

Witnesses to space shuttle launches are submerged by the roar of its engines and transfixed by the vision of power and flame lifting off like Elijah for the sky. They applaud. They shout. And they cry. And when the *Challenger* exploded, we all mourned. Something more than a rocket-borne boxcar was going up. It was the human spirit. A symbol of our loftiest visions was obliterated.

Our ancestors found signals of cosmic order in the glowing lights of heaven. They looked, as we do now, at the sky and tried to fashion from it a picture of the universe. Their tools were different from ours. And their interpretations of what they saw were different. Their cosmologies may seem quaint to us now, but I am not so sure our ideas about the Big Bang and the expanding universe—scientifically grounded ideas to which I subscribe—won't seem quaint to our descendants centuries from now.

Our ancient ancestors pondered the sky. We, on the other hand, don't just look at it. We go there. What was once a kingdom for gods whose gaze downward governed the earth is now a place we ourselves visit, a zone where space shuttles shuffle and where satellites circle and tell us many of the things we once believed only gods could know. By embracing the stories told by contemporary science—black holes, quasars, a runaway universe, planets parading to the tune of Newton's gravity—we have now put ourselves in the sky.

We've pressed a footprint on the moon, sent robots to watch the sunset from the icy deserts of Mars, and dispatched probes around the rings of Saturn, past the enigmatic moons of Uranus and Neptune, and off into interstellar space. There, those spacecraft will remain as monuments to cosmic order and to our journeys to the sky, monuments more permanent than anything our ancestors contrived.

But there is a continuity between us and our prehistoric ancestors. The path we were walking at Laetoli, in Tanzania, where we left footprints in the ash 3.7 million years ago, has taken us to the moon. The imprint of that boot sole on the Sea of Tranquility will be a record, for the eons, of that

The path creatures from earth have walked for 3.7 million years has led to the moon. The road and storytelling continue, beyond the blue horizon. (Left: Early hominid footprint at Laetoli, Tanzania. Photograph: T. White. Right: Apollo 11 *astronaut footprint on the moon's Sea of Tranquility. Photograph: NASA)*

time in the distant past when we took a step off the earth, touched down to steady ourselves, and pushed off again, beyond the blue horizon, for worlds unknown.

The blue horizon is, however, no more distant than the stories we tell. They draw us beyond the arena of everyday affairs and beckon us with the promise of transcendence and mystery. The allure of that unreachable border between earth and sky has propelled us into space. It is the cyclone to Oz, the step Through the Looking Glass, the stairway to the stars.

Bibliography

GENERAL MYTHOLOGY, FOLKLORE, AND ANCIENT RELIGION

d'Alviella, Count Goblet. *The Migration of Symbols*. 1894. Reprint. New York: University Books, Inc., 1956.

Armstrong, John. *The Paradise Myth*. London: Oxford University Press, 1969.

Barber, Richard. *A Companion to World Mythology*. New York: Delacorte Press, 1979.

Baring-Gould, Sabine. *Curious Myths of Middle Ages*. 1888. Reprint. New Hyde Park, New York: University Books, Inc., 1967.

Barnard, Mary. *The Mythmakers*. Athens, Ohio: Ohio University Press, 1966.

Berger, Pamela. *The Goddess Obscured*. Boston: Beacon Press, 1985.

Bettelheim, Bruno. *The Uses of Enchantment—the Meaning and Importance of Fairy Tales*. New York: Alfred A. Knopf, Inc., 1976.

Bierhorst, John, ed. *Four Masterworks of American Indian Literature*. New York: Farrar, Straus and Giroux, 1974.

Bierhorst, John, ed. *The Red Swan—Myths and Tales of the American Indians*. New York: Farrar, Straus and Giroux, 1976.

Brandon, S. G. F. *Creation Legends of the Ancient Near East*. London: Hodder and Stoughton Ltd., 1963.

Brandon, S. G. F. *The Judgment of the Dead*. London: Weidenfeld and Nicolson, 1967.

Bratton, Fred Gladstone. *Myths and Legends of the Ancient Near East*. New York: Thomas Y. Crowell Company, 1974.

Briggs, Katherine. *An Encyclopedia of Fairies*. New York: Pantheon Books, 1976.

Briggs, Katherine M., *The Fairies in Tradition and Literature*. 1967. Reprint. London: Routledge & Kegan Paul Limited, 1977.

Briggs, Katherine M. *The Personnel of Fairyland*. 1953. Reprint. Detroit, Michigan: Singing Tree Press, 1971.

Briggs, Katharine M. *The Vanishing People*. London: B. T. Batsford Ltd., 1978.

Budge, E. A. Wallis, *Amulets and Superstitions*. 1930. Reprint. New York: Dover Publications Inc., 1978.

Bulfinch, Thomas. *Bulfinch's Mythology*. Reprint. New York: Avenel Books, 1979.

Burland, C. A. *Myths of Life & Death*. New York: Crown Publishers, 1974.

Campbell, Joseph. *The Flight of the Wild Gander*. Chicago: Regnery Gateway, Inc., 1969.

Campbell, Joseph. *The Hero with a Thousand Faces*. Princeton, N.J.: Princeton University Press, 1949.

Campbell, Joseph. *Historical Atlas of World Mythology, Volume I: The Way of the Animal Powers:* New York: Alfred van der Marck Editions, 1983.

Campbell, Joseph. *Historical Atlas of World Mythology, Volume II: The Way of the Seeded Earth, Part I: The Sacrifice*. New York: Harper & Row, Publishers, Inc., 1988.

Campbell, Joseph. *Historical Atlas of World Mythology, Volume II: The Ways of the Seeded Earth, Part 2: Mythologies of the Primitive Planters: The Northern Americas*. New York: Harper & Row, Publishers, Inc., 1989.

Campbell, Joseph. *Historical Atlas of World Mythology, Volume II: The Way of the Seeded Earth, Part 3: Mythologies of the Primitive Planters: The Middle and Southern Americas*. New York: Harper & Row, Publishers, Inc., 1989.

Campbell, Joseph. *The Inner Reachers of Outer Space*. New York: Alfred van der Marck Editions, 1986.

Campbell, Joseph. *The Masks of God: Creative Mythology*. New York: Viking Press, 1968.

Campbell, Joseph. *The Masks of God: Occidental Mythology*. New York: Viking Press, 1964.

Campbell, Joseph. *The Masks of God: Oriental Mythology*. New York: Viking Press, 1962.

Campbell, Joseph. *The Masks of God: Primitive Mythology*. New York: Viking Press, 1959.

Campbell, Joseph. *The Mythic Image*. Princeton, N.J.: Princeton University Press, 1975.

Campbell, Joseph. *Transformations of Myth through Time*. New York: Harper & Row, Publishers, Inc., 1990.

Carlyon, Richard. *A Guide to the Gods*. New York: William Morrow and Company, Inc., 1982.

Carmer, Carl. *The Hurricane's Children*. New York: Farrar and Rinehart, Inc., 1937.

Cavendish, Richard. *The Black Arts*. New York: G. P. Putnam's Sons, 1967.

Cavendish, Richard, ed. *Legends of the World*. New York: Schocken Books, 1982.

Cavendish, Richard, ed. *Mythology: An Illustrated Ency-*

clopedia. New York: Rizzoli International Publications, 1980.
Chase, Richard. *Quest for Myth*. Baton Rouge: Louisiana State University Press, 1949.
Clark, Helen Archibald. *A Guide to Mythology*. Garden City, New York: Doubleday, Page & Company, 1913.
Cook, Roger. *The Tree of Life (Image for the Cosmos)*. New York: Avon Books, 1974.
Cotterell, Arthur. *A Dictionary of World Mythology*. New York: G. P. Putman's Sons, 1980.
Cotterell, Arthur. *The Macmillan Illustrated Encyclopedia of Myths & Legends*. New York: Macmillan Publishing Company, 1989.
Cox, George W. *The Mythology of the Aryan Nations*. London: Kegan Paul, Trench & Co., 1887.
Cruse, Amy. *The Book of Epic Heroes*. London: George G. Harrap & Company Ltd., 1935.
Darton, Frederick J. H. *Stories of Romance from the Age of Chivalry*. 1907. Reprint. New York: Arlington House, Inc., 1984.
Duffy, Maureen. *The Erotic World of Faery*. 1972. Reprint. New York: Avon Books, 1980.
Editor *et al. Encyclopedia of World Mythology*. New York: Galahad Books, 1975.
Editor *et al. New Larousse Encyclopedia of Mythology*. London: Hamlyn Publishing Group, 1959.
Editors of Time-Life Books. *The Book of Beginnings*. (The Enchanted World). Chicago: Time-Life Books Inc., 1986.
Editors of Time-Life Books. *Gods and Goddesses*. (The Enchanted World). Chicago: Time-Life Books Inc., 1986.
Eliade, Mircea. *Cosmos and History (The Myth of the Eternal Return)*. 1954. Reprint. New York: Harper & Row, 1959.
Eliade, Mircea. *The Forge and the Crucible*. 2d ed. Chicago: University of Chicago Press, 1978.
Eliade, Mircea. *The History of Religious Ideas. Vol. 1 (from the Stone Age to the Eleusinian Mysteries)*. Chicago: University of Chicago Press, 1978.
Eliade, Mircea. *The History of Religious Ideas. Vol. 2 (from Gautama Buddha to the Triumph of Christianity)*. Chicago: University of Chicago Press, 1982.
Eliade, Mircea. *The History of Religious Ideas. Vol. 3 (from Muhammad to the Age of Reforms)*. Chicago: University of Chicago Press, 1985.
Eliade, Mircea. *Myth and Reality*. 1963. Reprint. New York: Harper & Row, 1975.
Eliade, Mircea. *Myths, Dreams, and Mysteries*. 1960. Reprint. New York: Harper & Row, 1967.
Eliade, Mircea. *Patterns in Comparative Religion*. 1958. Reprint. New York: New American Library, 1974.
Eliade, Mircea. *From Primitives to Zen*. New York: Harper & Row, 1977.
Eliade, Mircea. *The Quest (History and Meaning in Religion)*. Chicago: University of Chicago Press, 1969.
Eliade, Mircea. *Rites and Symbols of Initiation (the Mysteries of Birth and Rebirth)* 1958. Reprint. New York: Harper & Row, 1975.
Eliade, Mircea. *The Sacred & the Profane (the Nature of Religion)*. 1959. Reprint. New York: Harcourt Brace Jovanovich, no date.
Eliade, Mircea. *Symbolism, the Sacred, & the Arts*. New York: The Crossroad Publishing Company, 1985.
Eliot, Alexander. *Myths*. New York: McGraw-Hill Book Company, 1976.
Ferm, Vergilius, ed. *Ancient Religions*. New York: Philosophical Library, 1950.
Fiske, John. *Myth and Myth-makers*. Boston: Houghton, Mifflin and Company, 1894.
Frankfort, H. *Kingship and the Gods*. Chicago: University of Chicago Press, 1948.
Frankfort, H. *et al. The Intellectual Adventure of Ancient Man*. Chicago: University of Chicago Press, 1946.
Frazer, Sir James George. *The Golden Bough* (13 volumes). New York: Macmillan and Company, 1936.
Frazer, Sir James George. *The Illustrated Golden Bough* (ed. Mary Douglas). Garden City, New York: Doubleday & Company, Inc., 1978.
Frazer, Lady. *Leaves from the Golden Bough*. New York: The Macmillan Company, 1924.
Gadon, Elinor W. *The Once and Future Goddess*. New York: Harper & Row, Publishers, Inc., 1989.
Gaster, Theodor H. *Thespis—Ritual, Myth, and Drama in the Ancient Near East*. Garden City, New York: Doubleday & Company, Inc., 1961.
Getty, Adele. *Goddess-Mother of Living Nature*. New York: Thames and Hudson, Inc., 1990.
Godwin, Joscelyn. *Athanasius Kircher—a Renaissance Man and the Quest for Lost Knowledge*. New York: Thames and Hudson, 1979.
Godwin, Joscelyn. *Robert Fludd—Hermetic Philosopher and Surveyor of Two Worlds*. Boulder, Colorado: Shambhala Publications, 1979.
Goodrich, Norma Lorre. *Myths of the Hero*. New York: The Orion Press, 1962.
Graves, Robert. *The White Goddess*. 1948. Reprint. New York: Vintage Books, no date.
Grimal, Pierre, ed. *Larousse World Mythology*. Seacaucus, New Jersey: Chartwell Books, 1973.
Grinsell, Leslie V. *Barrow, Pyramid and Tomb*. London: Thames and Hudson, 1975.
Guerber, H. A. *Myths and Legends of the Middle Ages*. London: George G. Harrap & Co. Ltd., 1909.

Guidoni, Enrico. *Primitive Architecture*. New York: Harry N. Abrams, Inc., Publishers, 1978.

Hamilton, Edith. *Mythology*. Boston: Little, Brown and Company, 1942.

Hamilton, Virginia. *In the Beginning—Creation Stories from Around the World*. San Diego, California: Harcourt Brace Jovanovich, Publishers, 1988.

Hartland, Edwin Sidney. *The Science of Fairy Tales*. New York: Scribner & Welford, 1891.

Heinberg, Richard. *Memories and Visions of Paradise*. Los Angeles: Jeremy P. Tarcher, Inc., 1989.

Henderson, Joseph L., and Oakes, Maud. *The Wisdom of the Serpent—the Myths of Death, Rebirth, and Resurrection*. New York: George Braziller, Inc., 1963.

Hooke, S. H. *The Labyrinth—Further Studies in the Relation between Myth and Ritual in the Ancient World*. London: Society for Promoting Christian Knowledge, 1935.

Hooke, S. H. *Myth and Ritual*. London: Oxford University Press, 1933.

Hooke, S. H. *Myth, Ritual, and Kingship*. Oxford: Oxford University Press, 1958.

Hultkrantz, Åke. *The Religions of the American Indians*. Berkeley and Los Angeles: University of California Press, 1979.

Huxley, Francis. *The Dragon—Nature of Spirit, Spirit of Nature*. New York: Collier Books, 1979.

Huxley, Francis. *The Eye—the Seer and the Seen*. New York: Thames and Hudson, 1990.

Huxley, Francis. *The Way of the Sacred*. Garden City, New York: Doubleday & Company, Inc., 1974.

Ions, Veronica. *The World's Mythology in Colour*. London: The Hamlyn Publishing Group Limited, 1974.

James, E. O. *The Ancient Gods*. New York: G. P. Putnam's Sons, 1960.

James, E. O. *The Cult of the Mother Goddess*. New York: Frederick A. Praeger, 1959.

James, E. O. *Seasonal Feasts and Festivals*. London: Thames and Hudson, 1961.

James, E. O. *The Tree of Life*. Leiden: E. J. Brill, 1966.

Jaynes, Julian. *The Origins of Consciousness and the Breakdown of the Bicameral Mind*. Boston: Houghton Mifflin Company, 1976.

Jennings, Philip S. *Medieval Legends*. New York: St. Martin's Press, Inc., 1983.

Jung, C. J. *The Archetypes and the Collective Unconscious*. Princeton, New Jersey: Princeton University Press, 1968.

Jung, C. J. *Man and His Symbols*. New York: Dell Publishing Company, Inc., 1968.

Jung, C. J. *Symbols of Transformation*. Princeton, New Jersey: Princeton University Press, 1967.

Kalweit, Holger. *Dreamtime & Inner Space*. Boston: Shambhala Publications. Inc., 1988.

Kearney, Michael. *World View*. Novato, California: Chandler & Sharp Publishers, Inc., 1984.

Keightley, Thomas. *The World Guide to Gnomes, Fairies, Elves and Other Little People*, 1880. Reprint. New York: Avenel Books, 1978.

Kelly, Walter K. *Curiosities of Indo-European Tradition and Folk-Lore*. London: Chapman & Hall, 1863.

Kirk, G. S. *Myth—Its Meaning & Functions in Ancient & Other Cultures*. Berkeley and Los Angeles: University of California Press, 1970.

Kramer, Samuel Noah, ed. *Mythologies of the Ancient World*. Garden City, New York: Doubleday & Company, Inc./Anchor Books, 1961.

Krickeberg, Walter; Trimborn, Hermann; Muller, Werner; and Zerries, Otto. *Pre-Columbian American Religions*. London: Weidenfeld and Nicolson, 1968.

Krupp, E. C. "Recasting the Past: Powerful Pyramids, Lost Continents, and Ancient Astronauts." *Science and the Paranormal* (George O. Abell and Barry Singer, eds.). New York: Charles Scribner's Sons, 1981, pp. 253–295.

Lang, Andrew. *Custom and Myth*. New York: Harper & Brothers, 1885.

Lang, Andrew. *Myth, Ritual, and Religion* (two volumes). London: Longmans, Green, and Co., 1887.

Leach, Maria, ed. *Funk & Wagnalls Standard Dictionary of Folklore, Mythology and Legend, Volumes One and Two*. New York: Funk & Wagnalls Company, 1949.

Leeming, David. *Mythology*. New York: Newsweek Books, 1976.

Lehmann, Arthur C., and Myers, James E., eds. *Magic, Witchcraft, and Religion—an Anthropological Study of the Supernatural*. Palo Alto, California: Mayfield Publishing Company, 1985.

Levi-Strauss, Claude. *From Honey to Ashes*. New York: Harper & Row, Publishers, Inc., 1973.

Levi-Strauss, Claude. *The Naked Man*. New York: Harper & Row, Publishers, Inc., 1973.

Levi-Strauss, Claude. *The Origin of Table Manners*. New York: Harper & Row, Publishers, Inc., 1978.

Levi-Strauss, Claude. *The Raw and the Cooked*. New York: Harper & Row, Publishers, Inc., 1969.

Lincoln, Bruce. *Myth, Cosmos, and Society—Indo-European Themes of Creation and Destruction*. Cambridge, Massachusetts: Harvard University Press, 1986.

Long, Charles H. *Alpha—the Myths of Creation*. New York: George Braziller, Inc., 1963.

L'Orange, H. P. *Studies in the Iconography of Cosmic Kingship*. New Rochelle, New York: Caratzas Brothers, Publishers, 1982.

Lovin, Robin W., and Reynolds, Frank E., eds. *Cosmogony and the Ethical Order.* Chicago: The University of Chicago Press, 1985.

Lowry, Shirley Park. *Familiar Mysteries.* New York: Oxford University Press, 1982.

Mackenzie, Donald A. *The Migration of Symbols.* New York: Alfred A. Knopf. 1926.

Mackenzie, Donald A. *Myths of Pre-Columbian America.* London: The Gresham Publishing Company Ltd, no date.

Maclagan, David. *Creation Myths—Man's Introduction to the World.* London: Thames and Hudson, 1977.

Maguire, Jack. *Creative Storytelling.* New York: McGraw-Hill Book Company, 1985.

Mercantante, Anthony S. *The Facts on File Encyclopedia of World Mythology and Legend.* New York: Facts on File, 1988.

Moncrieff, A. R. Hope. *Romance & Legend of Chivalry.* London: The Gresham Publishing Company Ltd, no date.

Munkur, Balaji. *The Cult of the Serpent.* Albany, New York: State University of New York Press, 1983.

Murray, Alexander S. *Manual of Mythology,* 1874. Reprint. New York: Tudor Publishing Company, no date.

Neumann, Erich. *The Great Mother.* Princeton, New Jersey: Princeton University Press, 1963.

Newall, Venetia. *An Egg at Easter.* London: Routledge & Kegan Paul plc Ltd, 1971.

Newberry, John Strong. *The Rainbow Bridge.* London: Macmillan and Co., Limited, 1934.

Olsen, Carl, ed. *The Book of the Goddess, Past and Present.* New York: The Crossroad Publishing Company, 1983.

Parker, Derek, & Parker, Julia. *The Immortals.* New York: McGraw-Hill Book Company, 1976.

Partridge, Eric. *Origins—A Short Etymological Dictionary of Modern English.* 1966. Reprint. New York: Greenwich House, 1983.

Pattazzoni, Raffaele. *The All-Knowing God.* London: Methuen & Co., Ltd, 1956.

Peet, Stephen D. *Prehistoric America, Volume Five, Myths and Symbols or Aboriginal Religions in America:* Chicago: Office of the American Antiquarian, 1905.

Perry, John Weir. *Lord of the Four Quarters—Myths of the Royal Father.* New York: George Braziller, 1966.

Philpot, Mrs. J. H. *The Sacred Tree.* London: Macmillan and Co., Limited, 1897.

Pritchard, James B., ed. *The Ancient Near East, volumes 1 and 2.* Princeton, New Jersey: Princeton University Press, 1958 and 1975.

Puhvel, Jaan. *Comparative Mythology.* Baltimore: The Johns Hopkins University Press, 1987.

Purce, Jill. *The Mystic Spiral—Journey of the Soul.* New York: Avon Books, 1974.

Raglan, Lord. *The Hero.* 1936. Reprint. New York: Vintage Books, Inc., 1956.

Raglan, Lord. *The Temple and the House.* New York: W. W. Norton & Company, Inc., 1964.

Rank, Otto. *The Myth of the Birth of the Hero and Other Writings.* New York: Vintage Books, Inc., 1959.

Robinson, Herbert Spencer, and Wilson, Knox. *Myths and Legends of All Nations.* Totowa, New Jersey: Rowman & Allenheld, 1976.

Savill, Sheila. *Pears Encyclopaedia of Myths and Legends, Volumes 1–4* ("The Ancient Near and Middle East and Classical Greece and Rome," "Western and Northern Europe and Central and Southern Africa," "The Orient," and "Oceania and Australia and the Americas"). London: Pelham Books, 1976.

Seboek, Thomas A., ed. *Myth—a Symposium* (Volume 5, Bibliographical and Special Series of the American Folklore Society). Philadelphia: American Folklore Society, 1955.

Seznec, Jean. *The Survival of the Pagan Gods.* New York: Pantheon Books, 1953.

Shapiro, Max S., and Hendrick, Rhoda A., eds. *Mythologies of the World: A Concise Encyclopedia.* Garden City, New York: Doubleday & Company, Inc., 1979.

Shepard, Paul, and Sanders, Barry. *The Sacred Paw—the Bear in Nature, Myth, and Literature.* New York: Viking Penguin Inc., 1985.

Spence, Lewis. *An Introduction to Mythology.* London: George G. Harrap & Company, Ltd., 1921.

Sproul, Barbara C. *Primal Myths—Creating the World.* New York: Harper & Row, Publishers, Inc., 1979.

Stewart. R. J. *The Elements of Creation Myth.* Longmead, Shaftsbury, Dorset: Element Books, 1989.

Stone, Merlin. *When God Was a Woman.* New York: The Dial Press, 1976.

Strenski, Ivan. *Four Theories of Myth in Twentieth-Century History.* Iowa City: University of Iowa Press, 1987.

Tatar, Maria. *The Hard Facts of the Grimms' Fairy Tales.* Princeton, New Jersey: Princeton University Press, 1987.

Thompson, Stith. *The Folktale.* New York: Holt, Rinehart and Winston, Inc., 1946.

Thompson, William Irwin. *The Time Falling Bodies Take to Light.* New York: St. Martin's Press, 1981.

Tyler, Edward B. *Primitive Culture* (two volumes). New York: G. P. Putnam's Sons, 1920.

Tyler, Edward B. *Researches in the Early History of Man-*

kind and the Development of Civilization. London: John Murray, 1878.

Watts, Alan W. *The Two Hands of God—the Myths of Polarity*. New York: George Braziller, Inc., 1963.

Williamson, John. *The Oak King, the Holly King, and the Unicorn*. New York: Harper & Row, Publishers, Inc., 1986.

Wilson, Peter Lamborn. *Angels*. New York: Pantheon Books, 1980.

Wosien, Maria-Gabriele. *Sacred Dance—Encounter with the Gods*. New York: Avon Books, 1974.

Yolen, Jane, ed. *Favorite Folktales from Around the World*. New York: Pantheon Books, 1986.

PREHISTORIC RELIGION

Bancroft, Anne. *Origins of the Sacred*. London: Arkana Paperbacks, 1987.

Briard, Jacques. *Myths et Symboles de l'Europe Preceltique —les Religions de l'Age du Bronze (2500–800 av. J.C.). Paris:* Les Editions Errance, 1987.

Burl, Aubrey. *Prehistoric Astronomy and Ritual*. Aylesbury, England: Shire Publications Ltd, 1983.

Burl, Aubrey. *Rites of the Gods*. London: J M Dent & Sons Ltd, 1981.

von Cles-Reden, Sibylle. *The Realm of the Great Goddess*. Englewood Cliffs, New Jersey: Prentice-Hall, Inc., 1962.

Crawford. O. G. S. *The Eye Goddess*. London: Phoenix House Ltd, 1957.

Dickson, D. Bruce. *The Dawn of Belief*. Tucson, Arizona: The University of Arizona Press, 1990.

Gelling, Peter, and Davidson, Hilda Ellis. *The Chariot of the Sun and Other Rites and Symbols of the Northern Bronze Age*. New York: Frederick A. Praeger, Inc., Publishers, 1969.

Gimbutas, Marija. *The Gods and Goddesses of Old Europe, 7000–3500 B.C.* London: Thames and Hudson, 1974.

Gimbutas, Marija. *The Language of the Goddess*. New York: Harper & Row, Publishers, Inc., 1989.

Gimbutas, Marija. "Perkunas/Perun—the Thunder God of the Balts and the Slavs," *The Journal of Indo-European Studies* 1, No. 4 (Winter, 1973): 466–478.

Hadingham, Evan. *Secrets of the Ice Age*. New York: Walker and Company, 1979.

James, E. O. *Prehistoric Religion*. New York: Frederick A. Praeger, 1957.

Levi-Strauss, Claude. *The Savage Mind*. Chicago: The University of Chicago Press, 1966.

Levy, G. Rachel. *Religious Conceptions of the Stone Age*. 1948. Reprint. New York: Harper & Row, 1963.

Maher, J. Peter. "*Haekmon:* '(Stone) Axe' and 'Sky' in Indo-European/Battle-Axe Culture," *The Journal of Indo-European Studies* 1, No. 4 (winter, 1973): 441–465.

Maringer, J. *The Gods of Prehistoric Man*. London: Wiedenfeld and Nicolson, 1960.

Marshack, Alexander. "Exploring the Mind of Ice Age Man," *National Geographic* 147, no. 1 (January, 1975): 64–89.

Solecki, Ralph S. *Shanidar, the First Flower People*. New York: Alfred A. Knopf, Inc., 1971.

SHAMANISM AND VISIONARY EXPERIENCES

Beyerstein, Barry L. "Neuropathology and the Legacy of Spiritual Possession," *The Skeptical Inquirer* XII, no. 3 (spring,1988): 248–262.

Blackmore, Susan. "Visions of the Dying Brain," *New Scientist*, no. 1611 (5 May 1988): 43–46.

Brodzky, Anne Trueblood; Danesewich, Rose; and Johnson, Nick, eds. *Stones, Bones and Skin—Ritual and Shamanic Art*. Toronto: The Society for Art Publications, 1977.

Davidson, H. R. Ellis., ed. *The Journey to the Other World*. Cambridge, England: D.S. Brewer Ltd., 1975.

dio Rios, Marlene Dobkin. *Hallucinogens: Cross-Cultural Perspectives*. Albuquerque: University of New Mexico Press, 1984.

Diószegi, V. *Tracing Shamans in Siberia*. Oosterhout, The Netherlands: Anthropological Publications, 1968.

Diószegi, V., and Hoppál, M., eds. *Shamanism in Siberia*. Budapest: Akadémiai Kiadó, 1978.

Dow, James. *The Shaman's Touch—Otomi Indian Symbolic Healing*. Salt Lake City: University of Utah Press, 1986.

Drury, Nevill. *The Elements of Shamanism*. Longmead, Shaftsbury, Dorset: Element Books, 1989.

Eliade, Mircea. *Shamanism (Archaic Techniques of Ecstasy)*. Princeton, New Jersey: Princeton University Press, 1964.

Furst, Peter T. *Hallucinogens and Culture*. San Francisco: Chandler & Sharp Publishers, 1976.

Furst, Peter T., ed. *Flesh of the Gods—the Ritual Use of Hallucinogens*. New York: Praeger Publishers, Inc., 1972.

Grim, John A. *The Shaman—Patterns of Siberian and Ojibway Healing*. Norman, Oklahoma: University of Oklahoma Press, 1983.

Grof, Stanislav, and Grof, Christina. *Beyond Death—the Gates of Consciousness*. London: Thames and Hudson, 1980.

Halifax, John. *Shamanic Voices—a Survey of Visionary Narratives*. New York: E. P. Dutton, 1979.

Halifax, Joan. *Shaman—the Wounded Healer*. New York: The Crossroad Publishing Company, 1982.

Harner, Michael J., ed. *Hallucinogens and Shamanism*. New York: Oxford University Press, Inc., 1973.

Hoppál, Mihály, and von Sadovszky, Otto, eds. *Shamanism Past and Present, Parts 1 and 2*. Budapest and Los Angeles/Fullerton: ISTOR Books, 1989.

Hufford, David J. *The Terror That Comes in the Night—an Experience-Centered Study of Supernatural Assault Traditions*. Philadelphia: University of Pennsylvania Press, 1982.

Larsen, Stephen. *The Shaman's Doorway*. Barrytown, New York: Station Hill Press, 1988.

Lewis, I. M. *Ecstatic Religion—a Study of Shamanism and Spirit Possession*. London: Routledge, 1989.

Loewe, Michael, and Blacker, Carmen, ed. *Divination and Oracles*. London: George Allen & Unwin Ltd., 1971.

Lommel, Andreas. *The World of the Early Hunters*. London: Evelyn, Adams and Mackay, 1967.

Nicholson, Shirley, ed. *Shamanism*. Wheaton, Illinois: The Theosophical Publishing House, 1987.

Nicholson, Shirley, ed. *Shamanism in Western North America*. 1938. Reprint. New York: Cooper Square Publishers, Inc., 1975.

Rutherford, Ward. *Shamanism—the Foundations of Magic*. Wellingborough, Northamptonshire: The Aquarian Press, 1986.

Siegel, Ronald K. "Life After Death," *Science and the Paranormal* (Abell, George O., and Singer, Barry, eds.). New York: Charles Scribner's Sons, 1981, pp. 159–184.

Siegel, R. K., and West, L. J. *Hallucinations*. New York: John Wiley & Sons, 1975.

Walker, Jr., Deward E., ed. *Systems of North American Witchcraft and Sorcery*. Moscow, Idaho: University of Idaho, 1970.

Wasson, R. Gordon. *Soma—Divine Mushroom of Immortality*. New York: Harcourt Brace Jovanovich, Inc., no date.

Wasson, R. Gordon. *The Wondrous Mushroom—Mycolatry in Mesoamerica*. New York: McGraw-Hill Book Company, 1980.

Wasson, R. Gordon; Hofmann, Albert; and Ruck, Carl A. P. *The Road to Eleusis*. New York: Harcourt Brace Jovanovich, Inc., 1978.

Wasson, R. Gordon; Kramrisch, Stella; Ott, Jonathan; and Ruck, Carl A. P. *Persephone's Quest: Entheogens and the Origin of Religion*. New Haven, Connecticut: Yale University Press, 1986.

CELESTIAL MYTHOLOGY, SKY LORE, AND ANCIENT CALENDARS

Abell, George O. "Moon Madness," *Science and Paranormal* (Abell, George O., and Singer, Barry, eds.). New York: Charles Scribner's Sons, 1981, pp. 95–104.

Allen, Richard Hickley. *Star Names—Their Lore and Meaning*. 1899. New York: Dover Publications, Inc., 1963.

anon. *The Kalendar & Compost of Shepherds*. London: Peter Davies, 1930.

Ashbrook, Joseph. "Astronomical Scrapbook: About an Astronomical Woodcut," *Sky and Telescope* (May, 1977), 356–358.

Ashe, Geoffrey. *The Ancient Wisdom*. London: Macmillan London Limited, 1977.

Belting, Natalia. *The Moon Is a Crystal Ball—Unfamiliar Legends of the Stars*. Indianapolis: The Bobbs-Merrill Company, Inc., 1952.

Blacker, Carmen, and Loewe, Michael, eds. *Ancient Cosmologies*. London: George Allen & Unwin Ltd, 1975.

Blake, John F. *Astronomical Myths*. London: Macmillan and Co., 1877.

Brown, Robert. *Researches into the Origin of the Primitive Constellations of the Greeks, Phoenicians and Babylonians, Volumes 1 and 2*. London: Williams and Norgate, 1900.

Davis, George A. "Why Did the Arabs Call Beta Persei 'al-Ghul'?" *Sky and Telescope* (February, 1957): 177.

de Santillana, Giorgio, and von Dechend, Hertha. *Hamlet's Mill*. Boston: Gambit, Incorporated, 1969.

Dolan, Edward F. *The Old Farmer's Almanac Book of Weather Lore*. Dublin, New Hampshire: Yankee Books, 1988.

Fagan, Cyril. *Zodiacs Old and New*. London: Anscombe & Company, 1951.

Fagan, Cyril. *Astrological Origins*. St. Paul, Minnesota: Llewellyn Publications, 1971.

Farrington, Oliver C. "The Worship and Folk-Lore of Meteorites," *Journal of American Folk-Lore* 13 (1900): 199–208.

Fernie, J. Donald. "Bloody Sirius," *American Scientist* 77 (September-October, 1989): 429–431.

von Franz, Marie-Louise. *Time—Rhythm and Repose*. London: Thames and Hudson, 1978.

Frazer, Sir James George. *The Worship of Nature, Volume 1* (*Volume 2* never published). London: Macmillan and Co., Limited, 1926.

Freier, George D. *Weather Proverbs*. Tucson, Arizona: Fisher Books, 1989.

de Freval, J. B. *The History of the Heavens*. London: J. Osborn, 1740.

Gingerich, Owen. "An Anachronism" (letter in "Peer Review"), *The Sciences*, no. 6 (November/December, 1984): 14.

Gingerich, Owen. letter to editor, *Scientific American* 52, no. 12 (December, 1976).

Gingerich, Owen. "The Origin of the Zodiac," *Sky and Telescope* (March, 1984): 218–220.

Gleadow, Rupert. *The Origin of the Zodiac*. New York: Castle Books, 1968.

Hadley, Eric, and Hadley, Tessa. *Legends of the Sun and Moon*. Cambridge, England: Cambridge University Press, 1983.

Harley, Timothy. *Moon Lore*. Reprint. Rutland, Vermont: Charles E. Tuttle Co.: Publishers, 1970.

Hawkes, Jacquetta. *Man and the Sun*. London: The Cresset Press, 1962.

Hazen, H. A. "The Origin and Value of Weather Lore," *Journal of American Folk-Lore* 13 (1900), 191–198.

Hentze, Carl. *Mythes et Symboles Lunaires*. Anvers: Editions "de Sikkel," 1932.

Herdeg, Walter. *The Sun in Art*. Zurich: Amstutz & Herdeg, Graphis Press, 1962.

Hunt, D. August. *The Road of the Sea*. Culver City, California: Labyrinthos, 1987.

Jablow, Alta, and Withers, Carl. *The Man in the Moon—Sky Tales from Many Lands*. New York: Holt, Rinehart and Winston, 1969.

Jobes, Gertrude, and Jobes, James. *Outer Space: Myths, Name Meanings, Calendars—from the Emergence of History to the Present Day*. New York: The Scarecrow Press, Inc., 1964.

Katzenstein, Ranee, and Savage-Smith, Emilie. *The Leiden Aratea—Ancient Constellations in a Medieval Manuscript*. Malibu, California: The J. Paul Getty Museum, 1988.

Krappe, Alexander Haggerty. "Les Péléiades," *Revue Archéoligique* 36 (Series 5, 1932): 77–93.

Krupp, E. C. "Celestial Wardrobe," *Griffith Observer* 47, No. 1 (January, 1983): 15–19.

Krupp, E. C. "Cultivating the Sky," *Griffith Observer* 47, no. 10 (October, 1983): 14–20.

Krupp, E. C. "The Dipper in Disguise," *Griffith Observer* 51, no. 12 (December, 1987): 1–18.

Krupp, E. C. "Facing the Sun," *Griffith Observer* 54, no. 1 (January, 1990): 1–13.

Krupp, E. C. "Moon Maids," *Griffith Observer* 52, no. 12 (December, 1988): 2–15.

Krupp, E. C. "Rayed Disks," *Griffith Observer* 53, no. 12 (December, 1989): 1–19.

Krupp, E. C. "Seven Sisters," *Griffith Observer* 55, no. 1 (January, 1991): 1–16.

Krupp, E. C. "Sky Riders," *Griffith Observer* 46, no. 6 (June, 1982): 17–20.

Kunitzsch, Paul, and Smart, Tim. *Short Guide to Modern Star Names and Their Derivations*. Wiesbaden, West Germany: Verlag Otto Harrassowitz, 1986.

Lalou, Etienne. *The Orion Book of the Sun*. New York: The Orion Press, Inc., 1960.

Lee, Albert. *Weather Wisdom*. Garden City, New York: Doubleday & Company, Inc., 1976.

Lehner, Ernst, and Lehner, Johanna. *Lore and Lure of Outer Space*. New York: Tudor Publishing, Company, 1964.

le Lionnais, François. *The Orion Book of Time*. New York: The Orion Press, Inc., 1960.

Lewis, Sir George Cornewall. *An Historical Survey of the Astronomy of the Ancients*. London: Parker, Son, and Bourn, West Strand, 1862.

Lockhart, Gary. *The Weather Companion*. New York: John Wiley & Sons, 1988.

Lum, Peter. *The Stars in Our Heaven—Myths and Fables*. New York: Pantheon Books, Inc., 1948.

Luomala, Katharine. *Oceanic, American Indian and African Myths of Snaring the Sun (Bernice P. Bishop Museum Bulletin 168)*. Honolulu: Bernice P. Bishop Museum, 1940.

Mercer, Samuel A. B. *Earliest Intellectual Man's Idea of the Cosmos*. London: Luzac & Co., 1957.

Miller, Roy Andrew. "Pleiades Perceived: Mul.Mul to Subaru," *American Oriental Society Journal* 108 (Jan./Mar., 1988): 1–25.

Nilsson, Martin P. *Primitive Time-reckoning*. Lund: C. W. K. Gleerup, 1920.

Olcott, William Tyler. *Star Lore of All Ages*. New York: G. P. Putnam's Sons, 1911.

Olcott, William Tyler. *Sun Lore of All Ages*. New York: G. P. Putnam's Sons, 1914.

O'Neil, W. M. *Time and the Calendars*. Sydney, Australia: Sydney University Press, 1975.

O'Neill, John. *The Night of the Gods, vol. 1*. London: Harrison & Sons and Bernard Quaritch. 1893.

Pecker, Jean-Claude. *The Orion Book of the Sky*. New York: The Orion Press, Inc., 1960.

Pennick, Nigel. *The Cosmic Axis*. Cambridge, England: Runestaff Publications, 1985.

Plunket, Emeline M. *Ancient Calendars and Constellations*. London: John Murray, 1903.

Porter, Jermain G. *The Stars in Song and Legend.* Boston: Ginn & Company, 1902.

Ridpath, Ian. *Star Tales.* New York: Universe Books, 1988.

Room, Adrian. *Dictionary of Astronomical Names.* London: Routledge, 1988.

Spinden, Herbert J. "Sun Worship," *The Smithsonian Report for 1939.* Washington, D.C.: Smithsonian Institution, 1940, pp. 447–469.

Staal, Julius D. W. *The New Patterns in the Sky—Myths and Legends of the Stars.* Blacksburg, Virginia: The McDonald & Woodward Publishing Company, 1988.

Toulson, Shirley. *The Winter Solstice.* London: Jill Norman & Hobhouse Ltd, 1981.

van der Waerden, B. L. "History of the Zodiac," *Archiv für Orientforschung* 16 (1953): 216–230.

van der Waerden, B. L. *Science Awakening II (the Birth of Astronomy).* Leyden: Noordhoff International Publishing, 1974.

Van Over, Raymond. *Sun Songs—Creation Myths from Around the World.* New York: Mentor Books/New American Library, 1980.

Varley, Desmond. *Seven, the Number of Creation.* London: G. Bell & Sons, 1976.

Vautier, Ghislaine, and McLeish, Kenneth. *The Shining Stars.* Cambridge, England: Cambridge University Press, 1981.

Vautier, Ghislaine, and McLeish, Kenneth. *The Way of the Stars.* Cambridge, England: Cambridge University Press, 1982.

Warren, William Fairfield. *The Earliest Cosmologies,* New York: Eaton & Mains, 1909.

Webb, E. J. *The Names of the Stars.* London: Nisbet & Co., Ltd., 1952.

Zerubavel, Eviatar. *The Seven Day Circle.* New York: The Free Press, 1985.

Zinner, Ernst. *Die Geschichte der Sternkunde,* Berlin: Verlag von Julius Springer, 1931.

Zinner, Ernst. *The Stars Above Us.* New York: Charles Scribner's Sons, 1957.

ARCHAEOASTRONOMY AND HISTORY OF ASTRONOMY

Aveni, Anthony F., ed. *Archaeoastronomy in the New World.* Cambridge: Cambridge University Press, 1982.

Aveni, Anthony F., ed. *Archaeoastronomy in Pre-Columbian America.* Austin: University of Texas Press, 1975.

Aveni, Anthony F., ed. *Native American Astronomy.* Austin: University of Texas Press, 1977.

Aveni, Anthony F., ed. *New Directions in American Archaeoastronomy.* Oxford: B. A. R., 1988.

Aveni, Anthony F., ed. *World Archaeoastronomy.* Cambridge: Cambridge University Press, 1989.

Aveni, Anthony F. *Empires of Time.* New York: Basic Books, Inc. Publishers, 1989.

Aveni, Anthony F. *Skywatchers of Ancient Mexico.* Austin: University of Texas Press, 1980.

Aveni, Anthony F., and Brotherston, Gordon. *Calendars in Mesoamerica and Peru and Native American Computations of Time.* Oxford: B. A. R., 1983.

Aveni, Anthony F., and Urton, Gary, eds. *Ethnoastronomy and Archaeoastronomy in the American Tropics.* New York: New York Academy of Sciences, 1982.

Baity, Elizabeth Chesley. "Archaeoastronomy and Ethnoastronomy So Far." *Current Anthropology* 14 (1973): 389–449.

Benson, Arlene, and Hoskinson, Tom, eds., *Earth and Sky—Papers from the Northridge Conference on Archaeoastronomy.* Thousand Oaks, California: Slo'w Press, 1985.

Berry, Arthur. *A Short History of Astronomy.* 1908. Reprint. New York: Dover Publications, Inc., 1968.

Brecher, Kenneth, and Feirtag, Michael, eds. *Astronomy of the Ancients.* Cambridge, Massachusetts: MIT Press, 1979.

Carlson, John B. "America's Ancient Skywatchers," *National Geographic Magazine* (March, 1990); 76–107.

Cornell, James. *The First Stargazers.* New York: Charles Scribner's Sons, 1981.

Dicks, D. R. *Early Greek Astronomy to Aristotle.* London: Thames and Hudson, 1970.

Dreyer, J. L. E. *A History of Astronomy from Thales to Kepler.* 1906. Reprint. New York: Dover Publications, Inc., 1956.

Hadingham, Evan. *Early Man and the Cosmos.* New York: Walker and Company, 1984.

Hawkins, Gerald S. *Beyond Stonehenge.* New York: Harper & Row, Publishers, Inc., 1973.

Hawkins, Gerald S. *Mindsteps to the Cosmos.* New York: Harper & Row, Publishers, Inc., 1983.

Heath, Sir Thomas. *Aristarchus of Samos, the Ancient Copernicus.* 1913. Reprint. New York: Dover Publications, Inc., 1981.

Heath, Thomas L. *Greek Astronomy.* London: J. M. Dent & Sons, Ltd, 1932.

Hodson, F. R., ed. "The Place of Astronomy in the Ancient World." *Philosophical Transactions of the Royal Society of London* 276, no. 1257; 1–276.

Krupp, E. C. "Ancient Watchers of the Sky," *1980 Science Year.* World Book Science Annual Chicago: World Book-Childcraft International, 1979, pp. 98–113.

Krupp, E. C. *Echoes of the Ancient Skies*. New York: Harper & Row, Publishers, Inc., 1983 (now only available in paperback from New American Library).

Krupp, E. C., ed. *Archaeoastronomy and the Roots of Science*. Boulder, Colorado/Washington, D.C.: Westview Press/American Association for the Advancement of Science, 1984.

Krupp, E. C., ed. *In Search of Ancient Astronomies*. Garden City, New York: Doubleday & Company, Inc., 1978.

Krupp, E. C. "Light and Shadow," *Griffith Observer* 54, no. 6 (June, 1983): 12–20.

Lloyd, G. E. R. *Early Greek Science: Thales to Aristotle*. New York: W. W. Norton & Company, Inc., 1970.

Lloyd, G. E. R. *Greek Science After Aristotle*. New York: W. W. Norton & Company, Inc., 1973.

Malville, J. McKim, and Putnam, Claudia. *Prehistoric Astronomy in the Southwest*. Boulder, Colorado: Johnson Publishing Company, 1989.

Marshack, Alexander. *The Roots of Civilization*. New York: McGraw-Hill Book Company, 1972.

Michell, John. *Secrets of the Stones—the Story of Astro-archaeology*. Harmondsworth, Middlesex, England: Penguin Books Ltd, 1977.

Neugebauer, O. *The Exact Sciences in Antiquity*. 2nd edition. 1957. Reprint. New York: Harper & Row, Publishers, Inc., 1962.

Neugebauer, O. *A History of Ancient Mathematical Astronomy, Part 2*. New York: Springer-Verlag, 1975.

O'Neil, W. M. *Early Astronomy from Babylon to Copernicus*. Sydney, Australia: Sydney University Press, 1986.

Panneokoek, A. *A History of Astronomy*. New York: Interscience Publishers, 1961.

Ruggles, C. L. N. *Records in Stone—Papers in Memory of Alexander Thom*. Cambridge: Cambridge University Press, 1988.

Sarton, George. *A History of Science, Volume 1 (Ancient Science Through the Golden Age of Greece)* and *Volume 2 (Hellenistic Science and Culture in the Last Three Centuries, B.C.)*. 1952 and 1959. Reprint. New York: W. W. Norton & Company, Inc., 1970.

Schiffman, Robert A., ed. *Visions of the Sky—Archaeological and Ethnological Studies of California Indian Astronomy* (Coyote Press Archives of California Prehistory No. 16). Salinas, California: Coyote Press, 1988.

Stephenson, F. R., and Walker, C. B. F. *Halley's Comet in History*. London: British Museum Publications Ltd, 1985.

Swarup, G.; Bag, A. K.; and Shukla, K. S. *History of Oriental Astronomy*. Cambridge: Cambridge University Press, 1987.

Thiel, Rudolf. *And There Was Light*. New York: Alfred A. Knopf, 1957.

Williamson, Ray A., ed. *Archaeoastronomy in the Americas*. Los Altos, California, and College Park, Maryland: Ballena Press and Center for Archaeoastronomy, 1981.

Williamson, Ray A. *Living the Sky—the Cosmos of the American Indian*. Boston: Houghton Mifflin Company, 1984.

JOURNALS

Archaeoastronomy. Bulletin of the Center for Archaeoastronomy: Center for Archaeoastronomy, Post Office Box X, College Park, Maryland, 20740.

Archaeoastronomy, Supplement to the *Journal for the History of Astronomy*, Science History Publications Ltd, Halfpenny Furze, Mill Lane, Chalfont St. Giles, Buckinghamshire, England, HP8 4NR, U.K.

ASTRONOMY AND METEOROLOGY

Abell, George O. *Exploration of the Universe*. 4th ed. New York: Holt, Rinehart and Winston, 1975.

Allen, David, and Allen, Carol. *Eclipse*. Sydney: Allen & Unwin Australia Pty Ltd, 1987.

Alter, Dinsmore; Cleminshaw, Clarence H.; and Phillips, John H. *Pictorial Astronomy*. 5th revised ed. New York: Harper & Row, Publishers, Inc., 1983.

anon. "The Seventh Star of the Pleiades," *Science* 86, sup 6, 23 July 1938. Reprinted in Corliss, William R. *Mysterious Universe: A Handbook of Astronomical Anomalies*. Glen Arm, Maryland: The Sourcebook Project, 1979.

Brewer, Bryan. *Eclipse*. Seattle: Earth View, Inc., 1978.

Boyer, Carl B. *The Rainbow—from Myth to Mathematics*. Princeton, New Jersey: Princeton University Press, 1987.

Burnham, Jr., Robert. *Burnham's Celestial Handbook*. 3 vols. New York: Dover Publications, Inc., 1977.

Chambers, George F. *The Story of Eclipses*. New York: D. Appleton and Company, 1913.

Chartrand, III, Mark R., and Wimmer, Helmut K. *Skyguide*. New York: Golden Press, 1982.

Cleminshaw, Clarence H. *The Beginner's Guide to the Skies*. New York: Thomas Y. Crowell, 1977.

Dickinson, Terence. *Exploring the Sky by Day*. Camden East, Ontario: Camden House Publishing, 1988.

Dickinson, Terence. *Exploring the Sky by Night*. Camden East, Ontario: Camden House Publishing, 1987.

Greenler, Robert. *Rainbows, Halos, and Glories*. Cambridge, England: Cambridge University Press, 1980.

Jacobs, Una. *Sun Calendar*. London: A & C Black (Publishers) Limited, 1983.
Krupp, E. C., and Krupp, Robin Rector. *The Big Dipper and You*. New York: William Morrow and Company, Inc., 1989.
Krupp, E. C., and Krupp, Robin Rector. *The Comet and You*. New York: Macmillan Publishing Company, 1985.
Long, Kim. *The Moon Book*. Boulder, Colorado: Johnson Publishing Company, 1988.
Meinel, Aden, and Meinel, Marjorie. *Sunsets, Twilights, and Evening Skies*. Cambridge, England: Cambridge University Press, 1983.
Minnaert, M. *The Nature of Light & Color in the Open Air*. New York: Dover Publications, Inc., 1954.
Moore, Patrick, ed. *The International Encyclopedia of Astronomy*. New York: Orion Books, 1987.
Olcott, William Tyler, and Putnam, Edmund W. *Field Book of the Skies*. New York: G. P. Putnam's Sons, 1936.
Pasachoff, Jay M. *Contemporary Astronomy*. 2nd edition. New York: Saunders College Publishing, 1981.
Peltier, Leslie C. *Leslie Peltier's Guide to the Stars*. Milwaukee: AstroMedia, 1986.
Zim, Herbert S., and Baker, Robert H. *Stars*. New York: Golden Press, 1975.

JOURNALS

Griffith Observer, 2800 East Observatory Road, Los Angeles, California, 90027.
Astronomy, 21027 Crossroads Circle, P.O. Box 1612, Waukesha, Wisconsin, 53187.
Mercury, Astronomical Society of the Pacific, 390 Ashton Avenue, San Francisco, California, 94112.
Sky and Telescope, 49 Bay State Road, Cambridge, Massachusetts, 02148.

GREECE AND ROME

Angus, S. *The Mystery-Religions*. 1928. Reprint. New York: Dover Publications, Inc., 1975.
Apollodorus. *The Library in Apollodorus—the Library, Volumes I and II*. Cambridge, Massachusetts: Loeb Classical Library, Harvard University Press, 1921.
Apollonius of Rhodes. *The Voyage of Argo*. Harmondsworth, England: Penguin Books Ltd, 1959.
Aratus. *The Phaenomena in Callimachus—Hymns and Epigrams, Lycophron, Aratus*. Cambridge, Massachusetts: Loeb Classical Library, Harvard University Press, 1921.
Beck, Roger. *Planetary Gods and Planetary Orders in the Mysteries of Mithras*. Leiden: E. J. Brill, 1988.
Bellingham, David. *An Introduction to Greek Mythology*. Seacaucus, New Jersey: Chartwell Books, 1989.
Brendel, Otto J. *Symbolism of the Sphere*. Leiden: E. J. Brill, 1977.
Brown, Robert. *Semitic Influence in Hellenic Mythology*. 1898. Reprint. Clifton, New Jersey: Reference Book Publishers, Inc., 1966.
Brumfield, Allaire Chandor. *The Attic Festivals of Demeter and Their Relation to the Agricultural Year*. Salem, New Hampshire: The Ayer Company, 1981.
Burkert, Walter. *Ancient Mystery Cults*. Cambridge, Massachusetts: Harvard University Press, 1987.
Burkert, Walter. *Structure and History in Greek Mythology and Ritual*. Berkeley and Los Angeles: University of California Press, 1979.
Campbell, Leroy A. *Mithraic Iconography and Ideology*. Leiden: E. J. Brill, 1968.
Condos, Theony. *Eratosthenes'* Katasterismoi. Los Angeles: University of Southern California, 1970 (Ph.D. thesis, translation and commentary).
Cumont, Franz. *After Life in Roman Paganism*. 1922. Reprint. New York: Dover Publications, Inc., 1959.
Cumont, Franz. *Astrology and Religion among the Greeks and Romans*. 1921. Reprint. New York: Dover Publications, Inc., 1960.
Cumont, Franz. *The Mysteries of Mithra*. 1902. Reprint. New York: Dover Publications, Inc., 1956.
Cumont, Franz. *Oriental Religions in Roman Paganism*. 1911. Reprint. New York: Dover Publications, Inc., 1956.
Evans, Arthur. "Mycenaean Tree and Pillar Cult and Its Mediterranean Relations," *The Journal of Hellenic Studies* 21 (1901): 100–204.
Farrar, F. A. *Old Greek Nature Stories*. New York: Thomas Y. Crowell & Company, Publishers, no date.
Ferguson, John. *The Religions of the Roman Empire*. Ithaca, New York: Cornell University Press, 1970.
Field, D. M. *Greek and Roman Mythology*. London: The Hamlyn Publishing Group Limited, 1977.
Flaceliere, Robert. *Greek Oracles*. London: Elek Books Ltd, 1976.
Fontenrose, Joseph. *Python—a Study of Delphic Myth and Its Origins*. 1959. Reprint Berkeley and Los Angeles: University of California Press, 1980.
Fox, William Sherwood. *The Mythology of All Races, Volume I—Greek and Roman*. Boston: Marshall Jones Company, 1916.
Frazer, Sir James George. *Studies in Greek Scenery, Legend and History*. London: Macmillan and Co., Limited, 1931.

Godolphin, F. R. B., ed. *Great Classical Myths.* New York: The Modern Library, 1964.

Godwin, Joscelyn. *Mystery Religions in the Ancient World.* London: Thames and Hudson Ltd, 1981.

Grant, Frederick C. *Hellenistic Religions.* Indianapolis: The Bobbs-Merrill Company, Inc., 1953.

Grant, Mary, trans. and ed. *The Myths of Hyginus.* Lawrence, Kansas: University of Kansas Publications, 1960.

Grant, Michael. *Myths of the Greeks and Romans.* Cleveland, Ohio: The World Publishing Company, 1962.

Grant, Michael. *Roman Myths.* New York: Dorset Press, 1984.

Grant, Michael, and Hazel, John. *Gods and Mortals in Classical Mythology: A Dictionary.* New York: Dorset Press, 1979.

Graves, Robert. *The Greek Myths.* New York: George Braziller, Inc., 1955.

Grimal, Pierre. *The Dictionary of Classical Mythology.* Oxford: Basil Blackwell Publisher, 1986.

Guerber, H. A. *The Myths of Greece and Rome.* London: George G. Harrap & Co. Ltd., 1907.

Halsberghe, Gaston H. *The Cult of Sol Invictus.* Leiden: E. J. Brill, 1972.

Harrison, Jane Ellen. *Epilegomena* and *Themis.* 1921 and 1927. Reprint. New Hyde Park, New York: University Books, Inc., 1966.

Herberger, Charles F. *The Riddle of the Sphinx—Calendric Symbolism in Myth and Icon.* New York: Vantage Press, 1979.

Herberger, Charles F. *The Thread of Ariadne—the Labyrinth of the Calendar of Minos.* New York: Philosphical Library, 1972.

Hesiod and Theognis. *Theogony/Works and Days and Elegies.* Harmondsworth, England: Penguin Books, 1973.

Homer. *The Iliad of Homer* (Richmond Lattimore, trans.). 1951. Reprint. Chicago: The University of Chicago Press, 1961.

Homer. *The Odyssey of Homer* (Richmond Lattimore, trans.). 1965. Reprint. New York: Harper & Row, Publishers, Inc., 1977.

Hulst, Cornelia Steketee. *Perseus and the Gorgon.* La Salle, Illinois: The Open Court Publishing Co., 1946.

Kerényi, C. *The Gods of the Greeks.* 1951. Reprint. New York: Thames and Hudson, 1982.

Kerényi, C. *The Heroes of the Greeks.* London: Thames and Hudson, 1959.

Kerényi, C. *The Religion of the Greeks and Romans.* New York: E. P. Dutton & Co., Inc., 1962.

Kirk, G. S. *The Nature of Greek Myth.* Harmondsworth, England: Penguin Books Ltd, 1974.

Lang, Andrew, trans. *The Homeric Hymns.* New York: Longmans, Green, and Co., 1899.

Mackenzie, Donald A. *Myths of Crete & Pre-Hellenic Europe.* London: The Gresham Publishing Company Ltd, no date.

Manilius. *Astronomica.* Cambridge, Massachusetts: Loeb Classical Library, Harvard University Press, 1977.

Meyer, Marvin W., ed. *The Ancient Mysteries.* New York: Harper & Row, Publishers, Inc., 1987.

Mikalson, Jon D. *The Sacred and Civil Calendar of the Athenian Year.* Princeton, New Jersey: Princeton University Press, 1975.

Moncrieff, A. R. Hope. *Classic Myth and Legend.* London: The Gresham Publishing Company Ltd, no date.

Morford, Mark P. O., and Lenardon, Robert J. *Classical Mythology.* New York: David McKay Company, Inc., 1977.

Nilsson, Martin P. *Greek Folk Religion.* 1940. Reprint. New York: Harper Torchbooks, 1961.

Nilsson, Martin P. *A History of Greek Religion.* Oxford: Oxford University Press, 1925.

Nilsson, Martin P. *Homer & Mycenae.* 1933. Reprint. Philadelphia: University of Pennsylvania Press, 1972.

Nilsson, Martin P. *The Minoan-Mycenaean Religion and Its Survival in Greek Religion.* Lund: C. W. K. Gleerup, 1968.

Nilsson, Martin P. *The Mycenaean Origin of Greek Mythology.* 1932. Reprint. Berkeley and Los Angeles: University of California Press, 1983.

Ovid. *The Fasti, Tristia, Pontic Epistles, Ibis, and Halieuticon.* London: Bell & Daldy, 1872.

Ovid. *Metamorphoses.* 1955. Reprint. Harmondsworth, England: Penguin Books Ltd, 1986.

Ovid. *Metamorphoses.* Oxford: Oxford University Press, 1987.

Perowne, Stewart. *Roman Mythology.* London: The Hamlyn Publishing Group Limited, 1969.

Pinsent, John. *Greek Mythology.* London: The Hamlyn Publishing Group Limited, 1969.

Richardson, Donald. *Greek Mythology for Everyone.* New York: Avenel Books, 1984.

Rose, H. J. *Ancient Greek Religion.* London: Hutchinson's University Library, 1946.

Rose, H. J. *Ancient Roman Religion.* London: Hutchinson's University Library, 1948.

Rose, H. J. *A Handbook of Greek Mythology.* 1928. Reprint. London: Methuen & Co. Ltd, 1985.

Schefold, Karl. *Myth and Legend in Early Greek Art.* New York: Harry N. Abrams, Inc., no date.

Schmidt, Joël. *Larousse Greek and Roman Mythology.* New York: McGraw-Hill Book Company, 1980.

Schwab, Gustav. *Gods and Heroes—Myths and Epics of Ancient Greece.* New York: Pantheon Books, 1974.
Scully, Vincent. *The Earth, the Temple, and the Gods—Greek Sacred Architecture,* New York: Frederick A. Praeger, Inc., Publishers, 1969.
Senior, Michael. *Greece and Its Myths.* London: Victor Gollancz Ltd, 1978.
Speidel, Michael P. *Mithras-Orion.* Leiden: E. J. Brill, 1980.
Spretnak, Charlene. *Lost Goddesses of Early Greece.* Boston: Beacon Press, 1984.
Stanford, W. B., and Luce, J. V. *The Quest for Ulysses.* New York: Praeger Publishers, Inc., 1974.
Stapleton, Michael. *A Dictionary of Greek and Roman Mythology.* New York: Bell Publishing Company, 1978.
Ulansey, David. *The Origins of the Mithraic Mysteries.* Oxford: Oxford University Press, 1989.
Vandenberg, Philipp. *The Mystery of the Oracles.* New York: The Macmillan Publishing Company, 1979.
van den Broek, R. *The Myth of the Phoenix—According to Classical and Early Christian Sources.* Leiden: E. J. Brill, 1972.
Vermaseren, Maarten J. *Cybele and Attis—the Myth and the Cult.* London: Thames and Hudson Ltd, 1977.
Ward. A. G. *The Quest for Theseus.* New York: Praeger Publishers, Inc., 1970.
Willetts, R. F. *Cretan Cults and Festivals.* London: Routledge and Kegan Paul, 1962.

THE BIBLE, JEWISH TRADITION, AND CHRISTIANITY

Beltz, Walter. *God and the Gods—Myths of the Bible.* Harmondsworth, England: Penguin Books Ltd, 1983.
Bible. American Standard Version. New York: Thomas Nelson & Sons, 1929.
Every, George. *Christian Mythology.* London: The Hamlyn Publishing Group Limited, 1970.
Farbridge, Maurice H. *Studies in Biblical and Semitic Symbolism.* 1923. Reprint. New York: KTAV Publishing House, Inc., 1970.
Frazer, Sir James George. *Folk-lore in the Old Testament.* New York: The Macmillan Company, 1927.
Ginzberg, Louis. *The Legends of the Jews* (seven volumes). Philadelphia: The Jewish Publication Society of America, 1937.
Goldstein, David. *Jewish Legends.* New York: Peter Bedrick Books, 1987.
Goodenough, Erwin R. *Jewish Symbols in the Greco-Roman Period.* Princeton, N.J.: Princeton University Press, 1988.
Graves, Robert, and Patai, Raphael. 1963. Reprint. *Hebrew Myths—the Book of Genesis.* New York: Greenwich House, 1983.
Gunkel, Hermann. *The Legends of Genesis.* New York: Schocken Books, Inc., 1964.
Hone, William, ed. *The Lost Books of the Bible.* 1926. Reprint. New York: Bell Publishing Company, 1979.
James, E. O. *Christian Myth and Ritual—a Historical Study.* 1933. Reprint. Gloucester, Massachusetts: Peter Smith, 1973.
Maunder, E. W. *The Astronomy of the Bible.* London: T. Sealey Clark & Co., Ltd., 1908.
McDannell, Colleen, and Lang, Bernhard. *Heaven—a History.* New Haven, Connecticut: Yale University Press, 1988.
Platt, J. Alden. *The Forgotten Books of Eden.* 1927. Reprint. New York: Bell Publishing Company, 1980.
Orchard, Thomas N. *The Astronomy of Milton's 'Paradise Lost.'* London: Longmans, Green and Co., 1896.
Orr, M. A. *Dante and the Early Astronomers.* London: Gall and Inglis, 1913.
Patai, Raphael. *Gates to the Old City—A Book of Jewish Legends.* New York: Avon Books, 1980.
Rappoport, Angelo S. *Ancient Israel—Myths and Legends.* Reprint. New York: Bonanza Books, 1987.

BRITAIN

Alexander, Marc. *British Folklore.* New York: Crescent Books, 1982.
Ashe, Geoffrey. *Mythology of the British Isles.* North Pomfret, Vermont: Trafalgar Square Publishing, 1990.
Ashe, Geoffrey. *The Quest for Arthur's Britain.* London: The Pall Mall Press Limited, 1968.
Barber, Richard. *King Arthur, Hero and Legend.* New York: St. Martin's Press, 1986.
Barber, Richard. *The Arthurian Legends—an Illustrated Anthology.* Totowa, New Jersey: Littlefield Adams & Company, 1979.
Branston, Brian. *The Lost Gods of England.* London: Thames and Hudson, 1974.
Briggs, Katharine. *British Folktales.* New York: Pantheon Books, 1977.
Ebbut, M. I. *Hero-Myths and Legends of the British Race.* New York: Farrar & Rinehart Publishers, no date.
Editor *et al. Folklore, Myths, and Legends of Britain.* London: The Reader's Digest Association Limited, 1977.
Green, Marian. *A Harvest of Festivals.* London: Longman Group Limited, 1980.
Green, Miranda. *The Gods of Roman Britain.* Aylesbury, Bucks, England: Shire Publications Ltd, 1983.

Hole, Christina. *A Dictionary of British Folk Customs.* 1976. Reprint. St. Albans, England: Granada Publishing Limited, 1978.

Hole, Christina. *English Folklore.* London: B. T. Batsford Ltd., 1940.

Hull, Eleanor. *Folklore of the British Isles.* London: Methuen & Co., Ltd., 1928.

Kightly, Charles. *The Customs and Ceremonies of Britain—an Encyclopaedia of Living Traditions.* New York: Thames and Hudson, Inc., 1986.

Kightly, Charles. *The Perpetual Almanack of Folklore.* London: Thames and Hudson Ltd, 1987.

Long, George. *The Folklore Calendar.* 1930. Reprint. East Ardsley, England: EP Publishing Limited, 1977.

Matthews, Caitlin. *Arthur and the Sovereignty of Britain.* London: Arkana, 1989.

Matthews, John. *The Elements of the Grail Tradition.* Longmead, Shaftsbury, Dorset: Element Books, 1990.

Matthews, John, and Matthews, Caitlin. *The Aquarian Guide to British and Irish Mythology.* Wellingborough, Northamptonshire: The Aquarian Press, 1988.

McCluskey, Stephen C. "The Mid-quarter Days and the Historical Survival of British Folk Astronomy," *Archaeoastronomy* (*JHA* Supplement, 13, 1989): S1-S19.

Owen, Gale R. *Rites and Religions of the Anglo-Saxons.* Totowa, New Jersey: Barnes & Noble Books, 1981.

Palmer, Geoffrey, and Lloyd, Noel. *A Year of Festivals—British Calendar Customs.* London: Frederick Warne & Co. Ltd, 1972.

Simpson, Jacqueline. *European Mythology.* New York: Peter Bedrick Books, 1987.

Spence, Lewis. *British Fairy Origins.* 1946. Reprint. Wellingborough, Northamptonshire: The Aquarian Press Limited, 1981.

Spence, Lewis. *The Fairy Tradition in Britain.* London: Rider and Company, 1948.

Stewart, Bob, and Matthews, John. *Legendary Britain.* London: Blandford Press, 1989.

Stone, Brian, trans. *Sir Gawain and the Green Knight.* Harmondsworth, England: Penguin Books, Ltd, 1959.

Weston, Jessie L. *From Ritual to Romance.* 1920. Reprint. Gloucester, Massachusetts, Peter Smith, 1983.

Westwood, Jennifer. *Albion, a Guide to Legendary Britain.* London: Grafton Books, 1985.

CELTS AND DRUIDS

Bellingham, David. *An Introduction of Celtic Mythology.* Seacaucus, New Jersey: Chartwell Books, 1990.

Cross, Tom Peete, and Slover, Clark Harris. *Ancient Irish Tales.* 1936. Reprint. New York: Barnes & Noble, 1969.

Curtin, Jeremiah. *Myths and Folk Tales of Ireland.* 1890. Reprint. New York: Dover Publications, Inc., 1975.

Davidson, H. R. Ellis. *Myths and Symbols in Pagan Europe—Early Scandinavian and Celtic Religions.* Syracuse, New York: Syracuse University Press, 1988.

Evans-Wentz, Walter Yeeling. *The Fairy-Faith in Celtic Countries.* New York: University Books, Inc., 1966.

Graves, Alfred Percival. *The Irish Fairy Book.* Reprint. New York: Greenwich House, 1983.

Green, Miranda. *The Gods of the Celts.* Totowa, New Jersey: Barnes and Noble Books, 1986.

Green, Miranda. *Symbol & Image in Celtic Religious Art.* London: Routledge, 1989.

Hicks, Ronald. "Astronomical Traditions of Ancient Ireland and Britain," *Archaeoastronomy* (Maryland) VIII (1985): 70–79.

Jacobs, Joseph, "Connla and the Fairy Maiden," *Celtic Fairy Tales.* 1892. Reprint. New York: Dover Publications, Inc., 1968, pp. 1–4.

Macalister, R. A. S. *Tara—a Pagan Sanctuary of Ancient Ireland.* New York: Charles Scribner's Sons, 1931.

MacBain, Alexander. *Celtic Mythology and Religion.* Stirling: Eneas Mackay, 1917.

MacCana, Proinsias. *Celtic Mythology.* London: The Hamlyn Publishing Group, 1970.

MacCulloch, John Arnott, and Máchal, Jan. *The Mythology of All Races, Volume III—Celtic and Slavic.* 1918. Reprint. New York: Cooper Square Publishers, Inc., 1964.

MacNeill, Máire. *The Festival of Lughnasa* (two volumes). 1962. Reprint. Dublin: University College, 1982.

McNeill, F. Marian. *The Silver Bough. Volume 1—Scottish Folk-lore and Folk-belief.* 1956. Reprint. Edinburgh: Canongate Publishing, 1989.

Owen, Trefor M. *Welsh Folk Customs.* 1959. Reprint. Llandysul, Dyfed (Wales): Gomer Press, 1987.

Rees, Alwyn, and Rees, Brinley. *Celtic Heritage—Ancient Tradition in Ireland and Wales.* New York: Grove Press, Inc., 1961.

Rhys, John. *Celtic Folklore: Welsh & Manx.* 1901. Reprint. New York: Benjamin Blom, Inc., 1972.

Rolleston, T. W. *Myths and Legends—the Celtic Race.* Boston: David D. Nickerson & Company Publishers, no date.

Ross, Anne. *Pagan Celtic Britain.* London: Routledge and Kegan Paul Limited, 1967.

Rutherford, Ward. *Celtic Mythology.* Wellingborough, Northamptonshire: The Aquarian Press, 1987.

Sharkey, John. *Celtic Mysteries—the Ancient Religion.* London: Thames and Hudson, 1975.

Squire, Charles. *Celtic Myth & Legend, Poetry & Romance.* London: The Gresham Publishing Company Limited, no date.

Webster, Graham. *Celtic Religion in Roman Britain.* Totowa, New Jersey: Barnes & Noble Books, 1987.

Wood-Martin, W. G. *Traces of the Elder Faiths of Ireland* (two volumes). London: Longmans, Green, and Co., 1902.

SCANDINAVIAN AND GERMANIC PEOPLES

Asbjørnsen, Peter Christen, and Moe Jørgen. *Norwegian Folktales.* New York: Pantheon Books, 1982.

Bauschautz, Paul C. *The Well and the Tree—World and Time in Early Germanic Culture.* Amherst, Massachusetts: The University of Massachusetts Press, 1982.

Branston, Brian. *Gods of the North.* London: Thames and Hudson, 1955.

Craigie, William Alexander. *The Religion of Ancient Scandinavia.* 1906. Reprint. Freeport, New York: Books for Libraries Press, 1969.

Crossley-Holland, Kevin. *The Norse Myths.* New York: Pantheon Books, 1980.

Davidson, H. R. Ellis. *Gods and Myths of Northern Europe.* Harmondsworth, England: Penguin Books Ltd, 1964.

Davidson, H. R. Ellis. *Myths and Symbols in Pagan Europe—Early Scandinavian and Celtic Religions.* Syracuse, New York: Syracuse University Press, 1988.

Davidson, H. R. Ellis. *Pagan Scandinavia.* New York: Frederick A. Praeger, Inc., Publishers, 1967.

Davidson, H. R. Ellis. *Scandinavian Mythology.* London: The Hamlyn Publishing Group Limited, 1969.

Dumézil, Georges. *Gods of the Ancient Northmen.* Berkeley and Los Angeles: University of California Press, 1973.

Grant, John. *An Introduction to Viking Mythology.* Seacaucus, New Jersey: Chartwell Books, Inc., 1990.

Grimm, Jacob. *Teutonic Mythology, Volumes 1–4.* 1883–1888. Reprint. Gloucester, Massachusetts: Peter Smith, 1976.

Guerber, H. A. *Myths of the Norsemen.* London: George G. Harrap & Company, no date.

MacCulloch, John Arnott. *The Mythology of All Races, Volume II—Eddic.* 1930. Reprint. New York: Cooper Square Publishers, Inc., 1964.

Mackenzie, Donald A. *Teutonic Myth and Legend.* London: The Gresham Publishing Company Ltd., no date.

Pennick, Nigel. *Hitler's Secret Sciences.* Suffolk, England: Neville Spearman Limited, 1981.

Reuter, O. S. *Skylore of the North.* 1936. Serialized in *Stonehenge Viewpoint,* (May–June, 1982), 20–25; (July–August, 1982), 13–19; (September–October, 1982), 21–26; and (November–December, 1982), 16–22.

Rydberg, Viktor. *Teutonic Mythology, Gods and Goddesses of the Northland, Volume II.* London: The Norroena Society, 1907.

Sturluson, Snorri. *The Prose Edda* (Jean I. Young, trans.). Berkeley and Los Angeles: University of California Press, 1964.

Taylor, Paul B., and Auden, W. H. Trans. *The Elder Edda—a Selection.* New York: Random House, 1967.

Tichenell, Elsa-Brita. *The Masks of Odin.* Pasadena, California: Theosophical University Press, 1985.

Turville-Petre, E. O. G. *Myth and Religion of the North.* New York: Holt, Rinehart and Winston, 1964.

CENTRAL AND EASTERN EUROPE

Beza, Marcu. *Paganism in Roumanian Folklore.* London: J. M. Dent & Sons Ltd., 1928.

Eliade, Mircea. *Zalmoxis, the Vanishing God.* Chicago: University of Chicago Press, 1972.

Dégh, Linda, ed. *Folktales of Hungary.* London: Routledge and Kegan Paul Limited, 1965.

Dömötör, Tekla. *Hungarian Folk Beliefs.* Bloomington, Indiana: Indiana University Press, 1982.

Georgieva, Ivanichka. *Bulgarian Mythology.* Sofia: Svyat Publishers, 1985.

Harding, Emily J., trans. *Slav Fairy Tales of the Slav Peasants and Herdsmen.* New York: A. L. Burt Company, Publishers, no date.

MacCulloch, John Arnott, and Máchal, Jan. *The Mythology of All Races, Volume III—Celtic and Slavic.* 1918. Reprint. New York: Cooper Square Publishers, Inc., 1964.

Mándoki, László. "Straw Path," *Acta Ethnographica—Academiae Scientiarum Hungaricae* 14 (1965): 117–139.

Ròheim, Gèza. *Hungarian and Vogul Mythology.* Locust Valley, New York: J. J. Augustin Publisher, 1954.

FINLAND AND SIBERIA

Bogoras, Waldemar. "Chukchee Mythology," *American Museum of Natural History Memoir, Volume 12, Part 1.* New York: American Museum of Natural History.

Bogoras, Waldemar. "Chuckchee Tales," *Journal of American Folk-Lore* 41 (1928): 298–371.

Curtin, Jeremiah. *A Journey in Southern Siberia*. London: Sampson Low, Marston & Company Limited, 1910.

Diószegi, V. *Popular Beliefs and Folklore Tradition in Siberia*. Budapest: Akadémiai Kiadò, 1968.

Friberg, Eino, trans. *The Kalevala—Epic of the Finnish People*. Helsinki: Otava Publishing Company Ltd., 1988.

Holmberg, Uno. *The Mythology of All Races: Finno-Ugric/Siberian. Vol IV*. Reprint. New York: Cooper Square Publishers, 1964.

Jochelson, Waldemar. "The Yakut," *Anthropological Papers of the American Museum of Natural History* 33, part 2 (1933): 35–225.

Levin, M. G., and Potapov, L. P., eds., *The Peoples of Siberia*. Chicago: The University of Chicago Press, 1964.

Lönnrot, Elias, compiler. *The Kalavala or Poems of the Kaleva District*. Cambridge, Massachusetts: Harvard University Press, 1963.

Nagishkin, Dmitri. *Folktales of the Amur—Stories from the Russian Far East*. New York: Harry N. Abrams, Inc., Publishers, 1980.

Pentikäinen, Juha Y. *Kalevala Mythology*. Bloomington, Indiana: Indiana University Press, 1989.

Winter, Jeanette. *The Girl and the Moon Man*. New York: Pantheon Books, 1984.

Zheleznova, Irina. *A Mountain of Gems—Fairy Tales of the Peoples of the Soviet Land*. Moscow: Raduga Publishers, 1975.

MESOPOTAMIA AND THE NEAR EAST

Ananikian, Mardiros, and Werner, Alice. *The Mythology of All Races, Volume VII—Armenian and African*. 1925. Reprint. New York: Cooper Square Publishers, Inc., 1964.

British Museum. *The Babylonian Legends of the Creation and the Fight between Bel and the Dragon*. London: British Museum, 1931.

Burrows, E. "The Constellation of the Wagon and Recent Archaeology," *Analecta Orientalia 12—Miscellanea Orientalia*. Rome: Pontificio Instituto Biblio, 1935, pp. 34–40.

Caquot, André, and Sznycer, Maurice. *Ugaritic Religion* (Iconography of Religions XV, 8). Leiden: E. J. Brill, 1980.

Dalley, Stephanie. *Myths from Mesopotamia*. Oxford: Oxford University Press, 1989.

Dawood, N. J. trans. *The Koran*. Harmondsworth, England: Penguin Books Ltd, 1961.

Gibson, J. C. L. *Canaanite Myths and Legends*. Edinburgh: T. & T. Clark Ltd, 1977.

Gray, John, *Near Eastern Mythology*. London: The Hamlyn Publishing Group Limited, 1969.

Hallo, William M., and van Dijk, J. J. A. *The Exaltation of Inanna*. New Haven, Connecticut: Yale University Press, 1968.

Hartner, Willy. "The Earliest History of the Constellations in the Near East and the Motif of the Lion-Bull Combat," *Journal of Near Eastern Studies XXIV*, nos. 1 & 2 (Jan.–Apr., 1965): 1–16.

Heidel, Alexander. *The Babylonian Genesis*. Chicago: The University of Chicago Press, 1951.

Heidel, Alexander. *The Gilgamesh Epic and Old Testament Parallels*. Chicago: The University of Chicago Press, 1946.

Hooke, S. H. *Babylonian and Assyrian Religion*. London: Hutchinson's University Library, 1953.

Hostetter, H. Clyde. "Inanna Visits the Land of the Dead: An Astronomical Interpretation," *Griffith Observer* 46, no. 2 (February, 1982): 9–15.

Hostetter, H. Clyde. "A Planetary Visit to Hades," *Archaeoastronomy* (Maryland), II, no. 4 (fall 1979): 7–10.

Jacobsen, Thorkild. *Toward the Image of Tammuz and Other Essays on Mesopotamian History and Culture*. Cambridge, Massachusetts: Harvard University Press, 1970.

Jacobsen, Thorkild. *The Treasures of Darkness—a History of Mesopotamian Religion*. New Haven, Connecticut: Yale University Press, 1976.

Jastrow, Morris. *Aspects of Religious Belief and Practice in Babylonia and Assyria*. 1911. Reprint. New York: Benjamin Blom, Inc., 1971.

Jastrow, Morris. *The Religion of Babylonia and Assyria*. Boston: Ginn & Company, 1898.

Jeremias, Alfred. *Handbuch der Altorientalischen Geisteskultur*. Leipzig: J. C. Hinrichs'sche Buchhandlung, 1913.

King, L. W. *Babylonian Religion and Mythology*. London: Kegan Paul, Trench, Trubner & Co., Ltd, 1899.

Kramer, Samuel Noah. *Sumerian Mythology*. 1944. Reprint. New York: Harper Torchbooks, 1961.

Krupp, E. C. "Astronomical Symbolism in Mesopotamian Religious Imagery," contributed paper, January 8, 1979, meeting of the American Astronomical Society, Mexico City. Abstract: *Archaeoastronomy Bulletin* (Maryland) II, no. 1 (November, 1978): 5.

Langdon, S. *Babylonian Menologies and the Semitic Calendars*. London: The British Academy, 1935.

Langdon, Stephen Herbert. *The Mythology of All Races, Volume V—Semitic*. 1931. Reprint. New York: Cooper Square Publishers, Inc., 1964.

Mackenzie, Donald A. *Myths of Babylonia & Assyria*. London: The Gresham Publishing Company Ltd, no date.

Obermann, Julian. *Ugaritic Mythology*. New Haven: Yale University Press, 1948.

Olmstead, A. T. "Babylonian Astronomy—Historical Sketch," *The American Journal of Semitic Languages and Literatures* LV, no. 2 (April, 1938): 113–129.

Oppenheim, A. Leo. "Man and Nature in Mesopotamian Civilization." *Dictionary of Scientific Biography, Volume XV, Supplement I, "Topical Essays"* (Charles Coulston Gillespie, ed.). New York: Charles Scribner's Sons, 1978, pp. 634–666.

Pallis, Svend Aage. *The Babylonian Akitu Festival (Historisk-Filologiske Meddelelser Volume 12)*. Copenhagen: Andr. Fred. Host & Son, 1926–27.

Rochberg-Halton, F. "Elements of the Babylonian Contribution to Hellenistic Astrology," *American Oriental Society Journal* 108 (January–March, 1988): 51–62.

Sandars, N. K., trans. *The Epic of Gilgamesh*. Harmondsworth, England: Penguin Books Ltd, 1960.

Sandars, N. K., trans. *Poems of Heaven and Hell from Ancient Mesopotamia*. Harmondsworth, England: Penguin Books Ltd, 1971.

Sayce, A. H. *Astronomy and Astrology of the Babylonians*. 1874. Reprint. San Diego: Wizards Bookshelf, 1981.

Sayce, A. H. *The Religions of Ancient Egypt and Babylonia*. Edinburgh: T. & T. Clark, 1902.

Segal, J. B. "The Sabian Mysteries—the Planet Cult of Ancient Harran," in *Vanished Civilizations of the Ancient World* (Edward Bacon, ed.). New York: McGraw-Hill Book Company, 1963, pp. 201–220.

Smith, George. *The Chaldean Account of the Genesis*. 1872. Reprint. Minneapolis, Minnesota: Wizards Book Shelf, 1977.

Spence, Lewis. *Myths and Legends of Babylonia and Assyria*. London: George G. Harrap & Company, Ltd., 1916.

Thompson, R. Campbell. *The Reports of the Magicians and Astrologers of Nineveh and Babylon in the British Museum, Volumes 1 and 2*. 1900. Reprint. New York: AMS Press Inc., 1977.

van der Waerden, B. L. "Babylonian Astronomy, II. The Thirty-six Stars," *Journal of Near Eastern Studies* 8 (1949): 6–26.

van der Waerden. B. L. "Mathematics and Astronomy in Mesopotamia." *Dictionary of Scientific Biography, Volume XV, Supplement I, "Topical Essays"* (Charles Coulston Gillespie, ed.). New York: Charles Scribner's Sons, 1978, pp. 667–680.

Weidner, Ernst F. *Handbuch der Babylonischen Astronomie*. Leipzig: J. C. Hinrichs'sche Buchhandlung, 1915.

Wolkstein, Diane, and Kramer, Samuel Noah. *Inanna, Queen of Heaven and Earth—Her Stories and Hymns from Sumer*. New York: Harper & Row, Publishers, Inc., 1983.

PERSIA

Fennelly, James M. "The Persepolis Ritual," *Biblical Archeologist* 43, no. 3 (summer, 1980): 135–162.

Hinnells, John R. *Persian Mythology*. London: The Hamlyn Publishing Group Limited, 1973.

Keith, A. Berriedale, and Carnoy, Albert J. *The Mythology of All Races, Volume VI—Indian and Iranian*. 1917. Reprint. New York: Cooper Square Publishers, Inc., 1964.

Masani, Sir Rustom. *Zoroastrianism—the Religion of the Good Life*. 1938. Reprint. New York: Collier Books, 1962.

Poebel, Arno. "The Names and the Order of the Old Persian and Elamite Months during the Achaemenian Period," *The American Journal of Semitic Languages and Literatures* LV, no. 2 (April, 1938): 130–141.

EGYPT

Allen, James P. *Genesis in Egypt—the Philosophy of Ancient Egyptian Creation Accounts*. New Haven, Connecticut: Yale Egyptological Seminar, Yale University, 1988.

Antoniadi, Eugene Michel. *L'Astronomie Egyptienne Depuis les Temps le Plus Recules*. Paris: Gauthiers-Villars, 1934.

Armour, Robert A. *Gods and Myths of Ancient Egypt*. Cairo: The American University in Cairo Press, 1986.

Badawy, Alexander. *A History of Egyptian Architecture, Vol. 1*. Giza, Egypt: Alexander Badawy, 1954.

Badawy, Alexander. *A History of Egyptian Architecture, The Empire (the New Kingdom)*. Berkeley and Los Angeles: University of California Press, 1968.

Badawy, Alexander. "The Stellar Destiny of Pharaoh and the So-Called Air-Shafts of Cheops' Pyramid," *Mitteilungen des Instituts fur Orientforschung*, Band X (1964): 189–206.

Barguet, Paul. *Le Temple d'Amon-Rê à Karnak*. Cairo: L'Institut Francais d'Archeologie Orientale du Caire, 1962.

Biot, J. B. *Recherches sur Plusieurs Points de l'Astronomie Egyptienne*. Paris: Chez Firmin Didot, Pere et Fils, 1823.

Bleeker, C. J. *Egyptian Festivals*. Leiden: E. J. Brill, 1967.

Bleeker, C. J. *Hathor and Thoth*. Leiden: E. J. Brill, 1973.

Boylan, Patrick. *Thoth, the Hermes of Egypt*. 1922. Reprint. Chicago: Ares Publishers, Inc., no date.

Breasted, James H. *The Dawn of Conscience*. New York: Charles Scribner's Sons, 1933.

Breasted, James H. *Development of Religion and Thought in Ancient Egypt*. 1912. Reprint. Philadelphia: University of Pennsylvania Press, 1972.

Brier, Bob. *Ancient Egyptian Magic*. New York: William Morrow and Company, Inc., 1980.

Brugsch, Heinrich. *Astronomical and Astrological Inscriptions on Ancient Egyptian Monuments (Thesaurus Inscriptionum Aegyptiacarum)*. 1883. English translation serialized in *Griffith Observer* by Griffith Observatory, 2800 East Observatory Road, Los Angeles, California, 90027, 1978–1980.

Budge, E. A. Wallis. *The Book of the Opening of the Mouth*. 1909. Reprint. New York: Arno Press, 1980.

Budge, E. A. Wallis. *The Egyptian Heaven and Hell*. 1906. Reprint. 3 vols. New York: AMS Press, 1976.

Budge, E. A. Wallis. *Egyptian Magic*. 1899. Reprint. London: Routledge & Kegan Paul, 1975.

Budge, E. A. Wallis. *Egyptian Religion*. 1900. Reprint. New York: Bell Publishing Company, no date.

Budge, E. A. Wallis. *From Fetish to God in Ancient Egypt*. 1934. Reprint. New York: Benjamin Blom, 1972.

Budge, E. A. Wallis. *The Gods of the Egyptians*. 1904. Reprint. 2 vols. New York: Dover Publications, Inc., 1969.

Budge, E. A. Wallis. *The Mummy*. 1925. Reprint. New York: Dover Publications, Inc., 1989.

Budge, E. A. Wallis. *Osiris and the Egyptian Resurrection*. 1911. Reprint. 2 vols. New York: Dover Publications, Inc., 1973.

Budge, E. A. Wallis. *Tutankhamen—Amenism, Atenism and Egyptian Monotheism*. 1923. Reprint. New York: Bell Publishing Company, no date.

Cerny, Jaroslav. *Ancient Egyptian Religion*. London: Hutchinson's University Library, 1952.

Clark, R. T. Rundle. *Myth and Symbol in Ancient Egypt*. London: Thames and Hudson, 1978.

Cole, John. *A Treatise on the Circular Zodiac of Tentyra*. London: Longmans & Co., 1824.

Cooke, Harold P. *Osiris, a Study of Myths, Mysteries and Religion*. London: C. W. Daniel Company, 1924.

David, A. Rosalie. *The Ancient Egyptians—Religious Beliefs and Practices*. London: Routledge & Kegan Paul, 1982.

David, Rosalie. *Cult of the Sun—Myth and Magic in Ancient Egypt*. London: J. M. Dent & Sons Ltd, 1980.

Davis, Virginia Lee. "Identifying Ancient Egyptian Constellations," *Archaeoastronomy* (*JHA* Supplement 9, 1985): S102–S104.

Davis, Virginia Lee. "Pathways to the Gods," *Ancient Egypt: Discovering Its Splendors* (Jules B. Bellard, ed.). Washington, D.C.: National Geographic Society, 1978, pp. 154–201.

Fairman, H. W. *The Triumph of Horus*. Berkeley and Los Angeles: University of California Press, 1969.

Faulkner, R. O. *The Ancient Egyptian Pyramid Texts*. Oxford: Oxford University Press, 1969.

Faulkner, R. O. *The Book of the Dead*. New York: The Limited Editions Club, 1972.

Faulkner, R. O. "The King and the Star-Religion in the Pyramid Texts," *Journal of Near Eastern Studies* XXV (1966): 153–161.

Frankfort, H. *Ancient Egyptian Religion*. New York: Columbia University Press, 1948.

Gingerich, Owen. "Ancient Egyptian Sky Magic," *Sky and Telescope* 65, no. 5 (May, 1983): 418–420.

Griffiths, J. Gwyn. *The Conflict of Horus & Seth*. Liverpool: Liverpool University Press, 1960.

Griffiths, J. Gwyn. *The Origins of Osiris and His Cult*. Leiden: E. J. Brill, 1980.

Hart, George. *A Dictionary of Egyptian Gods and Goddesses*. London: Routledge & Kegan Paul, 1986.

Hornung, Erik. *Conceptions of God in Ancient Egypt*. Ithaca, New York: Cornell University Press, 1982.

Hornung, Erik. *The Valley of the Kings—Horizon of Eternity*. New York: Timken Publishers, 1990.

Ions, Veronica. *Egyptian Mythology*. London: The Hamlyn Publishing Group Limited, 1968.

Jacq, C. *Egyptian Magic*. Warminster, Wiltshire, England: Aris & Phillips Ltd, 1985.

Kendall, Timothy. *Passing through the Netherworld—the Meaning and Play of* Senet, *an Ancient Egyptian Funerary Game*. Belmont, Massachusetts: Kirk Game Company, Inc., 1978.

Krupp, E. C. "Egyptian Astronomy: The Roots of Modern Timekeeping," *New Scientist* 85, (January 3, 1980): 24–27.

Krupp, E. C. "Great Pyramid Astronomy," *Griffith Observer*, 42, no. 3 (March, 1978), pp. 1–18.

Krupp, E. C. "Light in the Temples." *Records in Stone* (Clive Ruggles, ed.). Cambridge, England: Cambridge University Press, 1988, pp. 473–499.

Krupp, E. C. "The Sun Gods." *Fire of Life: The Smithsonian Book of the Sun* (Joe Goodwin *et al*, ed.). Washington, D.C.: Smithsonian Exposition Books, 1981, pp. 160–167.

Lockyer, J. Norman. *The Dawn of Astronomy*. London: Cassell and Company, 1894.

Mackenzie, Donald A. *Egyptian Myth and Legend.* London: The Gresham Publishing Company Ltd, no date.

Mercantante, Anthony S. *Who's Who in Egyptian Mythology.* New York: Clarkson N. Potter, Inc., 1978.

Mercer, Samuel A. B. *The Religion of Ancient Egypt.* London: Luzac & Co., 1949.

Morenz, Siegfried. *Egyptian Religion.* Ithaca, New York: Cornell University Press, 1973.

Müller, W. Max, and Scott, Sir James George. *The Mythology of All Races, Volume XII—Egyptian and Indo-Chinese.* Boston: Marshall Jones Company, 1918.

Neugebauer, O., and Parker, R. A. *Egyptian Astronomical Texts I. The Early Decans.* Providence, Rhode Island: Brown University Press, 1960.

Neugebauer, O., and Parker, R. A. *Egyptian Astronomical Texts II. The Ramesside Star Clocks.* Providence, Rhode Island: Brown University Press, 1964.

Neugebauer, O., and Parker, R. A. *Egyptian Astronomical Texts III. Decans, Planets, Constellations and Zodiacs.* 2 vols. Providence, Rhode Island: Brown University Press, 1969.

Otto, Eberhard. *Ancient Egyptian Art—the Cults of Osiris and Amon.* New York: Harry N. Abrams, Inc., no date.

Parker, Richard A. *The Calendars of Ancient Egypt.* Chicago: University of Chicago Press, 1950.

Parker, Richard A. "Egyptian Astronomy, Astrology, and Calendrical Reckoning." *Dictionary of Scientific Biography, Volume XV, Supplement I, "Topical Essays"* (Charles Coulston Gillespie, ed.). New York: Charles Scribner's Sons, 1978, pp. 706–727.

Petrie, W. M. Flinders. *Wisdom of the Egyptians,* vol. LXIII. London: Bernard Quaritch Ltd., 1940.

Piankoff, Alexandre. *The Litany of Re.* Princeton, New Jersey: Princeton University Press, 1964.

Piankoff, Alexandre. *Mythological Papyri.* Princeton, New Jersey: Princeton University Press, 1957.

Piankoff, Alexandre. *The Pyramid of Unas.* Princeton, New Jersey: Princeton University Press, 1968.

Piankoff, Alexandre. *The Shrines of Tut-Ankh-Amon.* Princeton, New Jersey: Princeton University Press, 1955.

Piankoff, Alexandre. *The Tomb of Ramesses VI.* Princeton, New Jersey: Princeton University Press, 1954.

Piankoff, Alexandre. *The Wandering of the Soul.* Princeton, New Jersey: Princeton University Press, 1974.

Plutarch. *Isis and Osiris* in *Moralia, Volume V.* Cambridge, Massachusetts: Loeb Classical Library, Harvard University Press, 1936.

Poole, Reginald Stuart. *Horae Aegyptiacae, or the Chronology of Ancient Egypt.* London: John Murray, 1851.

St. Clair, George. *Creation Records.* London: David Nutt, 1898.

Sayce, A. H. *The Religions of Ancient Egypt and Babylonia.* Edinburgh: T. & T. Clark, 1902.

Shorter, Alan W. *The Egyptian Gods.* London: Routledge & Kegan Paul, 1937.

Slosman, Albert. *L'astronomie selon les Egyptiens.* Paris: Editions Robert Laffont, 1983.

Spence, Lewis. *The Mysteries of Egypt.* Philadelphia: David McKay Company, no date.

Spence, Lewis. *Myths and Legends of Ancient Egypt.* London: George G. Harrap & Company, Ltd., 1915.

Te Velde, H. *Seth, God of Confusion.* Leiden: E. J. Brill, 1977.

Thomas, Angela P. *Egyptian Gods and Myths.* Aylesbury, Bucks, England: Shire Publications Ltd, 1986.

Trimble, Virginia, "Astronomical Investigation Concerning the So-Called Air-Shafts of Cheops' Pyramid," *Mitteilungen des Institut fur Orientforschung.* Band X (1964): 183–187.

Wainwright, Gerald A. *The Sky Religion in Egypt: Its Antiquity & Effects.* 1938. Reprint. Westport, Connecticut: Greenwood Press, 1971.

Watterston, Barbara. *The Gods of Ancient Egypt.* New York: Facts on File Publications, 1984.

AFRICA

Ananikian, Mardiros, and Werner, Alice. *The Mythology of All Races, Volume VII—Armenian and African.* 1925. Reprint. New York: Cooper Square Publishers, Inc., 1964.

Feldman, Susan, ed. *African Myths and Tales.* New York: Dell Publishing Company, Inc., 1963.

Griaule, M. *Conversations with Ogotemmêli—an Introduction to Dogon Religious Ideas.* Oxford: Oxford University Press, 1965.

Griaule, M., and Dieterlen, G. *The Pale Fox.* Chino Valley, Arizona: Continuum Foundation, 1986.

Huet, Michel. *The Dance, Art and Ritual of Africa.* New York: Pantheon Books, 1978.

Nyabongo, Akiki K. *Winds and Lights—African Fairy Tales.* New York: The Voice of Ethiopia, 1939.

Parrinder, Geoffrey. *African Mythology.* London: The Hamlyn Publishing Group Limited, 1967.

Radin, Paul, and Sweeney, James Johnson. *African Folktales and Sculpture.* New York: Pantheon Books, 1952.

Roberts, Allen F. " 'Comets Importing Change of Times and States': Ephemerae and Process among the Tabwa of Zaire," *American Ethnologist* (1982): 712–729.

Roberts, Allen F. "Passage Stellified: Speculation Upon Archaeoastronomy in Southeastern Zaire," *Archaeoastronomy* (Maryland) IV, no. 4 (October–December, 1981): 27–37.

Steffey, Philip C. "Some Serious Astronomy in the 'Sirius Mystery'." *Griffith Observer* 44, no. 9 (September, 1980): 10–20.

Steffey, Philip C. "Some Serious Astronomy in the 'Sirius Mystery' Part II, *Griffith Observer,* 46, no. 2 (February, 1982): 16–19; no. 3 (March, 1982): 16–20; no. 4 (April, 1982): 18–20; no. 5 (May, 1982): 19–20; no. 8 (August 1982): 19–20; no. 10 (October, 1982): 16–18; no. 11 (November, 1982): 8–11.

INDIA AND SOUTHEAST ASIA

Aiyangar, Narayan. *Ancient Hindu Mythology.* New Delhi: Deep & Deep Publications, 1983.

Bentley, John. *A Historical View of the Hindu Astronomy.* London: Smith, Elder, & Co., 1825.

Bernet Kempers, A. J. *Ageless Borobudur.* Servire/Wassenaar, 1976.

Covarrubias, Miguel. *Island of Bali.* New York: Alfred A. Knopf, 1942.

Daniélou, Alain. *The Gods of India.* New York: Inner Traditions International Ltd., 1985.

Dowson, John. *Hindu Mythology and Religion.* London: Asia Publishing House, 1989.

Dumarcay, Jacques. *Borobudur.* Oxford: Oxford University Press, 1978.

Eiseman, Jr., Fred B. *Bali: Sekala & Niskala. Volume 1: Essays on Religion, Ritual, and Art.* Berkeley, California: Periplus Editions, 1989.

Forman, Bedrich. *Borobudur—the Buddhist Legend in Stone.* London: Octopus Books Limited, 1980.

Getty, Alice. *The Gods of Northern Buddhism—Their History and Iconography.* 1928. Reprint. New York: Dover Publications, Inc., 1988.

Ghosh, Ekendranath. *Studies on Rigvedic Deities—Astronomical and Meteorological.* New Delhi: Cosmo Publications, 1983.

Gomez, Luis, and Woodward, Hiram W., Jr., eds. *Barabudur—History and Significance of a Buddhist Monument.* Berkeley, California: Asian Humanities Press, 1981.

Govinda, Lama Anagarika. *Psycho-cosmic Symbolism of the Buddhist Stupa.* Emeryville, California: Dharma Publishing, 1976.

Hackin, J. *Asiatic Mythology.* New York: Crescent Books, no date.

Hopkins, E. Washburn. *Epic Mythology.* 1915. Reprint. Delhi: Motilal Banarsidass, 1974.

Hostetter, H. Clyde. "The Days of the Scorpion," *Griffith Observer* 52, no. 3 (March, 1988): 6–10 and 16–19.

Ions, Veronica. *Indian Mythology.* London: The Hamlyn Publishing Group Limited, 1967.

Keith, A. Berriedale, and Carnoy, Albert J., *The Mythology of All Races, Volume VI—Indian and Iranian.* 1917. Reprint. New York: Cooper Square Publishers, Inc., 1964.

Lal, Kanwar. *Miracle of Konarak.* New York: Castle Books, Inc., 1968.

Macdonell, A. A. *Vedic Mythology.* 1898. Reprint. Delhi: Motilal Banarsidass, 1981.

Mackenzie, Donald A. *Indian Myth and Legend.* London: The Gresham Publishing Company Ltd, no date.

Mackenzie, Donald A. *Myths from Melanesia and Indonesia.* London: The Gresham Publishing Company Ltd, no date.

Michell, George. *The Hindu Temple.* New York: Harper & Row, Publishers, Inc., 1977.

Morgan, Thomas E. "The Burmese Era and Ancient Astronomy in Southeast Asia," *Archaeoastronomy* (Maryland) III, no. 2 (April–June, 1980): 20–21.

Müller, W. Max, and Scott, Sir James George. *The Mythology of All Races, Volume XII—Egyptian and Indo-Chinese.* Boston: Marshall Jones Company, 1918.

Mukherji, Kali Nath. *Popular Hindu Astronomy.* 1905. Reprint. Calcutta: Nirmal Mukherjea, 1969.

Narayan, R. K. *Gods, Demons, and Others.* New York: The Viking Press, 1964.

Nivedita, Sister, and Coomaraswamy, Ananda K. *Myths of the Hindus & Buddhists.* London: George G. Harrap & Company Ltd., 1913.

O'Flaherty, Wendy Doniger, trans., *Hindu Myths.* Harmondsworth, England: Penguin Books Ltd, 1975.

O'Flaherty, Wendy Doniger, trans., *The Rig Veda.* Harmondsworth, England: Penguin Books Ltd., 1981.

Pingree, David. "History of Mathematical Astronomy in India." *Dictionary of Scientific Biography, Volume XV, Supplement I, "Topical Essays"* (Charles Coulston Gillespie, ed.). New York: Charles Scribner's Sons, 1978, pp. 533–633.

Ramseyer, Urs. *The Art and Culture of Bali.* Oxford: Oxford University Press, 1977.

Sharma, Virendra Nath. "Model of Planetary Configuration in the *Mahabharata:* An Exercise in Archaeoastronomy," *Archaeoastronomy* IX (1986): 88–98.

Stewart, Joe D. "On Burmese Calendrics and Astronomy," *Archaeoastronomy* III, no. 3 (July–September, 1980): 17–20.

Tilak, Bal Gangadhar. *The Arctic Home in the Vedas*. Poona City, India: Messrs. Tilak Bros., 1956.

Tilak, Bal Gangadhar. *The Orion or Researches in the Antiquity of the Vedas*. Bombay: Mrs. Radhhabai Atmaram Sagoon, 1893.

Whittaker, Clio, ed. *An Introduction to Oriental Mythology*. Seacaucus, New Jersey: Chartwell Books, 1989.

Wilkins, W. J. *Hindu Mythology*. Calcutta: Thacker, Spink & Co., 1900.

CHINA, KOREA, AND JAPAN

Aston, W. G. *Shinto (The Way of the Gods)*. London: Longmans, Green and Co., 1905.

Aston, W. G., trans. *Nihongi—Chronicles of Japan from the Earliest Times to A.D. 697*. 1896. Reprint. Rutland, Vermont: Charles E. Tuttle Company, 1972.

Bodde, Derk. *Festivals in Classical China*. Princeton, New Jersey: Princeton University Press, 1975.

Bredon, Juliet, and Mitrophanow, Igor. *The Moon Year*. Shanghai: Kelly & Walsh, 1927.

Carpenter, Frances. *Tales of a Chinese Grandmother*. Garden City, New York: Doubleday & Company, Inc., 1946.

Carpenter, Frances. *Tales of a Korean Grandmother*. Garden City, New York: Doubleday & Company, Inc., 1947.

Carus, Paul. *Chinese Astrology*. 1907. Reprint. La Salle, Illinois: Open Court, 1974.

Chamberlain, Basil Hall, trans. *The Kojiki—Records of Ancient Matters*. 1920. Reprint. Rutland, Vermont: Charles E. Tuttle Company, 1982.

Christie, Anthony. *Chinese Mythology*. London: The Hamlyn Publishing Group Limited, 1968.

Cormack, Mrs. J. G. *Chinese Birthday, Wedding, Funeral & Other Customs*. Beijing: China Booksellers, Ltd., 1927.

Davis, F. Hadland. *Myths and Legends—Japan*. Boston: David D. Nickerson & Company, Publishers, no date.

Eberhard, Wolfram, ed. *Folktales of China*. London: Routledge and Kegan Paul Limited, 1965.

Editor *et al. Chinese Myths*. Shanghai: Juvenile and Children Publishing House, 1986.

Etter, Carl. *Ainu Folklore*. Chicago: Wilcox & Follett Company, 1949.

Ferguson, John C., and Anesaki, Masaharu. *The Mythology of All Races, Volume VIII—Chinese and Japanese*. 1928. Reprint. New York: Cooper Square Publishers, Inc., 1964.

Hackin, J. *Asiatic Mythology*. New York: Crescent Books, no date.

Heissig, Walter. *The Religions of Mongolia*. Berkeley and Los Angeles: University of California Press, 1970.

Krupp, E. C. "The Cosmic Temples of Old Beijing." *World Archaeoastronomy* (A. F. Aveni, ed.). Cambridge, England: Cambridge University Press, 1989, pp. 65–75.

Krupp, E. C. "The Mandate of Heaven," *Griffith Observer* 46, no. 6 (June, 1982): 8–17.

Krupp, E. C. "Shadows Cast for the Son of Heaven," *Griffith Observer* 46, no. 8 (August, 1982): 8–18.

Krupp, E. C. "Tombs That Touched the China Sky," *Griffith Observer* 46, no. 7 (July, 1982): 9–17.

Law, Joan, and Ward, Barbara E. *Chinese Festivals in Hong Kong*. Hong Kong: South China Morning Post Ltd., 1982.

Lim Sian-Tek. *Folk Tales from China*. New York: The John Day Company, 1944.

Li Xun. *Old Tales Retold*. Beijing: Foreign Languages Press, 1961.

Loewe, Michael. *Ways to Paradise, the Chinese Quest for Immortality*. London: George Allen & Unwin, 1979.

Mackenzie, Donald A. *Myths of China & Japan*. London: The Gresham Publishing Company Ltd, no date.

Morgan, Harry T. *Chinese Symbols and Superstitions*. South Pasadena, California: P. D. and Ione Perkins, 1942.

Needham, Joseph. *Science and Civilisation in China*, vol. 3, "Mathematics and the Sciences of the Heavens and the Earth." Cambridge: Cambridge University Press, 1959.

Piggott, Juliet. *Japanese Mythology*. London: The Hamlyn Publishing Group Limited, 1969.

de Saussure, Léopold. *Les Origines de l'Astronomie Chinoise*. Paris: Maisonneuve Freres, no date.

Schafer, Edward H. "Astral Energy in Medieval China," *Griffith Observer* 45, no. 7 (July, 1982): 18–20.

Schafer, Edward H. *Pacing the Void*. Berkeley and Los Angeles: University of California Press, 1977.

Seki, Keigo, ed. *Folktales of Japan*. London: Routledge and Kegan Paul Limited, 1963.

Staal, Julius D. W. *Stars of Jade—Astronomy and Star Lore of Very Ancient Imperial China*. Decatur, Georgia: Writ Press, 1984.

Tucci, Giuseppe. *The Religions of Tibet*. London: Routledge & Kegan Paul Ltd, 1980.

Walls, Jan, and Walls, Yvonne, ed. *Classical Chinese Myths*. Hong Kong: Joint Publishing Co., 1984.

Walters, Derek. *Chinese Astrology*. Wellingborough, Northamptonshire, England: The Aquarian Press, 1987.

Wechsler, Howard J. *Offerings of Jade and Silk—Ritual and Symbol in the Legitimation of the T'ang Dynasty*. New Haven, Connecticut: Yale University Press, 1985.
Werner, E. T. C. *Myths and Legends of China*. London: George G. Harrap & Co., Ltd., 1922.
Whittaker, Clio, ed. *An Introduction to Oriental Mythology*. Seacaucus, New Jersey: Chartwell Books, 1989.
Williams, C. A. S. *Outlines of Chinese Symbolism and Art Motives*. 1941. Reprint. Rutland, Vermont: Charles E. Tuttle Company, Inc., 1974.
Zong In-Sob. *Folk Tales from Korea*. London: Routledge & Kegan Paul, Ltd., 1952.

MICRONESIA, POLYNESIA, AND AUSTRALIA

Åkerblom, Kjell. *Astronomy and Navigation in Polynesia and Micronesia*. Publication 14 in the Monograph Series of The Ethnographical Museum, Stockholm. Stockholm: The Ethnographical Museum, 1968.
Alpers, Antony. *The World of the Polynesians*. 1970. Reprint. Oxford: Oxford University Press, 1987.
Andersen, Johannes C. *Myths and Legends of the Polynesians*. 1928. Reprint. Rutland, Vermont: Charles E. Tuttle Company, Inc., 1969.
Bates, Daisy. *Tales Told to Kabbarli*. Sydney: Angus and Robertson (Publishers) Pty Ltd., 1972.
Beckwith, Martha. *Hawaiian Mythology*. 1940. Reprint. Honolulu: University of Hawaii Press, 1970.
Berndt, Ronald M., and Berndt, Catherine H. *The Speaking Land*. Ringwood, Victoria: Penguin Books Australia Ltd, 1989.
Best, Elsdon. *The Astronomical Knowledge of the Maori*. 1955. Reprint. Dominion Museum Monograph No. 3. Wellington: Dominion Museum, 1978.
Best, Elsdon. *The Maori Division of Time*. Dominion Museum Monograph No. 4. Wellington: Dominion Museum, 1922.
Best, Elsdon. *Polynesian Voyagers*. Dominion Museum Monograph No. 5. Wellington: Dominion Museum, 1975.
Best, Elsdon. *Some Aspects of Maori Myth and Religion*. 1954. Reprint. Dominion Museum Monograph No. 1. Wellington: Dominion Museum, 1978.
Best, Elsdon. *Spiritual and Mental Concepts of the Maori*. 1954. Reprint. Dominion Museum Monograph No. 2. Wellington: Dominion Museum, 1978.
Cowan, James. *Legends of the Maori, Volume One*. 1930. Reprint. New York: AMS Press, 1977.
Dixon, Rolan Burrage. *The Mythology of All Races, Volume IX—Oceanic*, 1916. Reprint. New York: Cooper Square Publishers, Inc., 1964.
Gill, William Wyatt. *Myths and Songs from the South Pacific*. 1876. Reprint. New York: Arno Press, 1977.
Gladwin, Thomas. *East Is a Big Bird—Navigation and Logic on Puluwat Atoll*. Cambridge, Massachusetts: Harvard University Press, 1970.
Gonzalez, Marty Elizabeth. *The Archaeoastronomy and Ethnoastronomy of Easter Island*. Master of Arts thesis, California State University, Long Beach, 1984.
Goodenough, Ward H. *Native Astronomy in the Central Carolines*. Museum Monographs. Philadelphia: The University Museum, University of Pennsylvania, 1953.
Grey, George. *Polynesian Mythology*. 1855. Reprint. Christchurch, New Zealand: Whitcombe and Tombs Limited, 1965.
Hart, Roger, and Reed, A. W. *Maori Myth—the Supernatural World of the Maori*. Wellington, New Zealand: A. H. & A. W. Reed Ltd, 1977.
Henry, Teuira, *Ancient Tahiti*. Bernice P. Bishop Museum Bulletin 48. 1928. Reprint. Millwood, New York: Kraus Reprint, 1985.
Isaacs, Jennifer. *Australian Dreaming—40,000 Years of Aboriginal History*. Sydney: Lansdowne Press, 1980.
Johnson, Rubellite Kawena, and Mahelona, John Kaipo. *Na Inoa Hoku—a Catalogue of Hawaiian and Pacific Star Names*. Honolulu: Topgallant Publishing Company, Ltd., 1975.
Lewis, David. *The Voyaging Stars—Secrets of the Pacific Island Navigators*. New York: W. W. Norton & Company, Inc., 1978.
Lewis, David. *We, the Navigators*. Honolulu: The University Press of Hawaii, 1973.
Makemson, Maud Worcester. "Hawaiian Astronomical Concepts," *American Anthropologist* 40. (1938): 370–383.
Makemson, Maud Worcester. *The Morning Star Rises—an Account of Polynesian Astronomy*. New Haven: Yale University Press, 1941.
Massola, Aldo. *Bunjil's Cave—Myths, Legends and Superstitions of the Aborigines of South-East Australia*. Melbourne: Lansdowne Press, 1968.
Mountford, Charles P. *Nomads of the Australian Desert*. Melbourne: Rigby, 1976.
Parker, K. Langloh. *Australian Legendary Tales*. Sydney: Angus and Robertson, 1953.
Poignant, Roslyn. *Oceanic Mythology*. London: The Hamlyn Publishing Group Limited, 1967.
Reed, A. *Aboriginal Fables and Legendary Tales*. French Forest, NSW: Reed Books Pty Ltd, 1965.

Reed, A. *Aboriginal Legends—Animal Tales*. Sydney: A. H. & A. W. Reed Pty Ltd, 1978.

Reed. A. *Aboriginal Myths—Tales of the Dreamtime*. French Forest, NSW: Reed Books Pty Ltd, 1978.

Reed, A. W. *Fairy Tales from the Pacific Islands*. London: Frederick Muller Limited, 1969.

Reed, A. W. *Maori Legends*. Newton Abbot, Devon, England: David & Charles (Publishers) Ltd, 1972.

Reed, A. W. *Maori Myth*. Wellington: A. H. & A. W. Reed, 1977.

Reed, A. W. *Maori Tales of Long Ago*. Wellington, New Zealand: A. H. & A. W. Reed, no date.

Reed, A. W. *Myths and Legends of Australia*. Wellington: A. H. & A. W. Reed, 1965.

Reed, A. W. *Myths and Legends of Maoriland*. Wellington: A. H. & A. W. Reed, 1961.

Reed, A. W. *Myths and Legends of Polynesia*. Wellington: A. H. & A. W. Reed, 1974.

Reed, A. W. *Treasury of Maori Folklore*. Wellington: A. H. & A. W. Reed, 1963.

Reed, A. W., and Hames, Inez. *Myths and Legends of Fiji and Rotuma*. Wellington: A. H. & A. W. Reed, 1967.

Roberts, Ainslie, and Mountford, Charles P. *The Dawn of Time*. Adelaide, Australia: Rigby Limited, 1969.

Roberts, Ainslie, and Mountford, Charles P. *The Dreamtime*. Adelaide, Australia: Rigby Limited, 1965.

Roberts, Ainslie, and Mountford, Charles P. *The First Sunrise*. Adelaide, Australia: Rigby Limited, 1971.

Roberts, Ainslie, and Roberts, Dale. *Dreamtime Heritage*. Blackwood, South Australia: Art Australia, 1975.

Roberts, Ainslie, and Roberts, Dale. *Shadows in the Mist*. Blackwood, South Australia: Art Australia, 1989.

Smith, S. Percy, trans. *The Lore of the Whare-wananga; or Teachings of the Maori College on Religion, Cosmogony, and History, Part 1—Te Kauwae-runga, or 'Things Celestial.'* 1913. Reprint. New York: AMS Press, 1978.

Smith, W. Ramsay. *Myths and Legends of the Australian Aborigines*. New York: Farrar & Rinehart Publishers, no date.

Thomas, W. E. *Some Myths & Legends of the Australian Aborigines*. Melbourne: Whitcombe & Tombs Limited, 1923.

Thompson, Vivian L. *Hawaiian Myths of Earth, Sea, and Sky*. 1966. Reprint. Honolulu: University of Hawaii Press, 1988.

Westervelt, W. D. *Legends of Maui—a Demi-God of Polynesia and of His Mother Hina*. Honolulu: The Hawaiian Gazette Co., Ltd., 1910.

Williamson, Robert W. *Religious and Cosmic Beliefs of Central Polynesia, Volumes 1 and 2*, 1933. Reprint. New York: AMS Press Inc., 1977.

Williamson, Robert W. *Religion and Social Organization in Central Polynesia*. 1937. Reprint. New York: AMS Press, 1977.

NORTH AMERICAN INDIANS

Alexander, Hartley Burr. *The Mythology of All Races, Volume X—North American*. 1916. Reprint. New York: Cooper Square Publishers, Inc., 1964.

Alexander, Hartley Burr. *The World's Rim—Great Mysteries of the North American Indians*. Lincoln, Nebraska: University of Nebraska Press, 1953.

Beauchamp, W. M. "Onondaga Tale of the Pleiades," *Journal of American Folk-Lore* 13 (1900): 281–282.

Beck, Peggy V., and Walters, Anna L. *The Sacred—Ways of Knowledge, Sources of Life*. Tsaile (Navajo Nation), Arizona: Navajo Community College Press, 1977.

Benedict, Ruth. *Tales of the Cochiti Indians*. 1931. Reprint. Albuquerque, New Mexico: University of New Mexico Press, 1981.

Benedict, Ruth. *Zuni Mythology, Vol. II*. New York: Columbia University Press, 1935.

Benson, Arlene, and Buckskin, Floyd, "How the Seasons Began: An Ajumawi Narrative Involving Sun, Moon, North Star, and South Star," *Griffith Observer* 51, no. 7 (July, 1987): 1–11 and 15.

Bierhorst, John. *The Mythology of North America*. New York: William Morrow and Company, 1985.

Blackburn, Thomas C. *December's Child—a Book of Chumash Oral Narratives*. Berkeley and Los Angeles: University of California Press, 1975.

Boas, Franz. *Kwakiutl Ethnography* (edited and abridged edition). Chicago: The University of Chicago Press, 1966.

Brown, Joseph Epes, ed. *The Sacred Pipe—Black Elk's Account of the Seven Rites of the Oglala Sioux*. Norman, Oklahoma: University of Oklahoma Press, 1953.

Bruchac, Joseph. *Iroquois Stories—Heroes and Heroines, Monsters and Magic*. Trumansburg, New York: The Crossing Press, 1985.

Bruchac, Joseph. *Return of the Sun—Native American Tales from the Northeast Woodlands*. Freedom, California: The Crossing Press, 1989.

Buckstaff, Ralph N. "Stars and Constellations of Pawnee Sky Map," *American Anthropologist*, New Series 29 (1927): 279–285.

Bullchild, Percy. *The Sun Came Down—the History of the World as My Blackfeet Elders Told It*. San Francisco: Harper & Row Publishers, Inc., 1985.

Burland, Cottie. *North American Indian Mythology*. London: The Hamlyn Publishing Group Limited, 1965.

Burns, Louis F. *Osage Indian Customs and Myths*. Fallbrook, California: Ciga Press, 1984.

Carlson, John B., and Judge, James W., eds. *Astronomy and Ceremony in the Prehistoric Southwest*. Albuquerque: Maxwell Museum of Anthropology, 1987.

Ceci, Lynn, "Watchers of the Pleiades: Ethnoastronomy among Native Cultivators in Northeastern North America," *Ethnohistory* 25, No. 4 (fall, 1978): 301–317.

Chamberlain, Alex. "Some Items of Algonkian Folk-Lore," *Journal of American Folk-Lore* 13 (1900): 271–277.

Chamberlain, Von Del. "Astronomical Content of North American Plains Indian Calendars," *Archaeoastronomy* (*JHA* Supplement 6, 1984): S1-S54.

Chamberlain, Von Del. "Navajo Constellations in Literature, Art, Artifact and a New Mexico Rock Art Site," *Archaeoastronomy* (Maryland) VI (1983): 48–58.

Chamberlain, Von Del. *When the Stars Came Down to Earth: Cosmology of the Skidi Pawnee Indians of North America*. Los Altos, California: Ballena Press, 1982.

Cheney, Roberta Carkeek. *The Big Missouri Winter Count*. Happy Camp, California: Naturegraph Publishers, Inc., 1979.

Clark, Cora, and Williams, Texa Bowen. *Pomo Indian Myths and Some of Their Sacred Meanings*. New York: Vantage Press, Inc., 1954.

Clark, Ella E. *Indian Legends from the Northern Rockies*. Norman, Oklahoma: University of Oklahoma, Press, 1966.

Clark, Ella E. *Indian Legends of Canada*. 1960. Reprint. Toronto: McClelland and Stewart Limited, 1981.

Clark, Ella E. *Indian Legends of the Pacific Northwest*. Berkeley and Los Angeles. California: University of California Press, 1953.

Coffin, Tristram P., ed., *Indian Tales of North America*. Austin, Texas: University of Texas Press, 1961.

Coleman, Sister Bernardo Frogner, Ellen, and Eich, Estelle. *Ojibwa Myths and Legends*. Minneapolis: Ross & Haines, Inc., Publisher, 1961.

Connelley, William Elsey. *Indian Myths*. New York: Rand McNally & Company, 1928.

Cope, Leona. "Calendars of the Indians North of Mexico," *University of California Publications in American Archaeology and Ethnology* XVI, No. 4 (1919–1920): 119–176.

Curry, Jane Louise. *Back in the Beforetime—Tales of the California Indians*. New York: Margaret K. McElderry Books, 1987.

Curtin, Jeremiah. *Creation Myths of Primitive America*. Boston: Little, Brown & Co., 1898.

Curtin, Jeremiah. *Myths of the Modocs: Indian Legends of the Northwest*. 1912. Reprint. New York: Benjamin Blom, 1971.

Curtin, Jeremiah. *Seneca Indian Myths*. New York: E. P. Dutton & Company, 1922.

Curtis, Edward S. *The North American Indian, Volume 12—Hopi*. 1922. Reprint. New York: Johnson Reprint Corporation, 1978.

Curtis, Edward S. *The North American Indian, Volume 13—Hupa, Yurok, Karok, Wiyot, Tolowa and Tutuni, Shasta, Achomawi, and Klamath*. 1926. Reprint. New York: Johnson Reprint Corporation, 1980.

Curtis, Edward S. *The North American Indian, Volume 14—Kato, Wailuki, Yuki, Pomo, Wintun, Maidu, Miwok, and Yokuts*. 1924. Reprint. New York: Johnson Reprint Corporation, 1976.

Curtis, Edward S. *The North American Indian, Volume 15—Southern California Shoshoneans, Dieguеños, Plateau Shoshoneans, and Washo*. 1926. Reprint. New York: Johnson Reprint Corporation, 1978.

Curtis, Edward S. *The North American Indian, Volume 17—Tewa and Zuñi*. 1926. Reprint. New York: Johnson Reprint Corporation, 1978.

Curtis, Natalie, ed. *The Indians Book*. 1905. Reprint. New York: Bonanza Books, 1987.

Davis, Lee. *On This Earth: Hupa Land Domains, Images and Ecology on 'Deddeh Ninnisan,'* Ph.D. Thesis. Berkeley, California: University of California, Berkeley, 1988.

Dooling, D. M., ed. *The Sons of the Wind—the Sacred Stories of the Lakota*. New York: Parabola Books, 1984.

Dorsey, George A., and Voth, H. R. "The Oraibi Soyal Ceremony," *Field Columbian Museum Publication 55, Anthropological Series*, III, no. 1 (March, 1901).

Douglass, William Boone. "Notes on the Shrines of the Tewa and Other Pueblo Indians of New Mexico," *Proceedings of the Nineteenth International Congress of Americanists* (F. W. Hodge, ed.). Washington, D.C.: 1917, pp. 344–378.

Du Bois, Constance Goddard, "The Mythology of the Diegueños," *Journal of American Folk-Lore* 14 (1901): 181–185.

Du Bois, Constance Goddard, "Mythology of the Mission Indians," *Journal of American Folk-Lore* 19 (1906): 52–60.

Du Bois, Constance Goddard, "Mythology of the Mission Indians," *Journal of American Folk-Lore* 19 (1906): 145–164.

Du Bois, Constance Goddard, "The Religion of the Luiseño Indians of Southern California," *University of California Publications in American Archaeology and Ethnology* 8, no. 3 (June, 1908); 69–186.

Dutton, Bertha P. *Sun Father's Way—the Kiva Murals of Kuaua*. Albuquerque, New Mexico: The University of New Mexico Press, 1963.

Edmonds, Margot, and Clark, Ella E. *Voices of the Winds—Native American Legends*. New York: Facts on File, Inc., 1989.

Erdoes, Richard, and Ortiz, Alfonso, eds. *American Indian Myths and Legends*. New York: Pantheon Books, 1984.

Fewkes, J. W. "Sky-god Personations in Hopi Worship," *Journal of American Folk-Lore* 15 (1902): 14–32.

Fisher, Anne B. *Stories California Indians Told*. Berkeley, California: Parnassus Press, 1957.

Fletcher, Alice C. "Star Cult Among the Pawnee—a Preliminary Report," *American Anthropologist* 4 (1902): 730–736.

Fletcher, Alice C., and La Flesche, Francis. *The Omaha Tribe*. Twenty-seventh annual Report of the Bureau of American Ethnology 1905–1906. Washington, D.C.: Government Printing Office, 1911.

Frazer, Sir James George. *Native Races of America*. London: Percy Lund Humphries & Co., Ltd., 1939.

Frey, Rodney. *The World of the Crow Indians*. Norman, Oklahoma: University of Oklahoma Press, 1987.

Galloway, Patricia, ed. *The Southeastern Ceremonial Complex: Artifacts and Analysis*. Lincoln, Nebraska: University of Nebraska Press, 1989.

Gayton, A. H., and Newman, Stanley S. *Yakuts and Western Mono Myths (Anthropological Records* 5, no. 1). Berkeley and Los Angeles: University of California Press, 1940.

Gifford, Edward W., and Block, Gwendoline Harris. *California Indian Nights Entertainments*. Glendale, California: The Arthur H. Clark Company, 1930.

Goddard, Pliny Earle. "Life and Culture of the Hupa," *University of California Publications in American Archaeology and Ethnology* 1, no. 1 (September, 1903): 1–88.

Goldman, Irving. *The Mouth of Heaven—an Introduction to Kwakiutl Religious Thought*. New York: John Wiley & Sons, Inc., 1975.

Green, Jesse, ed., *Zuñi (Selected Writings of Frank Hamilton Cushing)*. Lincoln, Nebraska: University of Nebraska Press, 1979.

Griffin-Pierce, Trudy. "Ethnoastronomy in Navajo Sandpaintings of the Heavens," *Archaeostronomy* (Maryland) IX (1986): 62–69.

Grinnel, G. B. *Blackfoot Lodge Tales*. 1892. Reprint. Lincoln, Nebraska: University of Nebraska Press, 1962.

Grinnel, G. B. *By Cheyenne Campfires*. 1926. Reprint. Lincoln, Nebraska: University of Nebraska Press, 1971.

Grinnel, G. B. "Cheyenne Tales," *Journal of American Folk-Lore* 13 (1900): 161–183.

Hagar, Stansbury. "The Celestial Bear," *Journal of American Folk-Lore* 13 (1900): 92–103.

Haile, Berard. *Starlore Among the Navajo*. Santa Fe, New Mexico: William Gannon, 1977.

Holsinger, Rosemary. *Shasta Indian Tales*. Happy Camp, California: Naturegraph Publishers, Inc., 1982.

Hudson, Travis, ed. *Breath of the Sun—Life in Early California as Told by a Chumash Indian, Fernando Librado, to John P. Harrington*. Banning, Caifornia: Malki Museum Press, 1980.

Hudson, Travis, and Underhay, Ernest. *Crystals in the Sky: An Intellectual Odyssey Involving Chumash Astronomy, Cosmology and Rock Art*. Socorro, New Mexico: Ballena Press, 1978.

Hudson, Travis; Blackburn, Thomas; Curletti, Rosario; and Timbrook, Janice, eds., *The Eye of the Flute—Chumash Traditional History and Ritual*. Santa Barbara, California: Santa Barbara Museum of Natural History, 1977.

Hultkrantz, Ake. *Belief and Worship in Native America*. Syracuse, New York: Syracuse University Press, 1981.

Hultkrantz, Ake. *Native Religions of North America*. New York: Harper & Row, Publishers, Inc., 1987.

James, George Wharton. "A Saboba Origin-Myth," *Journal of American Folk-Lore* 15 (1902): 36–39.

Jenks, Albert Ernest. "The Bear Maiden," *Journal of American Folk-Lore* 15 (1902): 33–35.

Johnstone, Elizabeth Bayless, ed. *Bigfoot and Other Stories*. Visalia, California: Tulare County Department of Education, 1975.

Judson, Katharine Berry. *Myths and Legends of Alaska*. Chicago: A. C. McClurg & Co., 1911.

Judson, Katharine Berry. *Myths and Legends of British North America*. Chicago: A. C. McClurg & Co., 1917.

Judson, Katharine Berry. *Myths and Legends of California and the Old Southwest*. Chicago: A. C. McClurg & Co., 1916.

Kroeber, A. L. "Indian Myths of South Central California," *University of California Publications in American Archaeology and Ethnology* 4, no. 4 (May, 1907): 167–250.

Kroeber, A. L. "A Mission Record of the California Indians," *University of California Publications in American Archaeology and Ethnology* 8, no. 1 (May 1908): 1–227.

Kroeber, A. L. "Ethnography of the Cahuilla Indians," *University of California Publications in American Archaeology and Ethnology* 8, no. 2 (June, 1908): 29–68.

Kroeber, A. L. "Two Myths of the Mission Indians of

California," *Journal of American Folk-Lore* 19 (1906): 309–321.

Kroeber, A. L. *Yurok Myths*. Berkeley and Los Angeles, California: University of California Press, 1976.

Kroeber, A. L., and Gifford, E. W. *Karok Myths*. Berkeley and Los Angeles, California: University of California Press, 1980.

Kroeber, Theodora. *The Inland Whale*. Bloomington, Indiana: Indiana University Press, 1959.

Krupp, E. C. "Emblems of the Sky," *Ancient Images on Stone* (Van Tilburg, Jo Anne, ed.). Los Angeles: The Rock Art Archive, Institute of Archaeology, U.C.L.A., 1983.

Krupp, E. C. "Hiawatha in California," *The Astronomy Quarterly*, 7, no. 4 (1990).

Krupp, E. C., and Wubben, Bob. "When Things Are Divided in Half," *Rock Art Papers, Volume 7* (Ken Hedges, ed.). San Diego: San Diego Museum of Man, 1990, pp. 41–48.

Laird, Carobeth. *The Chemehuevis*. Banning, California: Malki Museum Press, 1976.

Laird, Carobeth. *Mirror and Pattern—George Laird's World of Chemehuevi Mythology*. Banning, California: Malki Museum Press, 1984.

Lankford, George E., ed. *Native Legends—Southeastern Legends: Tales from the Natchez, Caddo, Biloxi, Chickasaw, and Other Nations*. Little Rock, Arkansas: August House, 1987.

Latta, Frank, ed. *California Indian Folklore*. Shafter, California: F. F. Latta, 1936.

Levi-Strauss, Claude. *The Way of the Masks*. Seattle: University of Washington Press, 1982.

Linton, Ralph. "The Origin of the Skidi Pawnee Sacrifice to the Morning Star," *American Anthropologist*, New Series 28 (1926): 457–466.

Loeb, Edwin M. "The Creator Concept among the Indians of North Central California," *American Anthropologist*, New Series 28 (1926): 467–493.

Loeb, Edwin M. "Pomo Folkways," *University of California Publications in American Archaeology and Ethnology* 19, no. 2 (September, 1926): 152–339.

Logie, Alfred E. *Canadian Wonder Tales*. Chicago: Row, Peterson and Company, 1925.

Margolin, Malcolm, ed. *The Way We Lived—California Indian Reminiscences, Stories and Songs*. Berkeley, California: Heyday Books, 1981.

Marriott, Alice, and Rachlin, Carol K. *American Indian Mythology*. New York: Thomas Y. Crowell Company, 1968.

Marriott, Alice, and Rachlin, Carol K. *Plains Indian Mythology*, 1975. Reprint. New York: Meridian Classic/New American Library, 1985.

Masson, Marcelle. *A Bag of Bones—Legends of the Wintu Indians of Northern California*. Happy Camp, California: Naturegraph Company, 1966.

Mayo, Gretchen Will. *Earthmaker's Tales—North American Indian Stories About Earth Happenings*. New York: Walker and Company, 1989.

Mayo, Gretchen Will. *Star Tales—North American Indian Stories About the Stars*. New York: Walker and Company, 1987.

McCaskill, Don, ed. *Amerindian Cosmology*. Cosmos 4, Yearbook of the Traditional Cosmology Society. Brandon, Manitoba: The Canadian Journal of Native Studies, 1989.

McCluskey, Stephen C. "The Astronomy of the Hopi Indians," *Journal for the History of Astronomy* 8, 3 (October, 1977): 174–195.

McCluskey, Stephen C. "Calendars and Symbolism: Functions of Observation in Hopi Astronomy," *Archaeoastronomy* (*JHA* Supplement 15, 1990): S1–S16.

Merriam, C. Hart. *The Dawn of the World—Myths and Weird Tales Told by the Mewan Indians of California*. Cleveland, Ohio: The Arthur H. Clark Company, 1910.

Millman, Lawrence. *A Kayak Full of Ghosts—Eskimo Tales*. Santa Barbara, California: Capra Press, 1987.

Monroe, Jean Guard, and Williamson, Ray A. *They Dance in the Sky*. Boston: Houghton Mifflin Company, 1987.

Morriseau, Norval. *Legends of My People the Great Ojibway*. Toronto: McGraw-Hill Ryerson Limited, 1965.

Murie, James R. *Ceremonies of the Pawnee*. 1981. Reprint. Lincoln, Nebraska: University of Nebraska Press, 1989.

Neihardt, John G. *Black Elk Speaks*. Lincoln, Nebraska: University of Nebraska Press, 1961.

Newcomb, Franc Johnson; Fishler, Stanley; and Wheelwright, Mary C. "A Study of Navajo Symbolism," *Papers of the Peabody Museum of Archaeology and Ethnology, Harvard University* XXXII, no. 3 (1956): 1–100 (1978 reprint, Kraus Reprint Co., Millwood, New York).

Newell, Edythe W. *The Rescue of the Sun and Other Tales from the Far North*. Chicago: Albert Whitman & Company, 1970.

Norman, Howard, ed. *Northern Tales*. New York: Pantheon Books, 1990.

Olden, Sara Emilia. *Karoc Indian Stories*. San Francisco: Harr Wagner Publishing Co., 1923.

Ortiz, Alfonso. *The Tewa World*. Chicago: The University of Chicago Press, 1969.

Palmer, William R. *Why the North Star Stands Still and Other Indian Legends*. St. George, Utah: William I. Palmer, 1973.

Parker, Arthur C. *Seneca Myths & Folk Tales*. 1923. Reprint. Lincoln, Nebraska: University of Nebraska Press, 1989.

Parsons, Elsie Clews. *Pueblo Indian Religion, Vols. I and II*. Chicago: The University of Chicago Press, 1939.

Patencio, Chief Francisco. *Stories and Legends of the Palm Springs Indians*. Los Angeles: Times Mirror, 1943.

Pradt, George H. "Shakok and Miochin: Origin of Summer and Winter." *Journal of American Folk-Lore* 15 (1902): 88–90.

Reichard, Gladys A. "Literary Types and Dissemination of Myths," *Journal of American Folk-Lore* 34 (1921): 269–307.

Reichard, Gladys A. *Navaho Religion*. 1974. Reprint. Tucson, Arizona: The University of Arizona Press, 1983.

Ridington, Robin. "Images of Cosmic Union: Omaha Ceremonies of Renewal," *History of Religions* 28, No. 2 (November, 1988): 135–150.

Robinson, Gordon. *Tales of Kitamaat*. Kitimat, British Columbia: Northern Sentinel Press, 1956.

Russell, Frank. *The Pima Indians*. 1907. Reprint. Tucson, Arizona: The University of Arizona Press, 1980.

Sanderson, Grover C. *Peek-wa Stories—Ancient Indian Legends of California*. Galt, California: Grover C. Sanderson, 1986.

Sanger, Kay. *When the Animals Were People*. Banning, California: Malki Museum Press, 1983.

Saxton, Dean, and Saxton, Lucille. *Legends and Lore of the Papago and Pima Indians*. Tucson, Arizona: The University of Arizona Press, 1973.

Schlesier, Karl H. *The Wolves of Heaven—Cheyenne Shamanism, Ceremonies and Prehistoric Origins*. Norman, Oklahoma: University of Oklahoma Press, 1987.

Schmidt, W. *High Gods in North America*. Oxford: Oxford University Press, 1933.

Schoolcraft, Henry R. *The Myth of Hiawatha and Other Oral Legends, Mythologic and Allegoric, of the North American Indians*. 1856. Reprint. Millwood, New York: Kraus Reprint Co., 1977.

Scully, Vincent. *Pueblo—Mountain, Village, Dance*. New York: The Viking Press, Inc., 1975.

Shipek, Florence. *The Autobiography of Delfina Cuero*. Banning, California: Malki Museum Press, 1970.

Simmons, Leo W., ed. *Sun Chief*. New Haven, Connecticut: Yale University Press, 1942.

Smith, E. A. *Myths of the Iroquois*. 1883. Reprint. Oswheken, Ontario: Iroqrafts Ltd, 1989.

Spence, Lewis. *Myths and Legends—the North American Indians*. Boston: David D. Nickerson & Company, Publishers, no date.

Spier, Leslie. "Southern Diegueño Customs" *University of California Publications in American Archaeology and Ethnology* 20 (1923): 297–360.

Spier, Leslie. *Yuman Tribes of the Gila River*. 1933. New York: Dover Publications, Inc., 1978.

Stevenson, Matilda Coxe. *The Zuñi Indians: Their Mythology, Esoteric Societies, and Ceremonies*. Twenty-third Annual Report of the Bureau of American Ethnology 1901–1902. Washington, D.C.: Government Printing Office, 1904.

Swanton, John R. *Myths and Tales of the Southeastern Indians*. Bureau of American Ethnology Bulletin 88. Washington, D.C.: Government Printing Office, 1929.

Tedlock, Dennis, and Tedlock, Barbara, eds. *Teachings from the American Earth—Indian Religion and Philosophy*. New York: Liveright Publishing Corporation, 1975.

Thompson, Stith. *Tales of the North American Indians*. Cambridge, Massachusetts: Harvard University Press, 1929.

Titiev, Mischa. "Old Oraibi," *Papers of the Peabody Museum of American Archaeology and Ethnology, Harvard University* XXII, no. 1 (1944).

Tooker, Elisabeth. *The Iroquois Ceremonial of Midwinter*. Syracuse, New York: Syracuse University Press, 1970.

Tooker, Elisabeth, ed. *Native North American Spirituality of the Eastern Woodlands*. New York: Paulist Press, 1979.

Tozzer, Alfred M. "A Note on Star-Lore among the Navajos," *Journal of American Folk-Lore*, 21 (1908): 28–32.

Turner III, Frederick W., ed. *The Portable North American Indian Reader*. New York: Viking Press, 1973.

Tyler, Hamilton A. *Pueblo Gods and Myths*. Norman, Oklahoma: University of Oklahoma Press, 1954.

Underhill, Ruth. "A Papago Calendar Record," *The University of New Mexico Bulletin* 2, 5 (1938).

Vastokas, Joan M., and Vastokas, Romas K. *Sacred Art of the Algonkians*. Peterborough, Ontario: Mansard Press, 1973.

Voget, Fred W. *The Shoshoni-Crow Sun Dance*. Norman, Oklahoma: University of Oklahoma Press, 1984.

Voth, H. R. "The Traditions of the Hopi," *Field Columbian Museum Publications 96*, Anthropological Series Vol. VIII (March, 1905).

Walens, Stanley. *Feasting with Cannibals—an Essay on Kwakiutl Cosmology*. Princeton, New Jersey: Princeton University Press, 1981.

Waterman, T.T. "The Religious Practices of the Diegueño Indians," *University of California Publications in American Archaeology and Ethnology* 8, no. 6 (March, 1910): 271–358.

Weltfish, Gene. *The Lost Universe*. New York: Basic Books, 1965.

Wright, Barton, ed. *The Mythic World of the Zuni*. Albuquerque, New Mexico: University of New Mexico Press, 1988.

Young, Egerton R. *Algonquin Indian Tales*. New York: Fleming H. Revell, 1903.

Young, M. Jane. *Signs from the Ancestors—Zuni Cultural Symbolism and Perceptions of Rock Art*. Albuquerque: University of New Mexico Press, 1988.

Zeilik, Michael. "The Ethnoastronomy of the Historic Pueblos, 1: Calendrical Sun Watching," *Archaeoastronomy* (*JHA* Supplement 8,1985): S1–S24.

Zeilik, Michael. "The Ethnoastronomy of the Historic Pueblos, 2: Moon Watching," *Archaeoastronomy* (*JHA* Supplement 10, 1986): S1–S22.

Zeilik, Michael. "Sun Shrines and Sun Symbols in the U.S. Southwest," *Archaeoastronomy* (*JHA* Supplement 9, 1985): S86–S96.

Zolbrod, Paul G. *Diné bahané—the Navajo Creation Story*. Albuquerque: University of New Mexico Press, 1984.

MESOAMERICA

Alexander, Hartley Burr. *The Mythology of All Races, Volume XI—Latin-American*. 1920. Reprint. New York: Cooper Square Publishers, Inc., 1964.

Benitez, Fernando. *In the Magic Land of Peyote*. Austin: University of Texas Press, 1975.

Benson, Elizabeth P., ed. *Mesoamerican Sites and World-Views*. Washington, D.C.: Dumbarton Oaks Research Library and Collections, 1981.

Benson, Elizabeth P., and Griffin, Gillett G. *Maya Iconography*. Princeton, New Jersey: Princeton University Press, 1988.

Berrin, Kathleen, ed. *Art of the Huichol Indians*. New York: Harry N. Abrams, 1978.

Bierhorst, John. *The Mythology of Mexico and Central America*. New York: William Morrow and Company, Inc., 1990.

Boone, Elizabeth Hill, ed. *The Art and Iconography of Late Post-Classic Central Mexico*. Washington, D.C.: Dumbarton Oaks, 1982.

Boone, Elizabeth Hill, ed. *The Aztec Templo Mayor*. Washington, D.C.: Dumbarton Oaks Research Library and Collection, 1987.

Boone, Elizabeth Hill, ed. *Ritual Human Sacrifice in Mesoamerica*. Washington, D.C.: Dumbarton Oaks Research Library and Collection, 1984.

Bowditch, Charles P. *The Numeration, Calendar Systems and Astronomical Knowledge of the Mayas*. Cambridge, Massachusetts: Harvard University Press, 1910.

Broda, Johanna. "La Fiesta del Fuego Nuevo y el Culto Azteca de las Pleyades." *Homenaje a R. Girard: La Antropología Americanista en la Actualidad Tomo 2*. Mexico City: Editores Mexicanos Unidos, 1980, pp. 283–303.

Broda, Johanna; Carrasco, David; and Moctezuma, Eduardo Matos. *The Great Temple of Tenochtilan—Center and Periphery in the Aztec World*. Berkeley and Los Angeles: University of California Press, 1987.

Brundage, Burr Cartwright. *The Fifth Sun—Aztec Gods, Aztec World*. Austin, Texas: University of Texas Press, 1979.

Brundage, Burr Cartwright. *The Jade Steps—A Ritual Life of the Aztecs*. Salt Lake City: University of Utah Press, 1985.

Brundage, Burr Cartwright. *The Phoenix of the Western World—Quetzalcoatl and the Sky Religion*. Norman, Oklahoma: University of Oklahoma Press, 1982.

Brundage, Burr Cartwright. *A Rain of Darts—the Mexica Aztecs*. Austin, Texas: University of Texas Press, 1972.

Burland, C. A. *The Gods of Mexico*. New York: G. P. Putnam's Sons, 1967.

Burns, Allan F. *An Epoch of Miracles—Oral Literature of the Yucatec Maya*. Austin, Texas: University of Texas Press, 1983.

Carrasco, Davíd. *Quetzalcoatl and the Irony of Empire—Myths and Prophecies in the Azetec Tradition*. Chicago: The University of Chicago Press, 1982.

Carrasco, Davíd. *Religions of Mesoamerica*. New York: Harper & Row, Publishers, Inc., 1990.

Carrasco, Davíd, ed. *The Imagination of Matter*. Oxford: B.A.R., 1989.

Caso, Alfonso. *Los Calendarios Prehispanicos*. Mexico City: Universidad Nacional Autónoma de Mexico, 1967.

Coe, Michael. *Lords of the Underworld*. Princeton, New Jersey: Princeton University Press, 1978.

Coe, Michael. *The Maya Scribe and His World*. New York: The Grolier Society, 1972.

Coe, Michael. *Old Gods and Young Heroes*. Jerusalem: The Israel Museum, 1982.

Colby, Benjamin N., and Colby, Lore M. *The Daykeeper —the Life and Discourse of an Ixil Diviner*. Cambridge, Massachusetts: Harvard University Press, 1981.

Craine, Eugene R., and Reindorp, Reginald C., trans. and eds. *The Codex Pérez and The Book of Chilam Balam of Maní*. Norman, Oklahoma: University of Oklahoma Press, 1979.

de la Garza, Mercedes. *El Universo Sagrado de la Ser-*

piente entre los Mayas. Mexico City: Universidad Nacional Autónoma de Mexico, 1984.

Durán, Fray Diego. *The Aztecs (The History of the Indies of New Spain)*. 1588. New York: The Orion Press, Inc., 1964.

Durán, Fray Diego. *Book of the Gods and Rites* and *The Ancient Calendar*. Norman, Oklahoma: University of Oklahoma Press, 1971.

Edmonson, Munro S. *The Ancient Future of the Itza—the Book of Chilam Balam of Tizimin*. Austin, Texas: University of Texas Press, 1982.

Edmonson, Munro S. *The Book of the Year—Middle American Calendrical Systems*. Salt Lake City: University of Utah Press, 1988.

Edmonson, Munro S., trans. *Heaven Born Merida and Its Destiny—the Book Chilam Balam of Chumayel*. Austin, Texas: University of Texas Press, 1986.

Elzey, Wayne. "The Nahua Myth of the Suns," *Numen* XXIII, fasc. 2 (1976): 114–135.

Elzey, Wayne. "Some Remarks on the Space and Time of the 'Center' in Aztec Religion," *Estudios de Cultura Nahuatl, Vol. 12*. Mexico City: Instituto de Investigaciones Históricos/Universidad Nacional Autónoma de Mexico, 1976, pp. 315–334.

Fernández, Adela. *Pre-Hispanic Gods of Mexico*. Mexico City: Panorama Editorial, S.A., 1984.

Förstemann, Ernst. "Commentary on the Maya Manuscript in the Royal Public Library of Dresden," *Papers of the Peabody Museum of American Archaeology and Ethnology, Harvard University* IV, 2 (1906): 53–267.

Goetz, Delia, and Morley, Sylvanus G., trans. *Popol Vuh, the Sacred Book of the Ancient Quiché Maya*. Norman, Oklahoma: University of Oklahoma Press, 1950.

Gossen, Gary H. *Chamulas in the World of the Sun*. Cambridge, Massachusetts: Harvard University Press, 1974.

Heyden, Doris, and Villaseñor, Luis Francisco. *The Great Temple and the Aztec Gods*. Mexico City: Editorial Minutiae Mexicana, 1984.

Hunt, Eva. *The Transformation of the Hummingbird*. Ithaca, New York: Cornell University Press, 1977.

Kelley, David H. "Astronomical Identities of Mesoamerican Gods," *Archaeoastronomy* (*JHA* Supplement 2, 1980): S1–S54.

Kendall, Timothy. *Patolli, a Game of Ancient Mexico*. Belmont, Massachusetts: Kirk Game Company, 1980.

Kerr, Justin. *The Maya Vase Book, Volume 1*. New York: Kerr Associates, 1989.

Kerr, Justin. *The Maya Vase Book, Volume 2*. New York: Kerr Associates, 1990.

Kohler, Ulrich. " 'Sonnenstein' ohne Sonnengott," *Ethnologia Americana* 16/1, no. 91 (1979): 906–908.

Krupp, E. C. "An Aztec 'Calendar' Stone and Its Celestial Seal of Approval," *Griffith Observer* 45, no. 8 (July, 1981): 1–8.

Krupp, E. C. "The 'Binding of the Years,' the Pleiades, and the Nadir Sun," *Archaeoastronomy* (Maryland) 5, no. 1 (Jan.–Mar., 1982): 10–13.

Krupp, E. C. "The Observatory of Kukulcan," *Griffith Observer* 41, no. 9 (September, 1977): 1–20.

Krupp, E. C. "The Serpent Descending," *Griffith Observer* 41, no. 9, (September, 1982): 79–86.

La Barre, Weston. *The Peyote Cult*. New York: Schocken Books, 1969.

Landa, Friar Diego de. *Yucatan Before and After the Conquest (Relación de las cosas de Yucatan)*. 1566. New York: Dover Publications, Inc., 1978.

León-Portilla, Miguel. *Aztec Thought and Culture*. Norman, Oklahoma: University of Oklahoma Press, 1963.

Léon-Portilla, Miguel. *Mexico-Tenochtitlán: Su Espacio y Tiempo Sagrados*. Mexico City: Instituto Nacional de Antropologia e Historia, 1978.

Léon-Portilla, Miguel, ed. *Native Mesoamerican Spirituality*. New York: Paulist Press, 1980.

Léon-Portilla, Miguel. *Pre-Columbian Literatures of Mexico*. Norman, Oklahoma: University of Oklahoma Press, 1975.

León-Portilla, Miguel. *Time and Reality in the Thought of the Maya, Second Edition*. Norman, Oklahoma: University of Oklahoma Press: 1988.

Lounsbury, Floyd G. "Maya Numeration, Computation, and Calendrical Astronomy." *Dictionary of Scientific Biography, Volume XV, Supplement I, "Topical Essays"* (Charles Coulston Gillespie, ed.). New York: Charles Scribner's Sons, 1978, pp. 759–818.

Luckert, Karl W. *Olmec Religion*. Norman, Oklahoma: University of Oklahoma Press, 1976.

Makemson, Maud Worcester. *The Book of the Jaguar Priest (Book of Chilam Balam of Tizimin)*. New York: Henry Schuman, Inc., 1951.

McCluskey, Stephen C. "Maya Observations of Very Long Periods of Venus," *Journal for the History of Astronomy* 14, 2 (June, 1983): 92–101

Miller, Mary Ellen. *The Murals of Bonampak*. Princeton, New Jersey: Princeton University Press, 1986.

Morris, Walter F. "The Chiapas Maya Weavers' Vision of the Cosmos," *Archaeoastronomy* (Maryland) II, no. 3 (summer 1979): 8–10.

Myerhoff, Barbara G. *Peyote Hunt—the Sacred Journey of the Huichol Indians*. Ithaca, New York: Cornell University Press, 1974.

Nicholson, Henry B. "Religion in Pre-Hispanic Central Mexico," *Handbook of Middle American Indians, Volume*

10, Archaeology of Northern Mesoamerica, Part One (eds. Ekholm, Gordon F., and Bernal, Ignacio). Austin: University of Texas Press, 1971, pp. 395–446.

Nicholson, Irene. *Firefly in the Night.* London: Faber and Faber Limited, 1959.

Nicholson, Irene. *Mexican and Central American Mythology.* London: The Hamlyn Publishing Group Limited, 1967.

Pasztory, Esther. *Aztec Art.* New York: Harry N. Abrams, Inc., 1983.

Recinos, Adrián, and Goetz, Delia, trans., *The Annals of the Cakchiquels* and *Title of the Lords of Totonicapán.* Norman, Oklahoma: University of Oklahoma Press, 1953.

Robicsek, Francis, and Hales, Donald M. *The Maya Book of the Dead, the Ceramic Codex.* Charlottesville, Virginia: University of Virginia Art Museum, 1981.

Sahagùn, Fray Bernardino de, *Florentine Codex: General History of the Things of New Spain* (ed. Arthur J. O. Anderson and Charles E. Dibble). Santa Fe, New Mexico: The School of American Research and the University of Utah. *Book 1. The Gods,* 1970. *Book 2. The Ceremonies,* 1980. *Book 3. The Origin of the Gods,* 1978. *Books 4* and *5. The Soothsayers, the Omens,* 1976. *Book 7. The Sun, the Moon and Stars, and the Binding of the Years,* 1977.

Schele, Linda, and Freidel, David. *A Forest of Kings.* New York: William Morrow and Company, Inc., 1990.

Schele, Linda, and Miller, Jeffrey H. *The Mirror, the Rabbit, and the Bundle: "Accession" Expressions from the Classic Maya Inscriptions.* "Studies in Pre-Columbian Art & Archaeology Number Twenty-five." Washington, D.C.: Dumbarton Oaks Research Library and Collection, 1983.

Schele, Linda, and Miller, Mary Ellen. *The Blood of Kings—Dynasty and Ritual in Maya Art.* New York: George Braziller, Inc., 1986.

Séjourné, Laurette. *Burning Water: Thought and Religion in Ancient Mexico.* New York: Vanguard Press, 1956.

Séjourné, Laurette: *El Pensamiento Náhuatl Cifrado por los Calendarios.* Mexico City: Siglo Veintiuno Editores, SA, 1981.

Seler, Eduard. *Códice Borgia y Comentarios.* Mexico City: Fondo de Cultura Económica, 1963.

Seler, Eduard, *et al. Mexican and Central American Antiquities, Calendar Systems, and History (Bureau of American Ethnology Bulletin 28).* Washington, D.C.: Government Printing Office, 1904.

Soustelle, Jacques. *The Four Suns.* New York: Grossman Publishers, 1971.

Soustelle, Jacques. *El Universo de los Aztecas.* Mexico City: Fondo de Cultura Económica, 1979.

Spence, Lewis. *The Magic & Mysteries of Mexico.* Philadelphia: David McKay Company, no date.

Spence, Lewis. *The Myths of Mexico and Peru.* London: George G. Harrap & Company, Ltd., 1914.

Tedlock, Barbara. "Hawks, Meteorology and Astronomy in Quiché-Maya Agriculture," *Archaeoastronomy* (Maryland) VIII (1985): 80–88.

Tedlock, Barbara. *Time and the Highland Maya.* Albuquerque: University of New Mexico Press, 1982.

Tedlock, Dennis, trans. *Popol Vuh.* New York: Simon and Schuster, 1985.

Teeple, John E. "Maya Astronomy," *Contributions to American Anthropology and History* 1, no. 2 (Carnegie Institution of Washington, November, 1931): 29–116. Reprint. New York: Johnson Reprint Corporation, 1970.

Thompson, J. Eric S. *A Commentary on the Dresden Codex:* Philadelphia: American Philosophical Society, 1972.

Thompson, J. Eric S. *Maya Hieroglyphic Writing.* Norman, Oklahoma: University of Oklahoma Press, 1971.

Thompson, J. Eric S. *Maya History and Religion.* Norman, Oklahoma: University of Oklahoma Press, 1970.

Thompson, J. Eric S. *The Rise and Fall of Maya Civilization.* Norman, Oklahoma: University of Oklahoma Press, 1966.

Tichy, Franz, ed. *Space and Time in the Cosmovision of Mesoamerica (Lateinamerika Studien 10).* München: Wilhelm Fink Verlag, 1982.

Tompkins, Ptolemy. *This Tree Grows Out of Hell.* New York: HarperCollins Publishers, 1990

Townsend, Richard Fraser. *State and Cosmos in the Art of Tenochtitlán.* "Studies in Pre-Columbian Art and Archaeology Number Twenty." Washington, D.C.: Dumbarton Oaks, 1979.

Umberger, Emily. "The Structure of Aztec History," *Archaeoastronomy* (Maryland) IV, no. 4 (October–December, 1981): 10–17.

Villar, Maria Montoliu, *Cuando los Dioses Despertaron.* Mexico City: Universidad Nacional Autónoma de México, 1989.

Wicke, Charles R. "The Mesoamerican Rabbit in the Moon: An Influence from Han China?" *Archaeoastronomy* (Maryland) VII (1984): 46–55.

SOUTH AMERICA

Alexander, Hartley Burr. *The Mythology of All Races, Volume XI—Latin-American.* 1920. Reprint. New York: Cooper Square Publishers, Inc., 1964.

Bierhorst, John, ed. *Black Rainbow—Legends of the Incas*

and Myths of Ancient Peru. New York: Farrar, Straus and Giroux, 1976.

Bierhorst, John. *The Mythology of South America.* New York: William Morrow and Company, 1988.

Cobo, Father Bernabe. *History of the Inca Empire (Historia del Nuevo Mundo).* 1653. Austin, Texas: University of Texas Press, 1979.

Cobo, Father Bernabe. *Inca Religion & Customs (Historia del Nuevo Mundo).* 1653. Austin, Texas: University of Texas Press, 1990.

de Civrieux, Marc. *Watunna, an Orinoco Creation Cycle.* San Francisco: North Point Press, 1980.

de la Vega, Garcilaso. *The Incas (The Royal Commentaries of the Inca).* 1609. New York: The Orion Press, Inc., 1961.

Demarest, Arthur A. *Viracocha—the Nature and Antiquity of the Andean High God.* Cambridge, Massachusetts: Peabody Museum of Archaeology and Ethnology, Harvard University, 1981.

Hadingham, Evan. *Lines to the Mountain Gods—Nazca and the Mysteries of Peru.* New York: Random House, Inc., 1987.

Hugh-Jones, Stephen. *The Palm and the Pleiades.* Cambridge, England: Cambridge University Press, 1979.

Hyslop, John. *Inka Settlement Planning.* Austin, Texas: University of Texas Press, 1990.

Magaña, Edmundo. "Astronomîa de los Wayana de Surinam y Guayana francesa," *Journal of Latin American Lore* (U.C.L.A. Latin American Center) 13, no. 1 (1987): 47–71.

Magaña, Edmundo. "A Comparison Between Carib, Tukano/Cubeo and Western Astronomy," *Archaeoastronomy* (Maryland) V, no. 2 (April–June, 1982): 23–31.

Magaña, Edmundo. "South American Ethnoastronomy," *Myth and the Imaginary in the New World* (E. Magaña and P. Mason, eds.). Amsterdam: CEDLA/FORIS, 1986, pp. 399–426.

Magaña, Edmundo, and Jara, Fabiola. "Invention of the Sky," *Archaeoastronomy* (Maryland) VI (1983): 102–113.

Markham, Clements R. *Narratives of the Rites and Laws of the Yncas.* Reprint. New York: Burt Franklin, Publisher, no date.

Morrison, Tony. *Pathways to the Gods—the Mystery of the Andes Lines.* Salisbury, Wiltshire, England: Michael Russell (Publishing) Ltd, 1978.

Osborne, Harold. *South American Mythology.* London: The Hamlyn Publishing Group Limited, 1968.

Perrin, Michel. *The Way of the Dead Indians—Guajiro Myths and Symbols.* Austin: University of Texas Press, 1987.

Poma, Huamán. *Letter to a King (Nueva Corónica y Buen Gobierno).* 1567–1615. New York: E.P. Dutton, 1978.

Reichel-Dolmatoff, Gerardo. *Amazonian Cosmos.* Chicago: The University of Chicago Press, 1971.

Reichel-Dolmatoff, Gerardo: "Astronomical Models of Social Behavior Among Some Indians of Colombia," *Ethnoastronomy and Archaeoastronomy in the American Tropics.* New York: New York Academy of Sciences, 1982, pp. 165–181.

Reichel-Dolmatoff, Gerardo, *Beyond the Milky Way.* Los Angeles: U.C.L.A. Latin American Center Publications, 1978.

Reichel-Dolmatoff, Gerardo. "Brain and Mind in Desana Shamanism," *Journal of Latin American Lore* (U.C.L.A. Latin American Center) 7, no. 1 (1981): 73–98.

Reichel-Dolmatoff, Gerardo. "Desana Shamans' Rock Crystals and the Hexagonal Universe," *Journal of Latin American Lore* (U.C.L.A. Latin American Center) 5, no. 1 (1979): 117–128.

Reichel-Dolmatoff, Gerardo. "The Great Mother and the Kogi Universe: A Concise Overview," *Journal of Latin American Lore* (U.C.L.A. Latin American Center) 13, no. 1 (1987): 73–114.

Reichel-Dolmatoff, Gerardo. "The Loom of Life: A Kogi Principle of Integration," *Journal of Latin American Lore* (U.C.L.A. Latin American Center) 4, no. 1 (1978): 5–27.

Reichel-Dolmatoff, Gerardo. *The Shaman and the Jaguar.* Philadelphia: Temple University Press, 1975.

Reinhard, Johan. *The Nazca Lines—a New Perspective on Their Origin and Meaning.* Lima: Editorial Los Pinos E.I.R.L., 1985.

Roe, Peter G. *The Cosmic Zygote—Cosmology in the Amazon Basin.* New Brunswick, New Jersey: Rutgers University Press, 1982.

Roth, W. E. *An Inquiry into the Animism and Folk-Lore of the Guiana Indians.* Thirtieth Annual Report of the Bureau of American Ethnology 1908–1909. Washington, D.C.: Government Printing Office, 1915, pp. 103–386.

Spence, Lewis. *The Myths of Mexico and Peru.* London: George G. Harrap & Company, Ltd., 1914.

Sullivan, Lawrence E. *Icanchu's Drum—an Orientation to Meaning in South American Religions.* New York: Macmillan Publishing Company, 1988.

Taylor, Douglas. "Notes on the Star Lore of the Caribbees," *American Anthropologist* 48 (1946): 215–222.

Urton, Gary, ed. *Animal Myths and Metaphors in South America.* Salt Lake City: University of Utah Press, 1985.

Urton, Gary. *At the Crossroads of the Earth and Sky*. Austin: University of Texas Press, 1981.

Urton, Gary. "Calendrical Cycles and Their Projections in Pacariqtambo, Peru," *Journal of Latin American Lore* (U.C.L.A. Latin American Center) 12, no. 1 (1986): 45–64.

Urton, Gary. *The History of a Myth—Pacariqtambo and the Origin of the Inkas*. Austin, Texas: University of Texas Press, 1990.

Wilbert, Johannes. "Eschatology in a Participatory Universe: Destinies of the Soul among the Warao Indians of Venezuela." *Death and the Afterlife in Pre-Columbian America* (ed. Elizabeth P. Benson), Washington, D.C.: Dumbarton Oaks, 1973, pp. 163–189.

Wilbert, Johannes. *Folk Literature of the Warao Indians*. Los Angeles: U.C.L.A. Latin American Center, University of California, 1970.

Wilbert, Johannes, ed. *Folk Literature of the Selknam Indians*. Los Angeles: U.C.L.A. Latin American Center, University of California, 1975.

Wilbert, Johannes, ed. *Folk Literature of the Yamana Indians*. Los Angeles: U.C.L.A. Latin American Center, University of California, 1977.

Wilbert, Johannes, and Simoneau, Karin, eds. *Folk Literature of the Ayoreo Indians*. Los Angeles: U.C.L.A. Latin American Center, University of California, 1989.

Wilbert, Johannes, and Simoneau, Karin, eds. *Folk Literature of the Bororo Indians*. Los Angeles: U.C.L.A. Latin American Center, University of California, 1983.

Wilbert, Johannes, and Simoneau, Karin, eds. *Folk Literature of the Caduveo Indians*. Los Angeles: U.C.L.A. Latin American Center, University of California, 1989.

Wilbert, Johannes, and Simoneau, Karin, eds. *Folk Literature of the Chamacoco Indians*. Los Angeles: U.C.L.A. Latin American Center, University of California, 1987.

Wilbert, Johannes, and Simoneau, Karin, eds. *Folk Literature of the Chorote Indians*. Los Angeles: U.C.L.A. Latin American Center, University of California, 1985.

Wilbert, Johannes, and Simoneau, Karin, eds. *Folk Literature of the Gê Indians, Volume One*. Los Angeles: U.C.L.A. Latin American Center, University of California, 1978.

Wilbert, Johannes, and Simoneau, Karin, eds. *Folk Literature of the Gê Indians, Volume Two*. Los Angeles: U.C.L.A. Latin American Center, University of California, 1984.

Wilbert, Johannes, and Simoneau, Karin, eds. *Folk Literature of the Guajiro Indians, Volume One*. Los Angeles: U.C.L.A. Latin American Center, University of California, 1986.

Wilbert, Johannes, and Simoneau, Karin, eds. *Folk Literature of the Guajiro Indians, Volume Two*. Los Angeles: U.C.L.A. Latin American Center, University of California, 1986.

Wilbert, Johannes, and Simoneau, Karin, eds. *Folk Literature of the Mataco Indians*. Los Angeles: U.C.L.A. Latin American Center, University of California, 1982.

Wilbert, Johannes, and Simoneau, Karin, eds. *Folk Literature of the Mocoví Indians*. Los Angeles: U.C.L.A. Latin American Center, University of California, 1988.

Wilbert, Johannes, and Simoneau, Karin, eds. *Folk Literature of the Navaklé Indians*. Los Angeles: U.C.L.A. Latin American Center, University of California, 1987.

Wilbert, Johannes, and Simoneau, Karin, eds. *Folk Literature of the Tehuelche Indians*. Los Angeles: U.C.L.A. Latin American Center, University of California, 1984.

Wilbert, Johannes, and Simoneau, Karin eds. *Folk Literature of the Toba Indians, Volume One*, Los Angeles: U.C.L.A. Latin American Center, University of California, 1982.

Wilbert, Johannes, and Simoneau, Karin, eds. *Folk Literature of the Toba Indians, Volume Two*. Los Angeles: U.C.L.A. Latin American Center, University of California, 1989.

Wilbert, Johannes, and Simoneau, Karen, eds. *Folk Literature of the Yanomami Indians*. Los Angeles: U.C.L.A. Latin American Center, University of California, 1990.

Wilbert, Johannes and Simoneau, Karen, eds. *Folk Literature of the Yaruro Indians*. Los Angeles: U.C.L.A. Latin American Center, University of California, 1990.

Woodside, Joseph H. "Amahuaca Observational Astronomy," *Archaeoastronomy* (Maryland) III, no. 1 (winter 1980): 22–26.

Ziółkowski, Mariusz S., and Sadowski, Robert M., eds. *Time and Calendars in the Inca Empire*. Oxford; B.A.R., 1989.

Zuidema, R. Tom. *Inca Civilization in Cuzco*. Austin, Texas: University of Texas Press, 1990.

CHRISTMAS

Ashton, John. *A Righte Merrie Christmasse.* Reprint. New York: Benjamin Blom, 1968.

Editors of Time-Life Books. *The Book of Christmas.* (The Enchanted World). Chicago: Time-Life Books Inc., 1986.

Guthridge, Ian. *All About Christmas.* Port Melbourne, Australia; Medici Publications, 1988.

Harrison, Shirley. *Who Is Father Christmas?* North Pomfret, Vermont: David & Charles Inc., 1981.

Hildesheim, John of. *The Story of the Three Kings.* New York: The Metropolitan Museum of Art, 1955.

Jones, Charles W. *Saint Nicholas of Myra, Bari, and Manhattan.* Chicago; The University of Chicago Press, 1978.

Martin, Ernest L. *The Birth of Christ Recalculated!* Pasadena, California: Foundation of Biblical Research, 1978.

Miles, Clement A. *Christmas Customs and Traditions—Their History and Significance.* 1912. Reprint. New York: Dover Publications, Inc., 1976.

Mosley, John. *The Christmas Star.* Los Angeles: Griffith Observatory, 1987.

Nettel, Reginald, *Santa Claus.* Bedford, England: The Gordon Fraser Gallery Limited, 1957.

Samson, William. *A Book of Christmas.* New York: McGraw-Hill Book Company, 1968.

ASTROLOGY, NOSTRADAMUS, AND THE HARMONIC CONVERGENCE

Abell, George O. "Astrology," *Science and the Paranormal* (Abell, George O., and Singer, Barry, eds.). New York: Charles Scribner's Sons, 1981, pp. 70–94.

Argüelles, José. *The Mayan Factor: Path Beyond Technology.* Santa Fe, New Mexico: Bear & Company, 1987.

Carlson, Shawn. "A Double-blind Test of Astrology," *Nature* 318 (5 December 1985): 419–425.

Cazeau, Charles J. "Prophecy: The Search for Certainty," *The Skeptical Inquirer VII,* no. 1 (fall, 1982): 20–29.

Fraknoi, Andrew, "Perspective: Why Astrology Believers Should Feel Embarrassed," *San Jose Mercury News* (8 May 1988).

Hoebens, Piet Hein. "The Modern Revival of 'Nostradamitis'," *The Skeptical Inquirer* VII, no. 1 (fall, 1982): 38–45.

Kelly, Ivan. "The Scientific Case Against Astrology," *Mercury* IX, no. 6 (November-December, 1980): 135–142.

Kenton, Warren, *Astrology—the Celestial Mirror.* New York: Avon Books, 1974.

Lindsay, Jack. *Origins of Astrology.* New York: Barnes & Noble, Inc., 1971.

Mosley, John. "Nostradamus and the Doom of May," *Griffith Observer* 52, no. 5 (May, 1988): 2.

Randi, James. "Nostradamus: The Prophet for All Seasons," *The Skeptical Inquirer* VII, no. 1 (fall, 1982): 30–37.

Schultz, Ted, ed. *The Fringes of Reason: A Field Guide to New Age Frontiers, Unusual Beliefs & Eccentric Sciences.* New York: Harmony Books, 1989.

Tester, Jim. *A History of Western Astrology.* New York: Ballantine Books, 1987.

FLYING SAUCERS, UFOs, AND ALIEN ABDUCTIONS

Baker, Robert A. "The Aliens Among Us: Hypnotic Regression Revisited," *The Skepitcal Inquirer* XII, no. 2 (winter 1987–88): 148–162.

Bowen, Charles, *et al,* eds. *The Humanoids.* Chicago: Henry Regnery Company, 1969.

Curran, Douglas. *In Advance of the Landing: Folk Concepts of Outer Space.* New York: Abbeville Press, Inc., 1985.

Eberhart, George M. *UFO and the Extraterrestrial Contact Movement* (two volumes). Metuchen, New Jersey: The Scarecrow Press, Inc., 1986.

Ellis, Bill. "The Varieties of Alien Experience," *The Skeptical Inquirer* XII, no. 3 (Spring, 1988): 263–269.

Evans, Hilary. *The Evidence for UFOs.* Wellingborough, Northamptonshire: The Aquarian Press, 1983.

Evans, Hilary. *Gods, Spirits, Cosmic Guardians.* Wellingborough, Northamptonshire: The Aquarian Press, 1987.

Evans, Hilary. *Visions, Apparitions, Alien Visitors.* Wellingborough, Northamptonshire: The Aquarian Press, 1984.

Evans, Hilary, and Spencer, John, eds. *UFOs 1947–1987—the 40-Year Search for an Explanation.* London: Fortean Tomes, 1987.

Fuller, John G. *The Interrupted Journey.* New York: Berkeley Publishing Corporation, 1965.

Haines, Richard F., ed. *UFO Phenomena and the Behavioral Scientist.* Metuchen, New Jersey: The Scarecrow Press, Inc., 1979.

Hendry, Allan. *The UFO Handbook.* Garden City, New York: Doubleday & Company, Inc. 1979.

Hopkins, Budd. *Intruders—the Incredible Visitations at Copley Woods.* New York: Random House, Inc., 1987.

Hynek, J. Allen: *The UFO Experience*. Chicago: Henry Regnery Company, 1973.

Hynek, J. Allen, and Vallee, Jacques. *The Edge of Reality*. Chicago: Henry Regnery Company, 1975.

Jung, C.J. *Flying Saucers: A Modern Myth of Things Seen in the Sky*. New York: Harcourt, Brace and Company, 1959.

Klass, Philip J. *UFO Abductions—a Dangerous Game*. Buffalo, New York: Prometheus Books, 1988.

Lawson, Alvin H. " 'Alien' Roots: Six UFO Entity Types and Some Possible Earthly Ancestors," *MUFON Symposium Proceedings*, (1979): 152–175.

Lawson, Alvin H. "Hypnosis of Imaginary UFO" Abductees," unpublished manuscript, no date.

Lawson, Alvin H. "Perinatal Imagery in UFO Abduction Reports," *The Journal of Psychohistory* 12, no. 2, (fall, 1984): 211–239.

Lawson, Alvin H. "A Testable Hypothesis for the Origin of Fallacious Abduction Reports: Birth Trauma Imagery in CE-III Narratives," unpublished manuscript, 1981.

Leslie, Desmond, and Adamski, George. *Flying Saucers Have Landed*. New York: The British Book Centre, 1953.

Lloyd, Ann, ed. *There's Something Going on Out There*. London: Orbis Publishing, 1982.

Lucanio, Patrick. *Them or Us—Archetypal Interpretations of Fifties Alien Invasion Films*. Bloomington and Indianapolis: Indiana University Press, 1987.

Michell, John. *The Flying Saucer Vision*. 1967. Reprint. London: Abacus, 1974.

Persinger, Michael A., and Lafrenière, Gyslaine F. *Space-Time Transients and Unusual Events*. Chicago: Nelson-Hall, Inc., 1977.

Rimmer, John. *The Evidence of Alien Abductions*. Wellingborough, Northamptonshire: The Aquarian Press, 1981.

Sagan, Carl, and Page, Thornton. *UFO's—A Scientific Debate*. Ithaca, New York: Cornell University Press, 1972.

Saleh, Dennis. *Science Fiction Gold—Film Classics of the 50s*. New York: McGraw-Hill Book Company, 1979.

Sheaffer, Robert. *The UFO Verdict*. Buffalo, New York: Prometheus Books, 1980.

Strieber, Whitley. *Communion*. New York: Beech Tree Books/William Morrow and Company, Inc., 1987.

Strieber, Whitley. *Transformation*. New York: Beech Tree Books/William Morrow and Company, Inc., 1988.

Vallee, Jacques. *Confrontations*. New York: Ballantine Books, 1990.

Vallee, Jacques. *Dimensions*. Chicago: Contemporary Books, Inc., 1988.

Vallee, Jacques. *The Invisible College*. New York: E.P. Dutton, 1975.

Vallee, Jacques. *Messengers of Deception*. Berkeley, California: And/Or Press, 1979.

Vallee, Jacques. *Passport to Magonia (from Folklore to Flying Saucers)*. Chicago: Henry Regnery Company, 1969.

Index

References to illustrations are in italics.